Photosynthesis Bibliography

volume 6 1975

References no. 21505-25161 / AAR - ZUR

Editors Z. Šesták & Čatský

Dr. W. Junk b.v. - Publishers - The Hague 1980

Contributors:
Z. Šesták
J. Čatský
I. Tichá
J. Pospíšilová
J. Solárová
D. Hodáňová

ISBN 978-90-6193-045-7 ISBN 978-94-009-9164-4 (eBook)
DOI 10.1007/978-94-009-9164-4

PREFACE

The bibliography includes papers in all fields of photosynthesis research - from studies of model biochemical and biophysical systems of the photosynthesis mechanism to primary production studied by the so-called growth analysis. In addition to papers devoted entirely to photosynthesis, papers on other topics are included if they contain data on photosynthetic activity, photorespiration, chloroplast structure, chlorophyll and carotenoid synthesis and destruction, *etc.*, or if they contain valuable methodological information (measurement of selected environmental factors, leaf area, *etc.*). In many branches it has been very difficult to define the limits of interest for photosynthesis researchers. This problem has arisen *e.g.* in topics dealing with the transport of gases, where - in addition to the papers on CO_2 transfer - some papers on water vapour transfer are included, these being of general application. On the other hand, many papers dealing with the anatomy and physiology of stomata have been omitted, if the aspect of carbon dioxide or water vapour exchange has not been discussed.

This volume contains references to papers published in the year 1975, and, similarly to Vol. 5, also addenda including references published in the preceding period (*i.e.* 1966 - 1974). The numbers of these additional references are labelled with an asterisk in the list of references.

To maximize the value of the bibliography the references are arranged alphabetically by authors' names, and each volume is provided with three indexes. The Authors' Index contains all names of authors, co-authors and editors. The Subject Index covers only primary items chosen according to their interest for photosynthesis researchers. Starting with this volume, the Subject Index has been newly arranged and enlarged. It contains more details on the electron transport chain, carbon fixation pathways, gas exchange on leaf and canopy level, *etc.* In the Plant Index, only important crop plants and selected plant types and groups are indexed.

Cumulative indexes accompany Volumes 1, 5, and then every fifth volume, *i.e.* Volumes 10, 15, *etc.*

We have tried to cover fully the relevant papers which have appeared in the most important scientific periodicals and books. Articles published in local journals, mimeographed booklets, *etc.*, were chosen mostly from reprints and lists of publications received directly from the authors. Only abstracts published in regular journals were included.

Since some 3000 relevant papers are currently published every year and included in this bibliography, and since the majority of citations have been checked with the originals, collecting and preparing for publication of such a large amount of material would have been impossible without the collaboration of the authors of the relevant publications. The courtesy of those authors who have already supplied us with reprints is highly appreciated.

We acknowledge with thanks the cooperation of our colleagues from the Institute of Experimental Botany of the Czechoslovak Academy of Sciences in Prague, especially Mrs. **DRAHOMÍRA TEŽKÁ**, Mrs. **VLASTA FLORIÁNOVÁ**, Mrs. **LENKA KOLČABOVÁ**, Mrs. **MARIE MANDLOVÁ**, and Mrs. **MARTA ŠMÍDOVÁ** who helped in preparing the card material. Mr. **PETR ZÁZVORKA** and the librarian of the Institute, Mrs. **ZORA ZAWOYSKA** helped us with checking the references.

Dr. **Z. ŠESTÁK** and Dr. **J. ČATSKÝ**

Institute of Experimental Botany
Czechoslovak Academy of Sciences

Flemingovo n. 2
CS-160 00 PRAHA 6
Czechoslovakia

INSTRUCTIONS FOR USE

All references are arranged alphabetically according to the authors' names and the year of publication. They are numbered and these numbers are used in the indexes. In case of a book title, the number is preceded by B. An asterisk preceding the number denotes the reference published in the preceding period (1966 - 1974).

The references contain the original unshortened title of the paper (book). English, French and German titles are cited in the original language. Titles in other languages are supplemented with a translation in English (sometimes using the title of the respective English abstract or a shortened title with omitted deadweight words). Titles of Japanese, Chinese *etc.* papers are given in English translation only. The journals' names are abbreviated mainly according to the "Style Manual for Biological Journals" (Second Edition, Amer. Institute of Biological Sciences, Washington, D.C. 1964), *e.g.* :

Abhandlungen
Abstract
Abteilung
Academy
Acta
Africa
agricultural
Agriculture
Agronomy
Akademie (-emiya)
Algology
allgemeine
american
America
analytical
Anatomy
angewandte
Annals
annual
anorganisch (-nic)
applied
Arbeit
Archiv
Atmosphere
atomic
Australia
Beiheft
Belgique
Bericht
biochemical
Biochemistry
biokhimicheskiĭ
Biokhimiya
biological(-ogicheskiĭ)
Biology (-ogiya)
biophysical
Biophysics
Bodenkunde
bolgarskiĭ
botanical (-anicheskiĭ)
Botany
british
Bulletin
Canada
cellular (-ulaire)
central
chemical
Chemistry
chimicus
chinese
Chromatography
Commission
Communication
comparative
Comptes rendus
Conference
Congress
Contribution
Cytochemistry
Cytology
czechoslovak
Dendrology
Department
Deutschland
Disease
Dissertation
Doklady
Dopovidi
Ecology
Education
Embryology
Encyclopedia
Engineer
Enzymology
european
experimental
Experiment
Faculty
Federation
Fizika
Fiziologiya
Forestry
Forschung
Foundation
France
Gazette
general
genetical
Genetics
Gesellschaft
Giornale
helveticus
Histochemistry
Histology
Horticulture
hungaricus
Husbandry
imperial
Industry
inorganic
Institute
international
Investigation
italian
Izvestiya
Jahrbuch
japanese
Japan
Journal
Klasse
Laboratory
Landwirtschaft
Letters
Limnology
Magazin
marine
Mathematics
Microbiology
miscellaneous
molecular
Monograph
moskovskiĭ
Mycology
national
natural
Naturforschung
neerlandicus
Netherland
New Zealand
nuclear
Oceanography
Optics
organic
original
Otdelenie
Pathology
Pflanzen-
Philosophy
physical
Physics
physiological
Physiology
Phytopathology
Plant (-arum)
polish
Proceedings
Publication
Publishers
quantitative
Quarterly
Radiation
Radiobiology
Rasteniĭ
Recherche
Report
Research
Review
royal
russian
russkiĭ
scandinavicus
Science
Section
Series (-iya)
Society
sovetskiĭ
soviet
special
SSSR
Station
Supplement
Survey
Symposium
technical
Technology
Tijdschrift
Transaction
Travail (-aux)
tropical
Trudy
ukrainian
UK
US, USA
USSR
University
végétal
Virology
Virusforschung
Volume
Weekblad
Wetenschappen
Wissenschaft
Zeitschrift
Zeitung
Zentralblatt
Zhurnal

The numbers at the end of each reference of a journal article denote : volume (issue) : first page - last page, year of publication. The number of issue is given only in the journal where each issue is paginated separately.

Book titles are cited according to the title page, not to the book jacket or cover (if the names of the editors are not given on the title page, they are not cited in the reference). The publishing house, place and year of publication are included.

Brackets at the end of the reference give bibliographic details and explanations to the contents, not given in the original. The following abbreviations are used most often :

ab	abstract	Jap.	Japanese
Arm.	Armenian	Latv.	Latvian
Belorus.	Belorussian	Lithu.	Lithuanian
Bulg.	Bulgarian	Norweg.	Norwegian
Car	carotenoids	PC	paper chromatography
CC	column chromatography	PhAR	photosynthetically active radiation
Chin	Chinese		
Chl	chlorophyll	Pol.	Polish
Croat.	Croatian	Ps	photosynthesis
E	English	R	Russian
F	French	Roum.	Roumanian
G	German	Span.	Spanish
GC	gas chromatography	Swed.	Swedish
Georg.	Georgian	TLC	thin-layer chromatography
Hung.	Hungarian	Tr	transpiration
IRGA	infra-red gas analyser	Ukr.	Ukrainian
Ital.	Italian	Uz.	Uzbeg

The transliteration of Cyrillic characters is in accordance with the BSI-ASA/ /SC-Z39 draft table, *i.e.* :

Translit.	Cyrill.	Translit.	Cyrill.
a	а	p	п
b	б	r	р
ch	ч	s	с
d	д	sh	ш
e	е	shch	щ
ė	э	t	т
f	ф	ts	ц
g	г	u	у
i	и	v	в
ĭ	й	y	ы
k	к	ya	я
kh	х	yu	ю
l	л	z	з
m	м	zh	ж
n	н	"	ъ
o	о	'	ь

Several exceptions apply for Ukrainian, Belorussian and Serbian :

	Translit.	Cyrill.		Translit.	Cyrill.
Ukr.	y	и	Serbian :	ć	ћ
	i	і		dj	ђ
	ï	ї		dž	џ
Beloruss.				h	х
	ŭ	ў		j	j
				lj	љ
				nj	њ

Authors' names are presented in spelling used in the original paper. If this spelling does not correspond to the original spelling used by the author (*e.g.* Russiar papers of English authors), one spelling is referred to the other in the Authors' Index.

Printers' errors in the original papers are marked by underlining the respective words (letters).

21505 - **AARONSON, S., PATNI, N.J.** : The secretion and cell content of phosphatases by *Ochromonas danica* during the population cycle and in resting suspensions under some conditions of chemical and physical stress. - In: Les Cycles Cellulaires et leur Blocage chez Plusieurs Protistes. Coll. int. CNRS. Vol. 240. Pp. 137 - 144. Édit. CNRS, Paris 1975. [Chl.]

21506 - **ABILOV, Z.K., ABUTALYBOV, M.G., GASANOV, R.A.** : "Krasnoe padenie" kvantovogo vykhoda fluorestsentsii v protsesse dlitel'nogo zeleneniya ètiolirovannykh prorostkov. ["Red drop" in the quantum yield of fluorescence during prolonged greening of etiolated seedlings.] - Izv. Akad.Nauk azerb. SSR, Ser. biol. Nauk *1975* (3) : 3 - 7, 1975. [In R, ab: Azerb.]

21507 - **ACKER, S., DURANTON, J.** : Indépendance des formes spectrales de chlorophylle *a* et des holochromes chlorophylliens. - Biochim. biophys. Acta *387*: 279 - 287, 1975.

21508 - **ACKERSON, R.C., YOUNGNER, V.B.** : Responses of bermudagrass to salinity. - Agron. J. *67*: 678 - 681, 1975. [Ps.]

21509 - **ADABRA-MICHANOL, Y., MONÉGER, R., FRANCKE, B.** : Effets de la kinetine, à la lumière et à l'obscurité, sur les caroténoïdes et les chlorophylles des frondes de la *Spirodela polyrrhiza* (L.) SCHLEIDEN. - Physiol. vég. *13*: 619 - 635, 1975.

21510 - **ADAMS, M.S., ANDERSON, R., KOWAL, R.R.** : Temperature dependence of net photosynthesis in *Trientalis borealis* RAF. in Wisconsin. - Oecologia *18*: 199 - 207, 1975.

21511 - **ADEDIPE, N.O.** : Aspects of ^{14}C-sucrose translocation profiles in *Hibiscus esculentus* L. (okra). - In: **MARCELLE, R.** (ed.): Environmental and Biological Control of Photosynthesis. Pp.237 - 249. Dr.W.Junk b.v.Publ., The Hague 1975.

21512 - **ADEDIPE, N.O.** : Time-course of ^{14}C-sucrose translocation by successive leaves of *Hibiscus esculentus* L. (okra). - Biochem. Physiol. Pflanzen *167*: 419 - 426, 1975.

21513 - **ADEDIPE, N.O., ORMROD, D.P.** : Effects of light intensity on growth, and chlorophyll, carbohydrate and phosphorus contents of cowpea (*Vigna unguiculata* L.). - Biochem. Physiol. Pflanzen *167*: 301 - 309, 1975.

21514 - **ADOLFSEN, R., McCLUNG, J.A., MOUDRIANAKIS, E.N.** : Electrophoretic microheterogeneity and subunit composition of the 13S coupling factors of oxidative and photosynthetic phosphorylation. - Biochemistry *14*: 1727 - 1735, 1975.

21515 - **ADOLFSEN, R., MOUDRIANAKIS, E.N.** : Molecular microheterogeneity and subunit composition of the 13S coupling factors of oxidative and photosynthetic phosphorylation. - Fed. Proc. *33*: 1330, 1974.

21516 - **AIKAZYAN, V.Ts., NALBANDYAN, R.M.** : Plantacyanin from spinach. - FEBS Lett. *55*: 272 - 274, 1975.

21517 - **AIKIN, W.J., HANAN, J.J.** : Photosynthesis in the rose; effect of light intensity, water potential and leaf age. - J. amer. Soc. hort. Sci. *100*: 551 - 553, 1975.

21518 - **AIMI, R., SHIBASAKI, S.** : Diurnal change in bioelectric potential of *Phaseolus* plant in relation to the leaf movement and light condition. - Plant Cell Physiol. *16*: 1157 - 1162, 1975.

21519 - **AITKEN, A.** : Prokaryote-eukaryote relationships and the amino acid sequence of plastocyanin from *Anabaena variabilis*. - Biochem. J. *149*: 675 - 683, 1975.

21520 - **AKHMEDOV, Yu. K.** : [Effect of a group of cytokinin analogs on the level of chlorophylls and carotenoids.] - Uch. Zapiski, Minist. vyssh. sred. spets. Obrazovaniya azerb. SSR, Ser. biol. Nauk *1975* (2) : 66 - 70, 1975. [In Azerb., ab : R.]

21521 - **AKINOLA, J.O., WHITEMAN, P.C.** : Agronomic studies on pigeon pea (*Cajanus cajan* (L.) MILLSP.). II. Responses to sowing density. - Aust. J. agr. Res. *26*: 57 - 66, 1975. [Growth analysis.]

21522 - **AKINOLA, J.O., WHITEMAN, P.C.** : Agronomic studies on pigeon pea (*Cajanus cajan* (L.) MILLSP.). III. Responses to defoliation. - Aust. J. agr. Res. *26*: 67 - 79, 1975. [Growth analysis.]

21523 - **AKIYAMA, T., SAITO, Y., TAKEDA, T.** : [Dry matter production in corn plant. IV. Optical environment and light extinction coefficient of corn population.] - Proc. Crop Sci. Soc. Jap. *44* : 14 - 21, 1975. [In Jap., ab : E.]

21524 - **AKIYAMA, T., TAKEDA, T.** : [Studies on dry matter production in corn plant. III. Changes and interactions of plant factors constituting canopy photosynthesis according to the development of growth stages.] - Proc. Crop Sci. Soc. Jap. *44* : 7 - 13, 1975. [Growth analysis; in Jap., ab : E.]

21525 - **AKIYAMA, T., TAKEDA, T.** : [Studies on dry matter production in corn plant. V. Relation between the rate of leaf photosynthesis and dry matter production.] - Proc. Crop Sci. Soc. Jap. *44* : 269 - 274, 1975. [In Jap., ab :E.]

21526 - **AKIYAMA, T., TAKEDA, T.** : [Studies on dry matter production in corn plant. VI. Estimation of canopy photosynthesis by mathematical model.] - Proc. Crop Sci. Soc. Jap. *44* : 275 - 280, 1975. [In Jap., ab : E.]

21527 - **AKOBUNDU, I.O., DUKE, W.B., SWEET, R.D., MINOTTI, P.L.** : Basis for synergism of atrazine and alachlor combinations on Japanese millet. - Weed Sci. *23* : 43 - 48, 1975. [Ps.]

21528 - **AKOYUNOGLOU, G., MICHELINAKI-MANETA, M.** : Development of photosynthetic activity in flashed bean leaves. - In : **AVRON, M.**(ed.) : Proceedings of the Third International Congress on Photosynthesis. Vol. III. Pp. 1885 - 1896. Elsevier, Amsterdam - Oxford - New York 1975.

21529 - **AKSENOVICH, A.V.** : Svyaz' mezhdu aktivnost'yu karboangidrazy, intensivnost'yu fotosinteza i vodnym rezhimom yachmenya i pshenitsy. [Relation between carbonic anhydrase activity, photosynthetic rate and water relations in barley and wheat.] - Izv. sib. Otd. Akad. Nauk SSSR, Ser. biol. Nauk *1975* (3) : 106 - 110, 1975. [In R, ab : E.]

21530 - **AKULOVA, E.A., MURZAEVA, S.V., TAUKELEVA, Sh.N., RUZIEVA, R.Kh.** : Vliyanie IUK na fotofosforilirovanie izolirovannykh khloroplastov gorokha. [Effect of IAA on photophosphorylation in isolated pea chloroplasts.] - Biokhimiya *40* : 1205 - 1209, 1975. [In R, ab : E.]

21531 - **AKULOVA, E.A., MUZAFAROV, E.N., IVANOV, B.N., RUZIEVA, R.Kh., SHMELEVA, V.L.** : Kvertsetin-glikozid-kumarat - prirodnyǐ razobshchitel' fotofosforilirovaniya. [Quercetin-glucosidyl-*p*-coumarate, natural uncoupler of photophosphorylation.] Bioorg. Khim. *1* : 677 - 683, 1975. [In R, ab : E.]

21532 - **AKULOVICH, N.K., RASKIN, V.I., PARSHIKOVA, T.A., ORLOVSKAYA, K.I.** : Posledovatel'nost' vozniknoveniya spektral'nykh form khlorofillida pri stupenchatom vosstanovlenii protokhlorofillida ėtiolirovannykh list'ev. [Succession of spectral chlorophyllide forms during stepwise protochlorophyllide reduction in etiolated leaves.] - In : **SHLYK, A.A.** (ed.) : Biosintez i Sostoyanie Khlorofillov v Rastenii. Pp. 3 - 18, 243. Nauka i Tekhnika, Minsk 1975. [In R, ab:E.]

21533 - **ALAM, M.I., ISRAELSTAM, G.F.** : Photosynthesis and respiration of plants showing anomalous growth response to cyanide. - Z. Pflanzenphysiol. *75* : 25 - 30, 1975.

21534 - **ALBERTE, R.S., FISCUS, E.L., NAYLOR, A.W.** : The effects of water stress on the development of the photosynthetic apparatus in greening leaves. - Plant Physiol. *55* : 317 - 321, 1975.

21535 - **ALBERTE, R.S., HESKETH, J.D., BAKER, D.N.** : Aspects of predicting gross photosynthesis (net photosynthesis plus light and dark respiration) for an energy-metabolic balance in the plant. - In : **GATES, D.M., SCHMERL, R.B.** (ed.) : Perspectives of Biophysical Ecology. Pp. 87 - 98. Springer-Verlag, New York - Heidelberg - Berlin 1975.

21536 - **ALBERTE, R.S., NAYLOR, A.W.** : The role of cytokinins in chloroplast lamellar development. - Plant Physiol. *55* : 1079 - 1081, 1975.

21537 - **ALBERTE, R.S., THORNBER, J.P.** : The plasticity of the major pigmnet-protein complex of photosynthetic membranes. - Plant Physiol. *56* (Suppl.) : 10, 1975.

21538 - **ALDERFER, R.G.** : Photosynthesis in developing plant canopies. - In : **GATES, D.M., SCHMERL, R.B.** (ed.) : Perspectives in Biophysical Ecology. Pp. 227 - 238. Springer-Verlag, Berlin - Heidelberg - New York 1975.

*21539 - **ALEKSEEV, V.A.** : Ob izmerenii luchistoĭ ėnergii v svyazi s voprosami produktivnosti. [Measurement of radiant energy in relation to productivity problems.] - In : Soobshcheniya po Anatomii i Fiziologii Drevesnykh Rasteniĭ. Pp. 55 - 57. Leningrad. lesotekh. Akad. Kirova, Leningrad 1967. [In R, ab : E.]

*21540 - **ALEKSEEV, V.A.** : Vertikal'noe raspredelenie i sravnitel'naya otsenka fotosinteziruyushcheĭ massy khvoi. [Vertical distribution and comparative estimation of photosynthesizing active mass of needles.] - In : Struktura i Produktivnost' Elovykh Lesov Yuzhnoĭ Taĭgi. Pp. 162 - 170. Nauka, Leningrad 1973. [In R.]

*21541 - **ALEKSEEV, V.A., EFREMOV, D.F., MOROZOV, V.L., STEPANOVA, K.D., SHCHERBOVA, M.A.**: Rezhim solnechnoĭ radiatsii v fitotsenozakh Kamchatki. [Regime of solar radiation in phytocoenoses of Kamchatka.] - In : Biologicheskie Resursy Sushi Severa Dal'nego Vostoka. Vol.1. Pp. 237 - 248. Nauka, Vladivostok 1971. [In R.]

*21542 - **ALESHIN, A.D., KUMAKOV, V.A.** : Fotosinteticheskaya deyatel'nost' pobegov kushcheniya yarovoĭ pshenitsy. [Photosynthetic activity of suckers in spring wheat.] - In : Voprosy Povysheniya Produktivnosti Zernovykh Kul'tur. Pp. 48 - - 50, 316. Akad. Nauk SSSR, Irkutsk 1974. [In R.]

21543 - **AL-HASAN, R.H., COUGHLAN, S.J., PANT, A., FOGG, G.E.** : Seasonal variations in phytoplankton and glycollate concentrations in the Menai Straits, Anglesey. - J. mar. biol. Assoc. U.K. *55* : 557 - 565, 1975. [Chl.]

21544 - **ALIEV, D.A., AZIZOV, I.V.** : Fotokhimicheskaya aktivnost' izolirovannykh khloroplastov iz raznykh assimiliruyushchikh organov ozimoĭ pshenitsy. [Photochemical activity of isolated chloroplasts from different assimilatory organs of winter wheat.] - Izv. Akad. Nauk azerb. SSR, Ser. biol. Nauk *1975* (5) : 33 - 37, 1975. [In R, ab : Azerb.]

21545 - **ALIEV, K.A., MUZAFAROVA, S., RADZHABOV, H., KHOLMATOVA, M., ULUGBEKOVA, G., NASYROV, Yu.S.** : Chloroplast messenger RNA from etiolated pea seedlings. - In : **NASYROV, Yu. S., ŠESTÁK, Z.** (ed.) : Genetic Aspects of Photosynthesis. Pp. 93 - 103. Dr. W. Junk b.v. Publ., The Hague 1975.

*21546 - **ALLAKHVERDOV, B.L., POPOV, V.I., TAGEEVA, S.V.** : Primenenie metoda zamorozhennykh skolov dlya izucheniya ul'trastrukturnoĭ organizatsii fotosinteticheskikh membran. [Use of the freeze-etching method for studying the ultrastructural organization of photosynthetic membranes.] - In: Ėlektronnaya Mikroskopiya v Botanicheskikh Issledovaniyakh. Pp. 125 - 128. Petrozavodsk 1974.[In R.]

21547 - **ALLEN, J.F.** : A two-step mechanism for the photosynthetic reduction of oxygen by ferredoxin. - Biochem. biophys. Res. Commun. *66* : 36 - 43, 1975.

21548 - **ALLEN, J.F.** : Oxygen reduction and optimum production of ATP in photosynthesis. - Nature *256* : 599 - 600, 1975.

21549 - **ALLEN, L.H. Jr.** : Shade-cloth microclimate of soybeans. - Agron. J. *67* : 175 - - 181, 1975. [Stomatal diffusion resistance.]

21550 - **ALLEN, L.H. Jr., BOOTE, K.J., HAMMOND, L.C.** : Peanut stomatal diffusion resistance affected by soil water and solar radiation. - Soil Crop Sci. Soc. Florida Proc. *35* : 42 - 46, 1975.

21551 - **ALLEN, M.J., DAHLHOFF, J.A.** : Electrochemical studies on plant photosystems. II. Effects of catalase and peroxidase on thylakoid behaviour. - Bioelectrochem. Bioenerg. *2* : 177 - 183, 1975.

21552 - **ALLESSIO, M.L., TIESZEN, L.L.** : Leaf age effect on translocation and distribution of ^{14}C-photoassimilate in *Dupontia* at Barrow, Alaska. - Arctic alpine Res. *7* : 3 - 12, 1975.

21553 - **ALLESSIO, M.L., TIESZEN, L.L.** : Patterns of carbon allocation in an arctic tundra grass, *Dupontia fischeri* (*Graminae*), at Barrow, Alaska. - Amer. J. Bot. *62* : 797 - 807, 1975. [Photosynthates.]

21554 - **ALLEWELDT, G., DÜRING, H., WAITZ, G.** : Untersuchungen zum Mechanismus der Zuckereinlagerung in die wachsenden Weinbeeren. - Angew. Bot. *49* : 65 - 73, 1975.

21555 - **ALLEWELDT, G., HOFÄCKER, W.** : Einfluss von Umweltfaktoren auf Austrieb, Blüte, Fruchtbarkeit und Triebwachstum bei der Rebe. - Vitis *14* : 103 - 115. 1975.[Ps.]

21556 - **ALLMARAS, R.R., NELSON, W.W., VORHEES, W.B.** : Soybean and corn rooting in southwestern Minnesota : II. Root distributions and related water inflow. - Soil Sci. Soc. Amer. Proc. *39* : 771 - 777, 1975. [Growth analysis.]

21557 - **ALOFE, C.O., SCHRADER, L.E.** : Photosynthate translocation in tillered *Zea mays* following $^{14}CO_2$ assimilation. - Can. J. Plant Sci. *55* : 407 - 414, 1975.

21558 - **ALSCHER, R., SAUER, K., BASSHAM, J.A.** : Translational events linked to protochlorophyllide holochrome regeneration. - Plant Physiol. *56* (Suppl.) : 59, 1975.

21559 - **ALSOP, W.R., MORELAND, D.E.** : Effects of herbicides on the light-activated, magnesium-dependent ATPase of isolated spinach (*Spinacia oleracea* L.) chloroplasts. - Pesticide Biochem. Physiol. *5* : 163 - 170, 1975.

21560 - **ALTMAN, A., WAREING, P.F.** : The effect of IAA on sugar accumulation and basipetal transport of ^{14}C-labelled assimilates in relation to root formation in *Phaseolus vulgaris* cuttings. - Physiol. Plant. *33* : 32 - 38, 1975.

*21561 - **AL'ZHANOVA, R.M., GOLOVATYĬ, V.G.** : Vliyanie usloviĭ osveshcheniya yarovoĭ pshenitsy na nekotorye pokazateli ėnergeticheskogo obmena. [Effect of irradiance of spring wheat on some peculiarities of energetic metabolism.] - In : Voprosy Povysheniya Produktivnosti Zernovykh Kul'tur. Pp. 179 - 182, 325. Akad. Nauk SSSR, Irkutsk 1974. [In R.]

21562 - **AMBLER, R.P., BARTSCH, R.G.** : Amino acid sequence similarity between cytochrome *f* from a blue-green bacterium and algal chloroplasts. - Nature *235* : 285 - 288, 1975.

21563 - **AMBROSAŬ, A.L., SHCHUTSKAYA, V.V., PROKHARCHYK, R.A.** : Mekhanizmy parushėnnya funktsyĭ fotasintėtychnaga aparatu u raslin bul'by, pashkodzhanykh virusami. [Mechanisms of functional disturbance of the photosynthetic apparatus in sugar beet plants damaged by viruses.] - Vestsi Akad. Navuk belarus. SSR, Ser. biyal. Navuk *1975* (5) : 63 - 68, 1975. [In Belorus.]

21564 - **AMBROSOV, A.L., SHCHUTSKAYA, O.V.** : Vozdeĭstvie S i M virusov na nekotorye fiziologicheskie protsessy u rasteniĭ kartofelya. [Effect of S and M viruses on some physiological processes in potato plants.] - Dokl. Akad. Nauk belorus. SSR *19* : 279 - 281, 1975. [Chl, Car; in R.]

21565 - **AMERKHANOVA, M.B., BOGUSPAEV, K.K., SLYUSARENKO, A.G.** : Vydelenie khloroplastov i khloroplastnoĭ DNK iz list'ev gorokha. [Isolation of chloroplasts and chloroplast DNA from pea leaves.] - Biokhimiya *40* : 1312 - 1314, 1975. [In R, ab : E.]

21566 - **AMESZ, J.** : IV. Photosynthesis. Biophysical aspects. - Progr. Bot. *37* : 107 - - 120, 1975.

21567 - **AMESZ, J., de GROOTH, B.G.** : Light-induced absorbance changes due to Photosystems 1 and 2 in spinach chloroplasts at -50 °C. - Biochim. biophys. Acta *376* : 298 - 307, 1975.

21568 - **AMESZ, J., PULLES, M.P.J., de GROOTH, B.G., KERKHOF, P.L.M.** : Photoreactions of chloroplasts at -20 to -50 °C. - In : **AVRON, M.** (ed.) : Proceedings of the Third International Congress on Photosynthesis. Vol. I. Pp. 307 - 314. Elsevier, Amsterdam - Oxford - New York 1975.

21569 - **AMIRDZHANOV, A.G.** : Transpiratsionnyĭ raskhod vinogradnika v svyazi s ėnergeticheskim balansom rasteniĭ. [Transpiration expenditure of vineyard and energy balance of the plants.] - Fiziol. Rast. *22* : 98/ - 993, 1975. [In R, ab : E.]

21570 - **ANANYAN, A.A., EGIAZARYAN, A.G., SARKISYAN, E.M.** : Fiziologo-biokhimicheskaya kharakteristika sortov tomatov razlichnoĭ skorospelosti. [Physiological and biochemical characteristics of tomato cultivars ripening at different times.] - Biol. Zh. Armenii *28* (6) : 66 - 69, 1975. [Ps.]

21571 - **ANANYAN, A.A., TAROSOVA, E.O., EGIAZARYAN, A.G., AVETISYAN, S.A.** : Dinamika nakopleniya vitamina C i karotina v list'yakh i plodakh tomatov. [Dynamics of vitamin C and carotene accumulation in tomato leaves and fruits.] - Biol. Zh. Armenii *28* (3) : 93 - 96, 1975. [In R, ab : Arm.]

21572 - **ANDERSEN, W., WHITING, M., JOTKUR, S.** : Interaction between N_2-fixation, photosynthesis in blue-green algae. - Plant Physiol. *56* (Suppl.) : 34, 1975.

21573 - **ANDERSEN, W.R., GIBBS, M.** : Inhibition of CO_2 fixation in intact spinach chloroplasts by 3-phosphoglyceric acid. - Biochem. biophys. Res. Commun. *62* : 953 - 956, 1975.

21574 - **ANDERSON, J.M.** : Possible location of chlorophyll within chloroplast membranes.-Nature *253* : 536 - 537, 1975.

21575 - **ANDERSON, J.M.** : The molecular organization of chloroplast thylakoids. - Biochim. biophys. Acta *416* : 191 - 235, 1975.

21576 - **ANDERSON, L.E.** : Light modulation of the activity of carbon metabolism enzymes. - In : **AVRON, M.** (ed.) : Proceedings of the Third International Congress on Photosynthesis. Vol. II. Pp. 1393 - 1405. Elsevier, Amsterdam - Oxford - - New York 1975.

21577 - **ANDERSON, L.E.** : Ribulose-1,5-diphosphate carboxylase from *Rhodospirillum rubrum*. - In : **COLOWICK, S.P., KAPLAN, N.O.** (ed.) : Methods in Enzymology. Vol. 42. Pp. 457 - 461. Acad. Press, New York - San Francisco - London 1975.

21578 - **ANDERSON, L.E., AVRON, M.** : Membrane-bound vic-dithiol is involved in light--modulation of enzyme activity in chloroplasts. - Plant Physiol. *56* (Suppl.) : 82, 1975.

21579 - **ANDERSON, L.E., HEINRIKSON, R.L., NOYES, C.** : Chloroplast and cytoplasmic enzymes. Subunit structure of pea leaf aldolases. - Arch. Biochem. Biophys. *169* : 262 - 268, 1975.

21580 - **ANDERSON, L.E., SCHELLENTRAGER, R.C.** : Transfer of CO_2 from malate, 6-phosphogluconate or urea to ribulosediphosphate carboxylase in bound enzyme systems. - In : **AVRON, M.** (ed.) : Proceedings of the Third International Congress on Photosynthesis. Vol. II. Pp. 1449 - 1453. Elsevier, Amsterdam - Oxford - New York 1975.

21581 - **ANDERSON, M.C.** : Solar radiation and carbon dioxide in plant communities - conclusions. - In : **COOPER, J.P.** (ed.) : Photosynthesis and Productivity in Different Environments. Pp. 345 - 354. Cambridge Univ. Press, Cambridge - London - New York - Melbourne 1975.

21582 - **ANDERSON, O.R.** : The ultrastructure and cytochemistry of resting cell formation in *Amphora coffaeformis* (*Bacillariophyceae*). - J.Phycol. *11* : 272 - 281, 1975.[Chl.]

21583 - **ANDERSON, R.E., ANGER, G., PETERSSON, L., EHRENBERG, A., CAMMACK, R., HALL, D.O., MULLINGER, R., RAO, K.K.** : The iron electron-nuclear double resonance (ENDOR) of 4-Fe clusters in iron-sulfur proteins from *Chromatium* and *Clostridium pasteurianum*. - Biochim. biophys. Acta *376* : 63 - 71, 1975.

21584 - **ANDERSON, R.E., DUNHAM, W.R., SANDS, R.H., BEARDEN, A.J., CRESPI, H.L.** : On the nature of the iron sulfur cluster in a deuterated algal ferredoxin. - Biochim. biophys. Acta *408* : 306 - 318, 1975.

21585 - **ANDERSON, R.J., ROSS, R.T.** : Thermodynamics and kinetics of model photosynthetic reaction centers. - Biophys. J. *15* (2, Part 2) : 69a, 1975.

21586 - **ANDRE, C., VERCRUYSSE, A.** : Simultaneous determination of α- and β-carotene. Acta bot. neerl. *24* : 225 - 228, 1975.

21587 - **ANDREEVA, T.F., AVDEEVA, T.A., STEPANENKO, S.Yu.** : Vliyanie azotnogo pitaniya na aktivnost' glikolatoksidazy u rasteniĭ bobov i kukuruzy. [Effect of nitrogen nutrition on the activity of glycolate oxidase in broad-bean and maize plants.] - Fiziol. Rast. *22* : 553 - 557, 1975. [In R, ab : E.]

21588 - **ANDREO, C.S., VALLEJOS, R.H.** : Synergistic uncoupling of spinach chloroplasts by combinations of photophosphorylation inhibitors and carbonyl cyanide *p*-trifluoromethoxyphenylhydrazone. - Arch. Biochem. Biophys. *168* : 677 - 684, 1975.

21589 - **ANDREW, P.W., DELANEY, M.E., ROGERS, L.J., SMITH, A.J.** : Isolation and properties of a ferredoxin from *Anabaena flos-aquae*. - Phytochemistry *14* : 931 - 935, 1975.

21590 - **ANDREWS, A.K., SVEC, L.V.** : Photosynthetic activity of soybean pods at different growth stages compared to leaves. - Can. J. Plant Sci. *55* : 501 - 505, 1975.

21591 - **ANDREWS, P.T., JOHNSON, C.E., WALLBANK, B., CAMMACK, R., HALL, D.O., RAO, K. K.** : X-ray photoelectron spectra of iron-sulphur proteins. - Biochem. J. *149* : 471 - 474, 1975. [Ferredoxin.]

21592 - **ANDREWS, T.J., BADGER, M.R., LORIMER, G.H.** : Factors affecting interconversion between kinetic forms of ribulose diphosphate carboxylase-oxygenase from spinach. - Arch. Biochem. Biophys. *171* : 93 - 103, 1975.

21593 - **ANPILOGOVA, N.N.** : Nekotorye storony obmena veshchestv u rasteniĭ ozimykh pshenits i ottok assimilyatov v korni na rannikh fazakh razvitiya. [Some aspects of metabolism in winter wheat and efflux of photosynthates to roots during early stages of development.] - Byull. vses. Inst. Rastenievod. N. I. Vavilova *56* : 5 - 9, 1975. [In R.]

21594 - **ANSTIS, P.J.P., FRIEND, J., GARDNER, D.C.J.** : The role of xanthoxin in the inhibition of pea seedling growth by red light. - Phytochemistry *14* : 31 - 35, 1975. [Car.]

21595 - **ANTIA, N.J., BISALPUTRA, T., CHENG, J.Y., KALLEY, J.P.** : Pigment and cytological evidence for reclassification of *Nannochloris oculata* and *Monallantus salina* in the *Eustigmatophyceae*. - J. Phycol. *11* : 339 - 343, 1975.

21596 - **ANWAR, S.Y., REDDY, G.M.** : Effect of dimethyl sulphoxide in combination treatments with diethyl sulphate in the induction of chlorophyll mutations in *Oryza sativa* L. - Indian J. exp. Biol. *13* : 187 - 188, 1975.

21597 - **AOSHIMA, R., IRIYAMA, K., ASAI, H.** : Fluorescence properties of chlorophyll *a* and *b* monomolecular films at the air-water interface. - Biochim. biophys. Acta *406* : 362 - 369, 1975.

21598 - **APEL, K.** : Fractionation and characterization of the chloroplast membranes of *Acetabularia mediterranea*. - Protoplasma *83* : 168 - 169, 1975.

21599 - **APEL, K.** : Fractionation of the chloroplast membrane of *Acetabularia mediterranea*. - Plant Physiol. *56* (Suppl.) : 32, 1975.

21600 - **APEL, K., BOGORAD, L., WOODCOCK, C.L.F.** : Chloroplast membranes of the green alga *Acetabularia mediterranea*. I. Isolation of the Photosystem II. - Biochim. biophys. Acta *387* : 568 - 579, 1975.

21601 - **APEL, K., MILLER, K.R., BOGORAD, L., MILLER, G.** : Topography of the chloroplast membrane of *Acetabularia mediterranea*. - Plant Physiol. *56* (Suppl.) : 32, 1975.

21602 - **APEL, P., JANK, H.-W., LEHMANN, Ch.-O.** : Beschreibung des Wachstums von Weizenkaryopsen mit Hilfe einer Wachstumsfunktion. - Arch. Züchtungsforsch. *5* : 181 - 191, 1975.

21603 - **APPA RAO, S., JANA, M.K.** : Characteristics and inheritance of chlorophyll mutations in *Phaseolus mungo*. - Biol.Plant. *17* : 88 - 94, 1975.

*21604 - **ARGYROUDI-AKOYUNOGLOU, J., FELEKI, Z., AKOYUNOGLOU, G.** : [The effect of temperature on the enzymatic activity of RuDP-carboxylase.] - Chim. Chron. *2* : 280 - 284, 1970. [In Greek, ab : E.]

*21605 - **ARGYROUDI-AKOYUNOGLOU, J.H., AKOYUNOGLOU, G.** : Reconstitution of grana in high salt media. - In : **AVRON, M.** (ed.) : Proceedings of the Third International Congress on Photosynthesis. Vol. III. Pp. 2039 - 2048. Elsevier, Amsterdam - Oxford - New York 1975.

21606 - **ARGYROUDI-AKOYUNOGLOU, J.H., AKOYUNOGLOU, G.** : [Analysis of chlorophyll and other porphyrin "hydrolysates" in the amino acid analyzer.] - Chim. Chron. *2* : 276 - 279, 1970. [In Greek, ab : E.]

21607 - **ARMOND, P.A., ARNTZEN, C.J.** : Regulation of excitation energy distribution between Photosystems I and II in developing chloroplasts. - Biophys. J. *15* (2, Part 2) : 224a, 1975.

21608 - **ARMOND, P.A., ARNTZEN, C.J., DAVIS, D.J., GROSS, E.L.** : Correlations between chloroplast grana formation and onset of regulation of excitation energy distribution between photosystems I and II. - Plant Physiol. *56* (Suppl.) : 9, 1975.

21609 - **ARMSTRONG, W., WRIGHT, E.J.** : Radial oxygen loss from roots: The theoretical basis for the manipulation of flux data obtained by the cylindrical platinum electrode technique. - Physiol. Plant. *35* : 21 - 26, 1975. [Ps.]

*21610 - **ARNESON, R.D.** : *Pseudotetracystis*, a new chlorosarcinacean alga. - J. Phycol. *9* : 10 - 14, 1973. [Chloroplast.]

21611 - **ARNON, D.I., CHAIN, R.K.** : Regulation of ferredoxin-catalyzed photosynthetic phosphorylations. - Proc. nat. Acad. Sci. U S A *72* : 4961 - 4965, 1975.

*21612 - **ARNTZEN, C.J., ARMOND, P.A., VERNOTTE, C., BRIANTAIS, J.M.** : Surface localization of the photosystem II reaction center in chloroplast membranes. - Fed. Proc. *33* : 1329, 1974.

21613 - **ARNTZEN, C.J., BRIANTAIS, J.-M.** : Chloroplast structure and function. - In : **GOVINDJEE** (ed.) : Bioenergetics of Photosynthesis. Pp. 51 - 113. Academic Press, New York - San Francisco - London 1975.

21614 - **ARONOFF, S.** : The interaction of chlorophyll dimers with lutein. - Ann. New York Acad. Sci. *244* : 320 - 326, 1975.

21615 - **ARONOFF, S.** : The number of biologically possible porphyrin isomers. - Ann. New York Acad. Sci. *244* : 327 - 333, 1975.

21616 - **ARRON, G.P., BRADBEER, J.W.** : The use of two-dimensional polyacrylamide gel electrophoresis in the study of the biosynthesis of ribulosebisphosphate carboxylase. - In : **AVRON, M.** (ed.) : Proceedings of the Third International Congress on Photosynthesis. Vol. III. Pp. 2081 - 2088. Elsevier, Amsterdam - Oxford - New York 1975.

21617 - **ASADA, K., YOSHIKAWA, K., TAKAHASHI, M.-A., MAEDA, Y., ENMANJI, K.** : Superoxide dismutases from a blue-green alga *Plectonema boryanum*. - J. biol. Chem. *250* : 2801 - 2807, 1975.

21618 - **ASAMI, S., AKAZAWA, T.** : Biosynthetic mechanism of glycolate in *Chromatium* I. Glycolate pathway. - Plant Cell Physiol. *16* : 631 - 642, 1975.

21619 - **ASAMI, S., AKAZAWA, T.** : Biosynthetic mechanism of glycolate in *Chromatium*. II. Enzymic mechanism of glycolate formation by a transketolase system. - Plant Cell Physiol. *16* : 805 - 814, 1975.

21620 - **ASANOV, A.N.** : Kinetika vydeleniya kisloroda odnokletochnymi vodoroslyami. [Kinetics of oxygen evolution by unicellular algae.] - Probl. kosmich. Biol. *31* (Upravlenie fiziologicheskimi Protsessami i ikh Modelirovanie) : 293 - 310, 1975. [In R.]

21621 - **ASHENDEN, T.W., STEWART, W.S., WILLIAMS, W.** : Growth responses of sand dune populations of *Dactylis glomerata* L. to different levels of water stress. - J. Ecol. *63* : 97 - 107, 1975. [Growth analysis.]

21622 - **ASHWORTH, J.M.** : Adsorption and thin-layer chromatography. - In : **HALL, D.O., HAWKINS, S.E.** (ed.) : Laboratory Manual of Cell Biology. Pp. 188 - 189. English Universities Press, London 1975. [Chloroplast pigments.]

*21623 - **ASLING, Kh.S.** : Vliyanie na sl"nchevata energiya i vodata v"rkhu produktivnostta na rasteniyata. [Effect of solar energy and water on plant productivity.] - Fiziol. Rast. (Sofiya) *1* (2) : 103 - 114, 1974. [In Bulg., ab: E,R.]

*21624 - **ASLYNG, H.C.** : Evapotranspiration and plant production directly related to global radiation. - Nordic Hydrol. *5* : 247 - 256, 1974.

21625 - **ASSAF, R., LEVIN, I., BRAVDO, B.** : Effect of irrigation regimes on trunk and fruit growth rates, quality and yield of apple trees. - J. hort. Sci. *50* : 481 - 493, 1975.

21626 - **ATANASIU, L.** : Photosynthesis and respiration in sporophytes and gametophytes of two species of mosses. - Rev. roum. Biol. *20* : 44 - 48, 1975.

21627 - **ATKINS, C.A., CANVIN, D.T.** : Nitrate, nitrite and ammonia assimilation by leaves : Effect of inhibitors. - Planta *123* : 41 - 51, 1975. [Ps.]

21628 - **ATLAS, R.M., SCHOFIELD, E.** : Responses of the lichens *Peltigera aphthosa* and *Cetraria nivalis* and the alga *Nostoc commune* to sulfur dioxide, natural gas, and crude oil in Arctic Alaska. - Astarte *8* (2) : 53 - 60, 1975. [Ps.]

21629 - **AUCLAIR, D., GAUDILLÈRE, J.-P.** : Propriétés photosynthétiques de feuilles d'*Abies alba* (MILL.) et de *Picea abies* (L.). - Compt. rend. Acad. Sci. Paris, Sér. C *280* : 905 - 908, 1975.

21630 - **AUFHAMMER, W.**: Untersuchungen zur Beeinflussung des Assimilateaustausches zwischen Trieben verschiedener Ordnung durch physiologisch wirksame Substanzen bei der Sommergerste. - Z. Acker- Pflanzenbau *140* : 130 - 143, 1975.

21631 - **AUSLÄNDER, W., JUNGE, W.** : Neutral red, a rapid indicator for pH-changes in the inner phase of thylakoids. - FEBS Lett. *59* : 310 - 315, 1975.

21632 - **AUSLÄNDER, W., JUNGE, W.** : The electrical generator and the layer structure of the functional membrane in photosynthesis. - In : **AVRON, M.** (ed.) : Proceedings of the Third International Congress on Photosynthesis. Vol.I. Pp. 287 - 294. Elsevier, Amsterdam - Oxford - New York 1975.

21633 - **AUSSENAC, G., BOUCHON, J., DUCREY, M.** : La production photosynthétique forestière. - In : **COSTES, C.** (ed.) : Photosynthèse et Production Végétale. Pp. 217 - 242. Gauthier-Villars, Paris 1975.

21634 - **AUSTIN, R.B., EDRICH, J.** : Effects of ear removal on photosynthesis, carbohydrate accumulation and on the distribution of assimilated ^{14}C in wheat. - Ann. Bot. *39* : 141 - 152, 1975.

21635 - **AVARMAA, R.** : Lyuminestsentsiya bakteriorodopsina pri 77 i 4,2 K. [Bacteriorhodopsin luminescence at 77 and 4.2 K.] - Izv. Akad. Nauk est. SSR, Fiz., Mat. *24* : 351 - 353, 1975. [In R.]

21636 - **AVARMAA, R., REBANE, K.** : Fine-structured spectra of chlorophyll molecules in solid solutions. - Stud. biophys. *48* : 209 - 218, 1975.

21637 - **AVRATOVŠČUKOVÁ, N., FOUSOVÁ, S.** : Genetic variation of photosynthetic rate in leaf discs of *Zea mays* L. - In : **NASYROV, Yu.S., ŠESTÁK, Z.** (ed.) : Genetic Aspects of Photosynthesis. Pp. 343 - 347. Dr.W.Junk b.v. Publ., The Hague 1975.

21638 - **AVRON, M.** : The electron transport chain in chloroplasts. - In : **GOVINDJEE** (ed.) : Bioenergetics of Photosynthesis. Pp. 373 - 386. Academic Press, New York - San Francisco - London 1975.

21639 - **AVUNDZHYAN, Ė.S., KHACHATRYAN, B.S.** : Dinamika nakopleniya sakharov v list'yakh kartofelya v ontogeneze pri primenenii mikroėlementov. [Dynamics of sugar accumulation in potato leaves during ontogenesis using microelements.] - Biol.Zh. Armenii *28* (3) : 17 - 23, 1975. [In R.]

21640 - **AYERS, J.C. Jr., BARDEN, J.A.** : Net photosynthesis and dark respiration of apple leaves as affected by pesticides. - J. amer. Soc. hort. Sci. *100* : 24 - - 28, 1975.

21641 - **BABA, S., BABA, Y., KONISHI, T.** : A new combustion system for rapid preparation of biological samples for liquid scintillation counting. - Anal. Biochem. *66* : 243 - 252, 1975. [^{14}C analysis.]

21642 - **BABAYAN, R.S., GEVORKYAN, A.M., AĬRAPETYAN, R.B., SAAKYAN, M.A.** : Vliyanie streptomitsina na biosintez khlorofilla u prorostkov pshenitsy. [Effect of streptomycin on biosynthesis of chlorophyll in wheat seedlings.] - Fiziol. Rast. *22* : 484 - 489, 1975. [In R, ab : E.]

21643 - **BABCOCK, G.T., BLANKENSHIP, R.E., WARDEN, J.T., SAUER, K.** : EPR detection of the physiological donor to $P680^+$ in oxygen-evolving chloroplasts at room temperature. - Biophys. J. *15* (2, Part 2) : 247a, 1975.

21644 - **BABCOCK, G.T., SAUER, K.** : A rapid, light-induced transient in electron paramagnetic resonance signal II activated upon inhibition of photosynthetic oxygen evolution. - Biochim. biophys. Acta *376* : 315 - 328, 1975.

21645 - **BABCOCK, G.T., SAUER, K.** : The rapid component of electron paramagnetic resonance signal II: A candidate for the physiological donor to Photosystem II in spinach chloroplasts. - Biochim. biophys. Acta *376* : 329 - 344, 1975.

21646 - **BABCOCK, G.T., SAUER, K.** : Two electron donation sites for exogenous reductants in chloroplast photosystem II. - Biochim. biophys. Acta *396* : 48 - 62, 1975.

21647 - **BABUSHKIN, L.N.** : Transpiratsionnyĭ metod izmeneniya zharoustoĭchivosti rasteniĭ. [Transpiration method for determining heat resistance of plants.] - Fiziol. Rast. *22* : 647 - 653, 1975. [Ps; in R, ab : E.]

*21648 - **BABUSHKIN, L.N., KONSTANTINOV, P.T.** : Kontrol' ėffektivnosti ispol'zovaniya vody rasteniem putem sistematicheskikh izmeneniĭ intensivnosti fotosinteza i transpiratsii. [Monitoring the effectivity of water utilization by plant by means of systematic measurements of photosynthetic and transpiration rates.] - In : Tekhnologiya Vozdelivaniya Sel'skokhozyaĭstvennykh Kul'tur na Oroshaemykh Zemlyakh. Pp. 99 - 105. Shtiintsa, Kishinev 1975. - [In R.]

21649 - **BACCARINI MELANDRI, A., FABBRI, E., FIRSTATER, E., MELANDRI, B.A.** : Energy transduction in photosynthetic bacteria. VII. Inhibition of the coupling ATPase by *N*-ethylmaleimide related to the organized state of the membrane. - Biochim. biophys. Acta *376* : 72 - 81, 1975.

21650 - **BACCARINI MELANDRI, A., FABBRI, E., MELANDRI, B.A.** : Energy transduction in photosynthetic bacteria. VIII. Activation of the energy-transducing ATPase by inorganic phosphate. - Biochim. biophys. Acta *376* : 82 - 88, 1975.

*21651 - **BACHOFEN, R., LUTZ, H.U.** : Chromatophore membranes of the photosynthetic bacterium *Rhodospirillum rubrum* : Evidence for energy-rich bound phosphate. - In : **REID, E.** (ed.) : Methodological Developments in Biochemistry. Vol. 3. Pp. 233 - 240. Longman, London 1973.

21652 - **BADGER, M.R., ANDREWS, T.J., OSMOND, C.B.** : Detection in C_3, C_4 and CAM plant leaves of a low-$K_m(CO_2)$ form of RuDP carboxylase, having high RuDP oxygenase activity at physiological pH. - In : **AVRON, M.** (ed.) : Proceedings of the Third International Congress on Photosynthesis. Vol.II. Pp. 1421 - 1429. Elsevier, Amsterdam - Oxford - New York 1975.

21653 - **BAHL, J., MONÉGER, R.** : Composition lipidique des enveloppes d'étioplastes, d'étiochloroplastes et de chloroplastes de feuilles de Blé. - Compt. rend. Acad. Sci. Paris, Sér. D *281* : 1713 - 1716, 1975.

21654 - **BAHL, J., PHUNG-NHU-HUNG, S., LECHEVALLIER, D., MONÉGER, R.** : Effets de la lumière intermittente suivie ou non d'éclairement continu sur les lipides plastidiaux de feuilles étiolées de Blé. Comportement particulier du phosphatidylglycérol. - Physiol. vég. *13* : 115 - 124, 1975.

21655 - **BAIER, D., LATZKO, E.** : Properties and regulation of C-1-fructose-1,6-diphosphatase from spinach chloroplasts. - Biochim. biophys. Acta *396* : 141 - 147, 1975.

21656 - **BAIS, R.** : A rapid and sensitive radiometric assay for adenosine triphosphatase activity using Cerenkov radiation. - Anal. Biochem. *63* : 271 - 273, 1975.

*21657 - **BAKER, C.H., HORROCKS, R.D.** : A computer simulation of corn grain production. - Trans. ASAE *16* : 1027 - 1029, 1031, 1973.

21658 - **BAKER, C.H., HORROCKS, R.D., GOERING, C.E.** : Use of Gompertz function for predicting corn leaf area. - Trans. ASAE *18* : 323 - 326, 330, 1975.

21659 - **BAKER, E.F.I., KASSAM, A.H.** : Respiration by roselle (*Hibiscus sabdariffa*) in the Western State of Nigeria. - Trop. Agr. (Trinidad) *52* : 203 - 211, 1975. [Ps.]

*21660 - **BAKER, J.B., SWITZER, G.L., NELSON, L.E.** : Biomass production and nitrogen recovery after fertilization of young loblolly pines. - Soil Sci. Soc. Amer. Proc. *38* : 958 - 961, 1974.

21661 - **BAKER, N.R., HARDWICK, K.** : Biochemical and physiological aspects of leaf development in cocoa (*Theobroma cacao*). III. Changes in soluble sugar content and sucrose synthesizing capacity. - New Phytol. *75* : 519 - 524, 1975.

21662 - **BAKER, N.R., HARDWICK, K.** : A model for the development of photosynthetic units in cocoa leaves. - In : **AVRON, M.** (ed.) : Proceedings of the Third International Congress on Photosynthesis. Vol.III. Pp. 1897 - 1906. Elsevier, Amsterdam - Oxford - New York 1975.

21663 - **BAKER, N.R., HARDWICK, K., JONES, P.**: Biochemical and physiological aspects of leaf development in cocoa (*Theobroma cacao* L. II. Development of chloroplast ultrastructure and carotenoids. - New Phytol. *75* : 513 - 518, 1975.

21664 - **BAKER, N.R., STRASSER, R.J., BUTLER, W.L.** : Energy transfer between PSII and PS I in leaves grown in flashes. - Plant Physiol. *56* (Suppl.) : 46, 1975.

21665 - **BAKER, T.S., EISENBERG, D., EISERLING, F.A., WEISSMAN, L.** : The structure of form I crystals of D-ribulose-1,5-diphosphate carboxylase. - J. mol. Biol. *91* : 391 - 399, 1975.

*21666 - **BALANDREAU, J.P., DOMMERGUES, Y.R.** : Nycthermal variations in non-symbiotic nitrogen fixation in the rhizosphere. - Abstr. annu. Meet. amer. Soc. Microbiol. *73* : 7, 1973. [Ps.]

21667 - **BALDY, C.** : Importance de l'extinction de l'énergie lumineuse pour l'évolution des relations mutuelles de peuplements intraspécifiques de plantes fourragères. - Ann. Amélior. Plant. *25* : 25 - 27, 1975.

21668 - **BALTSCHEFFSKY, H.** : On the evolution of electron transport and phosphorylation in photosynthesis. - In : **AVRON, M.** (ed.) : Proceedings of the Third International Congress on Photosynthesis. Vol.III. Pp. 2061 - 2066. Elsevier, Amsterdam - Oxford - New York 1975.

21669 - **BALTSCHEFFSKY, M.** : The effect of dibromothymoquinone on light induced reactions in chromatophores from *Rhodospirillum rubrum*. - In : **AVRON, M.** (ed.): Proceedings of the Third International Congress on Photosynthesis. Vol.I. Pp. 799 - 806. Elsevier, Amsterdam - Oxford - New York 1975.

21670 - **BAMBERG, S.A., KLEINKOPF, G.E., WALLACE, A., VOLLMER, A.** : Comparative photosynthetic production of Mojave Desert shrubs. - Ecology *56* : 732 - 736, 1975.

21671 - **BAMBERGER, E.S., AVRON, M.** : Site of action of inhibitors of carbon dioxide assimilation by whole lettuce chloroplasts. - Plant Physiol. *56* : 481 - 485, 1975.

21672 - **BAMBERGER, E.S., EHRLICH, B.A., GIBBS, M.** : The glyceraldehyde 3-phosphate and glycerate 3-phosphate shuttle and carbon dioxide assimilation in intact spinach chloroplasts. - Plant Physiol. *55* : 1023 - 1030, 1975.

21673 - **BAMBERGER, E.S., EHRLICH, B.A., GIBBS, M.** : The d-glyceraldehyde-3-phosphate and 3-phosphoglycerate shuttle system and carbon dioxide assimilation in intact spinach chloroplasts. - In : **AVRON, M.** (ed.) : Proceedings of the Third International Congress on Photosynthesis. Vol. II. Pp. 1349 - 1362. Elsevier, Amsterdam - Oxford - New York 1975.

*21674 - **BANASIK, J.** : Ground flora production in a stand of *Tilio-Carpinetum*. - Bull. Acad. pol. Sci., Sér. Sci. biol. Cl.II. *21* : 593 - 599, 1973.

21675 - **BANERJI, D., KUMAR, N.** : Partial inhibition of the decay of Hill activity in isolated chloroplasts by kinetin. - Biochem. biophys. Res. Commun. *65* : 940 - - 944, 1975.

21676 - **BARABAL'CHUK, K.A.** : Prodolzhitel'nost' reaktsii fototaksisa khloroplastov kak pokazatel' svetochuvstvitel'nosti rasteniĭ. [Duration of phototactic reaction of chloroplasts as a characteristic of light sensitivity of plants.] - In : Obmen Veshchestv i Mineral'noe Pitanie Polivnykh Rasteniĭ. Pp. 46 - 56. Shtiintsa, Kishinev 1975. [In R, ab : E.]

21677 - **BARANINA, I.I.** : Sravnitel'noe izuchenie fotosinteza mutantov myagkoĭ ozimoĭ pshenitsy. [Comparative study of photosynthesis in soft winter wheat mutants.] - Izv. Akad. Nauk mold. SSR, Ser. biol. khim. Nauk *1975* (4) : 12 - 19, 89, 1975. [In R.]

21678 - **BARANOV, A.A., SAAKOV, V.S., CHUNAEV, A.S., BORSHCHEVSKAYA, T.N., KVITKO, K.V.** : Issledovanie reaktsiĭ khlorofilloobrazovaniya i svetozashchity u mutantov zelenykh vodorosleĭ metodom absorbtsionnoĭ spektrofotometrii. [Reactions of chlorophyll formation and light protection in mutants of green algae studied by absorption spectrophotometry.] - Fiziol. Rast. *22* : 702 - 711, 1975. [In R, ab : E.]

21679 - **BARBER, J., TELFER, A., MILLS, J., NICOLSON, J.**: Slow chlorophyll fluorescence changes in isolated intact chloroplasts: evidence for cation control. -

In : **AVRON, M.** (ed.) : Proceedings of the Third International Congress on Photosynthesis. Vol. I. Pp. 53 - 63. Elsevier, Amsterdam - Oxford - New York 1975.

21680 - **BARCKHAUS, R.H., WEINERT, H.** : Plastidenveränderungen in virus-infizierten Zellen in der Attraktionszone von *Sarracenia purpurea* L. - Protoplasma *84* : 101 - 108, 1975.

21681 - **BARICA, J.** : Collapses of algal blooms in prairie pothole lakes : their mechanism and ecological impact. - Ver. int. Verein. Limnol. *19* : 606 - 615, 1975. [Chl.]

21682 - **BARICA, J.** : Summerkill risk in prairie ponds and possibilities of its prediction. - J.Fish. Res. Board Can. *32* : 1283 - 1288, 1975.

21683 - **BARMORE, C.R.** : Effect of ethylene on chlorophyllase activity and chlorophyll content in calamondin rind tissue. - HortScience *10* : 595 - 596, 1975.

21684 - **BARNABAS, A.D., STEINKE, T.D.** : An autoradiographic study of the translocation of ^{14}C-labelled assimilates in *Eragrostis curvulla* (SCHRAD.) NEES at different stages of vegetative growth. - J. S. afr. Bot. *41* : 17 - 27, 1975.

21685 - **BAR-NUN, S., OHAD, I.** : Cytoplasmic and chloroplastic origin of chloroplast membrane proteins associated with PSII and PSI active centers in *Chlamydomonas reinhardi* Y1. - In: **AVRON, M.** (ed.) : Proceedings of the Third International Congress on Photosynthesis. Vol. III. Pp. 1627 - 1638. Elsevier, Amsterdam - Oxford - New York 1975.

21686 - **BAROOVA, S.R., HORVÁTH, I.** : Effect of the time of transplantation on dry--matter production and light-energy utilization in tomato. - Acta agron. Acad. Sci. hung. *24* : 99 - 104, 1975.

21687 - **BARR, R., CRANE, F.L.** : Chelaton inhibition of chloroplast photosystem II. - Plant Physiol. *56* (Suppl.) : 46, 1975.

21688 - **BARR, R., CRANE, F.L.** : The effect of chaotropic agents on photosynthetic reactions. - J. Bioenerg. *7* : 123 - 137, 1975.

21689 - **BARR, R., CRANE, F.L., GIAQUINTA, R.T.** : Dichlorophenylurea-insensitive reduction of silicomolybdic acid by chloroplast photosystem II. - Plant Physiol. *55* : 460 - 462, 1975.

21690 - **BARSKIĬ, E.L., BORISOV, A.Yu., IL'INA, M.D., SAMUILOV, V.D., FETISOVA, Z.G.** : Opredelenie kvantovogo vykhoda pervichnogo preobrazovaniya ėnergii pri fotosinteze. III. Ėksperimental'nye dannye dlya razlichnykh fotosinteziruyushchikh organizmov. [Determination of quantum yield of primary energy transformations in photosynthesis. III. Experimental data for different photosynthetic organisms.] - Mol. Biol. (Moskva) *9* : 275 - 282, 1975. [In R, ab : E.]

21691 - **BARSKY, E.L., BONCH-OSMOLOVSKAYA, E.A., OSTROUMOV, S.A., SAMUILOV, V.D., SKULACHEV, V.P.** : A study on the membrane potential and pH gradient in chromatophores and intact cells of photosynthetic bacteria. - Biochim. biophys. Acta *387* : 388 - 395, 1975.

21692 - **BARTA, A.L.** : Effect of nitrogen nutrition on distribution of photosynthetically incorporated $^{14}CO_2$ in *Lolium perenne*. - Can. J. Bot. *53* : 237 - 242, 1975.

21693 - **BARTOŠ, J., BERKOVÁ, E., ŠETLÍK, I.** : A versatile chamber for gas exchange measurements in suspensions of algae and chloroplasts. - Photosynthetica *9* : 395 - 406, 1975.

21694 - **BASIOUNY, F.M., BIGGS, R.H.** : Hill reaction and photosynthetic activity in relation to mineral element deficiencies. - Plant Physiol. *56* (Suppl.) : 4, 1975.

21695 - **BASNIZKI, J., EVENARI, M.** : The influence of a reflectant on leaf temperature and development of the globe artichoke (*Cynara scolymus* L.). - J. amer. Soc. hort. Sci. *100* : 109 - 112, 1975.

21696 - **BASSI, P.K., TREGUNNA, E.B.** : Carbon dioxide exchange during the photoperiodic control of flowering of *Xanthium pennsylvanicum* by red light. - Plant Physiol. *56* (Suppl.) : 74, 1975.

21697 - **BASZYŃSKI, T., PAŃCZYK, B., KRÓL, M., KRUPA, Z.** : The effect of nitrogen deficiency on some aspects of photosynthesis in maize leaves. - Z. Pflanzenphysiol. *74* : 200 - 207, 1975.

21698 - **BASZYŃSKI, T., TUKENDORF, A.** : Restoration of cyclic phosphorylation in heptane-extracted spinach chloroplasts by α-tocopherol. - FEBS Lett. *57* : 104 - -106, 1975.

21699 - **BATES, J.W., BROWN, D.H.** : The effect of seawater on the metabolism of some seashore and inland mosses. - Oecologia *21* : 335 - 344, 1975. [Chl, Ps.]

21700 - **BATIKYAN, G.G., GUKASYAN, L.A., SARKISYAN, S.A., ABRAMYAN, L.Kh.** : Ul'trastruktura khloroplastov list'ev pertsa s khlorofil'nymi izmeneniyami pri vozdeĭstvii nitrozometilmochevinoĭ. [Ultrastructure of chloroplasts of pepper leaves with chlorophyll changes induced by nitrosomethyl urea.] - Biol. Zh. Armenii *28* (5) : 14 - 16, 1975. [In R, ab : Arm.]

21701 - **BATLLE, A.M.Del.C., LLAMBIAS, E.B.C., WIDER de XIFRA, E., TIGIER, H.A.** : Porphyrin biosynthesis in the soybean callus tissue system - XV. The effect of growth conditions. - Int. J. Biochem. *6* : 591 - 606, 1975. [Chl.]

21702 - **BATT, T., WOOLHOUSE, H.W.** : Changing activities during senescence and sites of synthesis of photosynthetic enzymes in leaves of the Labiate, *Perilla frutescens* (L.) BRITT. - J. exp. Bot. *26* : 569 - 579, 1975.

21703 - **BATTERSBY, A.R., McDONALD, E.** : Biosynthesis of porphyrins, chlorins and corrins. - In : **SMITH, K.M.** (ed.) : Porphyrins and Metalloporphyrins. Pp. 61 - - 122. Elsevier, Amsterdam - Oxford - New York 1975. [Chl.]

*21704 - **BATZLI, G.O.** : Production, assimilation and accumulation of organic matter in ecosystems. - J. theor. Biol. *45* : 205 - 217, 1974.

21705 - **BAUER, H., LARCHER, W., WALKER, R.B.** : Influence of temperature stress on CO_2-gas exchange. - In : **COOPER, J.P.** (ed.) : Photosynthesis and Productivity in Different Environments. Pp. 557 - 586. Cambridge Univ. Press, Cambridge - London - New York - Melbourne 1975.

21706 - **BAUER, R., FRANCK, U.F.** : Light-induced variable photosystem II chlorophyll fluorescence as indicator for photosystem I activity. (Fluorescence behaviour of Hydrogenase containing green algae under aerobic and anaerobic conditions.)- In : **AVRON, M.** (ed.) : Proceedings of the Third International Congress on Photosynthesis. Vol. I. Pp. 171 - 182. Elsevier, Amsterdam - Oxford - New York 1975.

*21707 - **BAULD, J., BROCK, T.D.** : Algal excretion and bacterial assimilation in hot spring algal mats. - J. Phycol. *10* : 101 - 106, 1974. [Ps, Chl.]

21708 - **BAZZAZ, F.A., CARLSON, R.W., ROLFE, G.L.** : Inhibition of corn and sunflower photosynthesis by lead. - Physiol. Plant. *34* : 326 - 329, 1975.

21709 - **BAZZAZ, F.A., DUSEK, D., SEIGLER, D.S., HANEY, A.W.** : Photosynthesis and cannabinoid content of temperate and tropical populations of *Cannabis sativa*. - Biochem. System. Ecol. *3* : 15 - 18, 1975.

21710 - **BEALE, S.I., GOUGH, S.P., GRANICK, S.** : Biosynthesis of δ-aminolevulinic acid from the intact carbon skeleton of glutamic acid in greening barley. - Proc. nat. Acad. Sci. USA *72* : 2719 - 2723, 1975.

21711 - **BEARDALL, J., MORRIS, I.** : Effects of environmental factors on photosynthesis patterns in *Phaeodactylum tricornutum* (*Bacillariophyceae*). II. Effect of oxygen. - J. Phycol. *11* : 430 - 434, 1975.

*21712 - **BEARDEN, A.J., MALKIN, R.** : On the reversibility of the primary reaction of chloroplast photosystem I at low temperatures. - Fed. Proc. *33* : 1289, 1974.

21713 - **BEARDSELL, M.F., COHEN, D.** : Relationships between leaf water status, abscisic acid levels, and stomatal resistance in maize and sorghum. - Plant Physiol. *56* : 207 - 212, 1975.

21714 - **BECHER, B.M., CASSIM, J.Y.** : Effects of light perturbation on the absorption spectra of purple membrane from *Halobacteria halobium*. - Biophys. J. *15* (2, Part 2) : 65a, 1975.

21715 - **BECHER, B.M., CASSIM, J.Y.** : Effects of light perturbation on the circular dichroism of purple membrane from *Halobacteria halobium*. - Biophys. J. *15* (2, Part 2) : 66a, 1975.

21716 - **BECHER, B.M., CASSIM, J.Y.** : Improved isolation procedures for the purple membrane of *Halobacterium halobium*. - Prep. Biochem. *5* : 161 - 178, 1975.

21717 - **BECK, E., EICKENBUSCH, J.D., STEIGER, H.M.** : Characterisation of glycolate producing reactions in isolated spinach chloroplasts. - In: **AVRON, M.** (ed.) : Proceedings of the Third International Congress on Photosynthesis. Vol. II. Pp. 1329 - 1334. Elsevier, Amsterdam - Oxford - New York 1975.

*21718 - **BECKER, J.F., BRETON, J., GEACINTOV, N.E.** : The degree of fluorescence polarization of spinach chloroplasts and *Chlorella* oriented in a magnetic field. - Fed. Proc. *33* : 1462, 1974.

21719 - **BECKER, K., LICHTENTHALER, H.K.** : Formation of individual carotenoid components in *Raphanus* etioplasts in continuous far-red and white light. - Z. Pflanzenphysiol. *75* : 303 - 306, 1975.

21720 - **BEDDARD, G.S., PORTER, G., TREDWELL, C.J., BARBER, J.** : Fluorescence lifetimes in the photosynthetic unit. - Nature *258* : 166 - 168, 1975.

21721 - **BEDDARD, G.S., PORTER, G., WEESE, G.M.** : Model systems for photosynthesis. V. Electron transfer between chlorophyll and quinones in a lecithin matrix. - Proc. roy. Soc. London A *342*:317 - 325, 1975.

21722 - **BEER, S., SHOMER-ILAN, A., WAISEL, Y.** : Salt-stimulated phosphoenolpyruvate carboxylase in *Cakile maritima*. - Physiol. Plant. *34* : 293 - 295, 1975.

21723 - **BEHNKE, H.-D.** : Mikromorphologische Merkmale der Siebelement-Plastiden als ein Beitrag der Transmissionselektronmikroskopie zur Systematik der Samenpflanzen. - Ber. deut. bot. Ges. *88*:361 - 368, 1975.

21724 - **BEISENHERZ, W.W., KOTH, P.** : Der Einfluß von Chloramphenicol und Cycloheximid auf die Synthese von Ribulose-1,5-diphosphat-Carboxylase, NADP-abhängiger Glycerinaldehyd-3-phosphat-Dehydrogenase und Chlorophyll während der Leukoplasten-Chloroplasten-Transformation in Gewebekulturen von *Nicotiana tabacum*. - Z. Pflanzenphysiol. *75* : 201 - 210, 1975.

21725 - **BEKASOVA, O.D., EVSTIGNEEV, V.B.** : Spektral'nye i fotokhimicheskie svoĭstva allofikotsianina. [Spectral and photochemical properties of allophycocyanin.] - Izv. Akad. Nauk SSSR, Ser. biol. *1975* : 393 - 399, 1975. [In R, ab : E.]

21726 - **BEKASOVA, O.D., TSVETKOVA, A.M., EVSTIGNEEV, V.B.** : Ob izmeneniyakh pigmentnoĭ sistemy okeanicheskogo fitoplanktona. [Changes of the pigment system of ocean phytoplankton.] - In : Gidrofizicheskie i Opticheskie Issledovaniya v Indiĭskom Okeane. Pp. 235 - 242. Nauka, Moskva 1975. [In R.]

21727 - **BEKINA, R.M., LEBEDEVA, A.F., RUBIN, B.A.** : O svyazi fotookislitel'nykh prevrashcheniĭ organicheskikh kislot v khloroplastakh s fotosistemoĭ II. [Connection between photooxidative transformations of organic acids in chloroplasts and the photosystem II.] - Dokl. Akad. Nauk SSSR *224* : 960 - 963, 1975. [In R.]

21728 - **BEKINA, R.M., SHUVALOV, V.A., LYSENKO, G.G., MOSHENTSEVA, V.N., LEBEDEVA, A.F.** : Issledovanie mekhanizma fotookislitel'nykh prevrashcheniĭ organicheskikh kislot v khloroplastakh. [Mechanism of photooxidative transformations of organic acids in chloroplasts.] - Fiziol. Rast. *22* : 680 - 687, 1975. [In R, ab : E.]

21729 - **BELETSKIĬ, Yu.D., RAZORITELEVA, E.K.** : Ustoĭchivost' k N-nitrozo-N-metilmochevine tsitoplazmaticheskikh khlorofil'nykh mutantov podsolnechnika, indutsirovannykh ėtim agentom. [N-nitroso-N-methyl urea resistance of sunflower cytoplasmic chlorophyll mutants, induced by the agent.] - Genetika *11* (1) : 23 - 27, 1975. [In R, ab : E.]

21730 - **BELIKOV, P.S., MELEKHOV, E.I.** : Tipy krivykh vremennogo khoda fotosinteza i ikh zavisimost' ot sily i dlitel'nosti progreva lista. [Time curves of photosynthesis and their dependence on the intensity and duration of heating of the leaf.] - Fiziol. Rast. *22* : 466 - 470, 1975. [In R, ab : E.]

21731 - **BELYAEV, V.I., LENIN, A.I., PETIPA, T.S.** : Mathematical model of a pelagic ecosystem. - Mar. Biol. *31* : 1 - 6, 1975.

21732 - **BELYAEVA, M.G., LUKPANOV, Zh.L.** : O fiziologicheskikh osobennostyakh deĭstviya pestitsidov na ogurtsy. [Physiological peculiarities of pesticide effect on cucumbers.] - In : Trudy kazakh. nauch.-issled. Inst. Zashchity Rast. *13* (Zashchita Ovoshchnykh, Tekhnicheskikh i Kormovykh Kul'tur ot Vrediteleĭ, Bolezneĭ i Sornyakov) : 116 - 121, 1975. [Chl; in R.]

21733 - **BELYAEVA, O.B., KARNEEVA, N.V., STADNICHUK, I.N., LITVIN, F.F.** : Dinamika biosinteza nativnykh form khlorofilla ot nachal'nykh stadiĭ do zaversheniya protsessa zeleneniya ėtiolirovannykh list'ev. [Dynamics of biosynthesis of native chlorophyll forms from initial stages up to the completion of the greening process in etiolated leaves.] - Biokhimiya *40* : 951 - 961, 1975. [In R, ab : E.]

21734 - **BELYANIN, V.N., SPIROV, V.V., FURYAEV, E.A.** : Spektrometriya otdel'nykh kletok khlorelly. [Spectrophotometry of individual *Chlorella* cells.] - Biofizika *20* : 848 - 852, 1975. [In R, ab : E.]

21735 - **BELYANIN, V.N., VOLKOVA, E.K., TRUBACHEV, I.N., PAN'KOVA, I.M.** : Produktivnost' i biokhimicheskiĭ sostav khlorelly na razlichnykh urovnyakh osveshchennosti i azotnogo limitirovaniya. [Effect of illumination and nitrogen content on productivity and biochemical composition of *Chlorella*.] - Fiziol. Rast. *22* : 55 - 62, 1975. [In R, ab : E.]

21736 - **BEN-AMOTZ, A.** : Adaptation of the unicellular alga *Dunaliella parva* to a saline environment. - J. Phycol. *11* : 50 - 54, 1975. [Ps.]

21737 - **BEN-AMOTZ, A., ERBES, D.L., RIEDERER-HENDERSON, M.A., PEAVEY, D.G., GIBBS, M.**: H_2 metabolism in photosynthetic organisms. I. Dark H_2 evolution and uptake by algae and mosses. - Plant Physiol. *56* : 72 - 77, 1975.

21738 - **BEN-AMOTZ, A., GIBBS, M.** : H_2 evolution by preparations of *Chlamydomonas*, *Scenedesmus* and spinach. - Plant Physiol. *56* (Suppl.) : 9, 1975.

21739 - **BENEDICT, C.R.** : The comparative photosynthetic carbon metabolism of C_4 monocots and a marine monocot. - Plant Physiol. *56* : 27, 1975.

21740 - **BENEDICT, C.R., KOHEL, R.J.** : Export of ^{14}C-assimilates in cotton leaves. - Crop Sci. *15* : 367 - 372, 1975.

21741 - **BENEY, G., NIGON, V.** : Action of cycloheximide on δ-aminolevulinic acid and chlorophyll production in *Euglena gracilis*. - In : **AVRON, M.** (ed.) : Proceedings of the Third International Congress on Photosynthesis. Vol. III. Pp. 1801 - 1808. Elsevier, Amsterdam - Oxford - New York 1975.

21742 - **BENFIELD, E.F., HENDRICKS, A.C.** : Eutrophic gradient in Smith Mountain Lake, Virginia. - Virginia J. Sci. *26* : 20 - 26, 1975. [Ps,Chl.]

21743 - **BENGIS, C., NELSON, N.** : Purification and properties of the photosystem I reaction center from chloroplasts. - J. biol. Chem. *250* : 2783 - 2788, 1975.

21744 - **BEN-HAYYIM, G., DRECHSLER, Z., GOFFER, J., NEUMANN, J.** : Diaminobenzidine an electron donor to photosystem 1 and to photosystem 2 in chloroplasts. - Europe. J. Biochem. *52* : 135 - 141, 1975.

21745 - **BEN-HAYYIM, G., DRECHSLER, Z., NEUMANN, J.** : Evidence for sideness of the thylakoid membrane : studying photophosphorylation by photosystem I. - In : **AVRON, M.** (ed.) : Proceedings of the Third International Congress on Photosynthesis. Vol. II. Pp. 873 - 881. Elsevier, Amsterdam - Oxford - New York 1975.

21746 - **BEN-HAYYIM, G., NEUMANN, J.** : On the mechanism of action of silicomolybdic acid in chloroplasts. - FEBS Lett. *56* : 240 - 243, 1975.

21747 - **BENNETT, J., RADCLIFFE, C.** : Plastid DNA replication and plastid division in the garden pea. - FEBS Lett. *56* : 222 - 225, 1975.

21748 - **BENNETT, J.H., HILL, A.C.** : Interactions of air pollutants with canopies of vegetation. - In : **MUDD, J.B., KOZLOWSKI, T.T.** (ed.) : Responses of Plants to Air Pollution. Pp. 273 - 306. Academic Press, New York - San Francisco - London 1975. [Ps.]

*21749 - **BENNOUN, P.** : Caractérisation rapide de mutants photosynthétiques par leur cinétique d'induction de fluorescence. - Compt.rend. Acad. Sci. Paris, Sér. D *279* : 1915 - 1917, 1974.

21750 - **BENNOUN, P., BOUGES-BOCQUET, B.** : Inhibition of oxygen evolution following illumination of *Chlorella* cells with far-red light. - Biochim. biophys. Acta *408* : 180 - 185, 1975.

21751• - **BENNOUN, P., JUPIN, H.** : The relationship between thylakoid stacking, state I and state II phenomena in whole cells and the cation effects in chloroplasts of *Chlamydomonas reinhardi*. - In : **AVRON, M.** (ed.) : Proceedings of the Third International Congress on Photosynthesis. Vol. I. Pp. 163 - 169. Elsevier, Amsterdam - Oxford - New York 1975.

21752 - **BENNUN, A.** : A model mechanism for coupled phosphorylation. - In : **AVRON, M.** (ed.) : Proceedings of the Third International Congress on Photosynthesis. Vol. II. Pp. 1107 - 1120. Elsevier, Amsterdam - Oxford - New York 1975.

21753 - **BENNUN, A.** : A model pathway for coupled phosphorylation. - Plant Physiol. *56* (Suppl.) : 47, 1975.

21754 - **BENNUN, A.** : Hypothesis on the role of liganded states of proteins in energy transducing systems. - Biosystems *7* : 230 - 244, 1975.

21755 - **BEN-SHAUL, Y., SCHÖNFELD, M., NEUMANN, J.** : Photosynthetic reactions in the marine alga *Codium vermilara*. II. Structural studies. - Plant Physiol. *55* : 899 - 905, 1975.

21756 - **BEN-SHAUL, Y., SCHÖNFELD, M., NEUMANN, J.** : Photosynthetic reactions and structural studies in the marine alga *Codium vermilara*. - In : **AVRON, M.** (ed.) : Proceedings of the Third International Congress on Photosynthesis. Vol. II. Pp. 1533 - 1540. Elsevier, Amsterdam - Oxford - New York 1975.

21757 - **BERDYKULOV, Kh.A., BAKHRAMDZHANOVA, N.** : Zavisimost' intensivnosti fotosinteza stsenedesmusa ot temperatury i kontsentratsii uglekisloty. [Dependence of photosynthetic rate of *Scenedesmus* on temperature and CO_2 concentration.] - In : Vodorosli i Griby Sredneĭ Azii. Vol. 2. Pp. 103 - 107. FAN, Tashkent 1975. [In R.]

*21758 - **BERG, A.** : A short introduction to quantitative and qualitative chlorophyll methodologies. - Rapp. høyfjellsøkol. Forskningsstas. Finse, Norge *1974* (1) : 1 - 19, 1974.

*21759 - **BERG, A.** : A description and critical evaluation of methodology used in pigment analyses on alpine plants from Hardangervidda, Norway. - Rapp. høyfjellsøkol. Forskningsstas. Finse, Norge *1974* (1) : 20 - 35, 1974.

21760 - **BERG, A.** : Pigment structure of vascular plants, mosses and lichens at Hardangervidda, Norway. - In : **WIELGOLASKI, F.E.** (ed.) : Fennoscandian Tundra Ecosystems. Part I. Plants and Microorganisms. Pp. 217 - 224. Springer-Verlag, Berlin - Heidelberg - New York 1975.

21761 - **BERG, A., KJELVIK, S., WIELGOLASKI, F.E.** : Measurement of leaf areas and leaf angles of plants at Hardangervidda, Norway. - In : **WIELGOLASKI, F.E.** (ed.) : Fennoscandian Tundra Ecosystems. Part I. Plants and Microorganisms. Pp. 103 - - 110. Springer-Verlag, Berlin - Heidelberg - New York 1975.

21762 - **BERG, A., KJELVIK, S., WIELGOLASKI, F.E.** : Distribution of ^{14}C photosynthates in Norwegian alpine plants. - In : **WIELGOLASKI, F.E.** (ed.) : Fennoscandian Tundra Ecosystems. Part I. Plants and Microorganisms. Pp. 208 - 215. Springer-Verlag, Berlin - Heidelberg - New York 1975.

21763 - **BERG, S.P., KROGMANN, D.W.** : Mechanism of KCN inhibition of photosystem I. - J. biol. Chem. *250* : 8957 - 8962, 1975.

21764 - **BERG, S.P., KROGMANN, D.W.** : Evidence for copper removal as the mechanism by which KCN inhibits photosystem I. - Plant Physiol. *56* (Suppl.) : 47, 1975.

21765 - **BERGMANN, H., ROTH, D., JÄGER, E.** : Untersuchungen zur Eignung von Pflanzenwasserdefizitgrössen für die Beregnungssteuerung. - Arch. Acker- Pflanzenbau Bodenk. *19* : 107 - 120, 1975. [Ps.]

21766 - **BERING, C.L.Jr., DILLEY, R.A., CRANE, F.L.** : Inhibition of the membrane-bound

Mg^{++}-ATPase of chloroplasts by lipophilic chelators. - Biochem. biophys. Res. Commun. *63* : 736 - 741, 1975.

21767 - **BERKALOFF, C.** : Évolution de l'ultrastructure et de l'activité photosynthétique de l'algue verte *Protosiphon botryoides* au cours du reverdissement qui suit la levée de carence azotée. - J. Microscop. *24* : 365 - 376, 1975.

21768 - **BERKALOFF, C., DEROCHE, M.-E.** : Ultrastructure des membranes chloroplastiques progressivement extraites par des solvants de polarité croissante. - Compt. rend. Acad. Sci. Paris, Sér. D *280* : 439 - 442, 1975.

21769 - **BERKALOFF, C., KADER, J.C.** : Variations of the lipid composition during the formation of cysts in the green alga *Protosiphon botryoïdes*. - Phytochemistry *14* : 2353 - 2355, 1975. [Chl, Car.]

21770 - **BERKOVA, E., SHETLIK, I.** : Dinamika fotosinteza i reaktsii Khilla v kletochnykh tsiklakh *Scenedesmus quadricauda*. [Dynamics of photosynthesis and Hill reaction in cellular cycles of *Scenedesmus quadricauda*.] - In : **NECHAS, Ĭ.** (ed.) : Izuchenie Intensivnoĭ Kul'tury Vodorosleĭ. Vol.II. Pp. 37 - 57. Třeboň 1975. [In R.]

21771 - **BERLYN, M.B., ZELITCH, I.** : Photoautotrophic growth and photosynthesis in tobacco callus cells. - Plant Physiol. *56* : 752 - 756, 1975.

21772 - **BERMAN, T.** : Size fractionation of natural aquatic populations associated with autotrophic and heterotrophic carbon uptake. - Mar. Biol. *33* : 215 - 220, 1975.

21773 - **BERRY, J.A.** : Adaptation of photosynthetic processes to stress. - Science *188* : 644 - 650, 1975.

21774 - **BERRY, J.A., FORK, D.C., GARRISON, S.** : Mechanistic studies of thermal damage to leaves. - Carnegie Inst. Year Book *74* : 751 - 759, 1975.

21775 - **BERSENEVA, G.P., FINENKO, Z.Z.** : Kolichestvennoe opredelenie khlorofillov "*a*" i "*c*" v morskikh planktonnykh vodoroslyakh pri pomoshchi bumazhnoĭ khromatografii. [Quantitative determination of chlorophylls *a* and *c* in marine planctonic algae by paper chromatography.] - Okeanologiya *15* : 176 - 180, 1975.[In R, ab : E.]

21776 - **BERSENEVA, G.P., KRUPATKINA, D.K.** : Otnoshenie mezhdu skorost'yu fotosinteza i soderzhaniem khlorofilla "*a*" v tropicheskoĭ chasti Atlanticheskogo okeana. [Relation of photosynthetic rate and chlorophyll *a* content in the tropical part of Atlantic Ocean.] - In : Kompleksnye Issledovaniya v Mirovom Okeane. Pp. 359 - 360. Moskva 1975. [In R.]

21777 - **BERZBORN, R.J., JOHANSSON, B.C., BALTSCHEFFSKY, M.** : Immunological and fluorescence studies with the coupling factor ATPase from *Rhodospirillum rubrum*. - Biochim. biophys. Acta *396* : 360 - 370, 1975.

21778 - **BERZBORN, R.J., KOPP, F., MÜHLETHALER, K.**: Coupling factor 1 (CF_1), a mobile peripheral protein in the thylakoid surface of spinach chloroplasts. - In : **AVRON, M.** (ed.) : Proceedings of the Third International Congress on Photosynthesis. Vol. II. Pp. 809 - 820. Elsevier, Amsterdam - Oxford - New York 1975.

21779 - **BETHLENFALVAY, G., NORRIS, R.F.** : Phytotoxic action of desmedipham : influence of temperature and light intensity. - Weed Sci. *23* : 499 - 503, 1975.

21780 - **BEYELER, W., LUTZ, H.U., BACHOFEN, R.** : Membrane-bound phosphate as driving force for ATP synthesis in chromatophores of *Rhodospirillum rubrum*. - J. Bioenergetics *6* : 233 - 242, 1975.

21781 - **BHANDARI, M.C., SEN, D.N.** : Effect of growth retardants on seedling growth, excised cotyledons, chlorophylls, proteins and sugar contents in *Citrullus* species. - Biochem. Physiol. Pflanzen *167* : 135 - 140, 1975.

21782 - **BHOSALE, L.J., KARADGE, B.A.** : Photosynthetic efficiency and productivity of mangroves of Western India. - Bull. Dep. mar. Sci. Univ. Cochin *7* : 205 - 212, 1975.

21783 - **BHULLAR, B.S., DALY, J.M., REHFELD, D.W.** : Inhibition of dark CO_2 fixation and photosynthesis in leaf discs of corn susceptible to the host-specific toxin produced by *Helminthosporium maydis*, race T. - Plant Physiol. *56* : 1 - 7, 1975.

21784 - **BIANCHI, A., LAUDI, G.** : Fissazione di CO_2 in *Lathraea squamaria*. [CO_2 fixation in *Lathraea squamaria*.] - G. bot. ital. *109* : 59 - 63, 1975. [In Ital., ab : E.]

21785 - **BIDWELL, R.G.S., DURZAN, D.J.** : Some recent aspects of nitrogen metabolism. - In : **DAVIES, P.** (ed.) : Historical and Current Aspects of Plant Physiology. Pp. 152 - 225. Cornell Univ. Press, Ithaca, N.Y. 1975. [Ps.]

21786 - **BIDWELL, R.G.S., QUONG, E.C.K.** : Indoleacetic acid effect on the distribution of photosynthetically fixed carbon in the bean plant. - Biochem. Physiol. Pflanzen *168* : 361 - 370, 1975.

21787 - **BIELORAI, H., HOPMANS, P.A.M.** : Recovery of leaf water potential, transpiration, and photosynthesis of cotton during irrigation cycles. - Agron. J. *67* : 629 - 632, 1975.

*21788 - **BIERHUIZEN, J.F.** : Carbon dioxide supply and net photosynthesis. - Acta Horticult. *32* : 119 - 126, 1973.

21789 - **BIGGAR, J.W., YAUG, M.C.** : Simulation of biomass production by *Chlorella pyrenoidosa* and nitrogen and phosphorus depletion in a carbon limiting medium. - In : **VANSTEENKISTE, G.C.** (ed.) : Computer Simulation of Water Resources Systems. Pp. 503 - 510. North-Holland Publishing Company, Amsterdam 1975.

21790 - **BIL', K.Ya., KLIMOV, V.V., KARAPETYAN, N.V., KARPILOV, Yu.S.** : Aktivnost' i vozdeĭstvie dvukh fotosistem v khloroplastakh parenkhimnykh obkladok sosudistykh puchkov list'ev kukuruzy. [Activity and interaction between two photosystems in chloroplasts of parenchymal sheaths of vascular bundles of maize leaves.] - Dokl. Akad. Nauk SSSR *223* : 225 - 228, 1975. [In R.]

21791 - **BIL', K.Ya., OPARINA, L.A.** : Ul'trastrukturnye osobennosti kletok i plastid v kul'ture tkani pri geterotrofnom i avtotrofnom pitanii. [Ultrastructural peculiarities of cells and plastids during change from tissue culture heterotrophic nutrition to autotrophic nutrition.] - Bot. Zh. *60* : 1613 - 1617, 1975. [In R.]

21792 - **BILLECOCQ, A.** : Structure des membranes biologiques : localisation du sulfoquinovosyldiglycéride dans les diverses membranes des chloroplastes au moyen des anticorps spécifiques. - Ann. Immunol. (Inst. Pasteur) *126 C* : 337 - 352, 1975.

21793 - **BILLORE, S.K., MALL, L.P.** : Chlorophyll content as an ecological index of dry matter production. - J. indian bot. Soc. *54* : 75 - 77, 1975.

21794 - **BILLORE, S.K., MALL, L.P., DAWAR, K.** : Ecology of respiratory losses in teak and *Cymbopogon* grass with reference to deciduous forest and grassland. - Proc. Ind. nat. Sci. Acad. *41* (part B, 2) : 169 - 171, 1975. [Net primary production.]

21795 - **BILLORE, S.K., MALL, L.P., SINGH, V.P.** : Intercomparison of methods for primary production measurement in harvest technique. - Geobios *2* : 80 - 81, 1975.

21796 - **BILLOT, J.** : Étude des pigments des feuilles d'*Acalypha* : influence de la teneur en eau des solutions acétoniques sur l'extraction et l'état des anthocyanes et des chlorophylles. - Physiol. vég. *13* : 407 - 415, 1975.

21797 - **BINDER, A., GROMET-ELHANAN, Z.** : Depletion and reconstitution of photophosphorylation in chromatophore membranes of *Rhodospirillum rubrum*. - In : **AVRON, M.** (ed.) : Proceedings of the Third International Congress on Photosynthesis. Vol. II. Pp. 1163 - 1170. Elsevier, Amsterdam - Oxford - New York 1975.

21798 - **BINET, P.** : Remarques sur les possibilités d'amylogenèse des feuilles et des rhizomes d'*Aster tripolium* L. - Bull. Soc. bot. France *122* : 5 - 13, 1975.

21799 - **BIRD, A.F., LOVEYS, B.R.** : The incorporation of photosynthates by *Meloidogne javanica*. - J. Nematol. *7* : 111 - 113, 1975.

21800 - **BIRD, I.F., CORNELIUS, M.J., KEYS, A.J., KUMARASINGHE, S., WHITTINGHAM, C.P.** : The rate of metabolism by the glycollate pathway in wheat leaves during photosynthesis. - In : **AVRON, M.** (ed.) : Proceedings of the Third International Congress on Photosynthesis. Vol. II. Pp. 1291 - 1301. Elsevier, Amsterdam - Oxford - New York 1975.

21801 - **BISCOE, P.V., CLARK, J.A., GREGSON, K., McGOWAN, M., MONTEITH, J.L., SCOTT, R.K.** : Barley and its environment. I. Theory and practice. - J. appl. Ecol. *12* : 227 - 257, 1975. [Ps instrumentation.]

21802 - **BISCOE, P.V., GALLAGHER, J.N., LITTLETON, E.J., MONTEITH, J.L., SCOTT, R.K.** : Barley and its environment. IV. Sources of assimilate for the grain. - J. appl. Ecol. *12* : 295 - 318, 1975.

21803 - **BISCOE, P.V., SCOTT, R.K., MONTEITH, J.L.** : Barley and its environment. III. Carbon budget of the stand. - J. appl. Ecol. *12* : 269 - 293, 1975.

21804 - **BISHOP, D.G., NOLAN, W.G.** : Inhibition by dibromothymoquinone of photosynthetic electron transfer in chloroplasts of differing ultrastructure. - Arch. Biochem. Biophys. *168* : 594 - 600, 1975.

*21805 - **BISHOP, J.W.** : Effects of zooplankton on algae in Westhampton Lake. - Water Resources Res. Cent. Bull. *43* : 1 - 33, 1971. [Ps, Chl.]

21806 - **BISHOP, N.I., FRICK, M., JONES, L.W.** : Photohydrogen production in normal and mutant forms of various green algae. - Plant Physiol. *56* (Suppl.) : 9, 1975.

*21807 - **BISHOP, N.I., SICHER, R.** : The loss of photosystem II in an algal mutant lacking vitamin E. - Fed. Proc. *33* : 1329, 1974.

21808 - **BISKUPSKÝ, V.** : Primary production of the timber component of the biomass in the oak-hornbeam forest at Báb. - In : **BISKUPSKÝ, V.** (ed.) : Research Project Báb IBP Progress Report II. Pp. 39 - 45. Veda, Bratislava 1975.

21809 - **BISKUPSKÝ, V., OSZLÁNYI, J.** : The leaf area index of the timber component of the ecosystem at the research area of the IBP project at Báb. - In : **BISKUPSKÝ, V.** (ed.) : Research Project Báb IBP Progress Report II. Pp. 51 - 60. Veda, Bratislava 1975.

21810 - **BJÖRKMAN, O.** : Environmental and biological control of photosynthesis : inaugural address. - In : **MARCELLE, R.** (ed.) : Environmental and Biological Control of Photosynthesis. Pp. 1-16. Dr.W.Junk b.v. Publ., The Hague 1975.

21811 - **BJÖRKMAN, O.** : Thermal stability of the photosynthetic apparatus in intact leaves. - Carnegie Inst. Year Book *74* : 748 - 751, 1975.

21812 - **BJÖRKMAN, O., EHLERINGER, J.** : Comparison of the quantum yields for CO_2 uptake in C_3 and C_4 plants. - Carnegie Inst. Year Book *74* : 760 - 761, 1975.

21813 - **BJÖRKMAN, O., MOONEY, H.A., EHLERINGER, J.** : Comparison of photosynthetic characteristics of intact plants. - Carnegie Inst. Year Book *74* : 743 - 748, 1975.

21814 - **BLACK, C.C.** : Carbon assimilation pathways in higher plant cells and tissues.- In : **AVRON, M.** (ed.) : Proceedings of the Third International Congress on Photosynthesis. Vol. II. Pp. 1201 - 1208. Elsevier, Amsterdam - Oxford - New York 1975.

21815 - **BLACKBURN, T.H., KLEIBER, P., FENCHEL, T.** : Photosynthetic sulfide oxidation in marine sediments. - Oikos *26* : 103 - 108, 1975.

21816 - **BLANKENSHIP, R., McGUIRE, A., SAUER, K.** : Chemically induced dynamic electron polarization in chloroplasts at room temperature : Evidence for triplet state participation in photosynthesis. - Proc. nat. Acad. Sci. USA *72* : 4943 - 4947, 1975.

21817 - **BLANKENSHIP, R.E., BABCOCK, G.T., SAUER, K.** : Kinetic study of oxygen evolution parameters in Tris-washed, reactivated chloroplasts. - Biochim. biophys. Acta *387* : 165 - 175, 1975.

21818 - **BLANKENSHIP, R.E., BABCOCK, G.T., SAUER, K.** : Manganese and oxygen evolution in reconstituted chloroplasts. - Biophys. J. *15* (2, Pt.2) : 246 a, 1975.

21819 - **BLANKENSHIP, R.E., BABCOCK, G.T., WARDEN, J.T., SAUER, K.** : Observation of a new EPR transient in chloroplasts that may reflect the electron donor to photosystem II at room temperature. - FEBS Lett. *51* : 287 - 293, 1975.

*21820 - **BLANKENSHIP, R.E., SAUER, K.** : EPR study of the environment of manganese in Tris-washed chloroplasts. - Fed. Proc. *33* : 1255, 1974.

21821 - **BLASIE, J.K., TORRIANI, I., DUTTON, P.L.** : The X-ray structure of photosynthetic reaction center - phospholipid model membranes. - Fed. Proc. *33* : 1253, 1974.

21822 - **BLUM, A.** : Effect of the *Bm* gene on epicuticular wax and the water relations of *Sorghum bicolor* L. (Moench). - Isr. J. Bot. *24* : 50, 1975. [Radiation in canopy.]

21823 - **BLUM, A.** : Infrared photography for selection of dehydration-avoidant *Sorghum* genotypes. - Z. Pflanzenzücht. *75* : 339 - 345, 1975. [Leaf optical properties.]

21824 - **BOARDMAN, N.K.** : Trace elements in photosynthesis. - In : **NICHOLAS, D.J.D., EGAN, A.R.** (ed.) : Trace Elements in Soil-Plant-Animal Systems. Pp. 199 - 212. Academic Press, New York - London 1975.

21825 - **BOARDMAN, N.K., BJÖRKMAN, O., ANDERSON, J.M., GOODCHILD, D.J., THORNE, S.W.** : Photosynthetic adaptation of higher plants to light intensity : Relationship between chloroplast structure, composition of the photosystems and photosynthetic rates. - In : **AVRON, M.** (ed.) : Proceedings of the Third International Congress on Photosynthesis. Vol.III. Pp. 1809 - 1827. Elsevier, Amsterdam - Oxford - New York 1975.

21826 - **BÖCHER, T.W.** : Density determination in Arctic plant communities. - Phytocoenologia *2* : 73 - 86, 1975.

21827 - **BÖCHER, T.W.** : Structure of the multinodal photosynthetic thorns in *Prosopis kuntzei* HARMS. - Kong. dan. Vidensk. Selsk. Biol. Skr. *20* (8) : 1 - 43, 1975.

21828 - **BOCK, F. de, CARLIER, G.** : Stimulation spécifique, par les rayons X, de l'émission de gaz carbonique à l'obscurité des feuilles de *Pelargonium zonale* L. - Compt. rend. Acad. Sci. Paris, Sér. D *281* : 993 - 996, 1975.

B21829 - **BOGDANOV, N.I.** : Pervichnaya Produktsiya i Mikrobiologiya Kaǐrakkumskogo Vodokhranilishcha. [Primary Production and Microbiology of the Kairakum Water Reservoir.] - Donish, Dushanbe 1975. [In R.]

21830 - **BOGDANOVA, T.L., MENITSKAYA, I.M., PROTSENKO, D.F., KOMARENKO, N.I.** : K metodike polucheniya khlorofill-karotinovoǐ pasty iz sinezelenykh vodorosleǐ. [Method of producing chlorophyll-carotene paste from blue-green algae.] - Gidrobiol. Zh. *11* : 111 - 115, 1975. [In R.]

*21831 - **BOGDANOVIĆ, M.** : Effect of γ-radiation on chlorophyll formation in the dark in black pine cotyledons. - Period. Biol. *76* (1) : 40, 1974.

21832 - **BÖGER, P.** : Photosynthese in globaler Sicht. - Naturwiss. Rundschau *28* : 429 - - 435, 1975.

B21833 - **BÖGER, P.** : Photosynthese und Pflanzliche Produktivität. - Konstanzer Universitätsreden 76. Universitätsverlag GmbH, Konstanz 1975.

21834 - **BOGOMOLNI, R.A., KLEIN, M.P.** : Mobile charge carriers in photosynthesis. - Nature *258* : 88 - 89, 1975.

21835 - **BOGORAD, L.** : Genetic and evolutionary relationships between plastids and the nuclear-cytoplasmic system. - In : **NASYROV, Yu.S., ŠESTÁK, Z.** (ed.) : Genetic Aspects of Photosynthesis. Pp. 51 - 62. Dr. W. Junk b.v. Publ., The Hague 1975.

21836 - **BOGORAD, L.** : Phycobiliproteins and complementary chromatic adaption. - Annu. Rev. Plant Physiol. *26* : 369 - 401, 1975.

21837 - **BOGUSLAVSKIǏ, L.I., LOZHKIN, B.T., KISELEV, B.A.** : Generatsiya temnovogo potentsiala na bisloǐnykh lipidnykh membranakh, soderzhashchikh khlorofill. [Dark potential generation on bilayer lipid membranes containing chlorophyll.]- Dokl. Akad. Nauk SSSR *222* : 228 - 231, 1975. [In R.]

21838 - **BÖHME, H.** : Eine Methode zur schnellen Bestimmung des organischen Kohlenstoffs mit einem Naßoxydationsverfahren. - Acta hydrochim. hydrobiol. *3* : 327 - 332, 1975.

21839 - **BÖHME, H.** : Inhibition of cytochrome-b_6 oxidation by KCN. - FEBS Lett. *60* : 51 - 53, 1975.

21840 - **BOĬCHENKO, E.A.** : Aktivnost' zhelezosoderzhashcheĭ reduktazy uglekisloty rasteniĭ. [Activity of iron-containing reductase of carbon dioxide in plants.] - Fiziol. Rast. *22* : 770 - 775, 1975. [In R, ab : E.]

21841 - **BOKOVAYA, M.M., NILOVSKAYA, N.T.** : Ėndogennye ritmy fotosinteza posevov nekotorykh ovoshchnykh rasteniĭ. [Endogenous photosynthetic rhythms of sown cultures of some vegetables.]-In : Problemy Sozdaniya Biologo-Tekhnicheskikh Sistem Zhizneobespecheniya Cheloveka. Pp. 37 - 45. Nauka, Novosibirsk 1975. [In R.]

21842 - **BOLHÀR-NORDENKAMPF, H.R.** : Die Veränderungen des Chlorophyllgehaltes in ontogenetisch verschiedenen Blättern von *Phaseolus vulgaris* var. *nanus* L. nach Behandlung mit Atrazin. - Biochem. Physiol. Pflanzen *167* : 41 - 64, 1975.

21843 - **BOLTON, J.R.** : Electron transfer reactions in photochemistry and photobiology. - Polymer Preprints (Amer. chem. Soc.) *16* : 225 - 227, 1975. [Ps.]

21844 - **BOMSEL, J.L., SELLAMI, A.** : *In vivo* measurement of the rate of transfer of ∿P from adenylate through the chloroplast envelope. - In : **AVRON, M.** (ed.) : Proceedings of the Third International Congress on Photosynthesis. Vol. II. Pp. 1363 - 1367. Elsevier, Amsterdam - Oxford - New York 1975.

21845 - **BONDARENKO, V.I., ARTYUKH, A.D., MAKAROVA, A.Ya.** : Morozostoĭkost' i produktivnost' ozimoĭ pshenitsy v zavisimosti ot usloviĭ uvlazhneniya. [Frost resistance and productivity of winter wheat in dependence on conditions of moistening.] - Fiziol. Biokhim. kul't. Rast. *7* : 398 - 403, 1975. [Ps; in R, ab : E.]

21846 - **BONEN, L., DOOLITTLE, W.F.** : On the prokaryotic nature of red algal chloroplasts. - Proc. nat. Acad. Sci. USA *72* : 2310 - 2314, 1975.

21847 - **BONNEMAIN, J.L.** : Transport et distribution des produits de la photosynthèse.- In : **COSTES, C.** (ed.) : Photosynthèse et Production Végétale. Pp. 147 - 170. Gauthier-Villars, Paris 1975.

21848 - **BONNETT, H.T.** : Chloroplast uptake by isolated plant protoplasts. - Plant Physiol. *56* (Suppl.) : 38, 1975.

21849 - **BONTE, J., LOUGUET, P.** : Interrelations entre la pollution par le dioxyde de soufre et le mouvement des stomates chez le *Pelargonium* × *hortorum* : effets de l'humidité relative et de la teneur en gaz carbonique de l'air. - Physiol. vég. *13* : 527 - 537, 1975.

21850 - **BONZI, L.M., FABBRI, F.** : Chloroplast protrusions in *Arisarum proboscideum* (L) SAVI. - Caryologia *28* : 407 - 426, 1975.

21851 - **BONZON, M., BUIS, R., GREPPIN, H.** : Analyse factorielle de l'ultrastructure du chloroplaste d'épinard à l'état végétatif et floral. I. Jours courts de 8 heures (conditions végétatives). - Ber. schweiz. bot. Ges. *85* : 265 - 278, 1975.

21852 - **BONZON, M., BUIS, R., GREPPIN, H.** : Analyse factorielle de l'ultrastructure du chloroplaste d'épinard à l'état végétatif et floral. II. Transfert en photopériode continue (induction de la floraison). - Ber. schweiz. bot. Ges. *85* : 279 - 287, 1975.

21853 - **BOOYSEN, P. deV., NELSON, C.J.** : Leaf area and carbohydrate reserves in regrowth of tall fescue. - Crop Sci. *15* : 262 - 266, 1975. [Growth analysis.]

21854 - **BORCHERT, R.** : Endogenous shoot growth rhythms and indeterminate shoot growth in oak. - Physiol. Plant. *35* :152 - 157, 1975. [Leaf growth rhythms.]

21855 - **BORISEVICH, G.P., KONONENKO, A.A., VENEDIKTOV, P.S., VERKHOTUROV, V.N., RUBIN, A.B.** : Izmenenie spektrov pogloshcheniya karotinoidov v sukhikh plenkakh khromatoforov *Rhodospirillum rubrum* vo vneshnem ėlektricheskom pole. [Changes in absorption spectra of carotenoids in dried films of chromatophores of *Rhodospirillum rubrum* in an external electric field.] - Biofizika *20*: 250 - 253, 1975. [In R, ab : E.]

21856 - **BORISOV, A.Yu.** : Fotosinteziruyushchie organizmy kak ėffektivnye preobrazovateli svetovoĭ ėnergii v ėnergiyu ėlektrokhimicheskuyu. [Photosynthetic organisms as effective transformers of radiant energy into electrochemical energy.]

In : Metody Issledovaniya Fotokhimicheskikh Reaktsiĭ Fotosinteza *in Vitro* i *in Vivo*. Pp. 92 - 102. Pushchino 1975. [In R.]

21857 - **BORISOVA, I.G., BUDNITSKAYA, E.V.** : Lipoksigenaza khloroplastov. [Lipoxygenase of chloroplasts.] - Dokl. Akad. Nauk SSSR *225* : 439 - 441, 1975. [In R.]

21858 - **BORNEFELD, T., KAISER, W., KLEINSCHNITZ, W.** : Ein Gerät zur Probenentnahme in Ein-Sekunden-Abständen bei Photosyntheseexperimenten. - Z. Naturforsch. *30 C* : 680, 1975.

21859 - **BORNEFELD, T., SIMONIS, W.** : Correlations between phosphate uptake, photophosphorylation and metabolism in *Anacystis nidulans* as affected by carbonylcyanide M-chlorophenylhydrazone and chloramphenicol. - In : **AVRON, M.** (ed.) : Proceedings of the Third International Congress on Photosynthesis. Vol. II. Pp. 1557 - 1565. Elsevier, Amsterdam - Oxford - New York 1975.

21860 - **BÖRNER, T., SCHUMANN, B., KRAHNERT, S., PECHAUF, M., HERRMANN, F.H., KNOTH, R., HAGEMANN, R.** : Struktur und Funktion der genetischen Information in den Plastiden. XIII. Lamellarproteine bleicher Plastiden von Plastom- und Genmutanten von *Hordeum* und *Lycopersicon*. - Biochem. Physiol. Pflanzen *168* : 185 - 193, 1975.

21861 - **BORNKAMM, R.** : Water and the production of organic matter. - Ber. deut. bot. Ges. *88* : 197 - 203, 1975. [Chloroplast.]

21862 - **BORNKAMM, R., SALINGER, S., STREHLOW, H.** : Substanzproduktion und Inhaltsstoffe zweier Gräser in Rein- und Mischkultur. - Flora *164* : 437 - 448, 1975.

21863 - **BORZENKOVA, R.A., MOKRONOSOV, A.T.** : Formirovanie khloroplastov pri èksperimental'nykh vozdeĭstviyakh na beloksinteziruyushchuyu sistemu kletki. [Chloroplasts formation at experimental effect on protein synthesizing system of the cell.] - Nauch. Dokl. vyssh. Shkoly, biol. Nauki *18* (6) : 75 - 78, 1975. [In R.]

21864 - **BOTEY SERRA, J., FITE, D.G.** : Determinación de xantofilas por cromatografía líquida de alta presión. Aplicación al estudio de la deposición de xantofilas en la yema del huevo, como determinantes de su color. [Determination of xanthophylls by high pressure liquid chromatography. Application to the study of the deposition of xanthophylls in egg yolk as color determinants.] - Afinidad *32* : 249 - 255, 1975. [In Span., ab : E.]

21865 - **BOTHE, H.** : The occurrence of the pyruvate dehydrogenase complex and the pyruvate-ferredoxin oxidoreductase in blue-green bacteria. - Biochem. Soc. Trans. *3* : 376 - 377, 1975.

21866 - **BOUCHER, F., GINGRAS, G.** : The photogeneration of superoxide by isolated photoreaction center from *Rhodospirillum rubrum*. - Biochem. biophys. Res. Commun. *67* : 421 - 426, 1975.

21867 - **BOUGES-BOCQUET, B.** : Electron transport in Photosystem I in spinach chloroplasts. - Biochim. biophys. Acta *396* : 382 - 391, 1975.

21868 - **BOUGES-BOCQUET, B.** : Electron acceptors of photosystem II. - In : **AVRON, M.** (ed.) : Proceedings of the Third International Congress on Photosynthesis. Vol. I. Pp. 579 - 588. Elsevier, Amsterdam - Oxford - New York 1975.

21869 - **BOUMA, D.** : Effects of some metabolic phosphorus compounds on rates of photosynthesis of detached phosphorus-deficient subterranean clover leaves. - J. exp. Bot. *26* : 52 - 59, 1975.

21870 - **BOUNIAS, M.** : Modifications des relations quantitatives entre les acides aminés libres et les pigments photosynthétiques chez différent mutants chlorophylliens d'Orge et d'*Arabidopsis*. - Can. J. Bot. *53* : 708 - 718, 1975.

21871 - **BOURDU, R.** : Structure et développement des chloroplastes. - In : **COSTES, C.** (ed.) : Photosynthèse et Production Végétale. Pp. 1 - 32. Gauthier-Villars, Paris 1975.

21872 - **BOURDU, R., DELAY, C., DARMANADEN, J., HALPERN, S.** : Ontogénie et développement des structures foliaires et chloroplastiques de *Lactuca sativa*. - Physiol. vég. *13* : 265 - 289, 1975.

21873 - **BOURQUE, D.P., McMILLAN, P.N., CLINGENPEEL, W.J., NAYLOR, A.W.** : Ultrastructural effects of water stress on chloroplast development in Jack bean (*Canavalia ensiformis* [L.] DC). - Plant Physiol. *56* : 160 - 163, 1975.

21874 - **BOWES, G., M'BAKU, S.B., FRITZ, G.J.** : Photosynthetic CO_2 fixation and carbohydrate metabolism in cells isolated from *Digitaria decumbens* var Slenderstem.- Plant Physiol. *56* (Suppl.) : 25, 1975.

21875 - **BOWES, G., OGREN, W.L., HAGEMAN, R.H.** : pH dependence of the K_m (CO_2) of ribulose 1,5-diphosphate carboxylase. - Plant Physiol. *56* : 630 - 633, 1975.

21876 - **BOX, E.** : Quantitative evaluation of global primary productivity models generated by computers. - In : **LIETH, H., WHITTAKER, R.H.** (ed.) : Primary Productivity of the Biosphere. Pp. 265 - 283. Springer-Verlag, Berlin - Heidelberg - New York 1975.

21877 - **BOYD, C.E., SCARSBROOK, E.** : Influence of nutrient additions and initial density of plants on production of waterhyacinth *Eichhornia crassipes*. - Aquatic-Bot. *1* : 253 - 261, 1975. [Chl.]

21878 - **BOYER, J.S., McPHERSON, H.G.** : Physiology of water deficits in cereal crops.- Adv. Agron. *27* : 1 - 23, 1975. [Ps.]

21879 - **BOYER, P.D.** : Energy transduction and proton translocation by adenosine triphosphatases. - FEBS Lett. *50* : 91 - 94, 1975.

21880 - **BOYER, Y.** : Premières études sur les mécanismes de l'accélération de croissance induite par une sécheresse temporaire. - Compt. rend. Acad. Sci. Paris, Sér. D *280* : 283 - 286, 1975. [Chl.]

21881 - **BOYER, Y., PARCEVAUX, S. de, GUILLAUME, E.** : Étude de l'éclairage d'appoint en serre. I. Caractéristiques techniques de quelques dispositifs d'éclairage. - Oecol. Plant. *10* : 233 - 250, 1975. [PhAR.]

21882 - **BOYER, Y., PARCEVAUX, S. de, GUILLAUME, E.** : Étude de l'éclairage d'appoint en serre. II. Aspects écophysiologiques de la croissance du Lin en fonction des conditions d'éclairement. - Oecol. Plant. *10* : 251 - 266, 1975. [PhAR.]

21883 - **BOYLE, J.E., SMITH, D.C.** : Biochemical interactions between the symbionts of *Convoluta roscoffensis*. - Proc. roy. Soc. London B, Biol. Sci. *189* : 121 - 135, 1975. [Ps, Chl.]

21884 - **BRADBEER, J. W.** : A personal assessment of the state of knowledge of crassulacean acid metabolism (CAM) . - In : **MARCELLE, R.** (ed.) : Environmental and Biological Control of Photosynthesis. Pp. 369 - 371. Dr. W. Junk, b.v. Publ., The Hague 1975.

21885 - **BRADBEER, J.W., COCKBURN, W., RANSON, S.L.** : The labelling of the carboxyl carbon atoms of malate in *Kalanchoë crenata* leaves. - In : **MARCELLE, R.** (ed.): Environmental and Biological Control of Photosynthesis. Pp. 265 - 272. Dr. W. Junk, b.v. Publ., The Hague 1975.

21886 - **BRADY, C.J., TUNG, H.F.** : Rate of protein synthesis in senescing, detached wheat leaves. - Aust. J. Plant Physiol. *2* : 163 - 176, 1975. [Chl.]

21887 - **BRADY, R.A., GOLTZ, S.M., POWERS, W.L., KANEMASU, E.T.** : Relation of soil water potential to stomatal resistance of soybean. - Agron. J. *67* : 97 - 99, 1975.

21888 - **BRAND, J.J., CURTIS, V.A., TOGASAKI, R.K., SAN PIETRO, A.** : Partial reactions of photosynthesis in briefly sonicated *Chlamydomonas*. II. Photophosphorylation activities. - Plant Physiol. *55*:187 - 191, 1975.

21889 - **BRAND, J.J., MYERS, J.** : Pigment transformations in *Anacystis nidulans*. - Plant Physiol. *56* (Suppl.) : 10, 1975.

21890 - **BRÄNDLE, E.P.O., KÖTTER, R., ZETSCHE, K.** : Changes of pH in culture solution of *Acetabularia* caused by proton flux. - Protoplasma *83* : 173, 1975. [Ps.]

21891 - **BRANDT, A.B., ANTONINOVA, M.V., KISELEVA, M.I., SHARIPOV, K.A.** : Belkovo-uglevodnaya produktivnost' khlorelly i éffektivnost' ispol'zovaniya sveta razlichnogo spektral'nogo sostava v fotosinteze. [Protein-carbohydrate productivity of *Chlorella* and the efficiency of applying light of different spectral composition in photosynthesis.] - Biofizika *20* : 479 - 482, 1975. [In R, ab: E.]

21892 - **BRANDT, A.B., FINAKOV, G.Z.** : Udel'nyĭ fotosintez khlorelly v razlichnykh-oblastyakh spektra. [Specific photosynthesis of *Chlorella* in different spectral regions.] - Dokl. Akad. Nauk SSSR *221* : 990 - 992, 1975. [In R.]

21893 - **BRANDT, A.B., KISELEVA, M.I.** : O kvantovoĭ produktivnosti razlichnykh oblasteĭ spektra pri vyrashchivanii asinkhronnoĭ kul'tury khlorelly. [Quantum productivity of different spectrum regions during growing of asynchronous culture of *Chlorella*.] - Biofizika *20* : 859 - 861, 1975. [In R, ab : E.]

21894 - **BRANDT, A.B., KISELEVA, M.I., SHARIPOV, K.A.** : Vliyanie sveta razlichnogo spektral'nogo sostava na nakoplenie pigmentov, produktivnost' i ėffektivnost' ispol'zovaniya pogloshchennoĭ ėnergii v fotosinteze v techenie tsikla razvitiya khlorelly. [The effect of light of different spectral composition on the accumulation of pigments, productivity and efficiency of utilization of the absorbed energy in photosynthesis during the developmental cycle of *Chlorella*.]- Biofizika *20* : 862 - 866, 1975. [In R, ab : E.]

21895 - **BRANTON, D., BULLIVANT, S., GILULA, N.B., KARNOVSKY, M.J., MOOR, H., MÜHLETHALER, K., NORTHCOTE, D.H., PACKER, L., SATIR, B., SATIR, P., SPETH, V., STAEHLIN, L.A., STEERE, R.L., WEINSTEIN, R.S.** : Freeze-etching nomenclature. - Science *190* : 54 - 56, 1975. [Chloroplast.]

21896 - **BRAUM, J.G.** : Contribución del zooplancton a la formación de feopigmentos en el medio marino. [Contribution of zooplankton to the formation of phaeopigments in the sea.] - Bol. Inst. esp. Oceanogr. *200* : 1 - 18, 1975. [In Span., ab : E.]

21897 - **BRÄVDO, B., CANVIN, D.T.** : Evaluation of the rate of CO_2 production by photorespiration. - In : **AVRON, M.** (ed.) : Proceedings of the Third International Congress on Photosynthesis. Vol. II. Pp. 1277 - 1283. Elsevier, Amsterdam - Oxford - New York 1975.

21898 - **BRETON, J.** : Polarized light spectroscopy on oriented spinach chloroplasts : fluorescence emission and excitation spectra. - In : **AVRON, M.** (ed.) : Proceedings of the Third International Congress on Photosynthesis. Vol. I. Pp. 229 - 234. Elsevier, Amsterdam - Oxford - New York 1975.

21899 - **BRETON, J., ROUX, E.** : Examples d'utilisation des lasers dans les recherches sur la photosynthèse. - In : **JOUSSOT-DUBIEN, J.** (ed.) : Lasers in Physical Chemistry and Biophysics. Pp. 379 - 388. Elsevier, Amsterdam 1975.

21900 - **BRETON, J., ROUX, E., WHITMARSH, J.** : Dichroism of chlorophyll *a* absorption change at 700 nm using chloroplasts oriented in a magnetic field. - Biochem. biophys. Res. Commun. *64* : 1274 - 1277, 1975.

21901 - **BREUZÉ, G., PELRAS, A.** : Fluochronomètre destiné aux mesures de temps de déclin de fluorescence de l'ordre de la nanoseconde. - Détecteurs nucl. *55* : 518 - 526, 1975. [Chloroplast.]

21902 - **BREZEANU, A., TĂCINĂ, F., PAUCĂ-COMĂNESCU, M.** : The multiannual dynamics of the primary productivity of the herbeceous layer from some mixed fir-tree and beech-tree forests. - Rev. roum. Biol., Sér. Biol. vég. *20* : 161 - 177, 1975

21903 - **BRINKHUIS, B.H., TEMPEL, N.** : Photosynthesis in exposed saltmarsh macroalgae - J. Phycol. *11* (Suppl.) : 11, 1975.

21904 - **BRITTON, G., SINGH, R.K., GOODWIN, T.W., BEN-AZIZ, A.** : The carotenoids of *Rhodomicrobium vannielii (Rhodospirillaceae)* and the effect of diphenylamine on the carotenoid composition. - Phytochemistry *14* : 2427 - 2433, 1975.

21905 - **BRITZ, S.J.** : Circadian rhythm of chloroplast orientation in *Ulva*. - Plant Physiol. *56* (Suppl.) : 76, 1975.

21906 - **BRITZ, S.J.** : A model system to simulate chloroplast movement and accompanying transmittance changes in *Ulva*. - Carnegie Inst. Year Book *74* : 794 - 803, 1975.

21907 - **BRITZ, S.J.** : Inhibitor studies on the mechanism of chloroplast movement in *Ulva*. - Carnegie Inst. Year Book *74* : 803 - 805, 1975.

21908 - **BROCK, T.D.** : Effect of water potential on a *Microcoleus (Cyanophyceae)* from a desert crust. - J. Phycol. *11* : 316 - 320, 1975.

21909 - **BROCK, T.D** : Salinity and the ecology of *Dunaliella* from Great Salt Lake. - J. gen. Microbiol. *89* : 285 - 292, 1975. [Ps.]

21910 - **BROCK, T.D.** : The effect of water potential on photosynthesis in whole lichens and in their liberated algal components. - Planta *124* : 13 - 23, 1975.

21911 - **BROCK, T.D., REMINGTON, P., DOEMEL, W.A.** : Photochemical activity of single blue-green algal cells recorded by the use of nuclear track emulsion. - Photosynthetica *9* : 331 - 332, 1975.

21912 - **BRODA, E.** : Stand und Entwicklungstendenzen der Bioenergetik. - Physik.Blätter *31* (12) : 558 - 565, 1975.[Ps.]

21913 - **BRODA, E.** : The beginning of photosynthesis. - Origins Life *6* : 247 - 251, 1975.

B21914 - **BRODA, E.** : The Evolution of the Bioenergetic Processes. - Pergamon Press, Oxford - New York - Toronto - Sydney - Paris - Braunschweig 1975.

21915 - **BRODY, S.S.** : Surface properties of monomolecular films of reduced plastocyanin at a nitrogen-water interface. - Z. Naturforsch. *30 C* : 318 - 322, 1975.

21916 - **BROGÅRDH, T., JOHNSSON, A.** : Effects of magnesium, calcium and lanthanum ions on stomatal oscillations in *Avena sativa* L. - Planta *124* : 99 - 103, 1975. [Leaf resistance.]

21917 - **BROUERS, M.** : Optical properties of *in vitro* aggregates of protochlorophyllide in non-polar solvents. - Photosynthetica *9* : 304 - 310, 1975.

21918 - **BROUERS, M., SIRONVAL, C.** : Restoration of a $P_{657-647}$ form from $P_{645-638}$ in extracts of etiolated primary bean leaves. - Plant Sci. Lett. *4* : 175 - 181, 1975.

*21919 - **BROUWER, R.** : Dynamics of plant performance. - Acta Horticult. *32* : 31 - 49, 1973.[Growth analysis.]

21920 - **BROWN, A.S., FOSTER, J.A., VOYNOW, P.V., FRANZBLAU, C., TROXLER, R.F.** : Allophycocyanin from the filamentous cyanophyte, *Phormidium luridum*. - Biochemistry *14* : 3581 - 3588, 1975.

21921 - **BROWN, H.E., STEIN, E.R., SALDANA, G.** : Evaluation of *Brassica carinata* as a source of plant protein. - Agr. Food Chem. *23* : 545 - 547, 1975. [Car.]

21922 - **BROWN, J., ACKER, S., DURANTON, J.** : The difference in turnover rate between the chlorophyll *a* in the P700-chlorophyll *a*-protein and in the total chloroplast membranes. - Biochem. biophys. Res. Commun. *62* : 336 - 341, 1975.

21923 - **BROWN, J.A.** : Studies of the P700 chlorophyll *a*-protein complex. - Plant Physiol. *56* (Suppl.) : 86, 1975.

21924 - **BROWN, J.S.** : Metabolism and photocharacteristics of chlorophyll-protein complexes. - Carnegie Inst. Year Book *74* : 779 - 783, 1975.

21925 - **BROWN, J.S., ALBERTE, R.S., THORNBER, J.P.** : Comparative studies on the occurrence and spectral composition of chlorophyll-protein complexes in a wide variety of plant material. - In : **AVRON, M.** (ed.) : Proceedings of the Third International Congress on Photosynthesis. Vol. III. Pp. 1951 - 1962. Elsevier, Amsterdam - Oxford - New York 1975.

21926 - **BROWN, R.H., BROWN, W.V.** : Photosynthetic characteristics of *Panicum milioides*, a species with reduced photorespiration. - Crop Sci. *15* : 681 - 685, 1975.

21927 - **BROWN, W.V.** : Variations in anatomy, associations, and origins of Kranz tissue.- Amer. J. Bot. *62* : 395 - 402, 1975.

21928 - **BROWNING, G.** : Shoot growth in *Coffea arabica* L. I. Responses to rainfall when the soil moisture status and gibberellin supply are not limiting. - J. hort. Sci. *50* : 1 - 11, 1975. [Growth analysis.]

21929 - **BROWNING, G., FISHER, N.M.** : Shoot growth in *Coffea arabica* L. II. Growth flushing stimulated by irrigation. - J. hort. Sci. *50* : 207 - 218, 1975. [Growth analysis.]

21930 - **BRUCE, D.** : Evaluating accuracy of tree measurements made with optical instruments. - Forest Sci. *21* : 421 - 426, 1975.

21931 - **BRÜGGER, M., BOSCHETTI, A.** : Two-dimensional gel electrophoresis of ribosomal proteins from streptomycin-sensitive and streptomycin-resistant mutants of *Chlamydomonas reinhardi*. - Europe. J. Biochem. *58* : 603 - 610, 1975. [Chloroplast ribosomes.]

21932 - **BRULFERT, J., GUERRIER, D., QUIEROZ, O.** : Photoperiodism and enzyme rhythms : kinetic characteristics of the photoperiodic induction of Crassulacean acid metabolism. - Planta *125* : 33 - 44, 1975.

21933 - **BRÜNING, K., DRUMM, H., MOHR, H.** : On the role of phytochrome in controlling enzyme levels in plastids. - Biochem. Physiol. Pflanzen *168* : 141 - 156, 1975. [Ps.]

21934 - **BRUNKER, R.L.** : Autoinjector for the determination of picomolar quantities of ATP with a liquid scintillation counter. - Anal. Biochem. *63* : 418 - 422, 1975.

21935 - **BRUNOLD, C., SCHIFF, J.A.** : Assimilatory sulfate reduction in *Euglena gracilis* var. *bacillaris*. - Plant Physiol. *56* : 36, 1975.

21936 - **BRYLINSKY, M., MANN, K.H.** : The influence of morphometry and of nutrient dynamics on the productivity of lakes. - Limnol. Oceanogr. *20* : 666 - 667, 1975.

21937 - **BUCHANAN, B.B.** : Ferredoxin-activated fructose-1,6-diphosphatase systems of spinach chloroplasts. - In : **COLOWICK, S.P., KAPLAN, N.O.** (ed.) : Methods in Enzymology. Vol.42. Pp. 397 - 405. Academic Press, New York - San Francisco - London 1975.

*21938 - **BUCHANAN, B.B., KNAFF, D.B.** : Cytochrome *b* in photosynthetic sulfur bacteria. - Fed. Proc. *33* : 1578, 1974.

*21939 - **BUCHANAN, B.B., SCHÜRMANN, P.** : Ribulose 1,5-diphosphate carboxylase : a regulatory enzyme in the photosynthetic assimilation of carbon dioxide. - In : **HORECKER, B.L., STADTMAN, E.R.** (ed.) : Current Topics in Cellular Regulation. Vol. 7. Pp. 1 - 20. Academic Press, New York - London 1973.

21940 - **BUCHECKER, R., LIAAEN-JENSEN, S., EUGSTER, C.H.** : The identity of trollixanthin and trolliflor with neoxanthin. - Phytochemistry *14* : 797 - 799, 1975.

21941 - **BUCKE, C., COOMBS, J.** : Regulation of CO_2 assimilation in *Pennisetum purpureum*. - In : **AVRON, M.** (ed.) : Proceedings of the Third International Congress on Photosynthesis. Vol. II. Pp. 1567 - 1571. Elsevier, Amsterdam - Oxford - New York 1975.

21942 - **BUCKE, C., OLIVER, I.R.** : Location of enzymes metabolising sucrose and starch in the grasses *Pennisetum purpureum* and *Muhlenbergia montana*. - Planta *122* : 45 - 52, 1975.

21943 - **BUDD, T.W.** : Levels of carbonic anhydrase in response to light. - Plant Physiol. *56* : 72, 1975.

*21944 - **BUESA, R.J.** : Population and biological data on turtle grass (*Thalassia testudinum* KÖNIG, 1805) on the Northwestern Cuban shelf. - Aquaculture *4* : 207 - 226, 1974. [Production.]

21945 - **BUESA, R.J.** : Cocientes fisiológicos de plantas marinas. [Physiological quotients of marine plants.] - Res. Invest. (Cuba) *1975* (2) : 254, 1975. [Ps.]

21946 - **BUESA, R.J.** : Population biomass and metabolic rates of marine angiosperms on the northwestern Cuban shelf. - Aquat. Bot. *1* : 11 - 23, 1975. [Ps.]

21947 - **BUGAKOVA, A.N., PETRISHCHEVA, E.A., IL'YUSHCHENKO, T.E.** : Vliyanie nedostatka sery na vodnyĭ rezhim i intensivnost' fotosinteza gorokha i pshenitsy. [Effect of sulphur deficiency on water relations and photosynthetic rate in pea and wheat.] - Fiziol. Biokhim. kul't. Rast. *7* : 513 - 516, 1975. [In R, ab : E.]

21948 - **BUINOVA, M.G., EFIMOV, M.V.** : Vliyanie mikroėlementov na soderzhanie i sootnoshenie pigmentov v list'yakh rasteniĭ. [Effect of microelements on the content and ratio of pigments in plant leaves.]-Inform. Byull. Akad. Nauk SSSR, sibir. Filial, Ulan-Udė *10*(Mikroėlementy v Sibiri) : 91 - 106, 1975. [In R.]

21949 - **BUKHAR, I.E., MEDVEDEVA, T.N., ZHEKU, R.I.** : Soderzhanie khlorofilla, sukhikh veshchestv i ovodnennost' list'ev ozimoĭ pshenitsy v zavisimosti ot usloviĭ pitaniya. [Contents of chlorophyll, dry matter and water in winter wheat leaves in dependence on nutrition conditions.] - Izv. Akad. Nauk mold.SSR, Ser. biol. khim. Nauk *1975* (5) : 84 - 89, 94, 1975. [In R.]

21950 - **BULL, T. A.** : Row spacing and potential productivity in sugarcane. - Agron. J. *67*:421 - 423, 1975.

21951 - **BULL, T.A., GLASZIOU, K.T.** : Sugar cane. - In : **EVANS, L.T.** (ed.) : Crop Physiology. Pp. 51 - 72. Cambridge University Press, London - New York 1975.[Ps.]

21952 - **BUL'ON, V.V.** : Izmerenie produktsiĭ fitoplanktona metodom C^{14} s uchetom poter' mechenogo organicheskogo veshchestva v protsesse fil'tratsii. [Phytoplankton production measurements with ^{14}C under consideration for losses of labelled organic matter during filtration.]-Okeanologiya *15* : 503 - 507, 1975. [In R, ab : E.]

21953 - **BUL'ON, V.V.** : Produktsiya fitoplanktona ozera Baĭkal v raĭone pos. Listvennichnoe. [Phytoplankton production of the Lake Baikal in the region of the village Listvennichnoe.] - Dokl. Akad. Nauk SSSR *223* : 737 - 740, 1975. [In R.]

21954 - **BUNCE, N.J., HADLEY, M., MELLORS, A., SAFORD, W.E.** : A possible protective mechanism for plastoquinone-9 *in vivo*. - Photosynthetica *9* : 220 - 222, 1975.

21955 - **BUNT, J.S.** : Primary productivity of marine ecosystems. - In : **LIETH, H., WHITTAKER, R.H.** (ed.) : Primary Productivity of the Biosphere. Pp. 169 - 183. Springer-Verlag, Berlin - Heidelberg - New York 1975.

21956 - **BURGESS, M.D., COX, L.M.** : A bipolar analog integrator for use with net radiometers. - Agr. Meteorol. *15* : 385 - 391, 1975.

21957 - **BURKE, J.J., TRELEASE, R.N.** : Malate synthase and glycolate oxidase cytochemistry in cucumber microbodies. - Plant Physiol. *56* (Suppl.) : 87, 1975.

21958 - **BURKETT, R.D., TAYLOR, D.R.** : A bottle holder for *in situ* primary productivity studies. - Progr. Fish-Culturist *37* : 112, 1975.

21959 - **BUSCHMANN, C., LICHTENTHALER, H.K.** : Hill-reaction of chloroplasts from *Raphanus* seedlings grown with β-indoleacetic acid and kinetin. - In : **AVRON, M.** (ed.) : Proceedings of the Third International Congress on Photosynthesis. Vol. I. Pp. 753 - 756. Elsevier, Amsterdam - Oxford - New York 1975.

21960 - **BUSS, G.R., BARNES, D.K.** : Inheritance of a chlorotic seedling trait that affects zygote survival in diploid alfalfa. - Crop Sci. *15*:185 - 186, 1975.

21961 - **BUTA, J.G.** : Plant growth regulating activity of substituted phthalate esters.- Agr. Food Chem. *23* : 801 - 802, 1975. [Chl.]

21962 - **BUTLER, W.L., KITAJIMA, M.** : A tripartite model for chloroplast fluorescence.- In : **AVRON, M.** (ed.) : Proceedings of the Third International Congress on Photosynthesis. Vol. I. Pp. 13 - 24. Elsevier, Amsterdam - Oxford - New York 1975.

21963 - **BUTLER, W.L., KITAJIMA, M.** : Energy transfer in the photochemical apparatus of chloroplasts. - Biophys. J. *15* (2, Part 2) : 221 a, 1975.

21964 - **BUTLER, W.L., KITAJIMA, M.** : Fluorescence quenching in Photosystem II of chloroplasts. - Biochim. biophys. Acta *376* : 116 - 125, 1975.

21965 - **BUTLER, W.L., KITAJIMA, M.** : Energy transfer between photosystem II and photosystem I in chloroplasts. - Biochim. biophys. Acta *396* : 72 - 85, 1975.

21966 - **BYERRUM, R.U., BENSON, A.A.** : Effect of ammonia on photosynthetic rate and photosynthate release by *Amphidinium carterae (Dinophyceae)*. - J. Phycol. *11*: 449 - 452, 1975.

21967 - **BYKOV, O.D.** :Fiziologicheskie issledovaniya v oblasti fotosinteza (1965 - 1974 gg.). [Physiological studies in photosynthesis (1965 - 1974).] - Tr. prikl. Bot., Genet. Selektsii *56* (1) : 162 - 172, 1975. [In R.]

21968 - **BYKOV, O.D., LOKUTSIEVSKAYA, L.K., SAMOĬLOVA, L.A.** : Vliyanie izotopnogo sostava CO_2 na kinetiku fotosinteza C-3 i C-4 rasteniĭ. [Effect of isotope composition of CO_2 on photosynthesis kinetics in C_3 and C_4 plants.] - Byull. vses. Inst. Rastenievod. N.I. Vavilova *56* : 9 - 15, 1975. [In R.]

21969 - **BYSTROVA, M.I., PAKSHINA, E.V., KRASNOVSKIĬ, A.A.** : Izuchenie feofitinizatsii agregirovannykh form khlorofilla. [Pheophytinization of aggregated forms of chlorophyll.] - Biokhimiya *40* : 137 - 144, 1975. [In R, ab : E.]

21970 - **BYSTRYKH, E.E., MATORIN, D.N.** : Funktsional'naya aktivnost' fotosinteticheskogo apparata v ontogeneze rasteniĭ podsolnechnika. [Functional activity of the photosynthetic apparatus during ontogenesis of sunflower plants.] - Sel'skokhoz. Biol. *10* : 230 - 236, 1975. [In R, ab : E.]

21971 - **CABANYES, J. de, BOUTHELIER, V., MUZQUIZ, M.** : Metodo cromatografico de analise de carotenos. [Chromatographic method of analysis of carotenoids.] - An. Inst. nac. Invest. agr., Ser. gen. *1975* (3) : 81 - 129, 1975. [In Span., ab:E.]

21972 - **CADÉE, G.C.** : Primary production of the Guyana coast. - Neth. J. Sea Res. *9* : 128 - 143, 1975.

21973 - **CALDWELL, M.** : Primary production of grazing lands. - In : **COOPER, J.P.** (ed.): Photosynthesis and Productivity in Different Environments. Pp. 41 - 73. Cambridge University Press, Cambridge - London - New York - Melbourne 1975.

21974 - **CALÈ, M.T., FIGLIOLIA, A., IZZA, C., TOMBESI, L.** : Alcuni aspetti del metabolismo dei vegetali in funzione delle disponibilità idriche del suolo. [Some aspects of plant metabolism in relation to the water resources of soil.] - Ann. Ist. Sperim. Nutr. Piante *6* (6) : 3 - 9, 1975. [Ps; in Ital.,ab : E.]

21975 - **CALMES, J., VIGNES, D., VIALA, G.** : Influence de l'exposition (Sud ou Nord) sur le métabolisme glucidique et aminé des feuilles de Vigne vierge. - Oecol. Plant. *10* : 281 - 294, 1975.

21976 - **CALVAYRAC, R., LEDOIGT, G., DUBERTRET, G.** : Changements affèctant la composition macromoléculaire du plastidome d'*Euglena gracilis* Z., au cour de la croissance hétérotrophe. - Plant Sci. Lett. *5* : 365 - 374, 1975. [Ps, Chl.]

21977 - **CALVERT, A., SLACK, G.** : Effects of carbon dioxide enrichment on growth, development and yield of glasshouse tomatoes. I. Responses to controlled concentrations. - J. hort. Sci. *50* : 61 - 71, 1975.

*21978 - **CALVERT, H.E., DAWES, C.J.** : Comparative chloroplast ultrastructure in the genus *Caulerpa* LAMOUROUX, with special reference to phylogenetic relationships. - J. Phycol. *10* (Suppl.) : 4 - 5, 1974.

B21979 - **CALVIN, M.** : Photosynthesis as a Resource for Energy and Materials. - Rep. LBL-4239. Pp. 1 - 62. Lawrence Berkeley Lab., Univ. California, Berkeley 1975.

21980 - **CALVIN, M.** : Photosynthesis as a resource for energy and materials. - Kemia - Kemi *2* (2) : 46 - 57, 1975.

21981 - **CAMMACK, R.** : Effects of solvent on the properties of ferredoxins. - Biochem. Soc. Trans. *3* : 482 - 488, 1975.

21982 - **CAMMACK, R.** : Measurement of ATP synthesis by a radioisotope tracer technique: photosynthetic phosphorylation by chloroplasts. - In : **HALL, D.O., HAWKINS, S.E.** (ed.) : Laboratory Manual of Cell Biology. Pp. 36 - 38. The English University Press, London 1975.

21983 - **CAMMACK, R., EVANS, M.C.W.** : E.P.R. spectra of iron-sulphur proteins in dimethylsulphoxide solutions : Evidence that chloroplast Photosystem I particles contain 4Fe-4S centres. - Biochem. biophys. Res. Commun. *67* : 544 - 549, 1975.

21984 - **CAMP, R.R., WHITTINGHAM, W.F.** : Fine structure of chloroplasts in "green islands" and in surrounding chlorotic areas of barley leaves infected by powdery mildew. - Amer. J. Bot. *62* : 403 - 409, 1975.

21985 - **CAMPBELL, L.E., THIMIJAN, R.W., CATHEY, H.M.** : Spectral radiant power of lamps used in horticulture. - Trans. ASAE *18* : 952 - 956, 1975. [Ps.]

21986 - **CAMPILLO, A.J., HYER, R.C., KOLLMAN, V.H., SHAPIRO, S.L., SUTPHIN, H.D.** : Fluorescence lifetimes of α- and β-carotenes. - Biochim. biophys. Acta *387* : 533 - 535, 1975.

21987 - **CANTLEY, L.C. Jr., HAMMES, G.G.** : Characterization of nucleotide binding sites on chloroplast coupling factor 1. - Biochemistry *14* : 2968 - 2975, 1975.

21988 - **CANTLEY, L.C. Jr., HAMMES, G.G.** : Fluorescence energy transfer between ligand binding sites on chloroplast coupling factor 1. - Biochemistry *14* : 2976 - 2981, 1975.

21989 - **CANTLEY, L.C. Jr., HAMMES, G.G.** : Nucleotide binding sites on spinach chloroplast coupling factor 1. - Fed. Proc. *34* : 595, 1975.

21990 - **CAPBLANCQ, J., LAVANDIER, P.** : Production primaire et bilan de l'oxygène dissous dans un ruisseau des Pyrénées Centrales. - Ann. Limnol. *11* : 189 - 201, 1975.

21991 - **CARAPELLUCCI, P.A., MAUZERALL, D.** : Photosynthesis and porphyrin excited state redox reactions. - Ann. New York Acad. Sci. *244* : 214 - 238, 1975.

21992 - **CARELL, E.F.** : Modification by vitamin B_{12} of the cell cycle in *Euglena* : Detailed studies on the induction and recovery from B_{12} deficiency. - In : Les Cycles Cellulaires et leur Blocage chez Plusiers Protistes. (Coll. int. CNRS *240*). Pp. 321 - 329. Édit. CNRS, Paris 1975. [Chl, chloroplast.]

21993 - **CARITHERS, R.P., PARSON, W.W.** : Delayed fluorescence from *Rhodopseudomonas viridis* following single flashes. - Biochim. biophys. Acta *387* : 194 - 211, 1975.

21994 - **CARLSON, P.S., POLACCO, J.C.** : Plant cell cultures : genetic aspects of crop improvement. - Science *188* : 622 - 625, 1975.[Ps.]

21995 - **CARLSON, R.W., BAZZAZ, F.A., ROLFE, G.L.** : The effect of heavy metals on plants. II. Net photosynthesis and transpiration of whole corn and sunflower plants treated with Pb, Cd, Ni and Tl. - Environ. Res. *10* : 113 - 120, 1975.

21996 - **CARMELI, C., LIFSHITZ, Y.** : Pi-ATP exchange catalyzed by vesicles reconstituted from chloroplast lipids. - In : **AVRON, M.** (ed.) : Proceedings of the Third International Congress on Photosynthesis. Vol. II. Pp. 849 - 858. Elsevier, Amsterdam - Oxford - New York 1975.

21997 - **CARMELI, C., LIFSHITZ, Y., GEPSHTEIN, A.** : Control of proton translocation induced by ATPase activity in chloroplasts. - Biochim. biophys. Acta *376* : 249 - 258, 1975.

21998 - **CAROW, B.** : Ein Beitrag zur Blattflächenbestimmung. - Gartenbauwissenschaft *40* : 125 - 128, 1975.

21999 - **CARTER, J.L., GARRARD, L.A.** : Effect of 10 C-night temperature and leaf assimilate levels on photosynthesis of *Digitaria decumbens* STENT. - Soil Crop Sci. Soc. Florida Proc. *35* : 3 - 5, 1975.

22000 - **CASPERS, N.** : Untersuchungen über Individuendichte, Biomasse und kalorische Äquivalente des Makrobenthos eines Waldbaches. - Int. Rev. ges. Hydrobiol. *60* : 557 - 566, 1975.

22001 - **CASTELFRANCO, P.A., JONES, O.T.G.** : Protoheme turnover and chlorophyll synthesis in greening barley tissue. - Plant Physiol. *55* : 485 - 490, 1975.

22002 - **CATALANO, M., SCIANCALEPORE, V.** : I lipidi totali della grupa e degli organi fotosintetizzanti dell'olivo. Ricerche preliminari. [Total lipids of drupes and photosynthesizing organs of olive tree. Preliminary research.] - Riv. ital. Sostanze grasse *52* : 114 - 116, 1975. [In Ital., ab : E.]

22003 - **CATALANO, M., SCIANCALEPORE, V.** : La compozicione acidica dei lipidi neutri, glucolipidi e fosfolipidi estratti da foglie di olivo e da cloroplasti. Nota II. [Acidic composition of neutral lipids, glucolipids and phospholipids from olive tree leaves and chloroplasts. Note 2.] - Riv. ital. Sostanze grasse *52* : 276 - 280, 1975. [In Ital., ab : E.]

22004 - **CATLIN, P.B., OLSSON, E.A., BEUTEL, J.A.** : Reduced translocation of carbon and nitrogen from leaves with symptoms of Pearl Curl. - J. amer. Soc. hort. Sci. *100* : 184 - 187, 1975.

22005 - **ČATSKÝ, J., TICHÁ, I.** : A closed system for measurement of photosynthesis, photorespiration and transpiration rates. - Biol. Plant. *17* : 405 - 410, 1975.

22006 - **CAUBERGS, R., De GREEF, J.A.** : Behaviour of cotyledons in decapitated seedlings of *Phaseolus vulgaris* L. cv. Limburg. - Arch. int. Physiol. Biochim. *83* : 172 - 174, 1975. [Ps.]

22007 - **ČERNÝ, V.** : Dynamika růstu cukrovky při extrémních hustotách porostu. [Growth dynamics of sugar beet at extreme stand densities.] - Rost. Výroba (Praha) *21* : 37 - 46, 1975. [Dry matter production ; in Czech, ab : E, R.]

22008 - **CERVINKA, V., CHANCELLOR, W.J., COFFELT, R.J., CURLEY, R.G., DOBIE, J.B., HARRISON, B.D.** : Methods used in determining energy flows in California agriculture. - Trans. ASAE *18* : 246 - 251, 1975.

22009 - **CHABOT, J.F., CHABOT, B.F.** : Developmental and seasonal patterns of mesophyll ultrastructure in *Abies balsamea*. - Can. J. Bot. *53* : 295 - 304, 1975. [Peroxisome, chloroplast.]

22010 - **CHACKO MATHEW, RAMADASAN, A.** : Photosynthetic efficiency in relation to annual yield and chlorophyll content in the coconut palm. - J. Plantation Crops *3* : 26 - 28, 1975.

22011 - **CHAĬKA, M.T., SAVCHENKO, G.E.** : Fotoregulyatsiya biosinteza khlorofilla v protsesse razvitiya khloroplasta. [Photoregulation of chlorophyll biosynthesis and chloroplast development.] - In : Fotoregulyatsiya Metabolizma i Morfogeneza Rasteniĭ. Pp. 120 - 134, 251. Nauka, Moskva 1975. [In R.]

22012 - **CHAKRAVORTY, N.K., MUHSI, A.A.A.** : A study of the relationship between chlorophyll content, chlorophyll contributing factors and yield of different varieties of Aus paddy. - Bangladesh J. Bot. *4* (1-2) : 101 - 103, 1975.

22013 - **CHALKER, B.E., TAYLOR, D.L.** : Light-enhanced calcification, and the role of oxidative phosphorylation in calcification of the coral *Acropora cervicornis*.- Proc. roy. Soc. London *B 190* : 323 - 331, 1975. [Ps.]

22014 - **CHALMERS, D.J., CANTERFORD, R.L., JERIE, P.H., JONES, T.R., UGALDE, T.D.** : Photosynthesis in relation to growth and distribution of fruit in peach trees.- Aust. J. Plant Physiol. *2* : 635 - 645, 1975.

22015 - **CHALMERS, D.J., ENDE, B. van den** : Productivity of peach trees : factors affecting dry-weight distribution during tree growth. - Ann. Bot. *39* : 423 - 432, 1975.

22016 - **CHALONER, W.G., COLLINSON, M.E.** : Application of SEM to a sigillarian impression fossil. - Rev. Palaeobot. Palynol. *20* : 85 - 101, 1975. [Potential Ps.]

22017 - **CHAMPIGNY, M.L.** : Les phosphorylations photosynthétiques. - In : **COSTES, C.** (ed.) : Photosynthèse et Production Végétale. Pp. 73 - 87. Gauthier-Villars, Paris 1975.

22018 - **CHAMPIGNY, M.L., BISMUTH, E., LAVERGNE, D.** : Phosphoribulokinase and 3-phosphoglyceric acid kinase of spinach chloroplasts : Study of their participation in the photosynthetic carbon reduction cycle. - In : **AVRON, M.** (ed.) : Proceedings of the Third International Congress on Photosynthesis. Vol. II. Pp. 1441 - 1448. Elsevier, Amsterdam - Oxford - New York 1975.

22019 - **CHAMPIGNY, M.L., MOYSE, A.** : Effect of temperature on spinach chloroplast carboxylation activity. - Biochem. Physiol. Pflanzen *168* : 575 - 583, 1975.

22020 - **CHANCE, B., BALTSCHEFFSKY, M.** : Carotenoid and merocyanine probes in chromatophore membranes. - In : **EISENBERG, H., KATCHALSKI-KATZIR, E., MANSON, L.A.** (ed.) : Biomembranes. Vol. 7. Pp. 33 - 55. Plenum Publ. Comp., New York 1975.

22021 - **CHANCE, B., LEGALLAIS, V., SORGE, J., GRAHAM, N.** : A versatile time-sharing multichannel spectrophotometer, reflectometer and fluorometer. - Anal. Biochem. *66* : 498 - 514, 1975. [Cytochromes.]

22022 - **CHANCE, B., PORTE, M., HESS, B., OESTERHELT, D.** : Low temperature kinetics of H^+ changes of bacterial rhodopsin. - Biophys. J. *15* (2, Part 2) : 66 a, 1975.

22023 - **CHANG, C.S., JOHNSON, W.H.** : Moisture and weight distributions in bright leaf tobacco. - Tobacco Sci. *19* : 131 - 134, 1975. Tobacco *177* (21) : 67 - 70, 1975. [Leaf insertion, distribution on leaf area.]

22024 - **CHANG, C.W.** : Activation energy for thermal inactivation and K_m of carbonic anhydrase from the cotton plant. - Plant Sci. Lett. *4* : 109 - 113, 1975.

22025 - **CHANG, C.W.** : Carbon dioxide and senescence in cotton plants. - Plant Physiol. *55* : 515 - 519, 1975.

22026 - **CHANG, C.W.** : Extraction of starch and amylose from cotton leaves by a maceration technique. - Plant Physiol. *56* (Suppl.) : 78, 1975.

22027 - **CHANG, C.W.** : Carbonic anhydrase of the cotton plant. - Phytochemistry *14* : 119 - 121, 1975.

22028 - **CHANG, C.W.** : Fluorides. - In : **MUDD, J.B., KOZLOWSKI, T.T.** (ed.) : Responses of Plants to Air Pollution. Pp. 57 - 95. Academic Press, New York - San Francisco - London 1975. [Ps.]

22029 - **CHANG, C.W.** : Manometric method for determining bicarbonate content in plant leaves. - Anal. Biochem. *66* : 185 - 193, 1975.

22030 - **CHANG, N.K.** : [Determining cutting schedules for maximum yield and quality of orchardgrass and alfalfa.] - Seoul Univ. J. (B) *25* : 225 - 235, 1975.[Growth analysis; in Korean, ab : E.]

22031 - **CHANG, N.K.** : Theoretical studies on the kinetics of matter production and photosynthesis. - Seoul nat. Univ. Fac. Pap. Biol. Agr. Ser. (E) *4* : 11 - 21, 1975.

22032 - **CHAPMAN, D.J., LEECH, R.M.** : Identification of changes in photosynthetic amino acid metabolism in developing leaf tissue. - Plant Physiol. *56* (Suppl.) : 27, 1975.

22033 - **CHAPMAN, E.A., BAIN, J.M., GOVE, D.W.** : Mitochondria and chloroplast peripheral reticulum in the C_4 plants *Amaranthus edulis* and *Atriplex spongiosa*. - Aust. J. Plant Physiol. *2* : 207 - 223, 1975.

22034 - **CHAPMAN, S.B., HIBBLE, J., RAFAREL, C.R.** : Net aerial production by *Calluna vulgaris* on lowland heath in Britain. - J. Ecol. *63* : 233 - 258, 1975.

22035 - **CHAPMAN, S.B., HIBBLE, J., RAFAREL, C.R.** : Litter accumulation under *Calluna vulgaris* on a lowland heathland in Britain. - J. Ecol. *63* : 259 - 271, 1975. [Carbon budget.]

22036 - **CHARLES-EDWARDS, D.A.** : Efficiency and experiency in plant growth. - Ann. Bot. *39* : 161 - 162, 1975.

22037 - **CHARLES-EDWARDS, D.A., HO, L.C.** : Translocation and carbon metabolism in tomato leaves. - Ann. Bot. *40* : 387 - 389, 1975.

22038 - **CHARLES-EDWARDS, D.A., LUDWIG, L.J.** : A model of leaf carbon metabolism. - Ann. Bot. *39* : 819 - 829, 1975.

22039 - **CHARLES-EDWARDS, D.A., LUDWIG, L.J.** : The basis of expression of leaf photosynthetic activities. - In : **MARCELLE, R.** (ed.) : Environmental and Biological Control of Photosynthesis. Pp. 37 - 44. Dr. W. Junk b.v. Publ., The Hague 1975.

22040 - **CHARTIER, P.** : L'assimilation du CO_2 à l'échelle de la feuille et à celle de la culture. - In : **COSTES, C.** (ed.) : Photosynthèse et Production Végétale. Pp. 171 - 191. Gauthier-Villars, Paris 1975.

22041 - **CHARTIER, P., ČATSKÝ, J.** : Photosynthetic systems - conclusions. - In : **COOPER, J.P.** (ed.) : Photosynthesis and Productivity in Different Environments. Pp. 425 - 433. Cambridge University Press, Cambridge - London - New York - Melbourne 1975.

22042 - **CHATTERTON, N.J., HANNA, W.W., POWELL, J.B., LEE, D.R.** : Photosynthesis and transpiration of bloom and bloomless sorghum. - Can. J. Plant Sci. *55* : 641 - 643, 1975.

22043 - **CHAUDHARI, H.K., BARROW, J.R.** : Identification of cotton haploids by stomatal chloroplast-count technique. - Crop Sci. *15* : 760 - 763, 1975.

22044 - **CHEBOTAREVA, E.A., KUKUSHKIN, A.K.** : Issledovanie temperaturnoĭ zavisimosti intensivnosti dlinnovolnovoĭ fluorestsentsii list'ev vysshikh rasteniĭ. [Temperature dependence of the rate of long-wave fluorescence in leaves of higher plants.] - Biofizika *20* : 98 - 100, 1975. [In R, ab : E.]

22045 - **CHEKULAEVA, L.N., KOROLEV, Yu.N., TELEGIN, N.L., RIKHIREVA, G.T.** : Izuchenie obrazovaniya purpurnykh membran v protsesse kul'tivirovaniya solevykh bakteriĭ.

[Formation of purple membranes in the process of cultivation of salt bacteria.] - Biofizika *20* : 839 - 843, 1975. [In R, ab : E.]

22046 - **CHEN, C.-H., BERNS, D.S.** : Modification of intensity and direction of electron flow across bileaflet membranes. - Biophys. J. *15* (2, Part 2) : 241 a, 1975. [Chloroplast.]

22047 - **CHEN, C.-H., BERNS, D.S.** : Modification of intensity and direction of electron flow across bileaflet membranes. - Proc. nat. Acad. Sci. USA *72* : 3407 - 3411, 1975. [Biliproteins.]

22048 - **CHEN, C.W., ORLOB, G.T.** : Ecologic simulation for aquatic environments. - In : **PATTEN, B.C.** (ed.) : Systems Analysis and Simulation in Ecology. Vol. III. Pp. 476 - 588. Academic Press, New York - San Francisco - London 1975. [Chl.]

22049 - **CHEN, K., GRAY, J.C., WILDMAN, S.G.** : Fraction I protein and the origin of polyploid wheats. - Science *190* : 1304 - 1306, 1975.

22050 - **CHEN, K., KUNG, S.D., GRAY, J.C., WILDMAN, S.G.** : Polypeptide composition of fraction I protein from *Nicotiana glauca* and from cultivars of *Nicotiana tabacum*, including a male sterile line. - Biochem. Genet. *13* : 771 - 778, 1975.

22051 - **CHERMNYKH, L.N., CHUGUNOVA, N.G., KOSOBRYUKHOV, A.A., KARPILOVA, I.F.** : O vosstanovlenii aktivnosti fotosinteticheskogo apparata ogurtsov pri sozdanii optimal'nogo temperaturnogo rezhima v zone korneĭ. [Restoration of activity of the photosynthetic apparatus of cucumber at the optimum temperature relations in the root zone.] - Nauch. Dokl. vyssh. Shkoly, biol. Nauki *18* (11) : 72 - - 77, 1975. [In R.]

B22052 - **CHERNIGOVSKIĬ, V.N.** (ed.) : Problemy Kosmicheskoĭ Biologii. Tom 28. Eksperimental'nye Ėkologicheskie Sistemy Vklyuchayushchie Cheloveka. [Problems of Cosmic Biology. Vol. 28. Experimental Ecological Systems Including Man.] - Nauka, Moskva 1975. [Ps and productivity in systems algae - higher plants - - man; In R.]

22053 - **CHERNOV, Yu.J., DOROGOSTAISKAYA, E.V., GERASIMENKO, T.V., IGNATENKO, I.V., MATVEYEVA, N.V., PARINKINA, O.M., POLOZOVA, T.G., ROMANOVA, E.N., SCHAMURIN, V.F., SMIRNOVA, N.V., STEPANOVA, I.V., TOMILIN, B.A., VINOKUROV, A.A., ZALENSKY, O.V.** : Tareya. - In : **ROSSWALL, T., HEAL, O.W.** (ed.) : Ecological Bulletins (NFR 20. Structure and Function of Tundra Ecosystems). Pp. 159 - 182. Stockholm 1975. [Ps.]

22054 - **CHERNYAD'EV, I.I., TEREKHOVA, I.V., DOMAN, N.G., AL'BITSKAYA, O.N., GORONKOVA, O.I.** : Vliyanie pH sredy na skorost' assimilyatsii ugleroda i aktivnost' nekotorykh fermentov fotosinteza u spiruliny. [Effect of pH of the medium on the rate of carbon assimilation and the activity of some enzymes of photosynthesis in *Spirulina*.]- Fiziol. Rast. *22* : 903 - 909, 1975. [In R, ab : E.]

22055 - **CHERNYADYEV, I.I., GUSHCHINA, L.M., VAKLINOVA, S.G., DOMAN, N.G.** : Obtaining purified preparation of ribulose-diphosphate carboxylase from cells of phototrophic bacteria and algae. - Dokl. bolg. Akad. Nauk *28* : 821 - 824, 1975.

22056 - **CHEVALLIER, D.** : Effets d'un séjour à l'obscurité sur le pouvoir germinatif et sur l'évolution de l'appareil photosynthétique des spores de *Funaria hygrometrica*. - Physiol. Plant. *34*:216 - 220, 1975.

22057 - **CHIBISOV, A.K.** : Primenenie metoda impul'snogo fotoliza dlya issledovaniya okislitel'no-vosstanovitel'nykh fotoreaktsiĭ khlorofilla. [Use of the method of impulse photolysis for studying redox photoreactions of chlorophyll.] - In : Metody Issledovaniya Fotokhimicheskikh Reaktsiĭ Fotosinteza *in Vitro* i *in Vivo*. Pp. 102 - 116. Pushchino 1975. [In R.]

22058 - **CHIBISOV, A.K., ZAKHAROVA, N.I., SLAVNOVA, T.D., PESHKIN, A.F., KARYAKIN, A.V.**: Kratkovremennye svetoindutsirovannye izmeneniya pogloshcheniya khlorofill-belkovykh kompleksov 1-ĭ fotosistemy. [Short-time light-induced changes in the absorption of chlorophyll-protein complexes of the photosystem I.] - Dokl. Akad. Nauk SSSR *221* : 978 - 980, 1975. [In R.]

22059 - **CHIGNELL, C.F., CHIGNELL, D.A.** : A spin label study of purple membranes from *Halobacterium halobium*. - Biophys. J. *15* (2, Part 2) : 67a, 1975.

22060 - **CHIGNELL, C.F., CHIGNELL, D.A.** : A spin label study of purple membranes from *HAlobacterium halobium*. - Biochem. biophys. Res. Commun. *62* : 136 - 143, 1975.

22061 - **CHIKOV, V.I., NIKOLAEV, B.A.** : Vliyanie zasukhi na postfotosinteticheskie prevrashcheniya C^{14} v list'yakh bobov. [Effect of drought on post-photosynthetic transformations of ^{14}C in broad-bean leaves.] - Fiziol. Rast. *22* : 587 - 590, 1975. [In R, ab : E.]

22062 - **CHIMIKLIS, P., HEATH, R.** : Fluorescence transients of O_3 gassed *Chlorella*. - Plant Physiol. *56* (Suppl.) : 5, 1975.

22063 - **CHIN, P., BRODY, S.S.** : Monomolecular films of oxidized and reduced cytochrome *f*. - In : **AVRON, M.** (ed.) : Proceedings of the Third International Congress on Photosynthesis. Vol. I. Pp. 255 - 260. Elsevier, Amsterdam - Oxford - New York 1975.

22064 - **CHING, T.M., HEDTKE, S., GARAY, A.E.** : Energy state of wheat leaves in ammonium nitrate-treated plants. - Life Sci. *16* : 603 - 610, 1975. [Ps.]

22065 - **CHKHUBIANISHVILI, R.I.** : Morfologo-anatomicheskaya kharakteristika fotosinteziruyushchikh sistem buka vostochnogo sformirovavshikhsya v raznykh yarusakh drevostoya. [Morpho-anatomical characteristics of photosynthesizing systems of European beech formed in various layers of a stand.] - Soobshch. Akad. Nauk gruz. SSR *77* : 421 - 424, 1975. [In R, ab : E, Georg.]

22066 - **CHMORA, S.N., SLOBODSKAYA, G.A., NICHIPOROVICH, A.A.** : O vzaimosvyazi fotosinteza i fotodykhaniya u rasteniĭ s razlichnoĭ aktivnost'yu fotosinteticheskogo apparata. [Correlation between photosynthesis and photorespiration in plants with different activity of photosynthetic apparatus.] - Fiziol. Rast. *22* : 1101 - 1107, 1975. [In R, ab : E.]

22067 - **CHOE, H.T., NELSON, D.** : Light qualities, kinetin and chlorophyll synthesis as related to the growth response of *Pisum sativum*.- Plant Physiol. *56* (Suppl.): 30, 1975.

22068 - **CHOE, H.T., THIMANN, K.V.** : The metabolism of oat leaves during senescence. III. The senescence of isolated chloroplasts. - Plant Physiol. *55* : 828 - 834, 1975.

22069 - **CHOLLET, R., ANDERSON, L.L., HOVSEPIAN, L.C.** : The absence of tightly bound copper, iron, and flavin nucleotide in crystalline ribulose 1,5-bisphosphate carboxylase-oxygenase from tobacco. - Plant Physiol. *56* (Suppl.) : 26, 1975.

22070 - **CHOLLET, R., ANDERSON, L.L., HOVSEPIAN, L.C.** : The absence of tightly bound copper, iron and flavin nucleotide in crystalline ribulose 1,5-bisphosphate carboxylase-oxygenase from tobacco. - Biochem. biophys. Res. Commun. *64* : 97 - 107, 1975.

22071 - **CHOLLET, R., OGREN, W.L.** : Regulation of photorespiration in C_3 and C_4 species. - Bot. Rev. *41* : 137 - 179, 1975.

22072 - **CHOW, H.-C., SERLIN, R., STROUSE, C.E.** : The crystal and molecular structure and absolute configuration of ethyl chlorophyllide *a* dihydrate. A model for the different spectral forms of chlorophyll *a*. - J. amer. chem. Soc. *97* : 7230 - 7237, 1975.

22073 - **CHRELASHVILI, M.N., TAKAĬSHVILI, T.V., GAMKRELIDZE, L.M., DATIASHVILI, N.A.** : [Photosynthetic rate of some higher plants.] - Soobshch. Akad. Nauk gruz. SSR *77* : 161 - 163, 1975. [In Georg., ab : E, R.]

22074 - **CHRISTIE, E.K.** : Physiological responses of semiarid grasses. III. Growth in relation to temperature and soil water deficit. - Aust. J. agr. Res. *26* : 447- - 457, 1975. [Growth analysis.]

22075 - **CHRISTIE, E.K.** : Physiological responses of semiarid grasses. IV. Photosynthetic rates of *Thyridolepis mitchelliana* and *Cenchrus ciliaris* leaves. - Aust. J. agr. Res. *26* : 459 - 466, 1975.

22076 - **CHRISTIE, E.K., MOORBY, J.** : Physiological responses of semiarid grasses. I. The influence of phosphorus supply on growth and phosphorus absorption. - Aust. J. agr. Res. *26* : 423 - 436, 1975. [Growth analysis.]

22077 - **CHU, D.K., BASSHAM, J.A.** : Regulation of ribulose 1,5-diphosphate carboxylase by substrates and other metabolites. Further evidence for several types of binding sites. - Plant Physiol. *55* : 720 - 726, 1975.

22078 - **CHUA, N.-H., BENNOUN, P.** : Thylakoid membrane polypeptides of *Chlamydomonas reinhardtii* : Wild-type and mutant strains deficient in photosystem II reaction center. - Proc. nat. Acad. Sci. USA *72* : 2175 - 2179, 1975.

22079 - **CHUA, N.-H., MATLIN, K., BENNOUN, P.** : A chlorophyll-protein complex lacking in photosystem I mutants of *Chlamydonomas reinhardtii*. - J. Cell Biol. *67* : 361 - 377, 1975.

22080 - **CHUB, A.I.** : Izmenenie sostava assimilyatov u soi pod vliyaniem molibdena. [Changes in photosynthates composition in soybean induced by molybdenum.] - Fiziol. Biokhim. kul't. Rast. *7* : 607 - 610, 1975. [In R, ab : E.]

22081 - **CHUECA, A., LĂZARO, J.J., LŌPEZ GORGE, J., MAYOR, F.** : Fructosa-1,6-difosfatasa de cloroplastos de hojas de espinaca : purificacion del factor proteico ligado a dicha actividad. [Fructose-1,6-diphosphatase from spinach leaves chloroplasts : purification of the protein factor linked to this activity.] - An. Edafol. Agrobiol. *34* : 541 - 548, 1975. [In Span., ab : E.]

22082 - **CHUGUNOVA, N.G., BIL', K.Ya., CHERMNYKH, L.N.** : Struktura i fotosintez list'ev ogurtsov pri razlichnykh temperaturakh v kornevoĭ zone. [Structure and photosynthetic activity of cucumber leaves at different temperatures in root zone.] - Fiziol. Rast. *22* : 688 - 694, 1975. [In R, ab : E.]

22083 - **CIFFERRI, O., TIBONI, O., AMILENI, A.R.** : Effect of different antibiotics on the presence of a chloroplast-specific elongation factor in *Chlorella vulgaris*. - In : **PUISEUX-DAO, S.** (ed.) : Molecular Biology of Nucleocytoplasmic Relationships. Pp. 47 - 51. Elsevier, Amsterdam - Oxford - New York 1975.

22084 - **CIHA, A.J., BRUN, W.A.** : Stomatal size and frequency in soybeans. - Crop Sci. *15* : 309 - 313, 1975.

22085 - **CIMANOWSKI, J., NOVACKI, J., MILLIKAN, D.F.** : Effect of light on the development of powdery mildew on and some physiological and morphological changes in apple leaf tissue. - Phytoprotection *56* : 96 - 103, 1975. [Ps,Chl.]

22086 - **CLARK, J.B., LISTER, G.R.** : Photosynthetic action spectra of trees. I. Comparative photosynthetic action spectra of one deciduous and four coniferous tree species as related to photorespiration and pigment complements. - Plant Physiol. *55* : 401 - 406, 1975.

22087 - **CLARK, J.B., LISTER, G.R.** : Photosynthetic action spectra of trees. II. The relationship of cuticle structure to the visible and ultraviolet spectral properties of needles from four coniferous species. - Plant Physiol. *55* : 407 - 413, 1975.

22088 - **CLARKE, R.H., CONNORS, R.E., NORRIS, J.R., THURNAUER, M.C.** : Optically detected zero-field magnetic resonance studies of the photoexcited triplet state of the photosynthetic bacterium *Rhodospirillum rubrum*. - J. amer. chem. Soc. *97* : 7178 - 7179, 1975.

22089 - **CLAUSS, H.** : Photosynthese-Intensität und Stoffwechsel der *Acetabularia*-Zelle in Abhängigkeit von ihrem Entwicklungszustand. - Protoplasma *83* : 147 - 166, 1975.

22090 - **CLAYTON, R.K., PARSON, W.W.** : Excited states in the photochemistry of bacterial photosynthesis. - In : **AVRON, M.** (ed.) : Proceedings of the Third International Congress on Photosynthesis. Vol. I. Pp. 409 - 411. Elsevier, Amsterdam - Oxford - New York 1975.

22091 - **CLEMENT-METRAL, J.D.** : Direct observation of the rotation in a constant magnetic field of highly organized lamellar structures. - FEBS Lett. *50* : 257 - 260, 1975.

22092 - **CLEMENT-METRAL, J.D.** : Preparation of metabolically active protoplasts from the red alga *Porphyridium cruentum*. - In : **AVRON, M.** (ed.) : Proceedings of the Third International Congress on Photosynthesis. Vol.III. Pp. 2067 - 2071. Elsevier, Amsterdam - Oxford - New York 1975.

22093 - **CLEMENT-METRAL, J.D., FANICA-GAIGNIER, M.** : Regulation of bacteriochlorophyll synthesis in "low-iron" grown *Rhodopseudomonas spheroides* Y : Role of 5-amino levulinic synthetases. - In : **AVRON, M.** (ed.) : Proceedings of the Third International Congress on Photosyntesis. Vol.III. Pp. 2147 - 2152. Elsevier, Amsterdam - Oxford - New York 1975.

22094 - **CLEMENT-METRAL, J.D., FANICA-GAIGNIER, M.** : 5-aminolevulinic-acid synthetases from *Rhodopseudomonas spheroides* Y. Comparison of the purification and properties of enzymes extracted from bacteria grown in different iron concentrations. - Europe. J. Biochem. *59* : 73 - 77, 1975.

22095 - **CLEZY, P.S., FOOKES, C.J.R.** : Synthesis of the methyl ester of the magnesium--free derivative of chlorophyll c_2. - J. chem. Soc. chem. Commun. *17* : 707 - 708, 1975.

22096 - **CLIFFORD, P.E., LANGER, R.H.M.** : Pattern and control of distribution of ^{14}C-assimilates in reproductive plants of *Lolium multiflorum* LAM. var. *Westerwoldicum*. - Ann. Bot. *39* : 403 - 411, 1975.

22097 - **CLOUGH, B.F., MILTHORPE, F.L.** : Effects of water deficit on leaf development in tobacco. - Aust. J. Plant Physiol. *2* : 291 - 300, 1975. [Growth analysis.]

22098 - **COCKBURN, W., McAULAY, A.** : The pathway of malate synthesis in crassulacean acid metabolism. - In : **MARCELLE, R.** (ed.) : Environmental and Biological Control of Photosynthesis. Pp. 273 - 279. Dr. W. Junk b.v. Publ., The Hague 1975.

22099 - **COCKBURN, W., McAULAY, A.** : The pathway of carbon dioxide fixation in crassulacean plants. - Plant Physiol. *55* : 87 - 89, 1975.

*22100 - **COGDELL, R.J., BRUNE, D.C., CLAYTON, R.K.** : Effects of extraction and replacement of ubiquinone upon the photochemical activity of reaction centers and chromatophores from *Rhodopseudomonas spheriodes*. - FEBS Lett. *45* : 344 - 347, 1974.

22101 - **COGDELL, R.J., MONGER, T.G., PARSON, W.W.** : Carotenoid triplet states in reaction centers from *Rhodopseudomonas sphaeroides* and *Rhodospirillum rubrum*. - Biochim. biophys. Acta *408* : 189 - 199, 1975.

22102 - **COGGINS, C.W. Jr., HALL, A.E.** : Effects of light, temperature, and 2′,4′-dichloro-1-cyanoethanesulphonanilide on degreening of regreened "Valencia" oranges. - J. amer. Soc. hort. Sci. *100* : 484 - 487, 1975. [Chl.]

22103 - **COHEN, W.S., COHN, D.E., BERTSCH, W.** : Acceptor specific inhibition of photosystem II electron transport by uncoupling agents. - FEBS Lett. *49* : 350 - 355, 1975.

22104 - **COHEN, Y., JØRGENSEN, B.B., PADAN, E., SHILO, M.** : Sulphide-dependent anoxygenic photosynthesis in the cyanobacterium *Oscillatoria limnetica*. - Nature *257* : 489 - 492, 1975.

22105 - **COHEN, Y., PADAN, E., SHILO, M.** : Facultative anoxygenic photosynthesis in the cyanobacterium *Oscillatoria limnetica*. - J. Bacteriol. *123* : 855 - 861, 1975.

22106 - **COHN, D.E., COHEN, W.S., BERTSCH, W.** : Inhibition of Photosystem II by uncouplers at alkaline pH and its reversal by artificial electron donors. - Biochim. biophys. Acta *376* : 97 - 104, 1975.

22107 - **COHN, D.E., COHEN, W.S., LURIE, S., BERTSCH, W.** : Delayed light studies on photosynthetic energy conversion. IX. Evidence from msec emission for two distinct sites of action of treatments that inhibit on the oxidizing side of photosystem II. - In : **AVRON, M.** (ed.) : Proceedings of the Third International Congress on Photosynthesis. Vol. I. Pp. 65 - 74. Elsevier, Amsterdam - Oxford - New York 1975.

22108 - **COLE, D.F.** : Changes in leaf area and specific leaf weight of sugarbeet leaves during the growing season. - Crop Sci. *15* : 882 - 883, 1975.

22109 - **COLIJN, F., van BUURT, G.** : Influence of light and temperature on photosynthetic rate of marine benthic diatoms. - Marine Biol. *31* : 209 - 214, 1975.

22110 - **COLLINS, N., BROWN, R.H., MERRETT, M.J.** : Oxidative phosphorylation during glycollate metabolism in mitochondria from pnototrophic *Euglena gracilis*. - Biochem. J. *150* : 373 - 377, 1975. [Photorespiration.]

22111 - **COLLINS, N., MERRETT, M.J.** : The localization of glycollate-pathway enzymes in *Euglena*. - Biochem. J. *148* : 321 - 328, 1975.

22112 - **COLMAN, R.L., LAZENBY, A.** : Effect of moisture on growth and nitrogen response by *Lolium perenne*. - Plant Soil *42* : 1 - 13, 1975. [Growth analysis.]

22113 - **COLQUHOUN, A.J., HILLMAN, J.R.** : Endogenous abscisic acid and the senescence of leaves of *Phaseolus vulgaris* L. - Z. Pflanzenphysiol. *76* : 326 - 332, 1975. [Chl.]

22114 - **COLQUHOUN, A.J., HILLMAN, J.R., CREWE, C., BOWES, B.G.**: An ultrastructural study of the effects of abscisic acid on senescence of leaves of radish (*Raphanus sativus* L.). - Protoplasma *84* : 205 - 221, 1975. [Chloroplast.]

B22115 - Computer Simulation of a Cotton Production System. Users Manual. - Agr. Res. Serv. US Dep. Agr., New Orleans, Louisiana 1975.

22116 - **CONDE, L.F., KRAMER, P.J.** : The effect of vapor pressure deficit on diffusion resistance in *Opuntia compressa*. - Can. J.Bot. *53* : 2923 - 2926, 1975.

22117 - **CONDE, M.F., BOYNTON, J.E., GILLHAM, N.W., HARRIS, E.H., TINGLE, C.L., WANG, W.L.** : Chloroplast genes in *Chlamydomonas* affecting organelle ribosomes. Genetic and biochemical analysis of antibiotic-resistant mutants at several gene loci. - Mol. gen. Genet. *140* : 183 - 220, 1975.

22118 - **CONNOR, D.J.** : Growth, water relations and yield of wheat. - Aust. J. Plant Physiol. *2* : 353 - 366, 1975. [Growth analysis.]

22119 - **COOMBS, J., BALDRY, C.W.** : Metabolic regulation in C_4 photosynthesis : phosphoenol pyruvate carboxylase and 3C intermediates of the photosynthetic carbon reduction cycle. - Planta *124* : 153 - 158, 1975.

22120 - **COOMBS, J., MAW, S.L., BALDRY, C.W.** : Metabolic regulation in C_4 photosynthesis : the inorganic carbon substrate for PEP carboxylase. - Plant Sci. Lett. *4* : 97 - 102, 1975.

22121 - **COOPER, J.M., WILHM, J.** : Spatial and temporal variation in productivity, species diversity, and pigment diversity of periphyton in a stream receiving domestic and oil refinery effluents. - Southwest. Naturalist *19* : 413 - 428, 1975.

B22122 - **COOPER, J.P.** (ed.) : Photosynthesis and Productivity in Different Environments. - Cambridge University Press, Cambridge - London - New York - Melbourne 1975.

22123 - **COOPER, J.P.** : Control of photosynthetic production in terrestrial systems. - In : **COOPER, J.P.** (ed.) : Photosynthesis and Productivity in Different Environments. Pp. 593 - 621. Cambridge University Press, Cambridge - London - New York - Melbourne 1975.

22124 - **COPE, F.W.** : A review of the applications of solid state physics concepts to biological systems. - J. biol. Phys. *3* : 1 - 41, 1975. [Ps.]

22125 - **COPE, F.W.** : Kinetics of light emission by photosynthetic systems : Second order light decay kinetics means Elovich kinetics in a solid state reaction; 1.5 order light decay kinetics means second order reaction kinetics. - Bull. math. Biol. *37* : 79 - 83, 1975.

22126 - **CORKER, G.A., SHARPE, S.A.** : EPR detectable constituents in whole cell *Rhodospirillum rubrum*. - In : **AVRON, M.** (ed.) : Proceedings of the Third International Congress on Photosynthesis. Vol.I. Pp. 413 - 420. Elsevier, Amsterdam - Oxford - New York 1975.

22127 - **CORKER, G.A., SHARPE, S.A.** : Influence of light on the EPR-detectable, electron transport components in whole cell *Rhodospirillum rubrum*. - Photochem. Photobiol. *21* : 49 - 61, 1975.

22128 - **CORMACK, D.B., BATE, G.C.** : Methods of estimating leaf area from linear measurements of the *Macadamia integrifolia* cultivar Kakea. - Rhod. J. agr. Res. *13* : 45 - 53, 1975.

22129 - **CORMACK, D.B., BATE, G.C.** : Estimation of leaf area for some commercial macadamia cultivars. 1. Measurement of leaf parameters. - Rhod. J. agr. Res. *13* : 123 - 129, 1975.

22130 - **CORMIS, M. de** : Pollution de l'air et photosynthèse. - In : **COSTES, C.** (ed.) : Photosynthèse et Production Végétale. Pp. 259 - 283. Gauthier-Villars, Paris 1975.

22131 - **CORNIC, G., SARTI, A., LOUASON, G.** : L'effet Kok sur le *Lactuca sativa* cv. Romana. - Compt. rend. Acad. Sci. Paris, Sér. D *281* : 1837 - 1840, 1975.

22132 - **COSTE, B.** : Rôle des apports nutritifs minéraux rhodaniens sur la production organique des eaux du Golfe du Lion. - Tethys *6* : 727 - 740, 1974(1975).

B22133 - **COSTES, C.** (ed.) : Photosynthèse et Production Végétale. - Gauthier-Villars, Paris 1975.

22134 - **COSTES, C.** : Processus photochimiques et conversion énergétique de la lumière dans la feuille . - In : **COSTES, C.** (ed.) : Photosynthèse et Production Végétale. Pp. 33 - 72. Gauthier-Villars, Paris 1975.

22135 - **COSTES, C., BAZIER, R., BURGHOFFER, C., CARRAYOL, E., DEROCHE, M.E.** : Hydrogen bond breakdown in freeze-dried chloroplast lamellar proteins under various solvent treatments. - In : **AVRON, M.** (ed.) : Proceedings of the Third International Congress on Photosynthesis. Vol.III. Pp. 2049 - 2060. Elsevier, Amsterdam - Oxford - New York 1975.

22136 - **COULSON, C.L., HEATH, R.L.** : The interaction of peroxyacetyl nitrate (PAN) with the electron flow of isolated chloroplasts. - Atmos. Environ. *9* : 231 - 238, 1975.

22137 - **COURT, G.J.**: Photosynthesis and translocation studies of the symbiotic red algae *Janczewskia gardneri* and *Laurencia spectabilis (Rhodophyceae, Ceramiales.)* - J. Phycol. *11* (Suppl.) : 10, 1975.

22138 - **COVINGTON, W.W.** : Altitudinal variation of chlorophyll concentration and reflectance of the bark of *Populus tremuloides*. - Ecology *56* : 715 - 720, 1975.

22139 - **COX, G.S., HINCKLEY, T.M., ROBERTS, J., AUBUCHON, R., THOMPSON, D., DOUGHERTY, P., ASLIN, R., METCALF, C., CHAMBERS, J.** : Photosynthesis in white oak and sugar maple. - Res. Briefs, Sch. Forest., Fish. Wildlife, Univ. Missouri, Columbia 1975.

22140 - **COX, R.P.** : Oxygen evolution at sub-zero temperatures by chloroplasts suspended in fluid media. - FEBS Lett. *57* : 117 - 119, 1975.

22141 - **COX, R.P.** : The properties of cytochrome *f* and *P*700 in chloroplasts suspended in fluid media at sub-zero temperatures. - Europe. J. Biochem. *55* : 625 - 631, 1975.

22142 - **COX, R.P.** : The reduction of artificial electron acceptors at sub-zero temperatures by chloroplasts suspended in fluid media. - Biochim. biophys. Acta *387* : 588 - 598, 1975.

*22143 - **COYNE, P.I., KELLEY, J.J.** : Carbon dioxide partial pressures in arctic surface waters. - Limnol. Oceanogr. *19* : 928 - 938, 1974.

22144 - **COYNE, P.I., KELLEY, J.J.** : CO_2 exchange over the Alaskan arctic tundra : meteorological assessment by an aerodynamic method. - J. appl. Ecol. *12* : 587 - 611, 1975.

22145 - **CRAMER, W.A., HORTON, P.** : Recent studies on the chloroplast cytochrome *b*-559. - Photochem. Photobiol. *22* : 304 - 308, 1975.

22146 - **CRAMER, W.A., HORTON, P., WEVER, R.** : Proton-linked function of the chloroplast cytochrome *b*-559. - In : **QUAGLIARIELLO, E., PAPA, S., PALMIERI, F., SLATER, E.C., SILIPRANDI, N.** (ed.). : Electron Transfer Chains and Oxidative Phosphorylation. Pp. 31 - 36. North-Holland Publ. Comp., Amsterdam - Oxford; American Elsevier Publ. Comp., New York 1975.

22147 - **CRAN, D., DYER, A.F.** : The effect of a change in light quality on plastids of protonemata of *Dryopteris borreri*. - Plant Sci. Lett. *5* : 57 - 65, 1975.

22148 - **CRAUBNER, H.** : Hydration and partial specific volume of stroma-freed chloroplasts. - Colloid Polymer Sci. *253* : 713 - 719, 1975.

22149 - **CRAUBNER, H., KOENIG, F., SCHMID, G.H.** : Determination of the molecular weight and the hydrodynamic properties of a polypeptide from the thylakoid membrane by sedimentation, diffusion and binding measurements in dodecyl sulphate solutions. - Z. Naturforsch. *30 C* : 615 - 621, 1975.

22150 - **CRESSWELL, C.F., TEW, A.J., BAXTER, J.** : The influence of concentration and form of nitrogen supply on the carbon dioxide compensation point, photosynthetic rate, and enzymes associated with carbon dioxide exchange in selected C_4 photosynthetic plants. - In : **AVRON, M.** (ed.) : Proceedings of the Third International Congress on Photosynthesis. Vol. II. Pp. 1231 - 1248. Elsevier, Amsterdam - Oxford - New York 1975.

22151 - **CREUTZ, C., SUTIN, N.** : Reaction of tris(bipyridine)ruthenium(III) with hydroxide and its application in a solar energy storage system. - Proc. nat. Acad. Sci. USA *72* : 2858 - 2862, 1975.

22152 - **CREWS, C.E., VINES, H.M., BLACK, C.C. Jr.** : Postillumination burst of carbon dioxide in Crassulacean acid metabolism plants. - Plant Physiol. *55* : 652 - 657, 1975.

22153 - **CREWS, C.E., WILLIAMS, S.L., VINES, H.M.** : Characteristics of photosynthesis in peach leaves. - Planta *126* : 97 - 104, 1975.

22154 - **CROFTS, A.R., PRINCE, R.C., HOLMES, N.G., CROWTHER, D.** : Electrogenic electron transport and the carotenoid change in photosynthetic bacteria. - In : **AVRON, M.** (ed.) : Proceedings of the Third International Congress on Photosynthesis. Vol. II. Pp. 1131 - 1146. Elsevier, Amsterdam - Oxford - New York 1975.

22155 - **CRONBERG, G., GELIN, C., LARSSON, K.** : Lake Trummen restoration project. II. Bacteria, phytoplankton and phytoplankton productivity. - Verh. int. Verein. Limnol. *19* : 1088 - 1096, 1975.

22156 - **CROOKSTON, R.K., OZBUN, J.L.** : The occurrence and ultrastructure of chloroplasts in the phloem parenchyma of leaves of C_4 dicotyledons. - Planta *198* : 247 - 255, 1975.

22157 - **CROOKSTON, R.K., TREHARNE, K.J., LUDFORD, P., OZBUN, J.L.** : Response of beans to shading. - Crop Sci. *15* : 412 - 416, 1975. [Ps.]

22158 - **CROOME, R.L., TYLER, P.A.** : Phytoplankton biomass and primary productivity of Lake Leake and Tooms Lake, Tasmania. - Hydrobiologia *46* : 435 - 443, 1975.

22159 - **CROUSE, E., VANDREY, J., STUTZ, E.** : Comparative analyses of chloroplast and mitochondrial DNAs from *Euglena gracilis*. - In : **AVRON, M.** (ed.) : Proceedings of the Third International Congress on Photosynthesis. Vol. III. Pp. 1775 - 1786. Elsevier, Amsterdam - Oxford - New York 1975.

22160 - **CSATORDAY, K., LEHOCZKI, E., SZALAY, L.** : Energy transfer by chlorophyll *a* in detergent micelles. - Biochim. biophys. Acta *376* : 268 - 273, 1975.

22161 - **CSORBA, I., SZABAD, J., ERDEI, L., FAJSZI, Cs.** : Black lipid membranes (BLM) containing crystalline chlorophyll. - Photochem. Photobiol. *21* : 377 - 378, 1975.

22162 - **CURRY, R.B., BAKER, C.H., STREETER, J.G.** : SOYMOD I - a dynamic simulator of soybean growth and development. - Trans. ASAE *18* : 963 - 968, 974, 1975. [Ps.]

22163 - **CURRY, R.B., STREETER, J.G.** : An overview of SOYMOD, simulator of soybean growth and development. - In : Proceedings of 1975 Summer Computer Simulation Conferences. Pp. 954 - 960. San Francisco 1975. [Ps.]

22164 - **CURTIS, V.A., BRAND, J.J., TOGASAKI, R.K.** : Partial reactions of photosynthesis in briefly sonicated *Chlamydomonas*. I. Cell breakage and electron transport activities. - Plant Physiol. *55* : 183 - 186, 1975.

22165 - **CZARNOWSKI, M.** : Studies on ecophysiological method for estimation of photosynthetic production in leaves of chosen plant species. - Bull. Acad. pol. Sci., Sér. Sci. biol. *23* : 413 - 420, 1975.

22166 - **CZECZUGA, B.** : Carotenoids in three algae species from the Mediterranean Sea. - Nova Hedwigia *26* : 157 - 163, 1975.

22167 - **CZECZUGA, B.** : Spheroidenone - a dominating carotenoid in purple bacteria *Thiopedia rosea* WINOGR. (*Thiorodaceae*). - Bull. Acad. pol. Sci., Sér. Sci. biol. *23* : 181 - 184, 1975.

22168 - **CZECZUGA, B.** : The carotenoid content of galls produced by *Eriophyses tiliae* var. *rudis* NAL. *(Acarina)* on *Tilia cordata* MILL., leaves. - Marcellia *38* : 223 - 225, 1975.

*22169 - **CZECZUGA, B., GRĄDSKI, F.** : Phytoplankton primary production in some rivers of North-Eastern Poland. - Bull. Acad. pol. Sci., Sér. Sci. biol. *22* : 177 - 181, 1974.

22170 - **CZYGAN, F.-C.** : Farben und Farbstoffe. Einführung. - Ber. deut. bot. Ges. *88* : 3 - 5, 1975. [Chl, Car.]

22171 - **DADYKIN, V.P., POTAPOVA, A.D.** : O regulirovanii transpiratsionnogo raskhoda vlagi rasteniyami s pomoshch'yu antitranspirantov. [Regulation of transpiration loss of water by plants with the aid of antitranspirants.] - Izv. Akad. Nauk SSSR, Ser. biol. *1975* : 262 - 274, 1975. [Ps.]

22172 - **DALE, A.D., THOMSON, A.J., YOUNGSON, A.** : An automated adenosine triphosphatase assay. - Anal. Biochem. *67* : 332 - 335, 1975.

22173 - **DALGARN, M.C., WILSON, R.E.** : Net productivity and ecological efficiency of *Andropogon scoparius* growing in an Ohio Relict Prairie. - Ohio J. Sci. *75* : 194 - 197, 1975.

22174 - **DAM, R.J., KONGSLIE, K.F., GRIFFITH, O.H.** : Photoelectron quantum yields and photoelectron microscopy of chlorophyll and chlorophyllin. - Photochem. Photobiol. *22* : 265 - 268, 1975.

22175 - **DAMSZ, B.** : The ultrastructure of proplastids in the leaves of five species of Orchids. I. Organization of the inner membrane system and divisions of the proplastids. - Acta Soc. Bot. Pol. *44* : 501 - 509, 1975.

22176 - **DAMSZ, B.** : The ultrastructure of proplastids in the leaves of five species of Orchids. II. Intraplastid inclusions and DNA-containing areas. - Acta Soc. Bot. Pol. *44* : 511 - 518, 1975.

22177 - **DANFORTH, W.F.** : Phytoplankton productivity in Lake Michigan near Chicago. - J. Phycol. *11* (Suppl.) : 7, 1975.

22178 - **DANKANITS, E.** : Redukáló cukrok, szerves savak és szinezékek tanulmányozása egyes ribiszkefajtáknál. [Reducing sugars, organic acids and pigments in certain currant cultivars.] - Élelmiszervizsgálati Közlemények *21* : 48 - 53, 1975. [Chl, Car; in Hung., ab : E, F, G, R.]

22179 - **DANON, A., CAPLAN, S.R.** : Photophosphorylation and oxidative phosphorylation in *Halobacterium halobium*. - In : **AVRON, M.** (ed.) : Proceedings of the Third International Congress on Photosynthesis. Vol. III. Pp. 2163 - 2170. Elsevier, Amsterdam - Oxford - New York 1975.

22180 - **DAS, G.** : Changes in chloroplast structure during autospore formation in *Scenedesmus obtusiusculus*. - Protoplasma *84* : 175 - 180, 1975.

22181 - **DAS, G., RUNECKLES, V.C.** : Bisulphite-induced inactivation of growth and chlorophyll formation in *Chlorella pyrenoidosa*. - J. exp. Bot. *26* : 705 - 712, 1975.

22182 - **DAS, M., GOVINDJEE** : Action spectra of chlorophyll fluorescence in spinach chloroplast fractions obtained by solvent extraction. - Plant biochem. J. *2* (2) : 51 - 60, 1975.

22183 - **DAS, V.S.R., SANTAKUMARI, M.** : Stomatal behaviour towards four classes of herbicides as a basis of selectivity to certain weeds and crop plants. - Proc. indian Acad. Sci., Sect. B *82* : 108 - 116, 1975. [Ps.]

22184 - **DASH, M.C., HOTA, A.K., GURU, B.C., SENAPATI, B.K.** : Relationship on the community chlorophyll contents with reference to height and dry weight. - Geobios *2* : 126 - 127, 1975.

22185 - **DASHKEVICH, A.M.** : Uplyŭ khlarafosu i polikhlorpinenu na razvitstsë asimilyatsyĭnaga aparatu i gaspadarchuyŭ éfektyŭnasts' fotasintèzu bul'by. [Effect of chlorophos and polychlorpinene on the development of photosynthetic apparatus and the agricultural photosynthetic efficiency in potato.] - Vestsi Akad. Navuk belarus. SSR, Ser biyal. Navuk *1975* (6) : 45 - 48, 1975. [In Beloruss., ab : R.]

22186 - **DAVENPORT, D.C.** : Stomatal resistance from cuvette transpiration measurements. - Coll. agr. Res. Cent., Washington State Univ. Bull. *809* (Measurement of Stomatal Aperture and Diffusive Resistance) : 12 - 15, 1975.

22187 - **DAVENPORT, D.C., URIU, K., HAGAN, R.M.** : Antitranspirant effects on the water status of 'Manzanillo' olive trees. - J. amer. Soc. hort. Sci. *100* : 618 - 622, 1975. [Resistance to water vapour diffusion.]

22188 - **DAVIES, B.H.** : Carotenoids - aspects of biosynthesis and enzymology. - Ber. deut. bot. Ges. *88* : 7 - 25, 1975.

22189 - **DAVIES, W.J., KOZLOWSKI, T.T.** : Effect of applied abscisic acid and silicone on water relations and photosynthesis of woody plants. - Can. J. Forest. Res. *5* : 90 - 96, 1975.

22190 - **DAVIES, W.J., KOZLOWSKI, T.T.** : Stomatal responses to changes in light intensity as influenced by plant water stress. - Forest Sci. *21* : 129 - 133, 1975. [Stomatal resistance.]

22191 - **DAVIES, W.J., KOZLOWSKI, T.T.** : Effects of applied abscisic acid and plant water stress on transpiration of woody angiosperms. - Forest Sci. *21* : 191 - 195, 1975. [Stomatal resistance.]

22192 - **DAVIS, B., MERRETT, M.J.** : The glycolate pathway and photosynthetic competence in *Euglena*. - Plant Physiol. *55* : 30 - 34, 1975.

22193 - **DAVIS, D.J., ARMOND, P.A., GROSS, E.L., ARNTZEN, C.J.** : Development of excitation energy distribution regulation between Photosystems I and II during the final stages of chloroplast membrane differentiation. - Biophys. J. *15* (2, Part 2) : 223a, 1975.

*22194 - **DAVIS, D.J., GROSS, E.L.** : Ca^{+2} ion binding to pigment-protein complexes of spinach chloroplasts. - Fed. Proc. *33* : 1255, 1974.

22195 - **DAVIS, D.J., GROSS, E.L.** : Protein-protein interactions of light harvesting pigment protein from spinach chloroplasts. I. Ca^{2+} binding and its relation to protein association. - Biochim. biophys. Acta *387* : 557 - 567, 1975.

*22196 - **DAVIS, J.T., SPARKS, D.** : Assimilation and translocation patterns of carbon-14 in the shoot of fruiting pecan trees, *Carya illinoensis* KOCH. - J. amer. Soc. hort. Sci. *99* : 468 - 480, 1974.

*22197 - **DAVTYAN,V.A., KAZARYAN, V.V.** : Ob izmenenii korne-listovogo sootnosheniya i fiziologicheskoĭ aktivnosti list'ev odnoletnikh rasteniĭ pod deĭstviem pitatel'nogo rastvora razlichnoĭ kontsentratsii. [Changes in the leaf-root ratio and physiological activity of leaves of annual plants under the effect of nutrient solutions of various concentrations.] - Dokl. Akad. Nauk arm. SSR *45* (1) : 43 - 48, 1967. [Ps, Chl; In R, ab : Arm.]

22198 - **DEBRUNNER, P.G., SCHULZ, C.E., FEHER, G., OKAMURA, M.Y.** : Mossbauer study of reaction centers from *R. sphaeroides*. - Biophys. J. *15* (2, Part 2) : 226a, 1975.

22199 - **DE GREEF, J.A., CAUBERGS, R., MERTENS, R.** : Een zeer gevoelige spectrofotofluorometrische methode voor de detectie van kleine concentraties chlorofyl *b* in aanwezigheid van hoge concentraties aan chlorofyl *a*. [Spectrophotofluorometric method for detecting small concentrations of chlorophyll *b* in the presence of chlorophyll *a*.] - Arch. int. Physiol. Biochim. *83* : 955 - 957, 1975. [In Flem.]

22200 - **DE GREEF, J.A., VERBELEN, J.P.** : Plastid stroma crystals induced by 2,4-DNP treatment of *Phaseolus vulgaris* L. - In : **AVRON, M.** (ed.) : Proceedings of the Third International Congress on Photosynthesis. Vol. III. Pp. 2131 - 2137. Elsevier, Amsterdam - Oxford - New York 1975.

22201 - **DE GREEF, J.A., VERBELEN, J.P., MOEREELS, E.** : Preliminary studies on the site of primary action of light during de-etiolation processes of *Phaseolus* seedlings. - Arch. int. Physiol. Biochim. *83* : 957 - 958, 1975.

*22202 - **DELANEY, M.E., ROGERS, L.J.** : Studies on the sites of electron donation of diphenylcarbazide, quinol, and manganese ions to photosystem 2 by using 1,1,1-trichloro-2,2-*bis*(*p*-chlorophenyl)ethane (DDT) as an inhibitor. - Biochem. Soc. Trans. *1* : 726 - 729, 1973.

22203 - **DELEENS, E., LERMAN, J.C., NATO, A., MOYSE, A.** : Carbon isotope discrimination by the carboxylating reactions in C_3, C_4, and CAM plants. - In : **AVRON, M.**(ed.):

Proceedings of the Third International Congress on Photosynthesis. Vol. II. Pp. 1267 - 1276. Elsevier, Amsterdam - Oxford - New York 1975.

22204 - **DEMCHENKO, S.I.** : Vliyanie kontsentratsii i ėkspozitsii pri vozdeĭstvii N-nitrozo-N-metil-mochevinoĭ na semena *Arabidopsis thaliana* (L.).HEYNH. II. Zavisimost' chastoty khlorofil'nykh khimer v M_1 ot dozy mutagena. [Effect of concentration and exposure in treatment of *Arabidopsis thaliana* (L.) HEYNH. seeds by N-nitroso-N-methylurea. II. Dependence of chlorophyll chimeras frequency on the dosis of mutagens.]-In : Khimicheskie Supermutageny v Selektsii. Pp. 64 - 70. Nauka, Moskva 1975. [In R.]

22205 - **DEMCHENKO, S.I., AVETISOV, V.A., BUTENKO, R.G.** : Nature of chlorophyll chimeras in the M_1 of *Arabidopsis* by means of tissue culture. I. Production of regenerants from chlorophyll-deficient tissues. - In : **NASYROV, Yu.S., ŠESTÁK, Z.** (ed.) : Genetic Aspects of Photosynthesis. Pp. 209 - 215. Dr.W.Junk b.v. Publ., The Hague 1975.

22206 - **DENCHER, N., WILMS, M.** : Flash photometric experiments on the photochemical cycle of bacteriorhodopsin. - Biophys. Struct. Mech. *1* : 259 - 271, 1975.

22207 - **DEN ENGELSEN, D., DE KONING, B.** : Ellipsometry of black lipid membranes of egg lecithin and chloroplast extracts. - Photochem. Photobiol. *21* : 77 - 80, 1975. [Chl.]

22208 - **DENGLER, N.G., MacKAY, L.B.** : The leaf anatomy of beech, *Fagus grandifolia*. - Can. J. Bot. *53* : 2202 - 2211, 1975. [Internal surface of palisade and spongy parenchyma.]

22209 - **DePUIT, E.J., CALDWELL, M.M.** : Gas exchange of three cool semi-desert species in relation to temperature and water stress. - J. Ecol. *63* : 835 - 858, 1975.

22210 - **DePUIT, E.J., CALDWELL, M.M.** : Stem and leaf gas exchange of two arid land shrubs. - Amer. J. Bot. *62* : 954 - 961, 1975.

22211 - **DERA, J., HAPTER, R., MALEWICZ, B.** : Fluctuation of light in the euphotic zone and its influence on primary production. - Merentutkimuslait. julk. Havsforskningsint. Skr. *239* : 58 - 66, 1975.

*22212 - **DERERA, N.F., STOY, V.** : Varietal differences in photosynthetic efficiency of the awns. - In : **SEARS, E.R., SEARS, L.M.S.** (ed.) : Proceedings of the 4th International Wheat Genetics Symposium. Pp. 791 - 796. Missouri Agr. Exp. Sta., Columbia, Mo. 1973.

22213 - **DEROCHE, M.-E., CARRAYOL, E., COSTES, C.** : Étude de l'état de liaison des chlorophylles dans les lamelles chloroplastiques lyophilisées, par application de différentes familles de solvants. - Physiol. vég. *13* : 349 - 368, 1975.

22214 - **DESAI, T.S., SANE, P.V., TATAKE, V.G.** : Thermoluminescence studies on spinach leaves and *Euglena*. - Photochem. Photobiol. *21* : 345 - 350, 1975. [Ps.]

22215 - **DESMET, G., de RUYTER, A., RINGOET, A.** : Absorption and metabolism of CrO_4^{2-} by isolated chloroplasts. - Phytochemistry *14* : 2585 - 2588, 1975.

*22216 - **DETERS, D., NELSON, N.** : Approaches to the active site of chloroplast coupling factor I. - Fed. Proc. *33* : 1330, 1974.

22217 - **DETERS, D.W., RACKER, E., NELSON, N., NELSON, H.** : Partial resolution of the enzymes catalyzing photophosphorylation. XV. Approaches to the active site of coupling factor I. - J. biol. Chem. *250* : 1041 - 1047, 1975.

22218 - **DeVAULT, D., KUNG, M.C., HESS, B., OESTERHELT, D.** : Photolysis of bacterial rhodopsin. - Biophys. J. *15* (2, Part 2) : 67 a, 1975.

*22219 - **DeVAULT, D., SEIBERT, M.** : The rise time of P424 in *Chromatium* photosynthesis. - Fed. Proc. *33* : 1288, 1974.

22220 - **DEVAUX, J.** : Succession écologique, diversité spécifique et production primaire dans un lac oligotrophe d'Auvergne (France). - Verh. int. Verein. Limnol. *19* : 1165 - 1171, 1975.

22221 - **DÉVAY, M., FEHÉR, M.T.** : Temperature-induced changes in the plattern of chlorophyll complexes in wheat leaves. - Biochem. Physiol. Pflanzen *168* : 561 - 566, 1975.

22222 - **DÉVAY, M., RAJKI, S.** : Photosynthetic adaptation in the autumnization process. - In : **NASYROV, Yu.S., ŠESTÁK, Z.** (ed.) : Genetic Aspects of Photosynthesis. Pp. 357 - 362. Dr.W.Junk b.v. Publ., The Hague 1975.

22223 - **DEVIDÉ, Z.** : Biophysical investigation of chloroplast membranes isolated under sterile conditions.- Per. Biol. *77* : 73 - 74, 1975.

22224 - **DeYOE, D.R., BROWN, G.N.** : Characterization of western hemlock chloroplasts.- Plant Physiol. *56* (Suppl.) : 45, 1975.

*22225 - **DHALIWAL, A.S., CAMPIONE, A., CORDES, W.C.** : The effects of chloramphenicol on morphological and physiological characteristics of *Plectonema boryanum*. - J. Phycol. *10* (Suppl.) : 5, 1974. [Chl.]

22226 - **DICKMANN, D.I., GJERSTAD, D.H., GORDON, J.C.** : Developmental patterns of CO_2 exchange, diffusion resistance and protein synthesis in leaves of *Populus* × *euramericana*. - In : **MARCELLE, R.** (ed.) : Environmental and Biological Control of Photosynthesis. Pp. 171 - 181. Dr. W. Junk b.v. Publ., The Hague 1975.

22227 - **DICKMANN, D.I., GORDON, J.C.** : Incorporation of photosynthate into protein during leaf development in young *Populus* plants. - Plant Physiol. *56* (Suppl.): 60, 1975.

22228 - **DICKMANN, D.I., GORDON, J.C.** : Incorporation of ^{14}C-photosynthate into protein during leaf development in young *Populus* plants. - Plant Physiol. *56* : 23 - 27, 1975.

22229 - **DICKSON, R.E., LARSON, P.R.** : Incorporation of ^{14}C-photosynthate into major chemical fractions of source and sink leaves of cottonwood.- Plant Physiol. *56* : 185 - 193, 1975.

22230 - **DIERSTEIN, R., DREWS, G.** : Control of composition and activity of the photosynthetic apparatus of *Rhodopseudomonas capsulata* grown in ammonium-limited continuous culture. - Arch. Microbiol. *106* : 227 - 235, 1975.

22231 - **DILKS, T.J.K., PROCTOR, M.C.F.** : Comparative experiments on temperature responses of bryophytes : assimilation, respiration and freezing damage. - J. Bryol. *8* : 317 - 336, 1975.

*22232 - **DILLEY, R.A., DEAMER, D.** : Light-dependent chloroplast volume changes in chloride media. - J. Bioenerg. *2* : 33 - 38, 1971.

22233 - **DILLEY, R.A., GIAQUINTA, R.T.** : H^+ ion transport and energy transduction in chloroplasts. - In : **BRONNER, F., KLEINZELLER, A.** (ed.) : Current Topics in Membranes and Transport. Vol. 7. Pp. 49 - 107. Academic Press, New York - San Francisco - London 1975.

22234 - **DIMITROV, Kh.** : Izsledvaniya v"rkhu iziskvaniyata na topolite k"m svetlinata. [Light requirements of poplars.] - Gorskostop. Nauka *12* (3) : 3 - 15, 1975. [In Bulg., ab : E, R.]

22235 - **DINANT, M., AGHION, J.** : Chlorophylls attached to lipid and protein globules, absorption and fluorescence spectra, photo-oxidation. - Europe. J. Biochem. *52* : 515 - 520, 1975.

22236 - **DINER, B.** : Dependence of the turnover and deactivation reactions of photosystem II on the redox state of the pool A varied under anaerobic conditions. - In : **AVRON, M.** (ed.) : Proceedings of the Third International Congress on Photosynthesis. Vol. I. Pp. 589 - 601. Elsevier, Amsterdam - Oxford - New York 1975.

22237 - **DIRKS, W., RICHTER, G.** : Bedeutung von Blaulicht für die Synthese von RNA-Komponenten bei der Chloroplastendifferenzierung in isolierten Wurzeln (*Pisum sativum*). - Biochem. Physiol. Pflanzen *168* : 157 - 166, 1975.

22238 - **DI TORO, D.M., O'CONNOR, D.J., THOMANN, R.V., MANCINI, J.L.** : Phytoplankton-zooplankton-nutrient interaction model for Western Lake Erie. - In : **PATTEN, B.C.** (ed.) : Systems Analysis and Simulation in Ecology. Vol. III. Pp. 423 - 474. Academic Press, New York - San Francisco - London 1975.

22239 - **DITTRICH, P., HUBER, W.** : Carbon dioxide metabolism in members of the *Chlamydospermae*. - In: **AVRON, M.** (ed.) : Proceedings of the Third International Congress on Photosynthesis. Vol. II. Pp. 1573 - 1578. Elsevier, Amsterdam - Oxford - New York 1975.

22240 - **DODGE, A.D.** : Some mechanisms of herbicide action. - Sci. Prog. (Oxford) *62* : 447 - 466, 1975. [Ps, Car, chloroplast development.]

22241 - **DODGE, J.D.** : A survey of chloroplast ultrastructure in the *Dinophyceae*. - Phycologia *14* : 253 - 263, 1975.

*22242 - **DOLPHIN, D., MULJIANI, Z., ROUSSEAU, K., BORG, D.C., FAJER, J., FELTON, R.H.** : The chemistry of porphyrin π-cations. - Ann. New York Acad. Sci. *206* : 177 - 200, 1973. [Chl.]

22243 - **DÖRING, G.** : Further results on the photoactive chlorophyll a_{II} in photosynthesis. - Biochim. biophys. Acta *376* : 274 - 284, 1975.

22244 - **DÖRING, G.** : Investigations on the slow (200 μs) component of the photoactive chlorophyll reaction in system II of photosynthesis. - In : **AVRON, M.** (ed.) : Proceedings of the Third International Congress on Photosynthesis. Vol. I. Pp. 149 - 158. Elsevier, Amsterdam - Oxford - New York 1975.

*22245 - **DOROBANŢU, N., GĂINĂ, R.** : Dinamica fotosintezei şi respiraţiei la sfecla de zahăr în regim irigat şi neirigat. [Dynamics of photosynthesis and respiration in irrigated and non-irrigated sugar-beet.] - Lucrări şti. I.A.N.B. (Bucharest) Ser. A *17* : 38 - 44, 1974. [In Roum., ab : E.]

*22246 - **DOROKHOV, B.L., ZABRIYAN, D.P.** : Formirovanie listovoĭ poverkhnosti u podsolnechnika pri razlichnom mineral'nom pitanii. [Leaf area formation in sunflower under different mineral nutrition.]- Izv. Akad. Nauk mold. SSR, Ser. biol. khim. Nauk *1973* (5) : 24 - 26, 1973. [In R.]

22247 - **DOROKHOV, B.L., ZABRIYAN, D.P.** : Vliyanie mineral'nogo pitaniya na opticheskie svoĭstva list'ev podsolnechnika. [Effect of mineral nutrition on optical properties of sunflower leaves.] - Izv. Akad. Nauk mold. SSR, Ser. biol. khim. Nauk *1975* (1) : 14 - 20, 90, 1975. [In R.]

22248 - **DOUCHA, J., KUBÍN, Š.** : Die Synthese der Chlorophylle und die Veränderung ihreh spezifischen *in vivo*-Absorption im Laufe des Zellzyklus von *Scenedesmus quadricauda*. - In : **NECHAS, I.** (ed.) : Izuchenie Intensivnoĭ Kul'tury Vodorosleĭ. Vol. II. Pp. 59 - 72. Třeboň 1975.

22249 - **DOWNTON, W.J.S.** : The occurrence of C_4 photosynthesis among plants. - Photosynthetica *9* : 96 - 105, 1975.

22250 - **DOWNTON, W.J.S., HAWKER, J.S.** : Response of starch synthesis to temperature in chilling-sensitive plants. - In : **MARCELLE, R.** (ed.) : Environmental and Biological Control of Photosynthesis. Pp. 81 - 88. Dr. W. Junk b.v. Publ., The Hague 1975.

22251 - **DOWNTON, W.J.S., TÖRÖKFALVY, E.** : Effect of sodium chloride on the photosynthesis of *Aeluropus litoralis*, a halophytic grass. - Z. Pflanzenphysiol. *75* : 143 - 150, 1975.

22252 - **DOWNTON, W.J.S., TÖRÖKFALVY, E.** : Photosynthesis in developing asparagus plants. - Aust. J. Plant Physiol. *2* : 367 - 375, 1975.

22253 - **DRACHEV, L.A., KONDRASHIN, A.A., SAMUILOV, V.D., SKULACHEV, V.P.** : Generation of electric potential by reaction center complexes from *Rhodospirillum rubrum*. - FEBS Lett. *50* : 219 - 222, 1975.

22254 - **DRASKOVITS, R.M.** : Light intensity studies in beechwoods of different ages. - Acta bot. Acad. Sci. hung. *21* : 9 - 23, 1975.

22255 - **DREW, A.P., RUNNING, S.W.** : Comparison of two techniques for measuring surface area of conifer needles. - Forest Sci. *21* : 231 - 232, 1975.

*22256 - **DREW, E.A.** : Growth of a kelp forest at 60 metres in the straits of Messina. - Mem. Biol. mar. Oceanogr. N.S. *2* (6) : 135 - 157, 1972. [Ps.]

*22257 - **DREW, E.A.** : The biology and physiology of alga-invertebrate symbioses. III. *In situ* measurements of photosynthesis and calcification in some hermatypic corals. - J. exp. mar. Biol. Ecol. *13* : 165 - 179, 1973.

*22258 - **DREW, E.A.** : An ecological study of *Laminaria ochroleuca* PYL. growing below 50 metres in the Straits of Messina. - J. exp. mar. Biol. Ecol. *15* : 11 - 24, 1974. [Ps.]

22259 - **DREWS, G., DIERSTEIN, R., NIETH, K.F.** : The formation of the photosynthetic apparatus in cells of *Rhodopseudomonas capsulata*. - In : **AVRON, M.** (ed.) : Proceedings of the Third International Congress on Photosynthesis. Vol. III. Pp. 2139 - 2146. Elsevier, Amsterdam - Oxford - New York 1975.

22260 - **DREZNER, W.** : Wpływ stopnia krzewienia się roślin na akumulację biomasy organicznej u niektórych odmian zwyczajnej pszenicy ozimej. [Influence of tillering degree on the biomass accumulation in some winter wheat varieties.] - Acta agrobot. (Warszawa) *28* : 155 - 175, 1975. [Growth analysis; in Pol.,ab: E.]

22261 - **DRING, M.J., LÜNING, K.** : Induction of two-dimensional growth and hair formation by blue light in the brown alga *Scytosiphon lomentaria*. - Z. Pflanzenphysiol. *75* : 107 - 117, 1975. [Ps.]

22262 - **DROZDOVA, N.N., UMRIKHINA, A.V., PUSHKINA, E.M., KRASNOVSKIĬ, A.A.** : Obratimoe fotookislenie bakteriokhlorofillov *a* i *b* v vodnom rastvore detergenta. [Reversible photooxidation of bacteriochlorophylls *a* and *b* in detergent aqueous solution.] - Dokl. Akad. Nauk SSSR *225* : 1198 - 1201, 1975. [In R.]

22263 - **DUBÉ, P.A., STEVENSON, K.R., THURTELL, G.W., HUNTER, R.B.** : Effects of water stress on leaf respiration, transpiration rates in the dark and cuticular resistance to water vapor diffusion of two corn inbreds. - Can. J. Plant Sci. *55* : 565 - 572, 1975.

22264 - **DUBERTRET, G., AMBARD-BRETTEVILLE, F., CALVAYRAC, R.** : Étude de la formation des unités photosynthétiques au cours d'un cycle cellulaire chez *Euglena gracilis* Z par analyse de la fluorescence variable. - In : Les Cycles Cellulaires et leur Blocage chez Plusieurs Protistes. Coll. int. CNRS.Vol. 240. Pp. 109 - 113. Édit. CNRS, Paris 1975.

22265 - **DUBERTRET, G., JOLIOT, P.** : Formation of system II photosynthetic units during the greening of a dark-grown *Chlorella* mutant.- In : **AVRON, M.** (ed.) : Proceedings of the Third International Congress on Photosynthesis. Vol. III. Pp. 1877 - 1884. Elsevier, Amsterdam - Oxford - New York 1975.

22266 - **DUBEY, P.S., RAO, A.N.** : Amitrole toxicity to crops and its reversal by urea. - Curr. Sci. *44* : 247 - 248, 1975. [Chl.]

22267 - **DUBININA, G.A., GORLENKO, V.M.** : Novye nitchatye fotosinteziruyushchie zelenye bakterii s gazovymi vakuolyami. [New filamentous photosynthetic green bacteria with gas vacuoles.] - Mikrobiologiya *44* : 511 - 517, 1975. [In R, ab : E.]

22268 - **DUBINSKY, Z., BERMAN, T.** : Optical properties of Lake Kinneret and the light utilization efficiency of its phytoplankton. - Israel J. Bot. *24* : 53, 1975.

*22269 - **DUBINSKY, Z., ROTEM, J.** : Relations between algal populations and the pH of their media. - Oecologia *16* : 53 - 60, 1974. [Chl.]

22270 - **DUCKETT, J.G.** : An ultrastructural study of the differentiation of antheridial plastids in *Anthoceros laevis* L. - Cytobiologie *10* : 432 - 448, 1975.

22271 - **DUCREY, M.** : Utilisation des photographies hémisphériques pour le calcul de la perméabilité des couverts forestièrs au rayonnement solaire. I.- Analyse théorique de l'interception. - Ann. Sci. forest. *32* : 73 - 92, 1975.

22272 - **DUCREY, M.** : Utilisation des photographies hémisphériques pour le calçul de la perméabilité des couverts forestièrs au rayonnement solaire. II.- Étude expérimentale. - Ann. Sci. forest.*32* : 205 - 221, 1975.

22273 - **DUDA, M.** : Dry matter production of young plants in the forest profile used as phytometers. - In : **BISKUPSKÝ, V.** (ed.) : Research Project Báb IBP Progress Report II. Pp. 467 - 474. Veda, Bratislava 1975.

22274 - **DUGGAN, J.X., ANDERSON, L.E.** : Light-regulation of enzyme activity in *Anacystis nidulans* (RICHT.). - Planta *122* : 293 - 297, 1975. [Carbon metabolism enzymes.]

22275 - **DUJARDIN, E.** : Energy transfers between the protochlorophyll(ide) forms in lyophilized etiolated bean leaves. - Photosynthetica *9* : 283 - 287, 1975.

22276 - **DUJARDIN, E.** : Energy transfers from protochlorophyllide to chlorophyllide at room temperature in lyophilized bean leaves. - In : **AVRON, M.** (ed.) : Proceedings of the Third International Congress on Photosynthesis. Vol. I. Pp. 261 - 264. Elsevier, Amsterdam - Oxford - New York 1975.

22277 - **DUJARDIN, E., BONOTTO, S., SIRONVAL, C., KIRCHMANN, R.** : La différenciation-plastidale chez l'acétabulaire étudiée par l'émission de fluoréscence à 77 °K. - Plant Sci. Lett. *5* : 209 - 216, 1975.

22278 - **DUJARDIN, E., FOUASSIN, A., DEGRAEVE, H.** : Caractérisation des chlorophylles des huiles d'olive vierges par leur spectre de fluorescence et par chromatographie sur couche mince. - Rev. Fermentat. Indust. alim. *30* (3) : 63 - 65, 1975.

22279 - **DUKE, W.B., HAGIN, R.D., HUNT, J.F., LINSCOTT, D.L.** : Metal halide lamps for supplemental lighting in greenhouses : crop response and spectral distribution. - Agron. J. *67* : 49 - 53, 1975. [Crop growth.]

22280 - **DUMONT, I., KRUYEN, F., LEGROS, F.** : Fixation spécifique d'insuline à des chloroplastes isolés d'*Acetabularia mediterranea*. - Compt. rend. Séances Soc. Biol. *169* : 250 - 253, 1975.

22281 - **DUNAEVA, S.E.** : Razlichiya v ul'trastrukture khloroplastov rasteniĭ s raznym putem ugleroda pri fotosinteze. [Differences in chloroplast ultrastructure in plants with different carbon pathways in photosynthesis.] - Byull. vses. Inst. Rastenievod. N.I.Vavilova *56* : 22 - 28, 1975. [In R.]

22282 - **DUNCAN, W.G.** : Maize. - In : **EVANS, L.T.** (ed.) : Crop Physiology. Pp. 23 - 50. Cambridge University Press, London - New York 1975. [Ps.]

22283 - **DUNIWAY, J.M.** : Water relations in safflower during wilting induced by *Phytophthora* root rot. - Phytopathology *65* : 886 - 891, 1975. [Stomatal resistance.]

22284 - **DUNN, E.L.** : Environmental stresses and inherent limitations affecting CO_2 exchange in evergreen sclerophylls in mediterranean climates. - In : **GATES, D.M., SCHMERL, R.B.** (ed.) : Perspectives of Biophysical Ecology. Pp. 159 - 181. Springer-Verlag, Berlin - Heidelberg - New York 1975.

22285 - **DURAND, M.** : Fixation passive du calcium par les chloroplastes isolés de *Lupinus luteus* L. - Compt. rend. Acad. Sci. Paris, Sér. D *280* : 613 - 616, 1975.

22286 - **DURAND, M., GROUZIS, J.-P.** : Fixation du phosphate par les chloroplastes isolés de *Lupinus luteus* L. - Compt. rend. Acad. Sci. Paris, Sér. D *280* : 2753 - 2756, 1975.

22287 - **DURBIN, E.G., KRAWIEC, R.W., SMAYDA, T.J.** : Seasonal studies on the relative importance of different size fractions of phytoplankton in Narragansett Bay (USA). - Mar. Biol. *32* : 271 - 287, 1975. [Chl.]

22288 - **DUTTON, P.L., KAUFMANN, K.J., CHANCE, B., RENTZEPIS, P.M.** : Picosecond kinetics of the 1250 nm band of the *Rps. sphaeroides* reaction center : The nature of the primary photochemical intermediary state. - FEBS Lett. *60* : 275 - 280, 1975.

22289 - **DUTTON, P.L., PETTY, K.M., BONNER, H.S., MORSE, S.D.** : Cytochrome c_2 and reaction center of *Rhodopseudomonas spheroides* GA. membranes. Extinction coefficients, content, half-reduction potentials, kinetics and electric field alterations. - Biochim. biophys. Acta *387* : 536 - 556, 1975.

22290 - **DUYSEN, M.E., FREEMAN, T.P.** : Partial restoration of the high rate of plastid pigment development and the ultrastructure of plastids in detached water--stressed wheat leaves. - Plant Physiol. *55* : 768 - 773, 1975.

22291 - **DUYSENS, L.N.M., den HAAN, G.A., van BEST, J.A.** : Rapid reactions of photosystem 2 as studied by the kinetics of the fluorescence and luminescence of chlorophyll *a* in *Chlorella pyrenoidosa*. - In: **AVRON, M.** (ed.) : Proceedings of the Third International Congress on Photosynthesis. Vol. I. Pp. 1 - 12. Elsevier, Amsterdam - Oxford - New York 1975.

22292 - **DVIHALLY, S.T.** : Data on the O_2-household and primary production of the Nagyszéktő Lake near Kistelek (S. Hungary). - Ann. Univ. Sci. budapest. Rolando Eötvös nomin., Sect. biol. *17* : 5 - 20, 1975.

22293 - **DVIHALLY, Z.T.** : Schätzung der Primärproduktion im toten Donauarm von Tolna und im Wasser des Sees Nagyszéktő von Kistelek. - Symp. Biol. Hung. *15* : 97 - 102, 1975.

22294 - **DVIHALLY-TAMÁS, S.** : Primary production of the Hungarian Danube. - Verh. int. Verein. Limnol. *19* : 1717 - 1722, 1975.

22295 - **DWIVEDI, R.S.** : Studies on the efficiency of energy conversion in photosynthetic parts of wheat plants (*Triticum aestivum* L.). - Ann. Bot. *39* : 1077 - 1085, 1975.

22296 - **D'YACHKIN, I.I., BURLAKINA, A.V., BELYAKOVA, Z.P.** : Issledovanie opticheskikh svoĭstv list'ev v protsesse formirovaniya kachestva tabaka. [Leaf optical properties during formation of tobacco quality.]-Fiziol. Biokhim. kul't. Rast. *7* : 317 - 322, 1975. [In R, ab : E.]

22297 - **DYKYJOVÁ, D., PŘIBIL, S.** : Energy content in the biomass of emergent macrophytes and their ecological efficiency. - Arch. Hydrobiol. *75* : 90 - 108, 1975.

22298 - **DZNELADZE, A.A.** : Fotokhimicheskie reaktsii i fotosintez lista limona pri patologii. [Photochemical reactions and photosynthesis of lemon leaf at pathology.] - Tr. nauch.-issled. Inst. Zashchity Rast. gruz. SSR *27* : 198 - 208, 1975. [In R, ab : E.]

22299 - **DZNELADZE, A.A., KALICHAVA, G.S., MALANIYA, D.G.** : Ob ėnergomigratsii s fotoretseptorov zelenoĭ oblasti sveta (500 - 540 nm) na osnovnye pigmenty fotosinteza pri patologii. [Energy migration from photoreceptors of green region light (500 - 540 nm) to the main pigments of photosynthesis at pathology.] - Soobshch. Akad. Nauk gruz. SSR *80* (1) : 197 - 200, 1975. [In R, ab : E, Georg.]

22300 - **DZNELADZE, A.A., KALICHAVA, G.S., MALANIYA, D.G., KOSTENKO, A.I.** : Parametry generatsii ėnergii sveta list'yami vinograda pri khloroze. [Parameters of the light energy generation by the grape-vine leaf during chlorosis.] - Tr. nauch.-issled. Inst. Zashchity Rast. gruz. SSR *27* : 119 - 132, 1975. [In R, ab : E.]

22301 - **ECKARDT, F.E.** : Functioning of the biosphere at the primary production level - objectives and achievements. - In : **COOPER, J.P.** (ed.) : Photosynthesis and Productivity in Different Environments. Pp. 173 - 185. Cambridge Univ. Press, Cambridge - London - New York - Melbourne 1975.

22302 - **ECKARDT, F.E., HEIM, G., METHY, M., SAUVEZON, R.** : Interception de l'énergie rayonnante, échanges gazeux et croissance dans une fôret méditerranéenne à feuillage persistant (*Quercetum ilicis*). - Photosynthetica *9* : 145 - 156, 1975.

22303 - **EDMUNDS, L.N. Jr.**: Temporal differentiation in *Euglena* : Circadian phenomena in non-dividing populations and in synchronously dividing cells. - In : Les Cycles Cellulaires et leur Blocage chez Plusieurs Protistes. Coll. int. CNRS. Vol. 240. Pp. 53 - 67. Édit. CNRS, Paris 1975. [Ps, Chl, Car.]

22304 - **EDMUNDS, L.N. Jr., SAVILLO, R.L.** : Long-term *in vitro* culture of cell-free preparations of chloroplasts isolated from *Euglena gracilis* (Z). - In : **AVRON, M.** (ed.) : Proceedings of the Third International Congress on Photosynthesis. Vol. III. Pp. 1719 - 1730. Elsevier, Amsterdam - Oxford - New York 1975.

22305 - **EFIMTSEV, E.I., BOĬCHENKO, V.A., LITVIN, F.F.** : Fotoindutsirovannoe vydelenie vodoroda bakteriyami, vodoroslyami i vysshimi rasteniyami. [Photoinduced liberation of hydrogen by bacteria, algae and higher plants.] - Dokl. Akad. Nauk SSSR *220* : 986 - 989, 1975. [In R.]

22306 - **EFIMTSEV, E.I., BOĬCHENKO, V.A., LITVIN, F.F.** : Spektry deĭstviya fotosinteza i vydeleniya vodoroda u bakteriĭ, vodorosleĭ i vysshikh rasteniĭ. [Action spectra of photosynthesis and hydrogen liberation in bacteria, algae and higher plants.] - Dokl. Akad. Nauk SSSR *220* : 1238 - 1240, 1975. [In R.]

22307 - **EGAN, J.M. Jr., DORSKY, D., SCHIFF, J.A.** : Events surrounding the early development of *Euglena* chloroplasts. VI. Action spectra for the formation of chlorophyll, lag elimination in chlorophyll synthesis, and appearance of TPN-dependent triose phosphate dehydrogenase and alkaline DNase activities. - Plant Physiol. *56* : 318 - 323, 1975.

22308 - **EGGENBERG, P., ERISMANN, K.H.** : Tracerkinetic analysis of CO_2 assimilation pathway in the red alga *Porphyridium cruentum* in different spectral regions. - In : **AVRON, M.** (ed.) : Proceedings of the Third International Congress on Photosynthesis. Vol. II. Pp. 1547 - 1555. Elsevier, Amsterdam - Oxford - New York 1975.

22309 - **EGNÉUS, H.** : Effects of some chemicals on the oxygen evolution and oxygen uptake in isolated wheat chloroplasts irradiated without the addition of an oxidant. - Physiol. Plant. *33* : 203 - 213, 1975.

22310 - **EGNEUS, H., HEBER, U., MATTHIESEN, U., KIRK, M.** : Reduction of oxygen by the electron transport chain of chloroplasts during assimilation of carbon dioxide. - Biochim. biophys. Acta *408* : 252 - 268, 1975.

22311 - **EGOROVA, L.I.** : Posledeǐstvie kratkovremennogo progrevaniya list'ev na fotosintez. [After-effect of short-term heating on photosynthesis of leaves.] - Bot. Zh. *60* : 1000 - 1004, 1975. [In R.]

22312 - **EHARA, T., SHIHIRA-ISHIKAWA, I., OSAFUNE, T., HASE, E., OHKURO, I.** : Some structural characteristics of chloroplast degeneration in cells of *Euglena gracilis* Z during their heterotrophic growth in darkness. - J. Electron Microscopy *24* : 253 - 261, 1975.

22313 - **EHARA, Y., MISAWA, T.** : Occurrence of abnormal chloroplasts in tobacco leaves infected systemically with ordinary strain of cucumber mosaic virus. - Phytopathol. Z. *84* : 233 - 252, 1975.

22314 - **EHLERINGER, J.R., MILLER, P.C.** : A simulation model of plant water relations and production in the alpine tundra, Colorado. - Oecologia *19* : 177 - 193, 1975. [Ps.]

22315 - **EHLERINGER, J.R., MILLER, P.C.** : Water relations of selected plant species in the alpine tundra, Colorado. - Ecology *56* : 370 - 380, 1975. [Primary production.]

22316 - **EHRLER, W.L.** : Environmental and plant factors influencing transpiration of desert plants. - In : **HADLEY, N.F.** (ed.) : Environmental Physiology of Desert Organisms. Pp. 52 - 66. Dowden, Hutchinson & Ross, Stroudsburg 1975. [Resistances.]

22317 - **EICHENBERGER, W., BOSCHETTI, A.** : Light-induced greening in liquid cultures of streptomycin-bleached mutants of *Chlamydomonas reinhardi*. - FEBS Lett. *55* : 117 - 119, 1975.

22318 - **EICHHORN, M.** : Na^+- und NaCl-Effekte im CO_2-Fixierungsstoffwechsel der Pflanzen. - Biol. Rundsch. *13* : 122 - 123, 1975.

22319 - **EICHHORN, M., AUGSTEN, H.** : Der Einfluß von Bormangel und Borüberschuß auf *Scenedesmus obliquus* unter steady state-Bedingungen. - Biochem. Physiol. Pflanzen *167* : 87 - 96, 1975. [Ps.]

22320 - **EICKENBUSCH, J.D., SCHEIBE, R., BECK, E.** : Activated glykol aldehyde and ribulose diphosphate as carbon sources for oxidative glycolate formation in chloroplasts. - Z. Pflanzenphysiol. *75* : 375 - 380, 1975.

22321 - **EICKMEIER, W., ADAMS, M., LESTER, D.** : Two physiological races of *Tsuga canadensis* in Wisconsin. - Can. J. Bot. *53* : 940 - 951, 1975. [Ps.]

22322 - **EǏDEL'MAN, Z.M., POPOVA, O.F.** : Stanovlenie fotosinteticheskogo apparata pri zelenenii ètiolirovannykh prorostkov. [Establishment of photosynthetic apparatus in the process of greening of etiolated seedlings.] - Bot. Zh. *60* : 1019 - 1030, 1975. [In R.]

22323 - **EILATI, S.K., BUDOWSKI, P., MONSELISE, S.P.** : Carotenoid changes in the "Shamouti" orange peel during chloroplast-chromoplast transformation on and off the tree. - J. exp. Bot. *26* : 624 - 632, 1975.

22324 - **EISELE, R., ULLRICH, W.R.** : Stoichiometry between photosynthetic nitrate reduction and alkalinisation by *Ankistrodesmus braunii in vivo*. - Planta *123* : 117 - 123, 1975.

22325 - **EIZENGA, G.C., ROBLES, R.P., MILES, C.D.** : DCMU-insensitive quenching of fluorescence by ferricyanide with SMA or STA in normal and mutant maize chloroplasts. - Plant Physiol. *56* (Suppl.) : 47, 1975.

22326 - **ELGERSMA, O., VOORN, G.** : Deconvolution of absorption spectra at room temperature of chloroplast fragments. - In : **AVRON, M.** (ed.) : Proceedings of the Third International Congress on Photosynthesis. Vol. III. Pp. 1943 - 1949. Elsevier, Amsterdam - Oxford - New York 1975.

22327 - **ELIÁŠ, P.** : A contribution to the study of water relations of forest herbs. - Biológia (Bratislava) *30* : 771 - 779, 1975. [Chl.]

22328 - **ELKIN, L., PARK, R.B.** : Chloroplast fluorescence of C_4 plant. I. Detection with infrared color film. - Planta *127* : 243 - 250, 1975.

22329 - **ELKIN, L., PARK, R.B.** : Chloroplast fluorescence of C_4 plants. II. A photographic technique for obtaining relative fluorescence yields and spectra photographically. - Planta *127* : 187 - 199, 1975.

22330 - **ELKIN, L., PARK, R.B.** : Chloroplast fluorescence of C_4 plants. III. Fluorescence spectra and relative fluorescence yields of bundle-sheath and mesophyll chloroplasts. - Planta *127* : 37 - 47, 1975.

22331 - **ELLEN, J., SPIERTZ, J.H.J.** : The influence of nitrogen and Benlate on leaf-area duration, grain growth and pattern of N-, P-, and K-uptake of winter wheat (*Triticum aestivum*). - Z. Acker- Pflanzenbau *141* : 231 - 239, 1975.

22332 - **ELLENSON, J., SAUER, K.** : Electrophotoluminescence of chlroplasts. - Biophys. J. *15* (2, Part 2) : 221 a, 1975.

22333 - **ELLER, B.M.** : Die optische Eigenschaften der Blätter von *Rhododendron ferrugineum* L. and *Alnus viridis* (CHAIX) DC. - Ber. schweiz. bot. Ges. *85* : 25 - 30, 1975.

22334 - **ELLER, B.M., BRUNNER, U.** : Der Einfluß von Straßenstaub auf die Strahlungsabsorption durch Blätter. - Arch. Meteorol. Geophys. Bioklimatol., Ser. B *23*: 137 - 146, 1975.

22335 - **ELLIOT, W.M.** : Light-controlled leaf expansion in peas grown under different light conditions. - Plant Physiol. *55* : 717 - 719, 1975. [Chl.]

22336 - **ELLIS, J.** : Mitochondria and chloroplasts. - Nature *256* : 617 - 618, 1975.

22337 - **ELLIS, R., SPOONER, T., YAKULIS, R.** : Regulation of chlorophyll synthesis in the green alga *Golenkinia*. - Plant Physiol. *55* : 791 - 795, 1975.

22338 - **ELLIS, R.J.** : The synthesis of chloroplast membranes in *Pisum sativum*. - In : **TZAGALOFF, A.** (ed.) : Membrane Biogenesis. Pp. 247 - 278. Plenum Publ. Co., New York - London - Boston - Washington 1975.

22339 - **ELLIS, R.J.** : Inhibition of chloroplast protein synthesis by lincomycin and 2-(4-methyl-2,6-dinitroanilino)-*N*-methylpropionamide. - Phytochemistry *14* : 89 - 93, 1975.

22340 - **ELLSWORTH, R.K., CARNEY, C.F.** : Preparation of chlorin e_6 from microgram quantities of pheophorbide *a* and its absorption spectrum in diethyl ether. - Photosynthetica *9* : 333 - 336, 1975.

22341 - **ELLSWORTH, R.K., HERVISH, P.V.** : Biosynthesis of protochlorophyllide *a* from Mg-protoporphyrin IX *in vitro*. - Photosynthetica *9* : 125 - 139, 1975.

22342 - **ELLSWORTH, R.K., ST.PIERRE, L.A.** : Incorporation of the *S*-methyl group of methionine (*S*-methyl-^{14}C) into chlorophyll *a*. - Photosynthetica *9* : 340 - 342, 1975.

22343 - **EL NADI, A.H.** : Irrigation requirements of maize in a tropical environment.- Acta agron. Acad. Sci. hung. *24* : 423 - 429, 1975. [Growth analysis.]

22344 - **EL-SHARKAWI, H.M., MICHEL, B.E.** : Effects of soil salinity and air humidity on CO_2 exchange and transpiration of two grasses. - Photosynthetica *9* : 277 - 282, 1975.

22345 - **ELSTNER, E.F., HEUPEL, A.** : Lamellar superoxide dismutase of isolated chloroplasts. - Planta *123* : 145 - 154, 1975.

22346 - **ELSTNER, E.F., KONZE, J.R.** : Ethylene formation by isolated chloroplast lamellae in the dark. - Z. Naturforsch. *30 C* : 58 - 63, 1975.

22347 - **ELSTNER, E.F., STOFFER, C., HEUPEL, A.** : Determination of superoxide free radical ion and hydrogen peroxide as products of photosynthetic oxygen reduction. - Z. Naturforsch. *30 C* : 53 - 57, 1975.

22348 - **EMEL'YANOV, L.G., KUSHNIR, N.V., ANKUD, S.A.** : Vliyanie urovneĭ gruntovykh vod torfyanoĭ pochvy na vodoobmen i produktivnost' rasteniĭ v usloviyakh Poles'ya. [Effect of ground water of turf soil on water balance and productivity of plants in Poles'e.] - In : Pitanie i Obmen Veshchestv u Rasteniĭ. Pp. 171 - 179. Nauka i Tekhnika, Minsk 1975. [In R.]

22349 - **ENAMI, I., FUKUDA, I.** : Mechanisms of the acido- and thermo-phily of *Cyanidium caldarium* GEITLER I. Effects of temperature, pH and light intensity on the photosynthetic oxygen evolution of intact and treated cells. - Plant Cell Physiol. *16* : 211 - 220, 1975.

22350 - **ENAMI, I., NAGASHIMA, H., FUKUDA, I.** : Mechanisms of the acido- and thermo--phily of *Cyanidium caldarium* GEITLER II. Physiological role of the cell wall. - Plant Cell Physiol. *16* : 221 - 231, 1975. [Ps, Chl.]

22351 - **ENDO, M.** : [Studies on the daily change in fruit size of the Japanese pear. IV. Influence of shading on diurnal fluctuation of fruit.] - J. jap. Soc. hort. Sci. *43* : 347 - 358, 1975. [Ps.]

22352 - **ENYI, B.A.C.** : Effects of defoliation on growth and yield in groundnut (*Arachis hypogea*), cowpeas (*Vigna unguiculata*), soyabean (*Glycine max*) and green gram (*Vigna aurens*). - Ann. appl. Biol. *79* : 55 - 66, 1975.

22353 - **EPPLEY, R.W., SHARP, J.H.** : Photosynthetic measurements in the central North Pacific : The dark loss of carbon in 24-h incubations. - Limnol. Oceanogr. *20* : 981 - 987, 1975.

22354 - **ERABI, T., HIGUTI, T., KAKUNO, T., YAMASHITA, J., TANAKA, M., HORIO, T.** : Polarographic studies on ubiquinone-10 and rhodoquinone bound with chromatophores from *Rhodospirillum rubrum*. - J. Biochem. *78* : 795 - 801, 1975.

22355 - **ESAU, K.** : Crystalline inclusion in thylakoids of spinach chloroplasts. - J. Ultrastruct. Res. *53* : 235 - 243, 1975.

22356 - **ESHEL, Y., ILANI, S.** : Tolerance of cotton to five triazine herbicides. - Phytoparasitica *3* : 121 - 128, 1975. [Ps.]

22357 - **ESTRADA, M., VALLESPINÓS, F.** : Considaraciones estadísticas sobre algunos parâmetros oceanográficos en la región de afloramiento del NW de Africa. [Statistical considerations on some oceanographical parameters in the upwelling region of NW Africa.] - Res. Exp. cient. Buque oceanogr. "Cornide de Saavedra" *4* : 175 - 183, 1975. [Chl; in Span., ab : E.]

B22358 - **ETHERINGTON, J.R.** : Environment and Plant Ecology. - John Wiley & Sons, London - New York - Sydney - Toronto 1975. [Ps.]

22359 - **ETIENNE, A.L.** : Effects of carbonyl cyanide *m*-chlorophenyl hydrazone (CCCP) and of 3-(3,4 dichlorophenyl)-1,1-dimethylurea (DCMU) on photosystem II in the alga "*Chlorella pyrenoidosa*". - In : **AVRON, M.** (ed.) : Proceedings of the Third International Congress on Photosynthesis. Vol. I. Pp. 335 - 343. Elsevier, Amsterdam - Oxford - New York 1975.

22360 - **ETIENNE, A.L., LAVOREL, J.** : Triggered-luminescence in dark-adapted *Chlorella* cells and chloroplasts. - FEBS Lett. *57* : 276 - 279, 1975.

22361 - **ETTL, H.** : Die Teilung und Verformung des gegliederten Chromatophors von *Stigeoclonium stagnatile (Chlorophyceae)*. - Plant Syst. Evol. *124* : 179 - 186, 1975.

22362 - **ETTL, H., BŘEZINA, V.** : Teilungsverhalten der Chromatophoren in bezug auf die Mitose während des Lebenszyklus von *Diatoma hiemale* var. *mesodon*. - Plant Syst. Evol. *124* : 187 - 203, 1975.

22363 - **EVANS, E.H., CARR, N.G.** : Dark-light transitions with a heterotrophic culture of a blue-green alga. - Biochem. Soc. Trans. *3* : 373 - 376, 1975.

22364 - **EVANS, E.H., CARR, N.G.** : Dark to light transitions in *Chlorogloea fritschii*. - In : **AVRON, M.** (ed.) : Proceedings of the Third International Congress on Photosynthesis. Vol. III. Pp. 1861 - 1866. Elsevier, Amsterdam - Oxford - New York 1975.

22365 - **EVANS, L.T.** : Beyond photosynthesis - the role of respiration, translocation and growth potential in determining productivity. - In : **COOPER, J.P.** (ed.) : Photosynthesis and Productivity in Different Environments. Pp. 501 - 507. Cambridge University Press, Cambridge - London - New York - Melbourne 1975.

22366 - **EVANS, L.T.** : Photoperiodism and photosynthetic pathways. - Can. J. Bot. *53* : 590 - 591, 1975.

22367 - **EVANS, L.T.** : Crops and world food supply, crop evolution, and the origins of crop physiology. - In : **EVANS, L.T.** (ed.) : Crop Physiology. Pp. 1 - 22. Cambridge University Press, London - New York - Melbourne 1975. [Ps.]

22368 - **EVANS, L.T.** : The physiological basis of crop yield. - In : **EVANS, L.T.** (ed.) : Crop Physiology. Pp. 327 - 355. Cambridge University Press, London - New York - Melbourne 1975. [Ps.] -

22369 - **EVANS, L.T., WARDLAW, I.F., FISCHER, R.A.** : Wheat. - In : **EVANS, L.T.** (ed.) : Crop Physiology. Pp. 101 - 149. Cambridge University Press, London - New York - Melbourne 1975. [Ps.]

22370 - **EVANS, M.C.W.** : Isolation of chloroplasts from higher plants. - In : **HALL, D.O., HAWKINS, S.E.** (ed.) : Laboratory Manual of Cell Biology. Pp. 148 - 150. English Universities Press, London 1975.

22371 - **EVANS, M.C.W.** : Photosynthetic oxygen evolution and the relationship between electron transport and ATP synthesis in spinach chloroplasts. - In : **HALL, D.O., HAWKINS, S.E.** (ed.) : Laboratory Manual of Cell Biology. Pp. 151 - 153. English Universities Press, London 1975.

22372 - **EVANS, M.C.W.** : The partial purification and assay of ferredoxin from plants and bacteria. - In : **HALL, D.O., HAWKINS, S.E.** (ed.) : Laboratory Manual of Cell Biology. Pp. 183 - 185. English Universities Press, London 1975.

22373 - **EVANS, M.C.W.** : Iron-sulphur proteins in the photosynthetic electron-transport system of oxygen-evolving organisms. - Biochem. Soc. Trans. *3* : 492 - 495, 1975.

22374 - **EVANS, M.C.W., CAMMACK, R.** : The effect of the redox state of the bound iron--sulphur centres in spinach chloroplasts on the reversibility of P700 photooxidation at low temperatures. - Biochem. biophys. Res. Commun. *63* : 187 - 193, 1975.

22375 - **EVANS, M.C.W., CAMMACK, R., REEVES, S.C.** : Properties of the photochemical reaction centre of photosystem I in spinach chloroplasts. - In : **AVRON, M.** (ed.) : Proceedings of the Third International Congress on Photosynthesis. Vol. I. Pp. 383 - 388. Elsevier, Amsterdam - Oxford - New York 1975.

22376 - **EVANS, M.C.W., SIHRA, C.K., BOLTON, J.R., CAMMACK, R.** : Primary electron acceptor complex of photosystem I in spinach chloroplasts. - Nature *256* : 668 - 670, 1975.

22377 - **EVANS, N., GAMES, D.E., JACKSON, A.H., MATLIN, S.A.** : Applications of high--pressure liquid chromatography and field desorption mass spectrometry in studies of natural porphyrins and chlorophyll derivatives. - J. Chromatogr. *115* : 325 - 333, 1975.

22378 - **EVANS, T.A., KATZ, J.J.** : Evidence for 5- and 6-coordinated magnesium in bacteriochlorophyll *a* from visible absorption spectroscopy. - Biochim. biophys. Acta *396* : 414 - 426, 1975.

22379 - **EVENARI, M., BAMBERG, S., SCHULZE, E.-D., KAPPEN, L., LANGE, O.L., BUSCHBOM, U.** : The biomass production of some higher plants in Near-Eastern and American deserts. - In : **COOPER, J.P.** (ed.) : Photosynthesis and Productivity in Different Environments. Pp. 121 - 127. Cambridge University Press, Cambridge - London - New York - Melbourne 1975.

22380 - **EVENARI, M., SCHULZE, E.D., KAPPEN, L., BUSCHBOM, U., LANGE, O.L.** : Adaptive mechanisms in desert plants. - In : **VERNBERG, F.J.** (ed.) : Physiological Adaptation to the Environment. Pp. 111 - 129. Intext Educ. Publ., New York 1975. [Ps.]

22381 - **EVERS, A.K., WOLPERT, J.S., ERNST-FONBERG, M.L.** : An improved method for assay of carboxylation enzymes. - Anal. Biochem. *64* : 606 - 608, 1975.

22382 - **EVSTIGNEEV, V.B.** : Issledovanie fotokhimicheskikh reaktsiĭ fotosinteza *in vitro* i *in vivo*. [Studying photochemical reactions of photosynthesis *in vitro* and *in vivo*.] - In : Metody Issledovaniya Fotokhimicheskikh Reaktsiĭ Fotosinteza *in Vitro* i *in Vivo*. Pp. 3 - 11. Pushchino 1975. [In R.]

22383 - **EVSTIGNEEV, V.B.** : O vozmozhnoĭ roli "vspomogatel'nykh" fotokhimicheskikh reaktsiĭ v fotoregulyatsii protsessa fotosinteza. [Possible role of "accessory" photochemical reactions in photosynthesis regulation.]-In : Fotoregulyatsiya Metabolizma i Morfogeneza Rasteniĭ. Pp. 111 - 119, 251. Nauka, Moskva 1975. [In R.]

22384 - **EVSTIGNEEV, V.B., GAVRILOVA, V.A.** : O fotopotentsiale pri fotokhimicheskom vzaimodeĭstvii khlorofilla s durokhinonom. [Photopotential of photochemical interaction of chlorophyll with duroquinone.] - Biofizika *20* : 991 - 995, 1975. [In R, ab : E.]

22385 - **EVSTIGNEEV, V.B., GAVRILOVA, V.A.** : O vliyanii dièlektricheskoĭ postoyannoĭ sredy na fotookislenie khlorofilla *a* parabenzokhinonom. [The effect of dielectric constant of the medium on parabenzoquinone photooxidation of chlorophyll *a*.] - Biofizika *20* : 996 - 998, 1975. [In R, ab : E.]

22386 - **EVSTIGNEEV, V.B., SHKUROPATOV, A.Ya., STOLOVITSKIĬ, Yu.M.** : Fotogal'vanicheskiĭ èffekt v sloyakh khlorofilla i ftalotsianina magniya pri impul'snom osveshchenii. [Photogalvanic effect in solid chlorophyll and Mg-phthalocyanine films under pulse illumination.]-Stud. biophys. *49* : 27 - 42, 1975. [In R, ab : E.]

22387 - **EYLES, J.C.** : Effects of copper on two species of unicellular marine algae.- Annu. Rep. mar. Biochem. Unit *1974-75* : 16 - 17, 1975. [Ps.]

22388 - **FABBRI, F.** : Occurrence of stromacentre in chloroplasts of *Asplenium trichomanes* L. - Caryologia *28* : 539 - 548, 1975.

22389 - **FABRIS, G.L., HAMMER, U.T.** : Primary production in four small lakes in the Canadian Rocky Mountains. - Verh. int. Ver. theoret. angew. Limnol. *19* : 530 - 541, 1975.

22390 - **FAJER, J., BRUNE, D.C., DAVIS, M.S., FORMAN, A., SPAULDING, L.D.** : Primary charge separation in bacterial photosynthesis : Oxidized chlorophylls and reduced pheophytin. - Proc. nat. Acad. Sci. USA *72* : 4956 - 4960, 1975.

22391 - **FALK, H., HOORNAERT, G., ISENRING, H.-P., ESCHENMOSER, A.** : Über Enolderivate der Chlorophyllreihe. Darstellung von 13^2, 17^3-Cyclophäophorbid-enolen. - Helv. chim. Acta *58* : 2347 - 2357, 1975.

22392 - **FALKOWSKI, P.G., STONE, D.P.** : Nitrate uptake in marine phytoplankton : energy sources and the interaction with carbon fixation. - Mar. Biol. *32* : 77 - 84, 1975. [Ps.]

22393 - **FALUDI-DÁNIEL, Á.** : Pigment synthesis and photosynthetic activity in carotenoid deficient mutants of maize. - In : **NASYROV, Yu.S., ŠESTÁK, Z.**(ed.) : Genetic Aspects of Photosynthesis. Pp. 239 - 245. Dr. W. Junk b.v. Publ., The Hague 1975.

22394 - **FALUDI-DÁNIEL, Á., BRETON, J.** : A linear dichroism study using chloroplasts of various structure and pigment composition. - Photochem. Photobiol. *22* : 125 - 127, 1975.

22395 - **FALUDI-DÁNIEL, Á., DEMETER, S., HORVÁTH, G., JOÓ, F.** : Stacking capacity and chlorophyll forms of thylakoids in normal and mutant maize chloroplasts of different granum content. - In : **AVRON, M.** (ed.) : Proceedings of the Third International Congress on Photosynthesis. Vol. III. Pp. 1933 - 1942. Elsevier, Amsterdam - Oxford - New York 1975.

22396 - **FARAH, S.M.** : Effects of plant density and fertilization on the yield and quality of flue-cured tobacco in the Kenana area of the Sudan. - J. agr. Sci. *84*: 75 - 80, 1975. [Growth analysis.]

22397 - **FARINEAU, J.** : Photoassimilation of CO_2 by isolated bundle-sheath strands of

Zea mays. I. Stimulation of CO_2 assimilation by adding various intermediates of the photosynthetic cycle; evidence for a deficient photosystem II activity. - Physiol. Plant. *33* : 300 - 309, 1975.

22398 - **FARINEAU, J.** : Photoassimilation of CO_2 by isolated bundle-sheath strands of *Zea mays*. II. Role of malate as a source of CO_2 and reducing power for the photosynthetic activity of isolated bundle-sheaths cells of *Zea mays*. - Physiol. Plant. *33* : 310 - 315, 1975.

22399 - **FARINEAU, J.** : The effect of malate addition on $^{14}CO_2$ assimilation by isolated bundle-sheath cells of *Zea mays*. - In : **AVRON, M.** (ed.) : Proceedings of the Third International Congress on Photosynthesis. Vol. II. Pp. 1209 - 1217. Elsevier, Amsterdam - Oxford - New York 1975.

22400 - **FARINEAU, J., POPOVIC, R.** : Activation du photosystème II, par une courte période d'éclairement continu, chez les plastes de plantules de *Pinus jeffreyi* cultivées à l'obscurité. - Compt. rend. Acad. Sci. Paris, Sér. D *281* : 1317 - 1320, 1975.

22401 - **FARINEAU, N., ROUSSAUX, J.** : Influence de la 6-benzylaminopurine sur la différenciation plastidiale dans les cotylédons de concombre. - Physiol. Plant. *33* : 194 - 202, 1975.

22402 - **FARMER, R.E. Jr.** : Growth and assimilation rate of juvenile northern red oak : Effects of light and temperature. - Forest Sci. *21* : 373 - 381, 1975. [Growth analysis.]

22403 - **FARQUHAR, G.D.** : Dynamic stomatal behaviour. - Plant Physiol. *56* (Suppl.) : 21, 1975. [Stomatal resistance.]

22404 - **FARRAR, J.F.** : A method for investigating lichen growth rates and succession. - Lichenologist *6* : 151 - 155, 1975.

*22405 - **FAUSET, C.R., BROWN, A.P.** : The induction of chloroplast fluorescence *in vivo*. - Biochem. Soc. Trans. *1* : 897 - 899, 1973.

22406 - **FEDINA, I.S., VAKLINOVA, S.G.** : Light-induced uptake of oxygen from chloroplast fragments and from mesophyll cells and bundle sheath cells in maize. - Dokl. bolg. Akad. Nauk *28* : 1265 - 1268, 1975. [Ps enzymes.]

22407 - **FEDOROV, V.D., KORSAK, M.N.** : O fotosinteticheskom koėffitsiente i prizhiznennykh vydeleniyakh sine-zelenoĭ vodorosli *Anacysts nidulans*. [Photosynthetic coefficient and exudates of the blue-green alga *Anacystis nidulans*.] - Vest. mosk. Univ., Biol. Pochvoved. *30* (2) : 68 - 73, 1975. [In R, ab : E.]

22408 - **FEHER, G., HOFF, A.J., ISAACSON, R.A., ACKERSON, L.C.** : Endor experiments on chlorophyll and bacteriochlorophyll *in vitro* and in the photosynthetic unit. - Ann. New York Acad. Sci. *244* : 239 - 259, 1975.

22409 - **FEHÉR, M., DEVAY, M.** : Temperature-induced phase changes on the kinetics of photosynthetic electron transfer in winter and spring wheat chloroplasts. - Biochem. Physiol. Pflanzen *167* : 447 - 450, 1975.

22410 - **FEIERABEND, J.** : Developmental studies on microbodies in wheat leaves III. On the photocontrol of microbody development. - Planta *123* : 63 - 77, 1975.

22411 - **FEIGE, G.B.** : Beiträge zur Physiologie einheimischer Algen. 5. Einige Aspekte des photosynthetischen C-Metabolismus des Süßwasserrotalge *Audouinella violacea* (KÜTZ) HAMEL. - Z. Pflanzenphysiol. *75* : 339 - 345, 1975.

22412 - **FEIGE, G.B.** : Untersuchungen zur Ökologie und Physiologie der marinen Blaualgenflechte *Lichina pygmaea* AG. III. Einige Aspekte der photosynthetischen C-Fixierung unter osmoregulatorischen Bedingungen. - Z.Pflanzenphysiol. *77* : 1 - 15, 1975.

22413 - **FEKETE, G.** : Aerial environment and tolerance of *Polygonatum odoratum* (MILL.) DRUCE in natural communities. - Acta agron. Acad. Sci. hung. *24* : 89 - 97, 1975.

22414 - **FEKETE, G.** : Átültetési kísérletek *Polygonatum odoratum* ökotípusokkal különböző fénykörnyezetekben. II. A növekedési paraméterek módosulása. [Transplanting *Polygonatum odoratum* ecotypes under different irradiance. II. Modification of growth parameters.] - Bot. Közlem. *62* : 29 - 31, 1975. [Growth analysis; in Hung., ab : G.]

22415 - **FELLOWS, R.J., BOYER, J.S.** : Activity and *in vivo* and *in vitro* ultrastructure of chloroplasts of sunflower leaves having various water potentials. - Plant Physiol. *56* (Suppl.) : 61, 1975.

22416 - **FENNA, R.E., MATTHEWS, B.W.** : Chlorophyll arrangement in a bacteriochlorophyll protein from *Chlorobium limicola*. - Nature *258* : 573 - 577, 1975.

22417 - **FERRARI, I., BELLAVERE, C., CAMURRI, L., CATELLANI, M.** : Limnologia fisica e chimica e contenuti di clorofilla-*a* nel fitoplancton di un lago di montagna, il lago Santo Parmense. [Physical and chemical limnology and content of chlorophyll *a* in phytoplankton of the mountain lake Santo Parmense.] - Riv. Idrobiol. *14* (1-2) : 13 - 49, 1975. [In Ital., ab : E.]

22418 - **FERREE, D.C., HALL, F.R.** : Influence of Benomyl and oil on photosynthesis of apple leaves. - HortScience *10* : 128 - 129, 1975.

22419 - **FICK, G.W., LOOMIS, R.S., WILLIAMS, W.A.** : Sugar beet. - In : **EVANS, L.T.** (ed.) : Crop Physiology. Pp. 259 - 295. Cambridge University Press, Cambridge - London - New York 1975. [Ps production.]

22420 - **FIEDLER, U., HANSEN, E.H., RŮŽIČKA, J.** : Measurements of carbon dioxide with the air-gap electrode. Determination of the total inorganic and the total organic carbon contents in waters. - Anal. chim. Acta *74* : 423 - 435, 1975.

22421 - **FILIPPOV, G.L., VISHNEVSKIĬ, N.V., MAKSIMOVA, L.A.** : Osobennosti fotosinteticheskoĭ deyatel'nosti gibridov kukuruzy v usloviyakh orosheniya. [Peculiarities of photosynthetic activity of irrigated maize hybrids.] - Byull. vses. nauch.-issled. Inst. Kukuruzy *1975* (4) : 19 - 22, 1975. [In R.]

22422 - **FILIPPOVICH, I.I., ALINA, B.A., BEZSMERTNAYA, I.N., TONGUR, A.M., OPARIN, A.I.** : Relationship between the protein-synthesizing system and the structure of chloroplasts. - In : **NASYROV, Yu.S., ŠESTÁK, Z.** (ed.) : Genetic Aspects of Photosynthesis. Pp. 105 - 113. Dr. W. Junk b.v. Publ., The Hague 1975.

22423 - **FILIPPOVICH, I.I., NOZDRINA, V.N., OPARIN, A.I.** : Strukturnaya organizatsiya poliribosom khloroplastov v svyazi s biogenezom tilakoidov gran. [Structural organization of chloroplast polyribosomes in connection with biogenesis of grana thylakoids.] - Dokl. Akad. Nauk SSSR *225* : 223 - 226, 1975. [In R.]

22424 - **FILIPPOVICH, I.I., SVETLICHKIN, V.V., OPARIN, A.I.** : Izuchenie lokalizatsii sinteza RNK v tonkoĭ strukture khloroplastov. [Localization of RNA synthesis in fine structure of chloroplasts.] - Dokl. Akad. Nauk SSSR *225* : 1210 - 1212, 1975. [In R.]

22425 - **FIOLET, J.W.T., van der ERF-ter HAAR, L.** : Stimulation of the light-induced proton uptake by the uncoupler atebrin in isolated spinach chloroplasts. - In : **AVRON, M.** (ed.) : Proceedings of the Third International Congress on Photosynthesis. Vol. II. Pp. 1001 - 1011. Elsevier, Amsterdam - Oxford - New York 1975.

22426 - **FIOLET, J.W.T., van der ERF-ter HAAR, L., KRAAYENHOF, R., van DAM, K.** : On the stimulation of the light-induced proton uptake by uncoupling aminoacridine derivatives in spinach chloroplasts. - Biochim. biophys. Acta *387* : 320 - 334, 1975.

22427 - **FIOLET, J.W.T., VAN DE VLUGT, F.C.** : The pH indicator phenol red as an artificial electron acceptor in spinach chloroplasts. - FEBS Letters *53* : 287 - 291, 1975.

22428 - **FISCHER, K.S., WILSON, G.L.** : Studies of grain production in *Sorghum bicolor* (L. MOENCH). III. The relative importance of assimilate supply, grain growth capacity and transport system. - Aust. J. agr. Res. *26* : 11 - 23, 1975.

22429 - **FISCHER, K.S., WILSON, G.L.** : Studies of grain production in *Sorghum bicolor* (L.MOENCH). IV. Some effects of increasing and decreasing photosynthesis at different stages of the plant's development on the storage capacity of the inflorescence. - Aust. J. agr. Res. *26* : 25 - 30, 1975.

22430 - **FISCHER, K.S., WILSON, G.L.** : Studies of grain production in *Sorghum bicolor* (L. MOENCH). V. Effect of planting density on growth and yield. - Aust. J. agr. Res. *26* : 31 - 41, 1975.

22431 - **FISCHER, R.A.** : Yield potential in a dwarf spring wheat and the effect of shading. - Crop Sci. *15* : 607 - 613, 1975. [Growth analysis.]

22432 - **FISCHEROVÁ, H.** : Linkage relationships of recessive chlorophyll mutations in *Arabidopsis thaliana* (L.) HEYNH. - Biol. Plant. *17* : 182 - 188, 1975.

22433 - **FISHER, N.S.** : Chlorinated hydrocarbon pollutants and photosynthesis of marine phytoplankton : A reassessment. - Science *189* : 463 - 464, 1975.

22434 - **FISHER, R.W., MILLER, J.H.** : Growth regulation by ethylene in fern gametophytes. IV. Involvement of photosynthesis in overcoming ethylene inhibition of spore germination. - Amer. J. Bot. *62* : 1104 - 1111, 1975.

22435 - **FLEMING, A.A., PALMER, J.H.** : Variation in chlorophyll content of maize lines and hybrids. - Crop Sci. *15* : 617 - 620, 1975.

22436 - **FLINT, R.W., GOLDMAN, C.R.** : The effects of a benthic grazer on the primary productivity of the littoral zone of Lake Tahoe. - Limnol. Oceanogr. *20* : 935 - 944, 1975.

22437 - **FLOWER-ELLIS, J.G.K.** : Growth in populations of *Andromeda polifolia* on a subarctic mire. - In : **WIELGOLASKI, F.E.** (ed.) : Fennoscandian Tundra Ecosystems. Part I. Plants and Microorganisms. Pp. 129 - 134. Springer-Verlag, Berlin - Heidelberg - New York 1975. [Leaf area determination.]

22438 - **FLUHR, R., HAREL, E.** : Succinyl-CoA synthetase in greening maize leaves. - Phytochemistry *14* : 2157 - 2160, 1975. [Chl.]

22439 - **FLUHR, R., HAREL, E., KLEIN, S., MELLER, E.** : Control of δ-aminolevulinic acid and chlorophyll accumulation in greening maize leaves upon light-dark transitions. - Plant Physiol. *56* : 497 - 501, 1975.

22440 - **FLUHR, R., HAREL, E., KLEIN, S., NE'EMAN, E.** : The control of δ-aminolevulinic acid synthesis in greening maize leaves. - In : **AVRON, M.** (ed.) : Proceedings of the Third International Congress on Photosynthesis. Vol. III. Pp. 2097 - 2103. Elsevier, Amsterdam - Oxford - New York 1975.

22441 - **FOCKE, R.** : Der Vergleich verschiedener Lampentypen in Gewächshäusern und Kabinen und ihre Beziehung zum Ertrag bei Sommergerste und Weizen. - Arch. Züchtungsforsch. *5* : 133 - 147, 1975. [Ps.]

22442 - **FOGG, G.E.** : Biochemical pathways in unicellular plants. - In : **COOPER, J.P.** (ed.) : Photosynthesis and Productivity in Different Environments. Pp. 437 - 457. Cambridge University Press, Cambridge - London - New York - Melbourne 1975.

22443 - **FONG, F.K.** : Molecular symmetry and exciton interaction in photosynthetic primary events. - Appl. Phys. *6* : 151 - 166, 1975.

22444 - **FONG, F.K.** : Ester and keto carbonyl linkages in chlorophyll *a*, pyrochlorophyll *a*, and protochlorophyll *a*. - J. amer. chem. Soc. *97* : 6890 - 6892, 1975.

22445 - **FONG, F.K., KOESTER, V.J.** : Bonding interactions in anhydrous and hydrated chlorophyll *a*. - J. amer. chem. Soc. *97* : 6888 - 6890, 1975.

22446 - **FONTVIEILLE, D.** : Contribution à l'étude du role de l'agitation de l'eau sur la production primaire phytoplanctonique. - Ann. Limnol. *11* : 207 - 217, 1975.

22447 - **FORD, E.D.** : Competition and stand structure in some even-aged plant monocultures. - J. Ecol. *63* : 311 - 333, 1975. [Growth analysis.]

22448 - **FORD, M.A., THORNE, G.N.** : Effects of variation in temperature and light intensity at different times on growth and yield of spring wheat. - Ann. appl. Biol. *80* : 283 - 299, 1975.

22449 - **FORDE, B.J., WHITEHEAD, H.C.M., ROWLEY, J.A.** : Effect of light intensity and temperature on photosynthetic rate, leaf starch content and ultrastructure of *Paspalum dilatatum*. - Aust. J. Plant Physiol. *2* : 185 - 195, 1975.

22450 - **FORK, D.C., BROWN, J.S.** : The use of rapid light-induced absorbance changes of chlorophylls and carotenoids to distinguish between algae of different pigment composition. - J. Phycol. *11* (Suppl.) : 11, 1975.

22451 - **FORK, D.C., BROWN, J.S.** : A comparison of light-induced shifts in carotenoid absorption in representatives of different algal groups. - Carnegie Inst. Year Book *74* : 776 - 779, 1975.

22452 - **FORTI, G., ROSA, L., PUGGI, A., GARLASCHI, F.M.** : Effect of ammonium on photosynthesis in isolated chloroplasts. - In : **AVRON, M.** (ed.) : Proceedings of the Third International Congress on Photosynthesis. Vol. II. Pp. 1499 - 1505. Elsevier, Amsterdam - Oxford - New York 1975.

22453 - **FOSTER, J.M., IDSO, S.B.** : Light and assimilation number in a small desert, recharged-groundwater pond. - Oecologia *18* : 155 - 164, 1975.

22454 - **FOTT, J.** : Seasonal succession of phytoplankton in the fish pond Smyslov near Blatná, Czechoslovakia. - Arch. Hydrobiol. *46* (Suppl. - Algol. Stud. 12) : 259 - 279, 1975. [Chl.]

*22455 - **FOUSOVÁ, S., AVRATOVŠČUKOVÁ, N.** : Neaditivní složky dědičné proměnlivosti fotosyntetické aktivity terčíků a možnosti jejich detekce. [Nonadditive components of genetic variation in photosynthetic rate of leaf disks and possibilities for their detection.] - Acta Univ. Agr., Fac. agron. (Brno) *21* : 251 - 261, 1973. [In Czech, ab : E, G, R.]

22456 - **FOWLER, C.F.** : Measurement of H/e ratios in spinach chloroplast using flashing light. - Biophys. J. *15* (2, Part 2) : 249 a, 1975.

22457 - **FOX, J.L., MOYER, M.S.** : Effect of power plant chlorination on estuarine productivity. - Chesapeake Sci. *16* : 66 - 68, 1975.

22458 - **FRĄCKOWIAK, D.** : Properties of photosynthetic pigments and the primary processes of photosynthesis. - Pol. ekol. Stud. *1* : 17 - 26, 1975.

22459 - **FRĄCKOWIAK, D., BIAŁEK, G.** : Spectral and photochemical properties of β-carotene in deuterated solvents. - Bull. Acad. pol. Sci., Sér. Sci. math., astr., phys. *23* : 355 - 360, 1975.

22460 - **FRĄCKOWIAK, D., FIKSIŃSKI, K., GRABOWSKI, J.** : Low temperature absorption spectra of biliproteins. - Photosynthetica *9* : 185 - 191, 1975.

22461 - **FRĄCKOWIAK, D., JANUSZCZYK, L.** : Light scattering on suspension of sonicated and unsonicated algae. - Bull. Acad. pol. Sci., Sér. Sci. biol. *23* : 375 - 378, 1975.

22462 - **FRADKIN, L.I., KOLYAGO, V.M., MORDACHEVA, G.S., SAĬ, P.K.** : O vidovoĭ spetsifichnosti submembrannykh chastits khloroplastov. [Species peculiarities of submembrane chloroplast particles.] - In : **SHLYK, A.A.** (ed.) : Biosintez i Sostoyanie Khlorofillov v Rastenii. Pp. 161 - 182, 246. Nauka i Tekhnika, Minsk 1975. [In R, ab : E.]

22463 - **FRAGATA, M.** : Effects of aging on chlorophyll fluorescence and photosystem II electron transport in isolated chloroplasts. - Can. J. Bot. *53* : 2842 - 2845, 1975.

22464 - **FRALEIGH, P.C., WIEGERT, R.G.** : A model explaining successional change in standing crop of thermal blue-green algae. - Ecology *56* : 656 - 664, 1975.

22465 - **FRANCIS, G.W., STRAND, L.P., LIEN, T., KNUTSEN, G.** : Variations in the carotenoid content of *Chlamydomonas reinhardii* throughout the cell cycle. - Arch. Microbiol. *104* : 249 - 254, 1975.

22466 - **FRANZKE, C., GRUNERT, S., HASHEM, H.A.-A.** : Untersuchungen über den Einfluß verschiedener Adsorbentien auf die Inhaltsstoffe von Pflanzenöl. - Nahrung *19* : 707 - 713, 1975. [Chl, Car.]

22467 - **FREEMAN, T.P., DUYSEN, M.E.** : Effect of imposed water stress on development and ultrastructure of wheat chloroplasts. - Protoplasma *83* : 131 - 145, 1975.

22468 - **FRENKEL, C.** : Oxidative turnover of auxins in relation to the onset of ripening in Bartlett pear. - Plant Physiol. *55* : 480 - 484, 1975. [Chl.]

22469 - **FREYSSINET, G.** : Ribosomal protein synthesis during chloroplast development of dark-grown *Euglena gracilis*. - In : **AVRON, M.** (ed.) : Proceedings of the Third International Congress on Photosynthesis. Vol. III. Pp. 1731 - 1744. Elsevier, Amsterdam - Oxford - New York 1975.

22470 - **FRIČ, F.** : Translocation of ^{14}C-labelled assimilates in barley plants infected with powdery mildew (*Erysiphe graminis* f. sp. *hordei* MARCHAL). - Phytopathol. Z. *84* : 88 - 95, 1975.

22471 - **FRIČ, F., HORIČKOVÁ, B., HASPEL-HORVATOVIČ, E.** : Liquid scintillation counting of chlorophyll. - Int. J. appl. Radiat. Isotopes *26* : 509 - 514, 1975.

22472 - **FRICK, H.** : Phytochrome control of the lag phase of chlorophyll accumulation in *Lemna minor*. - Can. J. Bot. *53* : 2405 - 2410, 1975.

22473 - **FRICK, H., JONES, R.F.** : Inhibition of chlorophyll synthesis in *Lemna minor* by nalidixic acid. - Can. J. Bot. *53* : 2319 - 2324, 1975.

22474 - **FRIEDERICH, K.E., MOHR, H.** : Adenosine 5'-triphosphate content and energy charge during photomorphogenesis of the mustard seedling *Sinapis alba* L. - Photochem. Photobiol. *22* : 49 - 53, 1975.

22475 - **FRIEND, D.J.C.** : Adaptation and adjustment of photosynthetic characteristics of gametophytes and sporophytes of Hawaiian tree-fern (*Cibotium glaucum*) grown at different irradiances. - Photosynthetica *9* : 157 - 164, 1975.

22476 - **FRIEND, D.J.C.** : Light requirements for photoperiodic sensitivity in cotyledons of dark-grown *Pharbitis nil*. - Physiol. Plant. *35* : 286 - 296, 1975. [Chl.]

22477 - **FUHRHOP, J.-H., SMITH, K.M.** : Laboratory methods. - In : **SMITH, K.M.** (ed.) : Porphyrins and Metalloporphyrins. Pp. 757 - 869. Elsevier, Amsterdam - Oxford - New York 1975. [Chl.]

22478 - **FUJITA, Y.** : Further investigation of the light-induced cytochrome *b*-559 reaction in membrane fragments of the blue-green alga *Anabaena variabilis*. - Plant Cell Physiol. *16* : 1037 - 1048, 1975.

22479 - **FUKUYAMA, M., TAKEDA, T., OSHIRO, S.** : [Studies on the effects of oxygen concentration on the photosynthesis and the growth of crop plants. III. Effect of low oxygen concentration treatment for comparative long period on the growth in two row barley.] - Proc. Crop Sci. Soc. Jap. *44* : 1 - 6, 1975. [In Jap., ab : E.]

22480 - **FÜNFSCHILLING, J., WILLIAMS, D.F.** : Vibrationally resolved low temperature fluorescence spectra of porphin and chlorophylls *a* and *b* in organic glass matrices. - Photochem. Photobiol. *22* : 151 - 152, 1975.

*22481 - **GAASTRA, P.** : Energiebenutting door planten. [Energy utilization in plants.] - Bedrijfsontwikkeling *5* : 889 - 897, 1974. [In Hol.]

22482 - **GAENKO, O.N., DZHANUMOV, D.A., VESELOVSKIĬ, V.A., TARUSOV, B.N.** : Pervichnye fiziko-khimicheskie povrezhdeniya v khloroplastakh gorokha pri deĭstvii toksicheskikh kontsentratsiĭ NaCl. [Primary physico-chemical injuries in chloroplasts affected with toxic concentrations of NaCl.] - Sel'skokhoz. Biol. *10* : 849 - 853, 1975. [In R, ab : E.]

22483 - **GAFNI, A., HARDT, H., SCHLESSINGER, J., STEINBERG, I.Z.**: Circular polarization of fluorescence of chlorophyll in solution and in native structures. - Biochim. biophys. Acta *387* : 256 - 264, 1975.

22484 - **GALE, J.** : Water balance and gas exchange of plants under saline conditions. - In : **POLJAKOFF-MAYBER, A., GALE, J.** (ed.) : Plants in Saline Environments. Pp. 168 - 185. Springer-Verlag, Berlin - Heidelberg - New York 1975.

22485 - **GALE, J.** : The combined effect of environmental factors and salinity on plant growth. - In : **POLJAKOFF-MAYBER, A., GALE, J.** (ed.) : Plants in Saline Environments. Pp. 186 - 192. Springer-Verlag, Berlin - Heidelberg - New York 1975. [Ps.]

22486 - **GALE, J., KAPLAN, A., TAKO, T.** : Systematic errors in measurement of transpiration and photosynthesis by infrared gas analyses with varying oxygen : nitrogen ratios in the background gas. - J. exp. Bot. *26* : 702 - 704, 1975.

22487 - **GALKIN, V.I., KOSHKIN, V.A., BURENIN, V.I., BYKOV, O.D.** : Aktivnost' fotosinteticheskogo apparata svekly i ee zavisimost' ot faktorov vneshneĭ sredy i plodnosti. [Activity of photosynthetic apparatus in beet and its dependence on environmental factors and ploidy level.] - Byull. vses. Inst. Rastenievod. N.I.Vavilova *56* : 16 - 22, 1975. [In R.]

22488 - **GALLAGHER, J.L.** : The significance of the surface film in salt marsh plankton metabolism. - Limnol. Oceanogr. *20* : 120 - 123, 1975. [Ps.]

22489 - **GALLAHER, R.N., ASHLEY, D.A., BROWN, R.H.** : ^{14}C-photosynthate translocation in C_3 and C_4 plants as related to leaf anatomy. - Crop Sci. *15* : 55 - 59, 1975.

22490 - **GALLON, J.R., KURZ, W.G.W., LARUE, T.A.** : The physiology of nitrogen fixation by a *Gloeocapsa* sp. - In : **STEWART, W.D.P.** (ed.) : Nitrogen Fixation by Free--living Micro-organisms. Pp. 159 - 173. Cambridge University Press, London 1975. [Ps.]

22491 - **GALMICHE, J.M., GIRAULT, G.** : Restoration by silicotungstic acid of DCMU-inhibited photoreactions in spinach chloroplasts. - In : **AVRON, M.** (ed.) : Proceedings of the Third International Congress on Photosynthesis. Vol. I. Pp. 697 - 701. Elsevier, Amsterdam - Oxford - New York 1975.

22492 - **GAMALEĬ, Yu.V.** : Prodolzhitel'nost' zhizni khloroplastov v kletkakh mezofilla listopadnykh i vechnozelenykh rasteniĭ. [The age of chloroplasts in mesophyll of deciduous and evergreen plants.] - Tsitologiya *17* : 1243 - 1248, 1975. [In R, ab : E.]

22493 - **GANF, G.G.** : Photosynthetic production and irradiance-photosynthesis relationships of the phytoplankton from a shallow equatorial lake (Lake George, Uganda). - Oecologia *18* : 165 - 183, 1975.

22494 - **GANF, G.G., HORNE, A.J.** : Diurnal stratification, photosynthesis and nitrogen-fixation in a shallow, equatorial Lake (Lake George, Uganda). - Freshwater Biol. *5* : 13 - 39, 1975.

22495 - **GANTT, E.** : Phycobilisomes : Light-harvesting pigment complexes. - Bioscience *25* : 781 - 788, 1975.

22496 - **GANTT, E.** : Physiological and ultrastructural changes on nitrogen deprivation : disappearance and reappearance of phycobilisomes. - J. Phycol. *11* (Suppl.) : 14, 1975.

22497 - **GANTT, E., LIPSCHULTZ, C.A., ZILINSKAS, B.** : Evidence for a phycobilisome structural model. - Plant Physiol. *56* (Suppl.) : 10, 1975.

*22498 - **GARAY, A.S., LACZKÓ, I., CZÉGÉ, J., KOVÁCS, K.L., TOLVAJ, L., TÓTH, M.G., SZABÓ, M.** : Origin and biological role of molecular asymmetry. - Acta biol. Acad. Sci. hung. *24* : 137 - 156, 1973. [Ps.]

22499 - **GARCIA, A.F., DREWS, G., KAMEN, M.D.** : Electron transport in an *in vitro*-reconstituted bacterial photophosphorylating system. - Biochim. biophys. Acta *387* : 129 - 134, 1975.

22500 - **GARGAS, E.** : Nitrogen and phosphorus as growth limiting factors in a shallow fjord system. - Vatten *31* (1) : 56 - 63, 1975. [Primary production.]

22501 - **GARNIER, J., MAROC, J.** : Photooxidation of cytochrome *b*-559 in chloroplast fragments of non-photosynthetic mutants of *Chlamydomonas reinhardti*. - In : **AVRON, M.** (ed.) : Proceedings of the Third International Congress on Photosynthesis. Vol. I. Pp. 547 - 556. Elsevier, Amsterdam - Oxford - New York 1975.

22502 - **GARRARD, L.A., CARTER, J.L.** : Leaf assimilate level and photosynthesis of *Digitaria decumbens* STENT. - Soil Crop Sci. Soc. Florida Proc. *35* : 6 - 10, 1975.

22503 - **GARTSIDE, G.** : The energy cost of prospective fuels- with particular reference to fuels from "renewable" sources. - In : Feasibility of Alternative Renewable Resources. (Symp. Solar Energy Resources.) Pp. 41 - 54. Soc. Social Responsibility Sci., Canberra, A.C.T. 1975. [Ps.]

22504 - **GARWOOD, E.A., TYSON, K.C.** : The response of S 24 perennial ryegrass swards to irrigation. 2. Variation in soil- and plant-water status. - J. brit. Grassland Soc. *30* : 51 - 62, 1975. [Stomatal resistance.]

22505 - **GASH, J.H.C., STEWARD, J.B.** : The average surface resistance of a pine forest derived from Bowen ratio measurements. - Boundary-Layer Meteorol. *8* : 453 - 464, 1975.

22506 - **GASSMAN, M., DUGGAN, J.** : Chemical induction of porphyrin synthesis in higher plants. - In : **AVRON, M.** (ed.) : Proceedings of the Third International Congress on Photosynthesis. Vol. III. Pp. 2105 - 2113. Elsevier, Amsterdam - Oxford - New York 1975.

22507 - **GATES, D.M.** : Introduction : Biophysical ecology. - In : **GATES, D.M., SCHMERL, R.B.** (ed.) : Perspectives of Biophysical Ecology. Pp. 1 - 28. Springer-Verlag, Berlin - Heidelberg - New York 1975. [Ps, Chl, Car.]

22508 - **GAUSMAN, H.W., RODRIGUEZ, R.R., ESCOBAR, D.E.** :Ultraviolet radiation reflectance, transmittance, and absorptance by plant leaf epidermises. - Agron. J. *67* : 720 - 724, 1975.

22509 - **GAUSMAN, H.W., THOMAS, J.R., ESCOBAR, D.E., BERUMEN, A.** : Cotton leaf air volume and chlorophyll concentration affect reflectance of visible light. - J. Rio Grande Valley hort. Soc. *29* : 109 - 114, 1975.

22510 - **GAVAZZI, G., PICCARDO, C., MANZONI, L.** : A study of the effects of temperature and nutrients on the expression of chlorophyll mutants in maize. - Z. Pflanzenphysiol. *75* : 381 - 391, 1975.

22511 - **GAVIS, J., FERGUSON, J.F.** : Kinetics of carbon dioxide uptake by phytoplankton at high pH. - Limnol. Oceanogr. *20* : 211 - 221, 1975.

22512 - **GAVRILENKO, V.F., GRABOVSKAYA, M.I.** : Issledovanie porfirinovogo obmena v zavisimosti ot okislitel'no-vosstanovitel'nykh usloviĭ sredy. [Porphyrin exchange in dependence on oxidation-reduction conditions of environment.] - Nauch. Dokl. vyssh. Shkoly, biol. Nauki *18* (9) : 83 - 90, 1975. [Chl; in R.]

22513 - **GAVRILENKO, V.F., RUBIN, B.A., ZHIGALOVA, T.V.** : Izuchenie sopryagayushchikh faktorov v opytakh s razobshcheniem i rekonstruktsieĭ aktivno- i slabofosforiliruyushchikh sistem khloroplastov. [Coupling factors in the experiments with decoupling and reconstruction of the chloroplast phosphorylating systems of high and low activity.] - Fiziol. Rast. *22* : 445 - 452, 1975. [In R , ab:E.]

22514 - **GAZANCHYAN, R.M., ABILOV, Z.K., ALIEV, Z.Sh., GASANOV, R.A.** : Forms of chlorophyll and photochemical activities of photosystems 1 and 2 from grana and photosystem 1 from stroma lamellae isolated in different ways. - Photosynthetica *9* : 268 - 276, 1975.

22515 - **GEARING, P., van BAALEN, C., PARKER, P.L.** : Biochemical effects of technetium--99-pertechnetate on microorganisms. - Plant Physiol. *55* : 240 - 246, 1975.

22516 - **GEJ, B., BARANOWSKA, H., MAZUROWA, J., ROGOZIŃSKA, E.** : Wpływ nawożenia azotowego na niektóre właściwości fizjologiczne i morfologo-anatomiczną *Lolium multiflorum* LAM. (formy tetraploidalnej). [Effect of nitrogen nutrition on some physiological, morphological and anatomical properties of tetraploidal form of *Lolium multiflorum* LAM.] - Acta agrobot. *28* : 95 - 119, 1975. [In Pol., ab : E.]

22517 - **GELIN, C.** : Nutrients, biomass and primary productivity of nannoplankton in eutrophic Lake Vombsjön, Sweden. - Oikos *26* : 121 - 139, 1975.

22518 - **GEORGIEV, G., VENTO, H.** : Die Wirkung der Lichtintensität in Zusammenhang mit der Mineralernährung auf den Pigmentgehalt in Blättern von jungen Kaffeepflanzen. - Dokl. bolg. Akad. Nauk *28* : 825 - 828, 1975.

22519 - **GEORGIEV, M., SPASENOSKI, M.** : Influence of the soil moisture on the content of phosphorus in the leaves, stalks and peels, phosphorus fractions in the grain and the chlorophyll pigment in the leaves of winter wheat. - Acta bot. croat. *34* : 181, 1975.

22520 - **GEPSHTEIN, A., HOCHMAN, Y., CARMELI, C.** : Effect of the interaction between cation-ATP complexes and free cations on ATPase activity in *Chromatium* strain D chromatophores. - In : **AVRON, M.** (ed.) : Proceeding of the Third International Congress on Photosynthesis. Vol. II. Pp. 1189 - 1197. Elsevier, Amsterdam - Oxford - New York 1975.

22521 - **GERSTER, R., DIMON, B., PEYBERNES, A.** : The fate of oxygen in photosynthesis. - In : **AVRON, M.** (ed.) : Proceedings of the Third International Congress on Photosynthesis. Vol. II. Pp. 1589 - 1600. Elsevier, Amsterdam - Oxford - New York 1975.

22522 - **GERSTER, R., LORIMER, G.H., VENNESLAND, B.** : The extra O_2 evolved during nitrate utilization by *Chlorella*. - Plant Sci. Lett. *5* : 255 - 260, 1975.

22523 - **GEZELIUS, K.** : Extraction and some characteristics of ribulose-1,5-diphosphate carboxylase from *Pinus silvestris*. - Photosynthetica *9* : 192 - 200, 1975.

22524 - **GHOSH, K.K., JANA, P.K., MAITY, S.P.** : Influence of soil moisture tension and nitrogen on growth dynamics in wheat. - Ind. Agriculturist *19* : 51 - 57, 1975. [Growth analysis.]

22525 - **GIAQUINTA, R., DILLEY, R.A.** : A chloroplast membrane conformational change associated with water oxidation revealed by diazonium benzene sulfonate binding. - In : **AVRON, M.** (ed.) : Proceedings of the Third International Congress on Photosynthesis. Vol. II. Pp. 883 - 895. Elsevier, Amsterdam - Oxford - New York 1975.

22526 - **GIAQUINTA, R., ORT, D., DILLEY, R.** : Relationship between photosystem II energy transduction and membrane conformational changes. - Plant Physiol. *56* (Suppl.) : 46, 1975.

22527 **GIAQUINTA, R.T., DILLEY, R.A.** : A partial reaction in photosystem II : Reduction of silicomolybdate prior to the site of dichlorophenyldimethylurea inhibition. - Biochim. biophys. Acta *387* : 288 - 305, 1975.

22528 - **GIAQUINTA, R.T., ORT, D.R., DILLEY, R.A.** : The possible relationship between a membrane conformational change and photosystem II dependent hydrogen ion accumulation and adenosine 5'-triphosphate synthesis. - Biochemistry *14* : 4392 - 4396, 1975.

22529 - **GIBBONS, G.C., STRØBAEK, S., HASLETT, B., BOULTER, D.** : The N-terminal amino acid sequence of the small subunit of ribulose-1,5-diphosphate carboxylase from *Nicotiana tabacum*. - Experientia *31* : 1040 - 1041, 1975.

22530 - **GIBBS, M., LATZKO, E., LABER, L.J., HINES, G.** : Photosynthetic CO_2 assimilation in maize and spinach leaves and chloroplasts. - In : **NASYROV, Yu.S., ŠESTÁK, Z.** (ed.) : Genetic Aspects of Photosynthesis. Pp. 149 - 157. Dr. W. Junk b.v. Publ., The Hague 1975.

22531 - **GIESKES, W.W.C., KRAAY, G.W.** : The phytoplankton spring bloom in Dutch coastal waters of the North Sea. - Neth. J. Sea Res. *9* : 166 - 196, 1975. [Chl.]

22532 - **GIFFORD, R.M.** : Fuel requirements for growing plants. - In : Feasibility of Alternative Renewable Resources. (Symp. Solar Energy Resources.) Pp. 1 - 14. Soc. Social Responsibility Sci., Canberra 1975. [Ps.]

22533 - **GILLEN, L.A., BENEDICT, C.R.** : Ribulose-1,5-diP carboxylase in endosperms of germinating castor beans. - Plant Physiol. *56* (Suppl.) : 48, 1975.

22534 - **GILLER, Yu.E., VAKHIDOVA, L.R., ASOEVA, L.M., YUKHANANOVA, L.N., ABDULLAEVA, S.K., LIPKIND, B.I., KRASICHKOVA, G.V., YUSUPOVA, G.A.** : Synthetic pigment--protein-lipoid complexes - models of molecular organization and functional properties of the pigment system of the photosynthetic apparatus. - In : **NASYROV, Yu.S., ŠESTÁK, Z.** (ed.) : Genetic Aspects of Photosynthesis. Pp. 271 - 285. Dr. W. Junk b.v. Publ., The Hague 1975.

22535 - **GILLOTT, M.A., FLOYD, G.L., WARD, D.V.** : The role of sediment as a modifying factor in pesticide-algae interactions. - Environm. Entomol. *4* : 621 - 624, 1975. [Ps.]

22536 - **GIMMLER, H., SCHÄFER, G., HEBER, U.** : Low permeability of the chloroplast envelope towards cations. - In : **AVRON, M.** (ed.) : Proceedings of the Third International Congress on Photosynthesis. Vol. II. Pp. 1381 - 1392. Elsevier, Amsterdam - Oxford - New York 1975.

22537 - **GINZBURG, C.** : Dark CO_2 fixation in *Gladiolus* cormels and its regulation during the break of dormancy. - Plant Physiol. *56* : 51 - 55, 1975.

22538 - **GIRAUD, G.** : La production photosynthétique des Algues marines. - In : **COSTES, C.** (ed.) : Photosynthèse et Production Végétale. Pp. 243 - 257. Gauthier-Villars, Paris 1975.

22539 - **GIRAULT, G., GALMICHE, J.M., VERMEGLIO, A.** : CF_1 reconstitution of EDTA treated membrane fragments observed by means of photophosphorylation and 515 nm absorbance changes. - In : **AVRON, M.** (ed.) : Proceedings of the Third International Congress on Photosynthesis. Vol. II. Pp. 839 - 847. Elsevier, Amsterdam - Oxford - New York 1975.

22540 - **GIVAN, C.V.** : Glutamine synthesis and its relation to electron transport in pea chloroplasts. - Plant Physiol. *56* (Suppl.) : 66, 1975.

22541 - **GIVAN, C.V.** : Light-dependent synthesis of glutamine in pea-chloroplast preparations. - Planta *122* : 281 - 291, 1975.

22542 - **GJERSTAD, D.H., DICKMANN, D.I.** : Effect of leaf age on glycolate oxidase activity and photorespiration in *Populus* × *euramericana*. - Plant Physiol. *56* (Suppl.) : 26, 1975.

22543 - **GJESSING, Y.T., ØVSTEDAL, D.O.** : Energy budget and ecology of two vegetation types in Svalbard. - Astarte (J. arctic Biol.) *8* (2) : 83 - 92, 1975.

22544 - **GLAGOLEVA, T.A., REĬNUS, R.M., MOKRONOSOV, A.T., ZALENSKIĬ, O.V.** : Vliyanie kisloroda na fotosintez i fotodykhanie rasteniĭ pustyni yugo-vostochnykh Karakumov. [The effect of oxygen on photosynthesis and photorespiration in desert plants of south-east Karakums.] - Bot. Zh. *60* : 927 - 939, 1975. [In R, ab : E.]

22545 - **GLAZER, A.N., COHEN-BAZIRE, G.** : A comparison of cryptophytan phycocyanins.- Arch. Microbiol. *104* : 29 - 32, 1975.

22546 - **GLAZER, A.N., HIXSON, C.S.** : Characterization of R-phycocyanin. Chromophore content of R-phycocyanin and C-phycoerythrin. - J. biol. Chem. *250* : 5487 - 5495, 1975.

22547 - **GŁAŻEWSKI, S.** : Donory ^{14}C-asymilatów dla strąków i korzeni grochu. [The donors of ^{14}C-assimilates for pods and roots of pea plants.] - Pamiętnik puławski - Prace IUNG *64* : 189 - 208, 1975. [In Pol., ab : E, R.]

22548 - **GLIDEWELL, S.M., RAVEN, J.A.** : Measurement of simultaneous oxygen evolution and uptake in *Hydrodictyon africanum*. - J. exp. Bot. *26* : 479 - 488, 1975.

22549 - **GLOE, A., PFENNIG, N., BROCKMANN, H. Jr., TROWITZSCH, W.** : A new bacteriochlorophyll from brown-colored *Chlorobiaceae*. - Arch. Microbiol. *102* : 103 - 109, 1975.

22550 - **GLOOSCHENKO, V., GLOOSCHENKO, W.** : Effect of polychlorinated biphenyl compounds on growth of Great Lakes phytoplankton. - Can. J. Bot. *53* : 653 - 659, 1975. [Chloroplast.]

22551 - **GLOOSCHENKO, W.A.** : Nutrient-primary production relationships in central Lake Erie : A simple correlation approach. - Ohio J. Sci. *75* : 251 - 255, 1975. [Chl.]

22552 - **GLOVER, H., BEARDALL, J., MORRIS, I.** : Effects of environmental factors on photosynthesis patterns in *Phaeodactylum tricornutum (Bacillariophyceae)*. I. Effect of nitrogen deficiency and light intensity. - J. Phycol. *11* : 424 - 429, 1975.

22553 - **GLYNNE-JONES, E., MARSHALL, R., HANN, R.A., READ, G.** : The purification of chlorophyll *a* from spinach leaves by partition chromatography on Bio-Glas columns. - J. Chromatogr. *114* : 232 - 234, 1975.

22554 - **GNANAM, A., SEETHARAMA, N.** : Autonomy of organelles. - J. sci. ind. Res. (India) *34* : 375 - 385, 1975. [Chloroplast.]

22555 - **GNANARETHINAM, J.L.** : Quelques aspects biochimiques de la physiologie des radis (*Raphanus sativus* L.) traités par la simazine. I. Evolution des glucides. - Weed Res. *15* : 143 - 148, 1975. [Chl.]

22556 - **GOATLY, M.B., COOMBS, J., SMITH, H.** : Development of C_4 photosynthesis in sugar cane : Changes in properties of phosphoenolpyruvate carboxylase during greening. - Planta *125* : 15 - 24, 1975.

22557 - **GODZIEMBA-CZYŻ, J.** : Conformational changes in spinach (*Spinacia oleracea*) leaves chloroplasts *in vivo*. - Acta Soc. Bot. Pol. *44* : 277 - 287, 1975.

22558 - **GOEDHEER, J.C., KLEINEN HAMMANS, J.W.** : Efficiency of light conversion of the blue-green alga *Anacystis nidulans*. - Nature *256* : 333 - 335, 1975.

22559 - **GOEDHEER, J.C., SWART, J.** : Decay of delayed light with the diatom *Phaeodactylum tricornutum*. - Plant Sci. Lett. *4* : 335 - 341, 1975.

22560 - **GOL'D, V.M., GAEVSKIĬ, N.A., BELONOG, N.P.** : O roli fitokhroma v fotoindutsirovannykh konformatsionnykh izmeneniyakh khloroplastov gorokha na sinem i krasnom svetu. [Role of phytochrome in photoinduced conformation changes in pea chloroplasts under blue and red light.]- Nauch. Dokl. vyssh. Shkoly, biol. Nauki *18* (3) : 71 - 73, 1975. [In R.]

22561 - **GOL'DFELD, M.G., KHANGULOV, S.V., KHALILOV, R.I., TSAPIN, A.I.** : O putyakh vosstanovleniya reaktsionnykh tsentrov fotosistemy I fotosinteza. [Reaction centres of photosystem 1 reduction path.]-Biofizika *20* : 254 - 259, 1975. [In R, ab : E.]

22562 - **GOLDTHWAITE, J., BOGORAD, L.** : Ribulose-1,5-diphosphate carboxylase from leaf. - In : **COLOWICK, S.P., KAPLAN, N.O.** (ed.) : Methods in Enzymology. Vol. 42. Pp. 481 - 484. Academic Press, New York - San Francisco - London 1975.

22563 - **GOLOVKO, T.K.** : Izuchenie dykhaniya v svyazi s produktivnost'yu klevera (*Trifolim pratense*) v usloviyakh tsentral'nykh raĭonov Komi ASSR. [The role of respiration in productivity of *Trifolium pratense* grown in central regions of Komi ASSR.] - Bot. Zh. *60* : 1632 - 1638, 1975. [In R.]

22564 - **GOMÓLKA, B.** : Poglądy Jędrzeja Śniadeckiego na naturę procesu fotosyntezy. [Jędrzej Śniadecki's views on photosynthesis.] - Zesz. nauk. Univ. Jag. *395* : 49 - 93, 1975. [In Pol., ab : E.]

22565 - **GONCHARIK, M.N., MIKUL'SKAYA, S.A.** : Vliyanie ionov khlora i natriya na obrazovanie pigmentov i razmery plastid v list'yakh sakharnoĭ svekly. [Effect of chlorine and sodium ions on pigment formation and plastid size in sugar beet leaves.] - In : Fiziologo-biokhimicheskie Aspekty Rosta i Razvitiya Rasteniĭ. Pp. 28 - 36. Nauka i Tekhnika, Minsk 1975. [In R.]

22566 - **GONCHARIK, M.N., URBANOVICH, T.A.** : O vliyanii Cl^- na fotofosforilirovanie khloroplastov v usloviyakh nedostatochnogo vodoobespecheniya rasteniĭ. [Effect of Cl^- on photophosphorylation in chloroplasts from water deficient plants.] - In : Fiziologo-biokhimicheskie Aspekty Rosta i Razvitiya Rasteniĭ. Pp. 21 - 27. Nauka i Tekhnika, Minsk 1975. [In R.]

22567 - **GONCHAROVA, N.V., EVSTIGNEEV, V.B.** : Obrazovanie ATF v model'noĭ sisteme pri uchastii katalazy i perekisi vodoroda. [ATP formation in a model system induced by catalase and hydrogen peroxide.] - Dokl. Akad. Nauk SSSR *222* : 970 - 972, 1975. [In R.]

22568 - **GONCHAROVA, N.V., EVSTIGNEEV, V.B.** : O fotofosforilirovanii, sensibilizirovannom khlorofillom v adsorbirovannom sostoyanii. [Photophosphorylation sensitized by chlorophyll in the adsorbed state.] - Biokhimiya *40* : 622 - 628, 1975. [In R, ab : E.]

22569 - **GOODALL, D.W.** : Ecosystem modeling in the desert biome. - In : **PATTEN, B.C.** (ed.) : Systems Analysis and Simulation in Ecology. Vol. III. Pp. 73 - 94. Academic Press, New York - San Francisco - London 1975. [Production.]

22570 - **GORCHAKOVSKIĬ, P.L., KOROBEĬNIKOVA, V.P.** : Pervichnaya produktivnost' nekotorykh lugovykh soobshchestv yuzhnogo Urala. [Primary productivity of several meadow communities in the southern Urals.] - Ėkologiya *6* (3) : 5 - 17, 1975. [Caloric values; in R.]

22571 - **GORDON, M.E., LETHAM, D.S.** : Regulators of cell division in plant tissues. XXII. Physiological aspects of cytokinin-induced radish cotyledon growth. - Aust. J. Plant Physiol. *2* : 129 - 154, 1975. [Chl.]

B22572 - **GORYSHINA, T.K.** : Ėkologiya Travyanistykh Rasteniĭ Lesostepnoĭ Dubravy. [Ecology of Herbaceous Plants in a Forest-Steppe Oak Forest.] - Izdat. leningrad. Univ., Leningrad 1975. [Ps, Chl; in R.]

22573 - **GORYSHINA, T.K., ZABOTINA, L.N., PRUZHINA, E.G.** : Plastidnyĭ apparat travyanistykh rasteniĭ lesostepnoĭ dubravy v raznykh usloviyakh osveshchennosti. [Plastid apparatus of herbaceous plants of a forest-steppe oak forest under different illuminance.] - Ėkologiya *6* (5) : 15 - 22, 1975. [In R.]

22574 - **GOULD, J.M.** : Inhibition of Photosystem II-dependent phosphorylation in chloroplasts by mercurials. - Biochem. biophys. Res. Commun. *64* : 673 - 680, 1975.

22575 - **GOULD, J.M.** : Interactions of mercurials with the energy conservation sites in chloroplasts. - Fed. Proc. *34* : 596, 1975.

22576 - **GOULD, J.M.** : The phosphorylation site associated with the oxidation of exogenous donors of electrons to Photosystem I. - Biochim. biophys. Acta *387* : 135 - 148, 1975.

22577 - **GOULDEN, P.D., BROOKSBANK, P.** : Automated determinations of dissolved organic carbon in lake water. - Anal. Chem. *47* : 1943 - 1946, 1975.

22578 - **GOULDER, R.** : The effects of photosynthetically raised pH and light on some ciliated *Protozoa* in a eutrophic pond. - Freshwater Biol. *5* : 313 - 322, 1975.

B22579 - **GOVINDJEE** (ed.) : Bioenergetics of Photosynthesis. - Academic Press, New York - San Francisco - London 1975.

22580 - **GOVINDJEE, GOVINDJEE, R.** : Introduction to photosynthesis. - In : **GOVINDJEE** (ed.) : Bioenergetics of Photosynthesis. Pp. 1 - 50. Academic Press, New York - San Francisco - London 1975.

22581 - **GOVINDJEE, HAMMOND, J.H., SMITH, W.R., GOVINDJEE, R., MERKELO, H.** : Lifetime of the excised states *in vivo*. IV. Bacteriochlorophyll and bacteriopheophytin in *Rhodospirillum rubrum*. - Photosynthetica *9* : 216 - 219, 1975.

22582 - **GOVINDJEE, STEMLER, A.J., BABCOCK, G.T.** : A critical role of bicarbonate in the relaxation of reaction center II complex during oxygen evolution in isolated broken chloroplasts. - In : **AVRON, M.** (ed.) : Proceedings of the Third International Congress on Photosynthesis. Vol. I. Pp. 363 - 371. Elsevier, Amsterdam - Oxford - New York 1975.

22583 - **GRÄBER, P., WITT, H.T.** : Direct measurement of the protons pumped into the inner phase of the functional membrane of photosynthesis per electron transfer. - FEBS Lett. *59* : 184 - 189, 1975.

22584 - **GRACE, J., MALCOLM, D.C., BRADBURY, I.K.** : The effect of wind and humidity on leaf diffusive resistance in Sitka spruce seedlings. - J. appl. Ecol. *12* : 931 - 940, 1975.

22585 - **GRADYUSHKO, A.T., SOLOV'EV, K.N., TURKOVA, A.E., TSVIRKO, M.P.** : Fosforestsentsiya oktaėtilkhlorina, izobakteriooktaėtilkhlorina i ikh metallokompleksov. [Phosphorescence of octaethylchlorin, isobacteriooctaethylchlorin and their metallic complexes.] - Biofizika *20* : 602 - 607, 1975. [Chl analogues; in R, ab : E.]

22586 - **GRAEBER, P., WITT, H.T.** : Electrical potential difference, pH gradient and phosphorylation. On the relation between the transmembrane electrical potential difference, pH gradient and ATP formation in photosynthesis. - In : **AVRON, M.** (ed.) : Proceedings of the Third International Congress on Photosynthesis. Vol. I. Pp. 427 - 436. Elsevier, Amsterdam - Oxford - New York 1975.

22587 - **GRAEBER, P., WITT, H.T.** : The effect of temperature on flash induced transmembrane currents in chloroplasts of spinach. - In : **AVRON, M.** (ed.) : Proceedings of the Third International Congress on Photosynthesis. Vol. II. Pp. 951 - 956. Elsevier, Amsterdam - Oxford - New York 1975.

22588 - **GRAHL, H., WILD, A.** : Studies on the content of P700 and cytochromes in *Sinapis alba* during growth under two different light intensities. - In : **MARCELLE, R.** (ed.) : Environmental and Biological Control of Photosynthesis. Pp. 107 - 113. Dr. W. Junk b.v. Publ., The Hague 1975.

22589 - **GRANT, D.R.** : Comparison of evaporation from barley with Penman estimates. - Agr. Meteorol. *15* : 49 - 60, 1975. [Growth analysis.]

22590 - **GRAY, B.H., GANTT, E.** : Spectral properties of phycobilisomes and phycobiliproteins from the blue-green alga - *Nostoc* sp. - Photochem. Photobiol. *21* : 121 - 128, 1975.

22591 - **GRAY, E.A.** : Survival of *Escherichia coli* in stream water in relation to carbon dioxide and plant photosynthesis. - J. appl. Bacteriol. *39* : 47 - 54, 1975.

22592 - **GREENE, J.C., SOLTERO, R.A., MILLER, W.E., GASPERINO, A.F., SHIROYAMA, T.** : The relationship of laboratory algal assays to measurements of indigenous phytoplankton in Long Lake, Washington. - In : **MIDDLEBROOKS, E.J., FALKENBORG, D.H., MALONEY, T.E.** (ed.) : Biostimulation and Nutrient Assessment. Pp. 93 - 126. Utah State Univ., Logan, Utah 1975. [Chl, productivity.]

22593 - **GREENWOOD, E.A.N., FARRINGTON, P., BERESFORD, J.D.** : Characteristics of the canopy, root system and grain yield of a crop of *Lupinus angustifolius* cv. Unicrop. - Aust. J. agr. Res. *26* : 497 - 510, 1975.

22594 - **GREGORY, P., BRADBEER, J.W.** : Plastid development in primary leaves of *Phaseolus vulgaris*. Development of plastid adenosine triphosphatase activity during greening. - Biochem. J. *148* : 433 - 438, 1975.

22595 - **GREGORY, R.P.F.** : Evidence that circularly dichroic chlorophyll forms *a*-682 and *a*-710 are oriented at right angles to the thylakoid membranes of whole chloroplasts, and that the circular dichroism is light-dependent. - Biochem. J. *148* : 487 - 497, 1975.

22596 - **GREGORY, R.P.F., RAPS, S.** : The relation of light-harvesting chlorophyll to the chlorophyll-protein complexes CP1 and CP2. - In : **AVRON, M.** (ed.) : Proceedings of the Third International Congress on Photosynthesis. Vol. III. Pp. 1977 - 1982. Elsevier, Amsterdam - Oxford - New York 1975.

22597 - **GREGSON, K., BISCOE, P.V.** : Barley and its environment. II. Strategy for computing. - J. appl. Ecol. *12* : 259 - 267, 1975. [Measurements in canopy : data sampling and processing.]

22598 - **GRESHAM, C.A., SINCLAIR, T.R., WUENSCHER, J.E.** : A ventilated diffusion porometer for measurement of the stomatal resistance of pine fascicles. - Photosynthetica *9* : 72 - 77, 1975.

22599 - **GRIFFITHS, M., EDMONDSON, W.T.** : Burial of oscillaxanthin in the sediment of Lake Washington. - Limnol. Oceanogr. *20* : 945 - 952, 1975.

22600 - **GRIFFITHS, W.T.** : Some observations on chlorophyll(ide) synthesis by isolated etioplasts. - Biochem. J. *146* : 17 - 24, 1975.

22601 - **GRIFFITHS, W.T.** : Characterization of the terminal stages of chlorophyll(ide) synthesis in etioplast membrane preparations. - Biochem. J. *152* : 623 - 635, 1975.

22602 - **GRIFFITHS, W.T., JONES, O.T.G.** : Magnesium 2,4-divinylphaeoporphyrin a_5 as a substrate for chlorophyll biosynthesis *in vitro*. - FEBS Lett. *50* : 355 - 358, 1975.

22603 - **GRIME, J.P., HUNT, R.** : Relative growth-rate: its range and adaptive significance in a local flora. - J. Ecol. *63* : 393 - 422, 1975. [Growth analysis.]

22604 - **GRIMME, L.H.** : Biologische und biotechnologische Beiträge zur Welternährungslage. - Naturwiss. Rundschau *28* (5) : 149 - 153, 1975. [Ps.]

22605 - **GRIMME, L.H., BOARDMAN, N.K.** : Bound cytochrome-553 in photosystem I subchloroplast fragments from *Chlorella*. - In : **AVRON, M.** (ed.) : Proceedings of the Third International Congress on Photosynthesis. Vol. III. Pp. 2115 - 2124. Elsevier, Amsterdam - Oxford - New York 1975.

22606 - **GRINENKO, V.V., BELETSKAYA, D.K.** : Radiatsionnyĭ rezhim krony grushi v zavisimosti ot formy i podvoya. [Radiation regime of a pear-tree crown in dependence on the shape and rootstock.] - Fiziol. Biokhim. kul't. Rast. *7* : 68 - 74, 1975. [In R, ab : E.]

22607 - **GRINTAL', A.R.** : Temperaturnaya adaptatsiya fotosinteza *Laminaria saccharina* (L.) LAM. [Adaptation of photosynthesis to temperature in *Laminaria saccharina* (L.) LAM.] - Bot. Zh. *60* : 256 - 265, 1975. [In R.]

22608 - **GRISHINA, G.S., MALISHEVSKIĬ, S., FRANKEVICH, A., POSKUTA, Yu., VOSKRESENSKAYA, N.P.** : Sravnitel'noe deĭstvie krasnogo i sinego sveta na metabolizm C^{14} u kukuruzy pri razlichnykh kontsentratsiyakh kisloroda. [Comparative effect of red and blue light on ^{14}C metabolism in maize at different oxygen concentrations.] - Fiziol. Rast. *22* : 27 - 33, 1975. [In R, ab : E.]

22609 - **GRODZINSKI, B., COLMAN, B.** : The effect of osmotic stress on the oxidation of glycolate by the blue-green alga *Anacystis nidulans*. - Planta *124* : 125 - 133, 1975. [Ps.]

22610 - **GRODZINSKIĬ, A.M.** : Raboty N.G. Kholodnogo o vozdushnom pitanii rasteniĭ i sovremennaya allelopatiya. [Publications of N.G. Kholodnyĭ on the air nutrition of plants and the contemporary allelopathy.] - In : Fiziologo-biokhimicheskie Osnovy Vzaimodeĭstviya Rasteniĭ v Fitotsenozakh. Vol. 6. Pp. 3 - 8. Naukova Dumka, Kiev 1975. [Ps; in R.]

22611 - **GRODZINSKIĬ, A.M., BOGDAN, G.P.** : O vliyanii allelopaticheskogo faktora na aktivnost' adenozintrifosfatazy u rasteniĭ. [Effect of the allelopathic factor on the activity of adenosinetriphosphatase in plants.] - In : Fiziologo--biokhimicheskie Osnovy Vzaimodeĭstviya Rasteniĭ v Fitotsenozakh. Vol.6. Pp. 15 - 19. Naukova Dumka, Kiev 1975.[In R.]

22612 - **GROMET-ELHANAN, Z.** : Effect of aurovertin on energy conversion reactions in *Rhodospirillum rubrum* chromatophores. - In : **AVRON, M.** (ed.) : Proceedings of the Third International Congress on Photosynthesis. Vol. I. Pp. 791 - 797. Elsevier, Amsterdam - Oxford - New York 1975.

22613 - **GROMET-ELHANAN, Z., LEISER, M.** : Postillumination adenosine triphosphate synthesis in *Rhodospirillum rubrum* chromatophores. II. Stimulation by a K^+ diffusion potential. - J. biol. Chem. *250* : 90 - 93, 1975.

22614 - **GRONEBAUM-TURCK, K., MATHÉ, P.** : Der Einfluß von Fluorverbindungen auf die Chloroplastenpigmente von Pappel-, Holunder- und Fliederblättern im Freiland bei verschiedener Belastung. - Europe. J. Forest Pathol. *5* : 183 - 184, 1975. [Chl, Car.]

22615 - **GROSS, E.L., WYDRZYNSKI, T., VanderMEULEN, D., GOVINDJEE** : Monovalent and divalent cation-induced changes in chlorophyll *a* fluorescence and chloroplast structure. - In : **AVRON, M.** (ed.) : Proceedings of the Third International Congress on Photosynthesis. Vol. I. Pp. 345 - 361. Elsevier, Amsterdam - Oxford - New York 1975.

22616 - **GROSS, J.A., STROZ, R.J., BRITTON, G.** : The carotenoid hydrocarbons of *Euglena gracilis* and derived mutants. - Plant Physiol. *55* : 175 - 177, 1975.

22617 - **GROVES, R.H.** : Growth and development of five populations of *Themeda australis* in response to temperature. - Aust. J. Bot. *23* : 951 - 963, 1975. [Growth analysis.]

22618 - **GRUMBACH, K.H., LICHTENTHALER, H.K.** : Kinetic of lipoquinone and pigment synthesis during red light-induced thylakoid formation in etiolated barley seedlings. - Z. Naturforsch. *30 C* : 337 - 341, 1975.

22619 - **GRUMBACH, K.H., LICHTENTHALER, H.K.** : Photooxidation of the plastohydroquinone-9 pool in plastoglobuli during onset of photosynthesis. - In : **AVRON, M.** (ed.) : Proceedings of the Third International Congress on Photosynthesis. Vol. I. Pp. 515 - 523. Elsevier, Amsterdam - Oxford - New York 1975.

22620 - **GUARDINO, V., AVERNA, V.** : Lipids in photosynthetic tissues of plants with Hatch-Slack metabolic pathway. I : Fatty acids composition of pigweed (*Amaranthus retroflexus* L.) leaf and chloroplast lipids. - Riv. ital. Sostanze grasse *52* : 198 - 202, 1975.

22621 - **GUDKOV, N.D., STOLOVITSKIĬ, Yu.M., EVSTIGNEEV, V.B.** : Impul'snaya fotoprovodimost' rastvorov khlorofilla i ego analogov. II. Vliyanie polyarnosti sredy na vykhod ion-radikalov pri fotookislenii khlorofilla *a*. [Impulse photoconductivity of chlorophyll solutions and its analogs. II. Effect of medium polarity on yield of ion radicals during chlorophyll *a* photooxidation.] - Biofizika *20* : 214 - 218, 1975. [In R, ab : E.]

22622 - **GUDKOV, N.D., STOLOVITSKIĬ, Yu.M., EVSTIGNEEV, V.B.** : Impul'snaya fotoprovodimost' rastvorov khlorofilla i ego analogov. III. Fotoprovodimost' geksanolovykh rastvorov khlorofilla *a* v prisutstvii *n*-benzokhinona, kineticheskie i temperaturnye izmereniya. [Impulse photoconductivity of chlorophyll solutions and its analogs. III. Photoconductivity of hexanol solutions of chlorophyll *a* in the presence of *n*-benzoquinone. Kinetic and temperature measurements.] - Biofizika *20* : 807 - 811, 1975. [In R, ab : E.]

22623 - **GUÉRIN-DUMARTRAIT, E., LECLERC, J.-C., HOARAU, J.** : Composition pigmentaire et activité photosynthétique de *Porphyridium* sp. cultivé sur milieu carencé en fer à partir d'un inoculum synchronisé. - In : Les Cycles Cellulaires et leur Blocage chez Plusieurs Protistes. Coll. int. CNRS. Vol. 240. Pp. 95 - 99. Édit. CNRS, Paris 1975.

22624 - **GUERN, M., BOURDU, R., ROUX, M.** : Polyploidie et appareil photosynthétique chez l'*Hippocrepis comosa* L. - Photosynthetica *9* : 40 - 51, 1975.

22625 - **GUIKEMA, J.A., YOCUM, C.F.** : Analysis of site II electron flow and photophosphorylation. - Plant Physiol. *56* (Suppl.) : 45, 1975.

22626 - **GUILLOTIN, J., REISS-HUSSON, F.** : Cytoplasmic and outer membranes separation in *Rhodopseudomonas sphaeroides*. - Arch. Microbiol. *105* : 269 - 275, 1975.

22627 - **GUILLOT-SALOMON, T., TUQUET, C.** : Modifications de l'ultrastructure des plastes de feuilles d'Orge étiolées soumises à l'action d'éclairs répétés. - Compt. rend. Acad. Sci. Paris, Sér. D *280* : 1685 - 1688, 1975.

22628 - **GULAYA, N.K., TYUTEN'KOVA, N.L.** : Fotosintez planktona i destruktsiya v Kapchagaĭskom vodokhranilishche na tret'em godu ego sushchestvovaniya. [Plankton photosynthesis and degradation in the Kapchagaĭ reservoir in the third year of its existence.] - Gidrobiol. Zh. *11* (3) : 43 - 46, 1975. [In R.]

22629 - **GUN-AAZHAV, T., KUKUSHKIN, A.K., SOLNTSEV, M.K.** : Izmeneniya vykhoda fluorestsentsii zelenogo lista pod deĭstviem sveta, pogloshchaemogo fotosistemoĭ 1 i fotosistemoĭ 2. [The change in fluorescence yield of a green leaf under the action of light absorbed by photosystem 1 and photosystem 2.] - Biofizika *20* : 260 - 265, 1975. [In R, ab : E.]

22630 - **GUPTA, P.K., YASHVIR** : Induced mutations in foxtail millet (*Setaria italica* BEAUV.) I. Chlorophyll mutations induced by gamma rays, EMS and DES. - Theor. appl. Genet. *45* : 242 - 249, 1975.

22631 - **GURICHEVA, N.P., DEMINA, O.M., KOZLOVA, G.I., NOMOKONOV, L.I., STEPANOVA, K.D.**: Produktivnost' lugovykh soobshchestv. [Productivity of meadow communities.] - In : Resursy Biosfery (Itogi sovetskikh issledovaniĭ po Mezhdunarodnoĭ biologicheskoĭ programme). Vol. 1. Pp. 96 - 127. Nauka, Leningrad 1975. [In R, ab : E.]

22632 - **GURINOVICH, G.P.** : Ėlementarnye fotoprotsessy v molekulakh i metody ikh izucheniya. [Elementary photoprocesses in molecules and methods of their study.] - In : Metody Issledovaniya Fotokhimicheskikh Reaktsiĭ Fotosinteza *in Vitro* i *in Vivo*. Pp. 11 - 53. Pushchino 1975. [In R.]

22633 - **GURINOVICH, G.P., LOSEV, A.P., SARZHEVSKAYA, M.V.** : Izuchenie struktury produktov fotokhimicheskogo gidrirovaniya khlorofilla i ego proizvodnykh. [Structure of products of photochemical hydration of chlorophyll and its derivatives.] - Dokl. Akad. Nauk belorus. SSR *19* : 1129 - 1131, 1975. [In R.]

22634 - **GUTELMACHER, B.L.** : Relative significance of some species of algae in plankton primary production. - Arch. Hydrobiol. *75* : 318 - 328, 1975.

22635 - **GUTIERREZ, M., HUBER, S.C., KU, S.B., KANAI, R., EDWARDS, G.E.** : Intracellular localization of carbon metabolism in mesophyll cells of C_4 plants. - In : **AVRON, M.** (ed.) : Proceedings of the Third International Congress on Photosynthesis. Vol. II. Pp. 1219 - 1230. Elsevier, Amsterdam - Oxford - New York 1975.

22636 - **GUTSER, R.** : Carotinbildung in Möhren unter dem Einfluß einer Behandlung mit Harnstoffderivaten, Carbamaten und Amiden. - Landw. Forsch. 31/II. Sonderheft (Stand und Leistung agrikulturchemischer und agrarbiologischer Forschung XXIX) : 235 - 244, 1975.

22637 - **HÄDER, D.-P.** : The effect of inhibitors on the electron flow triggering photo-

-phobic reactions in *Cyanophyceae*. - Arch. Microbiol. *103* : 169 - 174, 1975.

22638 - **HAEDER, H.E.** : Einfluß chloridischer und sulfatischer Ernährung auf Assimilation und Assimilatverteilung in Kartoffelpflanzen. - Landwirtsch. Forsch. Kongreßband *1975* (Sonderheft 32/1) : 122 - 131, 1975.

22639 - **HAEDER, H.E., MENGEL, K.** : Einfluß der Lichtintensität bei variierter Kaliumernährung auf CO_2-Assimilation und Ertragsbildung bei Sommerweizen. - Z. Pflanzenern. Bodenkunde *1975* : 573 - 582, 1975. [Ps.]

22640 - **HAEHNEL, W.** : Light-induced absorbance changes of plastocyanin *in situ* and its functional role in chloroplasts. - In : **AVRON, M.** (ed.) : Proceedings of the Third International Congress on Photosynthesis. Vol. I. Pp. 557 - 568. Elsevier, Amsterdam - Oxford - New York 1975.

22641 - **HAFF, L.** : Detection of messenger-like RNA transcribed from chloroplast DNA in etiolated and greened maize seedlings. - Plant Physiol. *56* (Suppl.) : 32, 1975.

22642 - **HÄGELE, W., DRISSLER, F.** : Photosynthese (2. Teil). - Physik unserer Zeit *6* (1) : 2 - 14, 1975.

22643 - **HAGEMANN, R., HERRMANN, F., BÖRNER, T.** : The use of plastid and gene mutants of higher plants in studying the genetic control of plastid functions. - In : **NASYROV, Yu.S., ŠESTÁK, Z.** (ed.) : Genetic Aspects of Photosynthesis. Pp. 115 - 118. Dr. W. Junk b.v. Publ., The Hague 1975.

22644 - **HAGER, A.** : Die reversiblen, lichtabhängigen Xanthophyllumwandlungen im Chloroplasten. - Ber. deut. bot. Ges. *88* : 27 - 44, 1975.

22645 - **HAGER, H.** : Die forstliche Produktion in Wechselbeziehung zu den atmosphärischen Standortsfaktoren. - Allgem. Forstzeit. *86* : 327 - 328, 1975. [CO_2 fluxes.]

22646 - **HAINES, E.B., DUNSTAN, W.M.** : The distribution and relation of particulate organic material and primary productivity in the Georgia Bight, 1973 - 1974. - Estuar. coast. mar. Sci. *3* : 431 - 441, 1975.

22647 - **HAISMAN, D.R., CLARKE, M.W.** : The interfacial factor in the heat-induced conversion of chlorophyll to pheophytin in green leaves. - J. Sci. Food Agr. *26* : 1111 - 1126, 1975.

22648 - **HALES, B.J.** : Dependency of rate of biological electron transport on protein motion and availability of phonon levels. - Biophys. J. *15* (2, Part 2) : 113a, 1975.

22649 - **HALL, A.E., BJÖRKMAN, O.** : Model of leaf photosynthesis and respiration. - In : **GATES, D.M., SCHMERL, R.B.** (ed.) : Perspectives of Biophysical Ecology. Pp. 55 - 72. Springer-Verlag, Berlin - Heidelberg - New York 1975.

22650 - **HALL, A.E., CAMACHO-B, S.E., KAUFMANN, M.R.** : Regulation of water loss by citrus leaves. - Physiol. Plant. *33* : 62 - 65, 1975. [Stomatal resistance.]

22651 - **HALL, A.E., KAUFMANN, M.R.** : Regulation of water transport in the soil-plant--atmosphere continuum. - In : **GATES, D.M., SCHMERL, R.B.** (ed.) : Perspectives of Biophysical Ecology. Pp 187 - 202. Springer-Verlag, Berlin - Heidelberg - New York 1975. [Ps.]

22652 - **HALL, A.E., KAUFMANN, M.R** : Stomatal response to environment with *Sesamum indicum* L. - Plant Physiol. *55* : 455 - 459, 1975. [Ps.]

22653 - **HALL, A.E., YERMANOS, D.M.** : Leaf conductance and leaf water status of sesame strains in hot, dry climates. - Crop Sci. *15* : 789 - 793, 1975. [Stomatal resistance.]

22654 - **HALL, C.A.S., MOLL, R.** : Methods of assessing aquatic primary productivity. - In : **LIETH, H., WHITTAKER, R.H.** (ed.) : Primary Productivity of the Biosphere. Pp. 19 - 53. Springer-Verlag, Berlin - Heidelberg - New York 1975.

22655 - **HALL, D.O.** : Kinetics of oxygen utilisation by yeast and *E. coli* using an oxygen electrode. - In : **HALL, D.O., HAWKINS, S.E.** (ed.) : Laboratory Manual of Cell Biology. Pp. 162 - 165. English Universities Press, London 1975. [Ps.]

22656 - **HALL, D.O.** : Dye reduction by isolated chloroplasts. - In : **HALL, D.O., HAW-**

KINS, S.E. (ed.) : Laboratory Manual of Cell Biology. Pp. 176 - 178. English Universities Press, London 1975.

22657 - **HALL, D.O.** : Phosphorylation and electron transport in photosystems I and II of chloroplasts. - In : **QUAGLIARIELLO, E., PAPA, S., PALMIERI, F., SLATER, E. C., SILIPRANDI, N.** (ed.) : Electron Transfer Chains and Oxidative Phosphorylation. Pp. 411 - 416. North-Holland Publ. Comp., Amsterdam - Oxford - New York 1975.

22658 - **HALL, D.O., CAMMACK, R., RAO, K.K., EVANS, M.C.W., MULLINGER, R.** : Ferredoxins, blue-green bacteria and evolution. - Biochem. Soc. Trans. *3* : 361 - 368, 1975.

22659 - **HALL, D.O., RAO, K.K., CAMMACK, R.** : The iron sulphur proteins : structure, function and evolution of a ubiquitous group of proteins. - Sci. Prog. *62* : 285 - 317, 1975.

22660 - **HALL, D.O., RAO, K.K., MULLINGER, R.** : Biological functions of iron-sulphur proteins. - Biochem. Soc. Trans. *3* : 472 - 479, 1975.

22661 - **HALL, F.R., FERREE, D.C.** : Influence of twospotted spider mite populations on photosynthesis of apple leaves. - J. econom. Entomol. *68* : 517 - 520, 1975.

22662 - **HALL, R., ALI, A., BUSCH, L.V.** : *Verticillium* wilt of *Chrysanthemum* : development of wilt in relation to leaf diffusive resistance and vascular conductivity. - Can. J. Bot. *53* : 1200 - 1205, 1975. [Stomatal conductance.]

22663 - **HALL, S.M., HILLMAN, J.R.** : Correlative inhibition of lateral bud growth in *Phaseolus vulgaris* L. Timing of bud growth following decapitation. - Planta *123* : 137 - 143, 1975. [Photosynthate transport.]

22664 - **HALLDAL, P., NORDSTRÖM, B., ÖQUIST, G.** : Quantacorrected spectrometer for recording different types of photobiological, photochemical, and spectrophotometric measurements. - Physiol. Plant. *35* : 5 - 10, 1975.

22665 - **HÄLLGREN, J.E., HUSS, K.** : Effects of SO_2 on photosynthesis and nitrogen fixation. - Physiol. Plant. *34* : 171 - 176, 1975.

22666 - **HALLIWELL, B.** : Hydroxylation of *p*-coumaric acid by illuminated chloroplasts. The role of superoxide. - Europe. J. Biochem. *55* : 355 - 360, 1975.

22667 - **HALSEY, Y.D., BYERS, B.** : A large photoreactive particle from *Chromatium vinosum* chromatophores. - Biochim. biophys. Acta *387* : 349 - 367, 1975.

22668 - **HALVA, E., LESÁK, J.** : Studium produkční efektivnosti krátkodobých pícních společenstev. [Production efficiency of temporary clover-grass associations.] - Acta Univ. Agr. Fac. agron. *23* : 471 - 489, 1975. [In Czech, ab : E, G, R.]

22669 - **HAMLIN, L., TILLBERG, J.-E., SUNDBERG, I.** : Effects of azide on photophosphorylation in isolated spinach chloroplasts. - Physiol. Plant. *34* : 296 - 299, 1975.

22670 - **HAMPP, R., SANKHLA, N., HUBER, W.** : Effect of EMD-IT-5914 on chlorophyll synthesis in leaves of *Pennisetum typhoides* seedlings. - Physiol. Plant. *33* : 53 - 57, 1975.

22671 - **HAMPP, R., SCHNABL, H.** : Effect of alluminium ions on $^{14}CO_2$-fixation and membrane system of isolated spinach chloroplasts. - Z. Pflanzenphysiol. *76* : 300 - 306, 1975.

22672 - **HAMPP, R., ZIEGLER, H.** : Lichtabhängige Neusynthese von δ-Aminolaevulinsäure--Dehydratase in isolierten *Avena*-Etioplasten. - Planta *124* : 255 - 260, 1975.

22673 - **HANIGK, H., LICHTENTHALER, H.K.** : The kinetic of thylakoid prenyl lipid formation and oxygen evolution in *Scenedesmus* cultures grown under photo-autotrophic and photo-heterotrophic conditions. - In : **AVRON, M.** (ed.) : Proceedings of the Third International Congress on Photosynthesis. Vol. III. Pp. 2021 - 2028. Elsevier, Amsterdam - Oxford - New York 1975.

22674 - **HÄNSEL, H.** : Nachweis von multiplen Chlorophyllmutationen bei Gerste und ein Modell zur Abschätzung einer "optimalen" Mutationsrate unter Berücksichtigung multipler Mutationen. - Biológia (Bratislava) *30* : 161 - 173, 1975.

22675 - **HÄNSEL, H., SIMON, W., EHRENDORFER, K.** : Selektion auf zwei quantitative Charaktere nach Neutronen- und EMS-Behandlung bei Sommergerste. Ein angewandtes

Mikromutations-Experiment. - Biológia (Bratislava) *30* : 401 - 411, 1975. [Chl.]

22676 - **HANSEN, G.K.** : A dynamic continuous simulation model of water state and transportation in the soil-plant-atmosphere system. I. The model and its sensitivity. - Acta Agr. scand. *25* : 129 - 149, 1975. [Growth analysis.]

22677 - **HANSEN, J., MØLLER, I.** : Percolation of starch and soluble carbohydrates from plant tissue for quantitative determination with anthrone. - Anal. Biochem. *68* : 87 - 94, 1975.

22678 - **HANSEN, P.** : Produktion, fordeling og udnyttelse af fotosyntater i æbletræer. [Production, distribution and utilization of photosynthates in apple trees.] - Tidsskrift Planteavl *79* : 133 - 170, 1975. [In Dan., ab : E.]

22679 - **HANSEN, T.A., SEPERS, A.B.J., van GEMERDEN, H.** : A new purple bacterium that oxidizes sulfide to extracellular sulfur and sulfate. - Plant Soil *43* : 17 - 27, 1975.

22680 - **HANSEN, V.** : On the correlation between measured and estimated values of radiation balance. - Meld. Norges Landbrukshøgsk. *54* (23) : 1 - 9, 1975.

22681 - **HANSON, T.L.** : Needle biomass of lodgepole pine from tree dimensions. - Trans. ASAE *18* : 491 - 492, 496, 1975.

22682 - **HARBOUR, J.R., BOLTON, J.R.** : Superoxide formation in spinach chloroplasts : electron spin resonance detection by spin trapping. - Biochem. biophys. Res. Commun. *64* : 803 - 807, 1975.

22683 - **HARDT, H., MALKIN, S., EPEL, B.** : Fluorescence and triggered luminescence of mutants of *Chlamydomonas reinhardi* with lesions associated with photosystem II. - In : **AVRON, M.** (ed.) : Proceedings of the Third International Congress on Photosynthesis. Vol. I. Pp. 75 - 82. Elsevier, Amsterdam - Oxford - New York 1975.

22684 - **HARDY, R.W.F., HAVELKA, U.D.** : Nitrogen fixation research : a key to world food ? - Science *188* : 633 - 643, 1975. [Ps and nitrogen fixation.]

22685 - **HARDY, R.W.F., HAVELKA, U.D.** : Photosynthate as a major factor limiting nitrogen fixation by field-grown legumes with emphasis on soybeans. - In : **NUTMAN, P.S.** (ed.) : Symbiotic Nitrogen Fixation in Plants. IBP Vol. 7. Pp. 421 - 439. Cambridge University Press, Cambridge - London - New York - Melbourne 1975.

22686 - **HARI, P., LUUKKANEN, O., PELKONEN, P., SMOLANDER, H.** : Comparisons between photosynthesis and transpiration in birch. - Physiol. Plant. *33* : 13 - 17, 1975.

22687 - **HARLIN, M.M., CRAIGIE, J.S.** : The distribution of photosynthate in *Ascophyllum nodosum* as it relates to epiphytic *Polysiphonia lanosa*. - J. Phycol. *11* : 109 - 113, 1975.

22688 - **HARRIS, D.A., SLATER, E.C.** : Tightly-bound nucleotides of coupling ATPases - structural and functional aspects. - In : **QUAGLIARIELLO, E., PAPA, S., PALMIERI, F., SLATER, E.C., SILIPRANDI, N.** (ed.) : Electron Transfer Chains and Oxidative Phosphorylation. Pp. 379 - 384. North-Holland Publ. Comp., Amsterdam - Oxford - New York 1975.

22689 - **HARRIS, D.A., SLATER, E.C.** : Tightly bound nucleotides of the energy-transducing ATPase of chloroplasts and their role in photophosphorylation. - Biochim. biophys. Acta *387* : 335 - 348, 1975.

22690 - **HARTIG, P.R., SAUER, K.** : Fluorescent labeled coupling factor : identification of an ATP induced conformational change. - Plant Physiol. *56* (Suppl.) : 46, 1975.

22691 - **HARTMANN, E.** : Influence of light on the bioelectric potential of the bean (*Phaseolus vulgaris*) hypocotyl hook. - Physiol. Plant. *33* : 266 - 275, 1975. [Ps, Chl.]

22692 - **HASE, E.** : Recent progress in the physiology of *Chlorella*. - In : **TOKIDA, J., HIROSE, H.** (ed.) : Advance of Phycology in Japan. Pp. 170 - 180. Dr. W. Junk b.v. Publ., The Hague 1975. [Chloroplast.]

22693 - **HASE, T., WADA, K., MATSUBARA, H.** : A minor component of ferredoxin from *Aphanothece sacrum* cells. - J. Biochem. (Tokyo) *78* : 605 - 610, 1975.

22694 - **HASHIMOTO, H., MURAKAMI, S.** : Dual character of lipid composition of the envelope membrane of spinach chloroplasts. - Plant Cell Physiol. *16* : 895 - 902, 1975.

22695 - **HASHWA, F.** : Thiosulfate metabolism in some red phototrophic bacteria. - Plant Soil *43* : 41 - 47, 1975.

22696 - **HASPELOVÁ-HORVATOVIČOVÁ, A.** : Zmeny asimilačných farbív ako indikátory odolnosti rastlín po poškodení. [Changes in assimilatory pigments as factors indicating the resistance of plants after damage.] - Acta Inst. bot. Acad. Sci. slov., Ser. B *1* : 141 - 151, 1975. [In Slovak, ab : E, R.]

22697 - **HATCH, M.D.** : C_4-pathway photosynthesis in *Portulaca oleracea* and the significance of alanine labelling. - Planta *125* : 273 - 279, 1975.

22698 - **HATCH, M.D., KAGAWA, T., CRAIG, S.** : Subdivision of C_4-pathway species based on differing C_4 acid decarboxylating systems and ultrastructural features. - Aust. J. Plant Physiol. *2* : 111 - 128, 1975.

22699 - **HATCH, M.D., SLACK, C.R.** : Pyruvate, P_i dikinase from leaves. - In : **COLOWICK, S.P., KAPLAN, N.O.** (ed.) : Methods in Enzymology. Vol. 42. Pp. 212 - 219. Academic Press, New York - San Francisco - London 1975.

22700 - **HATTERSLEY, P.W., WATSON, L.** : Anatomical parameters for predicting photosynthetic pathways of grass leaves : The "maximum lateral cell count" and the "maximum cells distant count". - Phytomorphology *25* : 325 - 333, 1975.

22701 - **HAUN, J.R.** : Potato growth-environment relationships. - Agr. Meteorol. *15* : 325 - 332, 1975.

22702 - **HAUPT, W., TRUMP, K.** : Lichtorientierte Chloroplastenbewegung bei *Mougeotia* : Die Größe des Phytochromgradienten steuert die Bewegungsgeschwindigkeit. - Biochem. Physiol. Pflanzen *168* : 131 - 140, 1975.

22703 - **HAUPT, W., ÜBEL, H.** : Zum Mechanismus der Phytochromwirkung bei der Chloroplastenbewegung von *Mougeotia*. - Z. Pflanzenphysiol. *75* : 165 - 171, 1975.

22704 - **HAUSKA, G.** : Biochemical aspects of electron transport and phosphorylation in chloroplasts. - Ber. deut. bot. Ges. *88* : 303 - 318, 1975.

22705 - **HAUSKA, G.A.** : The effect of uncouplers on artificial donor shuttles for photosystem I. - In : **AVRON, M.** (ed.) : Proceedings of the Third International Congress on Photosynthesis. Vol. I. Pp. 689 - 696. Elsevier, Amsterdam - Oxford - New York 1975.

22706 - **HAUSKA, G., OETTMEIER, W., REIMER, S., TREBST, A.** : Energy conservation in photoreductions by photosystem I. Shuttles of artificial electron donors for photosystem I across the thylakoid membrane. - Z. Naturforsch. *30 C* : 37 - 45, 1975.

22707 - **HAUSKA, G., TREBST, A., KÖTTER, C., SCHULZ, H.** : 1,2,3-thiadiazolyl-phenyl-ureas, new inhibitors of photosynthetic and respiratory energy conservation. - Z. Naturforsch. *30 C* : 505 - 510, 1975.

22708 - **HAVEMAN, J., LAVOREL, J.** : Identification of the 120 μs phase in the decay of delayed fluorescence in spinach chloroplasts and subchloroplast particles as the intrinsic back reaction. The dependence of the level of this phase on the thylakoids internal pH. - Biochim. biophys. Acta *408* : 269 - 283, 1975.

22709 - **HAVEMAN, J., MATHIS, P., VERMEGLIO, A.** : Light-induced absorption changes in the near-ultraviolet of the primary electron acceptor of Photosystem II at liquid nitrogen temperature. - FEBS Lett. *58* : 259 - 261, 1975.

22710 - **HAWCROFT, D.M.** : Investigation of the Calvin cycle using radioactive carbon dioxide. - In : **HALL, D.O., HAWKINS, S.E.** (ed.) : Laboratory Manual of Cell Biology. Pp. 39 - 43. English Universities Press, London 1975.

22711 - **HAWCROFT, D.M., FRIEND, J.** : Stimulation by added lipids of photophosphorylation associated with photosystem I. - In : **AVRON, M.** (ed.) : Proceedings of the Third International Congress on Photosynthesis. Vol. I. Pp. 603 - 608. Elsevier, Amsterdam - Oxford - New York 1975.

22712 - **HAWKE, J.C., LEESE, B.M., LEECH, R.M.** : Lipid biosynthesis by intact mesophyll and bundle sheath chloroplasts from maize. - Phytochemistry *14* : 1733 - 1736, 1975.

22713 - **HAWKER, J.S., HATCH, M.D.** : Sucrose-phosphate phosphohydrolase (sucrose phosphatase) from plants. - In : **COLOWICK, S.P., KAPLAN, N.O.** (ed.) : Methods in Enzymology. Vol. 42. Pp. 341 - 347. Academic Press, New York - San Francisco - London 1975.

22714 - **HAY, R.L., WOODS, F.W.** : Distribution of carbohydrates in deformed seedling root systems. - Forest Sci. *22* : 263 - 267, 1975.

22715 - **HAYASHIYA, K.** : [Polyhedral disease virus-inactivating substance of the silkworm. Photochemical reaction related to red fluorescent protein production.] - Kagaku to Seibutsu *13* : 634 - 635, 1975. [In Jap.]

22716 - **HAYDOCK, K.P., SHAW, N.H.** : The comparative yield method for estimating dry matter yield of pasture. - Aust. J. exp. Agr. anim. Husb. *15* : 663 - 670, 1975.

22717 - **HEAD, W.D., CARPENTER, E.J.** : Nitrogen fixation associated with the marine macroalga *Codium fragile*. - Limnol. Oceanogr. *20* : 815 - 823, 1975. [Ps.]

B22718 - **HEATH, O.V.S.** : Stomata. Oxford Biology Reader. Vol 37. - Oxford University Press, London 1975.

22719 - **HEATH, R.L.** : Ozone. - In : **MUDD, J.B., KOZLOWSKI, T.T.** (ed.): Responses of Plants to Air Pollution. Pp. 23 - 55. Academic Press, New York - San Francisco - London 1975. [Stomatal resistance.]

22720 - **HEATHCOTE, P., HALL, D.O.** : Noncyclic photophosphorylation in photosystem II and photosystem I : stoichiometry and photosynthetic control. - In : **AVRON, M.** (ed.) : Proceedings of the Third International Congress on Photosynthesis. Vol. I. Pp. 463 - 471. Elsevier, Amsterdam - Oxford - New York 1975.

22721 - **HEBER, U.** : Energy transfer within leaf cells. - In : **AVRON, M.** (ed.) : Proceedings of the Third International Congress on Photosynthesis. Vol. II. Pp. 1335 - 1348. Elsevier, Amsterdam - Oxford - New York 1975.

22722 - **HEBER, U., KIRK, M.R.** : Efficiency of coupling between phosphorylation and electron transport in intact chloroplasts. - In : **AVRON, M.** (ed.) : Proceedings of the Third International Congress on Photosynthesis. Vol. II. Pp. 1041 - 1046. Elsevier, Amsterdam - Oxford - New York 1975.

22723 - **HEBER, U., KIRK, M.R.** : Aufnahme und Umsatz exportierter Photosyntheseprodukte durch die Chloroplasten während der Assimilation von CO_2. - Biochem. Physiol. Pflanzen *168* : 211 - 223, 1975.

22724 - **HEBER, U., KIRK, M.R.** : Flexibility of coupling and stoichiometry of ATP formation in intact chloroplasts. - Biochim. biophys. Acta *376* : 136 - 150, 1975.

22725 - **HEDLEY, C.L., HARVEY, D.M.** : The involvement of CO_2 uptake in the flowering behaviour of two varieties of *Antirrhinum majus*. - In : **MARCELLE, R.** (ed.) : Environmental and Biological Control of Photosynthesis. Pp. 149 - 160. Dr. W. Junk b.v. Publ., The Hague 1975.

22726 - **HEDLEY, C.L., HARVEY, D.M.** : Variation in the photoperiodic control of flowering of two cultivars of *Antirrhinum majus* L. - Ann. Bot. *39* : 257 - 263, 1975. [Ps.]

22727 - **HEDLEY, C.L., ROWLAND, A.O.** : Changes in the activities of some respiratory and photosynthetic enzymes during the early leaf development of *Antirrhinum majus* L. - Plant Sci. Lett. *5* : 119 - 126, 1975.

22728 - **HEICHEL, G.H., FRINK, C.R.** : Anticipating the energy needs of American agriculture. - J. Soil Water Conservation *30* : 48 - 53, 1975. [Ps.]

22729 - **HEIZMANN, P.** : Control of chloroplast rRNA synthesis in *Euglena*. - In : **AVRON, M.** (ed.) : Proceedings of the Third International Congress on Photosynthesis. Vol. III. Pp. 1745 - 1754. Elsevier, Amsterdam - Oxford - New York 1975.

22730 - **HEIZMANN, P., HOWELL, S.H.** : Detection of generalized and RUDP-carboxylase (large subunit) specific messenger activity in *Chlamydomonas reinhardi* RNA preparations. - Plant Physiol. *56* (Suppl.) : 69, 1975.

22731 - **HELDER, R.J.** : Polar potassium transport and electrical potential difference across the leaf of *Potamogeton lucens* L. - Proc. konінkl. nederl. Akad. Wetenschappen (Amsterdam) Ser. C *78* : 189 - 197, 1975. [Bicarbonate uptake.]

22732 - **HELDT, H.W., FLIEGE, R., LEHNER, K., MILOVANCEV, M., WERDAN, K.** : Metabolite movement and CO_2 fixation in spinach chloroplasts. - In : **AVRON, M.** (ed.) : Proceedings of the Third International Congress on Photosynthesis. Vol. II. Pp. 1369 - 1379. Elsevier, Amsterdam - Oxford - New York 1975.

22733 - **HELLINGWERF, K.J., MICHELS, P.A.M., DORPEMA, J.W., KONINGS, W.N.** : Transport of amino acids in membrane vesicles of *Rhodopseudomonas spheroides* energized by respiratory and cyclic electron flow. - Europe. J. Biochem. *55* : 397 - 406, 1975.

*22734 - **HELLKVIST, J., RICHARDS, G.P., JARVIS, P.G.** : Vertical gradients of water potential and tissue water relations in Sitka spruce trees measured with the pressure chamber. - J. appl. Ecol. *11* : 637 - 667, 1974. [Growth analysis.]

22735 - **HEMPHILL, J.K., VENKETESWARAN, S.** : Callus growth and pigment analysis of three soybean phenotypes (*Glycine max.* (L.) MERRILL.). - Plant Physiol. *56* (Suppl.) : 37, 1975.

22736 - **HENNIES, H.-H.** : Die Sulfitreduktase aus *Spinacia oleracea* - ein Ferredoxin--abhängiges Enzym. - Z. Naturforsch. *30 C* : 359 - 362, 1975.

22737 - **HENRIQUES, F., PARK, R.B.** : Further chemical and morphological characterization of chloroplast membranes from a chlorophyll *b*-less mutant of *Hordeum vulgare*. - Plant Physiol. *55* : 763 - 767, 1975.

22738 - **HENRIQUES, F., VAUGHAN, W., PARK, R.** : High resolution gel electrophoresis of chloroplast membrane polypeptides. - Plant Physiol. *55* : 338 - 339, 1975.

22739 - **HENRY, E.W.** : Peroxidases in tobacco abscission zone tissue. III. Ultrastructural localization in thylakoids and membrane-bound bodies of chloroplasts. - J. Ultrastructure Res. *52* : 289 - 299, 1975.

22740 - **HENRY, E.W.** : Polyphenol oxidase activity in thylakoids and membrane-bound granular components of *Nicotiana tabacum* chloroplasts. - J. Microscop. Biol. cell. *22* : 109 - 116, 1975.

22741 - **HENRY, E.W.** : Ultrastructural localization of polyphenol oxidase in chloroplasts of *Brassica napus* L. cv. Zephyr. - Plant Physiol. *56* (Suppl.) : 32, 1975.

22742 - **HERBLAND, A., PAGES, J.** : L'adénosine triphosphate (ATP) dans le dome de Guinée. Distribution verticale et signification écologique. - Cah. ORSTOM, Sér. Océanogr. *13* : 163 - 169, 1975.

22743 - **HERMAN, E.M.** : Turnover of the pyrenoid and the presence of tubules in the chloroplasts of a marine *Cachonia (Pyrrhophyta)*. - J. Phycol. *11* (Suppl.) : 8, 1975.

22744 - **HERMAN, E.M., SWEENEY, B.M.** : Circadian rhythm of chloroplast ultrastructure in *Gonyaulax polyedra*, concentric organization around a central cluster of ribosomes. - J. Ultrastructure Res. *50* : 347 - 354, 1975.

22745 - **HERODEK, S., TAMÁS, G.** : Phytoplankton production in lake Balaton. - Symp. Biol. Hung. *15* : 29 - 34, 1975.

22746 - **HERODEK, S., TAMÁS, G.** : The primary production of phytoplankton in the Keszthely Basin of Lake Balaton in 1973 - 1974. - Ann. Inst. Biol. (Tihany) *42* : 175 - 190, 1975.

22747 - **HERRMANN, K.** : Über den Carotin-(Provitamin A-) Gehalt der Gemüse- und Obstarten. - Ernährungs-Umschau *22* (2) : 45 - 49, *22* (3) : 75 - 77, 1975.

22748 - **HERVO, G., PAILLOTIN, G., THIERY, J., BREUZE, G.** : Détermination de différentes durées de vie de fluorescence manifestées par la chlorophylle *a in vivo*. - J. Chim. phys. *72* : 761 - 766, 1975.

22749 - **HESKETH, J.D., McKINION, J.M., JONES, J.W., BAKER, D.N., LANE, H.C., THOMPSON, A.C., COLWICK, R.F.** : Problems in building computer models for photosynthesis and respiration. - In : **MARCELLE, R.** (ed.) : Environmental and Biological Control of Photosynthesis. Pp. 53 - 60. Dr. W. Junk b.v. Publ., The Hague 1975.

22750 - **HESLOP-HARRISON, J.** : Crops, commodities & energy capture. - Biologist *22* (2) : 60 - 67, 1975.

22751 - **HEVESI, J., MOLNÁR, M.** : Time stability of detergent (micelle) systems containing organic dyes. - Acta Univ. szeged.-Acta phys. chem., nova Ser. *21* : 31 - 36, 1975. [Ps models.]

22752 - **HEWITT, H.G., AYRES, P.G.** : Changes in CO_2 and water vapour exchange rates in leaves of *Quercus robur* infected by *Microsphaera alphitoides* (powdery mildew). - Physiol. Plant Pathol. *7* : 127 - 137, 1975.

22753 - **HEYN, M.P., BAUER, P.-J., DENCHER, N.A.** : A natural CD label to probe the structure of the purple membrane from *Halobacterium halobium* by means of exciton coupling effects. - Biochem. biophys. Res. Commun. *67* : 897 - 903, 1975.

22754 - **HEYSER, W., LEONARD, O., HEYSER, R., FRITZ, E., ESCHRICH, W.** : The influence of light, darkness, and lack of CO_2 on phloem translocation in detached maize leaves. - Planta *122* : 143 - 154, 1975. [Photosynthate translocation.]

22755 - **HICKMAN, J.C., PITELKA, L.F.** : Dry weight indicates energy allocation in ecological strategy analysis of plants. - Oecologia *21* : 117 - 121, 1975.

22756 - **HICKMAN, M., KLARER, D.M.** : The effect of the discharge of thermal effluent from a power station on the primary productivity of an epiphytic algal community. - Brit. phycol. J. *10* : 81 - 91, 1975.

22757 - **HIEKEL, H.-G., BÖHM, H.** : Zur Anwendung elektronischer Volumenmessungen für Produktivitätsuntersuchungen an Flüssigkeitskulturen einzelliger Grünalgen. - Arch. Hydrobiol. *46* (Suppl.-Algol. Stud. 12) : 280 - 288, 1975.

22758 - **HIEKEL, H.G., BÖHM, H.** : Zur Anwendung elektronischer Volumenmessungen für Produktivitätsuntersuchungen an Flüssigkeitskulturen einzelliger Grünalgen. - In : **NECHAS, Ĭ.** (ed.) : Izuchenie Intensivnoĭ Kul'tury Vodorosleĭ. Vol. II. Pp. 225 - 232. Třeboň 1975.

22759 - **HIGUTI, T., ERABI, T., KAKUNO, T., HORIO, T.** : Role of ubiquinone-10 in electron transport system of chromatophores from *Rhodospirillum rubrum*. - J.Biochem. (Tokyo) *78* : 51 - 56, 1975.

22760 - **HILDEBRAND, E., DENCHER, N.** : Two photosystems controlling behavioural responses of *Halobacterium halobium*. - Nature *257* : 46 - 48, 1975.

22761 - **HINCKLEY, T.M., DOUGHERTY, P.M., FISHER, J.A., BENCI, J.F.** : Relationship between xylem pressure potential, vapor pressure deficit, soil moisture and stomatal conductance. - Plant Physiol. *56* (Suppl.) : 74, 1975.

22762 - **HINCKLEY, T.M., SCHROEDER, M.O., ROBERTS, J.E., BRUCKERHOFF, D.N.** : Effect of several environmental variables and xylem pressure potential on leaf surface resistance in white oak. - Forest Sci. *21* : 201 - 211, 1975. [Stomatal resistance.]

22763 - **HINDÁK, F.** : The phototrophic edaphon of the oak-hornbeam forest at Báb. - In : **BISKUPSKÝ, V.** (ed.) : Research Project Báb, IBP Progress Report II. Pp. 177 - 183. Veda, Bratislava 1975. [Chl.]

22764 - **HINDE, R., SMITH, D.C.** : The role of photosynthesis in the nutrition of the mollusc *Elysia viridis*. - Biol. J. Linnean Soc. *7* : 161 - 171, 1975.

22765 - **HINDS, W.T.** : Energy and carbon balances in cheatgrass : an essay in autecology. - Ecol. Monogr. *45* : 367 - 388, 1975.

22766 - **HIPKINS, M.F., BARBER, J.** : Analysis of kinetics and temperature sensitivity of delayed fluorescence from uncoupled spinach chloroplasts. - In : **AVRON, M.** (ed.) : Proceedings of the Third International Congress on Photosynthesis. Vol. I. Pp. 101 - 114. Elsevier, Amsterdam - Oxford - New York 1975.

22767 - **HIROSE, H.** : Photoreactive pigments of algae and algal phylogeny. - In : **TOKIDA, J., HIROSE, H.** (ed.) : Advance of Phycology in Japan. Pp. 52 - 65. Dr. W. Junk b. v. Publ., The Hague 1975.

22768 - **HIROSE, T.** : Relations between turnover rate, resource utility and structure of some plant populations : A study in the matter budgets. - J. Fac. Sci. Univ. Tokyo, Sect. III. Bot. *11* (11) : 355 - 407, 1975.

22769 - **HIROTA, O., TAKEDA, T., SAITO, Y.** : [Studies on the utilization of solar radiation by crop stands. I. A photosynthetically active solarimeter devised to use in leaf canopy.] - Proc. Crop Sci. Soc. Jap. *44* : 357 - 363, 1975. [In Jap., ab : E.]

22770 - **HO, L.C., NICHOLS, R.** : The role of phloem transport in the translocation of sucrose along the stem of carnation cut flowers. - Ann. Bot. *39* : 439 - 446, 1975. [Photosynthate translocation.]

22771 - **HO, L.C., REES, A.R.** : Aspects of translocation of carbon in the tulip. - New Phytol. *74* : 421 - 428, 1975.

22772 - **HOAD, G.V.** : Effect of osmotic stress on abscisic acid levels in xylem sap of sunflower (*Helianthus annuus* L.). - Planta *124* : 25 - 29, 1975. [Photosynthate translocation.]

22773 - **HOCHMAN, A., CARMELI, C.** : Photosynthetic electron transport in *Rhodopseudomonas capsulata*. - In : **AVRON, M.** (ed.) : Proceedings of the Third International Congress on Photosynthesis. Vol. I. Pp. 777 - 789. Elsevier, Amsterdam - Oxford - New York 1975.

22774 - **HOCHMAN, A., FRIDBERG, I., CARMELI, C.** : The location and function of cytochrome c_2 in *Rhodopseudomonas capsulata* membranes. - Europe. J. Biochem. *58* : 65 - 72, 1975.

22775 - **HODÁŇOVÁ, D.** : Chlorophyll, leaf area-dry weight indexes and PhAR attenuation in developing sugar beet canopy. - Photosynthetica *9* : 211 - 215, 1975.

22776 - **HODÁŇOVÁ, D.** : Specific leaf weight and photosynthetic rate in sugar beet leaves of different age. - Biol. Plant. *17* : 314 - 317, 1975.

22777 - **HODDINOTT, J., SWANSON, C.A.** : Carbohydrate translocation in bean plants : light and sink effects. - Plant Physiol. *56* (Suppl.) : 18, 1975.

22778 - **HODKINSON, I.D.** : Energy flow and organic matter decomposition in an abandoned beaver pond ecosystem. - Oecologia *21* : 131 - 139, 1975. [Ps.]

22779 - **HOFAECKER-KLETT, I., BERINGER, H.** :The influence of temperature and nitrogen fertilization on fatty acids in leaves of rye-grass (*Lolium Perenne*). - Z. Pflanzenern. Bodenkunde *1975* (2) : 147 - 151, 1975. [Chl.]

22780 - **HOFER, I., STRASSER, R.J., SIRONVAL, C.** : The polypeptide patterns of membrane fractions from normal and primary thylakoids of bean leaves. - Photosynthetica *9* : 246 - 254, 1975.

22781 - **HOFER, I., STRASSER, R.J., SIRONVAL, C.** : The polypeptide pattern of fractions isolated from normal thylakoids and from primary thylakoids before and after induction of the ability to evolve oxygen. - In : **AVRON, M.** (ed.) : Proceedings of the Third International Congress on Photosynthesis. Vol. III. Pp. 1685 - 1690. Elsevier, Amsterdam - Oxford - New York 1975.

22782 - **HOFFMANN, E., LENZ, F.** : Effect of callus formation and growth regulators on $^{14}CO_2$-uptake of *Citrus* leaves. - Gartenbauwissenschaft *40* : 35 - 36, 1975.

22783 - **HOFFMANN, E., MIX, G., LENZ, F.** : Der Stärkegehalt der Chloroplasten bei fruchttragenden und nicht fruchttragenden Auberginen- und Erdbeerpflanzen. - Angew. Bot. *49* : 115 - 121, 1975.

B22784 - **HOFFMANN, P.** : Photosynthese. (Wiss.Taschenb. 158). - Akademie-Verlag, Berlin 1975.

22785 - **HOFFMANN, P.** : Regulative Wechselbeziehungen zwischen Photosynthese und Atmung bei CCC-behandelten Primärblättern von *Triticum aestivum* L. - Biochem. Physiol. Pflanzen *168* : 553 - 560, 1975.

22786 - **HOFFMANN, P., SCHWARZ, Zs.** : Characterization of regulative interactions between the autotrophic and heterotrophic system in *Phaseolus vulgaris* and *Triticum aestivum* seedlings. - In : **MARCELLE, R.** (ed.) : Environmental and Biological Control of Photosynthesis. Pp. 191 - 200. Dr. W. Junk b.v. Publ., The Hague 1975.

22787 - **HOFSTRA, G., HESKETH, J.D.** : The effects of temperature and CO_2 enrichment on photosynthesis in soybean. - In : **MARCELLE, R.** (ed.) : Environmental and Biological Control of Photosynthesis. Pp. 71 - 80. Dr. W. Junk b.v. Publ., The Hague 1975.

22788 - **HØJERSLEV, N.** : A spectral light absorption meter for measurements in the sea. - Limnol. Oceanogr. *20* : 1024 - 1034, 1975.

22789 - **HOLE, C.C., DODGE, A.D.** : Inter-organ control of chlorophyll synthesis in primary leaves of dark-grown *Phaseolus aureus*. - Physiol. Plant. *34* : 22 - 25, 1975.

22790 - **HOLLISTER, T.A., WALSH, G.E., FORESTER, J.** : Mirex and marine unicellular algae : accumulation, population growth and oxygen evolution. - Bull. environ. Contamination Toxicol. *14* : 753 - 759, 1975. [Ps.]

22791 - **HOLMES, K.S., DARLEY, W.M.** : Physiological responses of benthic algal productivity in the salt marsh. - J. Phycol. *11* (Suppl.) : 10, 1975. [Ps.]

22792 - **HOLT, D.A., BULA, R.J., MILES, G.E., SCHREIBER, M.M., PEART, R.M.** : Environmental physiology, modeling and simulation of alfalfa growth : 1. Conceptual development of SIMED. - Indiana agr. Exp. Sta. Res. Bull. *907* : 1- 26, 1975.

22793 - **HOLUB, Z.** : Štúdium vplyvu fluórových exhalátov na rastliny. Časť A. Účinky NaF na fyziologické procesy listov *Triticum vulg*. Časť B. Analytické stanovenie fluóru v rastlinách. [Effect of fluorine exhalations on plants. A. Effects of sodium fluoride on physiological processes of leaves of *Triticum vulgare*. B. Analytical determination of fluorine in plants.] - Quaest. geobiol. *18* : 45 - 139, 1975. [Ps, Chl; in Slovak, ab : E, F, G, R.]

22794 - **HOLZAPFEL, C., BAUER, R.** : Computer simulation of primary photosynthetic reactions - compared with experimental results on O_2-exchange and chlorophyll fluorescence of green plants. - Z. Naturforsch. *30 C* : 489 - 498, 1975.

22795 - **HOLZAPFEL, C., HAUG, A.** : Time course of microsecond delayed light emission from *Scenedesmus obliquus* at intermittent illumination. - Photochem. Photobiol. *21* : 209 - 211, 1975.

22796 - **HOMANN, P.H.** : Carbon dioxide exchange of young tobacco leaves in light and darkness. - In : **MARCELLE, R.** (ed.) : Environmental and Biological Control of Photosynthesis. Pp. 183 - 190. Dr. W. Junk b.v. Publ., The Hague 1975.

22797 - **HOPEN, H.J., OEBKER, N.F.** : Mulch effects on ambient carbon dioxide levels and growth of several vegetables. - HortScience *10* : 159 - 161, 1975.

22798 - **HOPF, F.R., WHITTEN, D.G.** : Photochemistry of porphyrins and metalloporphyrins. - In : **SMITH, K.M.** (ed.) : Porphyrins and Metalloporphyrins. Pp. 667 - 700. Elsevier, Amsterdam - Oxford - New York 1975. [Chl.]

22799 - **HOPKINS, W.G., HAYDEN, D.B., WALDEN, D.B.** : Analysis of greening in virescent mutants of maize by *in vivo* spectrophotometry. - Can. J. Bot. *53* : 2720 - 2724, 1975.

22800 - **HORAK, A., ZALIK, S.** : Development of photoreductive activity in plastids of a virescens mutant of barley. - Can. J. Bot. *53* : 2399 - 2404, 1975.

22801 - **HORAK, A., ZALIK, S.** : Photophosphorylation of α,β-methylene adenosine 5'-diphosphate an analogue of ADP, by spinach chloroplasts. - In : **AVRON, M.** (ed.) : Proceedings of the Third International Congress on Photosynthesis. Vol. II. Pp. 1067 - 1072. Elsevier, Amsterdam - Oxford - New York 1975.

22802 - **HORECKER, B.L.** : Fructose bisphosphate aldolase from spinach. - In : **COLOWICK, S.P., KAPLAN, N.O.** (ed.) : Methods in Enzymology. Vol. 42. Pp. 234 - 239. Academic Press, New York - San Francisco - London 1975.

22803 - **HORI, T., UEDA, R.** : The fine structure of algal chloroplasts and algal phylogeny. - In : **TOKIDA, J., HIROSE, H.** (ed.) : Advance of Phycology in Japan. Pp. 11 - 42. Dr. W. Junk b.v. Publ., The Hague 1975.

22804 - **HORIČKOVÁ, B.** : Vplyv postreku Karathane na zmeny v obsahu chlorofylu *a* v jačmeni napadnutom *Erysiphe graminis* f. sp. *hordei* MARCHAL. [Influence of the spray with the fungicide Karathane upon the chlorophyll *a* content in barley infected by powdery mildew (*Erysiphe graminis* f. sp. *hordei* MARCHAL).] - Acta Inst. bot. Acad. Sci. slov., Ser. B *1* : 403 - 411, 1975. [In Slovak, ab : E, R.]

22805 - **HORIČKOVÁ, B., HASPELOVÁ-HORVATOVIČOVÁ, A.** : Zmeny obsahu chlorofylov v listoch jačmeňa napadnutého hubou *Erysiphe graminis* F. sp. *hordei* MARCHAL po postreku ochrannou látkou. [Changes in chlorophyll content in the leaves of barley infected with the fungus *Erysiphe graminis* f. sp. *hordei* MARCHAL after spraying with a protective substance.] - Pol'nohospodárstvo *21* : 963 - 971, 1975. [In Slovak, ab : E, R.]

22806 - **HORIE, M., KAMIYAMA, K., MIKOSHIBA, K.** : Growth pattern and climate. 3.5.1. Developmental analysis by a statistical method. - In : **MURATA, Y.** (ed.) : JIBP Synthesis. Vol.11. Pp. 122 - 135. Tokyo University Press, Tokyo 1975. [Growth analysis.]

22807 - **HORNER, R.A.** : Biology of the phytoplankton in Prudhoe Bay, Alaska. - J. Phycol. *11* (Suppl.) : 20, 1975. [Chl.]

22808 - **HOROVITZ, C.T., WEGMANN, K.** : Response of *Chlorella* to scandium and potassium at various pH. - Physiol. Plant. *33* : 113 - 117, 1975. [Ps, Chl.]

22809 - **HORTON, P., BÖHME, H., CRAMER, W.A.** : On the pathway and mechanism of the cytochrome *b*-559 photoreactions. - In : **AVRON, M.** (ed.) : Proceedings of the Third International Congress on Photosynthesis. Vol.I. Pp. 535 - 545. Elsevier, Amsterdam - Oxford - New York 1975.

22810 - **HORTON, P., CRAMER, W.A.** : Acid-base induced redox changes of the chloroplast cytochrome *b-559*. - FEBS Lett. *56* : 244 - 247, 1975.

22811 - **HORTON, P., CRAMER, W.A.** : Light-induced turnover of chloroplast cytochrome *b*-559 at low pH. - Biophys. J. *15* (2, Part 2) : 250 a, 1975.

22812 - **HORTON, P., CRAMER, W.A.** : Light-induced turnover of chloroplast cytochrome *b*-559 in the presence of N-methylphenazonium methosulphate. - Biochim. biophys. Acta *396* : 310 - 319, 1975.

22813 - **HORTON, P., LEECH, R.M.** : The effect of adenosine 5'-triphosphate on the Shibata shift and on associated structural changes in the conformation of the prolamellar body in isolated maize etioplasts. - Plant Physiol. *55* : 393 - 400, 1975.

22814 - **HORTON, P., LEECH, R.M.** : The effect of ATP on the photoconversion of protochlorophyllide in isolated etioplasts of *Zea mays*. - Plant Physiol. *56* : 113 - 120, 1975.

22815 - **HORVÁTH, G., GARAB, G.I., HALÁSZ, N., FALUDI-DÁNIEL, Á.** : Maturation of thylakoids in mesophyll and bundle sheath chloroplasts of maize. - In : **AVRON, M.** (ed.) : Proceedings of the Third International Congress on Photosynthesis. Vol. III. Pp. 1925 - 1932. Elsevier, Amsterdam - Oxford - New York 1975.

22816 - **HORVÁTH, G., GARAB, G.I., MUSTÁRDY, L.A., HALÁSZ, N., FALUDI-DÁNIEL, Á.** : The development of thylakoids and photochemical properties of mesophyll and bundle sheath chloroplasts of greening maize leaves. - Plant Sci. Lett. *5* : 239 - 244, 1975.

22817 - **HORVÁTH, I., KÁLMÁN, F., TITLYANOV, E.A.** : The influence of monochromatic blue and red light on the electron microscope structure of chloroplasts. - Acta bot. Acad. Sci. hung. *21* : 273 - 278, 1975.

22818 - **HORVÁTH, I., SZALAY, L.** : Photosynthetic production and wavelength-dependent energy migration. - Acta biochim. biophys. Acad. Sci. hung. *10* : 123 - 128, 1975.

22819 - **HORWITZ, B.A., SAMISH, Y.B.** : Light-stimulated bioelectric response in *Spirodela oligorrhiza* and its relation to photosynthesis. - Z. Pflanzenphysiol. *76* : 182 - 189, 1975.

22820 - **HOSHINA, S., KAJI, T., NISHIDA, K.** : Photoswelling and light-inactivation of isolated chloroplasts. I. Change in lipid content in light-aged chloroplasts. - Plant Cell Physiol. *16* : 465 - 474, 1975.

22821 - **HOSHINA, S., NISHIDA, K.** : Photoswelling and light-inactivation of isolated chloroplasts II. Functional and structural changes in isolated chloroplasts under the influence of lysolecithin. - Plant Cell Physiol. *16* : 475 - 484, 1975.

B22822 - **HOSHINO, M.** : Studies on the Tropical Forage Crop in Thailand. - Trop. Agr. Res. Center, Min. Agr. Forest. Japan 1975. [Growth analysis.]

22823 - **HOSHINO, M., ONO, S., SIRIKIRATAYANOND, N.** : Dry matter production of tropical grasses and legumes in Thailand. - Jap. agr. Res. Quart. *9* : 240 - 243, 1975.

22824 - **HOSOI, K., SOE, G., KAKUNO, T., HORIO, T.** : Effects of pH indicators on various activities of chromatophores of *Rhodospirillum rubrum*. - J. Biochem. (Tokyo) *78* : 1331 - 1346, 1975.

22825 - **HOUGH, R.A., WETZEL, R.G.** : The release of dissolved organic carbon from submersed aquatic macrophytes : Diel, seasonal, and community relationships. - Verh. internat. Verein. Limnol. *19* : 939 - 948, 1975. [Primary production, photorespiration.]

22826 - **HOUSLEY, T.L., SCHRADER, L.E., SETTER, T.L.** : Transport of assimilates following pulse labelling in nonnodulated and nodulated soybeans. - Plant Physiol. *56* (Suppl.) : 18, 1975.

22827 - **HOVENKAMP-OBBEMA, R.** : Effect of added aminolaevulinic acid upon synthesis of chlorophyll in *Euglena gracilis*. - Z. Pflanzenphysiol. *75* : 1 - 5, 1975.

22828 - **HOWARD, R.J., GAYLER, K.R., GRANT, B.R.** : Products of photosynthesis in *Caulerpa simpliciuscula (Chlorophyceae)*. - J. Phycol. *11* : 463 - 471, 1975.

22829 - **HRUŠKA, L., JANÍČEK, J., BEDNÁŘOVÁ, E.** : Productivity of main field crops in southern Moravia. - Rost. Výroba (Praha) *21* : 809 - 816, 1975. [Growth analysis.]

22830 - **HUANG, C.-Y., BOYER, J.S., VANDERHOEF, L.N.** : Acetylene reduction (nitrogen fixation) and metabolic activities in soybean having various leaf and nodule water potentials. - Plant Physiol. *56* : 222 - 227, 1975. [Ps.]

22831 - **HUANG, C.-Y., BOYER, J.S., VANDERHOEF, L.N.** : Limitation of acetylene reduction (nitrogen fixation) by photosynthesis in soybean having low water potentials. - Plant Physiol. *56* : 228 - 232, 1975.

22832 - **HUBER, F.** : Respiratorischer Kohlenstoffverbrauch alpiner Zwergstrauchbestände. - Verhandl. Ges. Ökol. (Wien) *1975* : 31 - 35, 1975. [Net production and respiration.]

22833 - **HUBER, S., EDWARDS, G.** : The effect of oxygen on CO_2 fixation by mesophyll protoplast extracts of C_3 and C_4 plants. - Biochem. biophys. Res. Commun. *67* : 28 - 34. 1975.

22834 - **HUBER, S.C., EDWARDS, G.E.** : Effect of DBMIB, DCMU and antimycin A on cyclic and noncyclic electron flow in C_4 mesophyll chloroplasts. - FEBS Lett. *58* : 211 - 214, 1975.

22835 - **HUBER, S.C., EDWARDS, G.E.** : Inhibition of phosphoenolpyruvate carboxylase from C_4 plants by malate and aspartate. - Can. J. Bot. *53* : 1925 - 1933, 1975.

22836 - **HUBER, S.C., EDWARDS, G.E.** : An evaluation of some parameters required for the enzymatic isolation of cells and protoplasts with CO_2 fixation capacity from C_3 and C_4 grasses. - Physiol. Plant. *35* : 203 - 209, 1975.

22837 - **HUBER, S.C., EDWARDS, G.E.** : C_4 photosynthesis : Light-dependent CO_2 fixation by mesophyll cells, protoplasts, and protoplast extracts of *Digitaria sanguinalis*. - Plant Physiol. *55* : 835 - 844, 1975.

22838 - **HUBER, S.C., EDWARDS, G.E.** : Regulation of oxaloacetate, aspartate, and malate formation in mesophyll protoplast extracts of three types of C_4 plants. - Plant Physiol. *56* : 324 - 331, 1975.

22839 - **HUBER, S.C., EDWARDS, G.E.** : Regulation of oxalacetate, aspartate and malate formation in mesophyll protoplast extracts of several C_4 plants. - Plant Physiol. *56* (Suppl.) : 27, 1975.

22840 - **HUBER, S.C., GUTIERREZ, M., EDWARDS, G.E.** : An evaluation of some parameters required for the isolation of cells and protoplasts with photosynthetic capacity. - Plant Physiol. *56* (Suppl.) : 74, 1975.

22841 - **HUDOBOVÁ, E., DUDA, M.** : Dynamics of changes in pigment content during vegetation in *Pulmonaria officinalis* ssp. *maculosa* and *Cornus mas* in an oak-hornbeam ecosystem. - In : **BISKUPSKÝ, V.** (ed.) : Research Project Báb IBP Progress Report II. Pp. 185 - 203. Veda, Bratislava 1975.

22842 - **HUDOCK, M.O., TOGASAKI, R.K., LIEN, S., HOSEK, M., SAN PIETRO, A.** : Uniparentally inherited mutation affecting light induced phosphorylation (LIP) in *Chlamydomonas reinhardi*. - Plant Physiol. *56* (Suppl.) : 10, 1975.

22843 - **HUGUET, J.-G., PRADE, J.-L.** : La chlorose du prunier d'Ente en Aquitaine. - Compt. rend. Séances Acad. Agr. France *61* : 208 - 217, 1975.

22844 - **HUKKERI, S.B., SHARMA, A.K., NIMBOLE, N.N., BASANTANI, H.T.** : Stress-day index for timing of irrigation for potato. - Indian J. agr. Sci. *45* : 515 - 523, 1975. [Growth analysis.]

22845 - **HUME, D.J.** : Translocation of ^{14}C-labelled photosynthetic assimilates in cassava (*Manihot esculenta* CRANTZ). - Ghana J. agr. Sci. *8* : 69 - 75, 1975.

*22846 - **HUMPHREY, G.F.** : Effects of carbon dioxide and phosphate supplied during growth, on phosphorus content and photosynthetic rates of some unicellular marine algae. - J. mar. biol. Assoc. India *16* : 358 - 366, 1974.

22847 - **HUMPHREY, G.F.** : Phosphorus deficiency in algae. - Annu. Rep. mar. Biochem. Unit *1974-75*: 22, 1975. [Ps.]

22848 - **HUMPHREY, G.F.** : The photosynthesis : respiration ratio of some unicellular marine algae. - J. exp. mar. Biol. Ecol. *18* : 111 - 119, 1975.

22849 - **HUNT, I.V., FRAME, J., HARKESS, R.D.** : Potential productivity of red clover varieties in S.W. Scotland. - J. brit. Grassland Soc. *30* : 209 - 216, 1975.

22850 - **HUNT, R., STRIBLEY, D.P., READ, D.J.** : Root/shoot equilibria in cranberry (*Vaccinium macrocarpon* AIT.). - Ann. Bot. *39* : 807 - 810, 1975. [Ps.]

22851 - **HUSZÁR, J., PAULECH, C.** : Vplyv huby *Erysiphe cichoracearum* DC. na intenzitu fotosyntézy a obsah chlorofylov v listoch tabaku. [Effect of *Erysiphe cichoracearum* DC. on photosynthetic rate and chlorophyll content in tobacco leaves.] - Acta Inst. bot. Acad. Sci. slov., Ser. B *1* : 79 - 90, 1975. [In Slovak, ab : E, R.]

*22852 - **HUTH, W.** : Das Verhalten einiger Enzyme des Kohlenhydratstoffwechsels in Kartoffel-X-Virus-kranken Tabakpflanzen. - Phytopathol. Z. *77* : 117 - 124, 1973. [Chl.]

22853 - **HUTSON, K.G., ROGERS, L.J.** : Two plant-type ferredoxins in light-grown and dark-grown cells of the blue-green bacterium *Nostoc* strain MAC. - Biochem. Soc. Trans. *3* : 377 - 379, 1975.

22854 - **HUZULÁK, J., ELIÁŠ, P.** : Within-crown pattern of ecophysiological features in leaves of *Acer campestre* and *Carpinus betulus*. - Folia geobot. phytotax. *10* : 337 - 350, 1975. [Growth analysis.]

22855 - **IANCU, M.** : Influența conținutului da apă din sol, a factorilor climatici și a încărcăturii cu fructe a pomilor asupra creșterii unor organe vegetative la măr. [Influence of the soil water content, of the climatic factors and of the fruit load of trees upon the growth of some of the apple vegetative organs.] - Lucrările științ. Inst. Cercetări Pitești *4* : 173 - 194, 1975. [Growth analysis; in Roum., ab : E, F, R.]

22856 - **ICHIKAWA, T., INOUE, Y., SHIBATA, K.** : Characteristics of thermoluminescence bands of intact leaves and isolated chloroplasts in relation to the water-splitting activity in photosynthesis. - Biochim. biophys. Acta *408* : 228 - 239, 1975.

22857 - **ICHIKAWA, T., INOUE, Y., SHIBATA, K.** : Delayed light emission and variable fluorescence from intermittently illuminated wheat leaves under continuous illumination related to activation of the latent water-splitting system. - Plant Sci. Lett. *4* : 369 - 376, 1975.

22858 - **IDSO, S.B., FOSTER, J.M.** : An analytical study of three characteristic forms of light-forced primary production in aquatic ecosystems. - Oecologia *18* : 145 - 154, 1975.

22859 - **IHLENFELDT, M.J.A., GIBSON, J.** : CO_2 fixation and its regulation in *Anacystis nidulans (Synechococcus)*. - Arch. Microbiol. *102* : 13 - 21, 1975.

22860 - **IIJIMA, T., MARUYAMA, H.** : [Processing adaptability of determinate type tomato varieties (4) On the methods of carotenoid determination in tomato fruits.] - J. Fac. Agr. Shinshu Univ. *12* : 29 - 36, 1975. [In Jap., ab : E.]

22861 - **IIZUKA, S., NAKASHIMA, T.** : [Response of red tide organisms to sulphide.] - Bull. Plankton Soc. Jap. *22* : 27 - 32, 1975. [Ps; in Jap., ab : E.]

22862 - **IKEGAMI, I., KATOH, S.** : Enrichment of Photosystem I reaction center chlorophyll from spinach chloroplasts. - Biochim. biophys. Acta *376* : 588 - 592, 1975.

22863 - **IKENAGA, T., MATUO, M., OHASHI, H.** : [Studies on the physiology and ecology of *Amaranthus viridis*. 2. The effect of nutritional conditions on the growth and chlorophyll content in *Amaranthus viridis*.] - Zasso Kenkyu *20* : 156 - 160, 1975. [In Jap., ab : E.]

22864 - **IKUSHIMA, T.** : Action of furylfuramide (AF-2) on chloroplasts of *Euglena gracilis*. - Annu. Rep. Res. Reactor Inst. Kyoto Univ. *8* : 83 - 85, 1975.

22865 - **ILLYES, Gh., NEAMŢU, G., BODEA, C.** : Cercetări chemotaxonomice la plante inferioare. I. Pigmenţii carotenoidici şi clorofilieni din *Cymbella cymbiformis* KÜTZ. [Chemotaxonomic investigations on lower plants. I. Carotenoids and chlorophylls in *Cymbella cymbiformis* KÜTZ.] - Stud. Cerc. Biochim. *18* (2) : 109 - 113, 1975. [In Roum., ab : E.]

22866 - **ILMAVIRTA, V.** : Diel periodicity in the phytoplankton community of the oligotrophic lake Pääjärvi, southern Finland. II. Late summer phytoplanktonic biomass. - Ann. bot. fenn. *12* : 37 - 44, 1975.

22867 - **ILMAVIRTA, V.** : Dynamics of phytoplanktonic production in the oligotrophic lake Pääjärvi, southern Finland. - Ann. bot. fenn. *12* : 45 - 54, 1975.

22868 - **IMAI, H., FUKUYAMA, M., YAMADA, Y.** : Comparative studies on the photosynthesis of higher plants. V. Differences in the rate of sugar formation from radioactive compounds among C_4-plants. - Soil Sci. Plant Nutr. *21* : 253 - 261, 1975.

22869 - **IMAI, H., IWAI, S., YAMADA, Y.** : Comparative studies on the photosynthesis of higher plants. IV. Further studies on the photosynthetic sugar formation pathway in C_4-plants. - Soil Sci. Plant Nutr. *21* : 13 - 19, 1975.

22870 - **IMAMALIEV, A.I., RAKHMANKULOV, S.A., AZIZKHODZHAEV, A.** : Vliyanie gibridizatsii na strukturu i funktsiyu fotosinteticheskogo apparata khlopchatnika. [Effect of hybridization on the structure and function of photosynthesizing apparatus in cotton.] - Fiziol. Rast. *22* : 923 - 928, 1975. [In R , ab : E.]

22871 - **IMURA, T., FURUTSUKA, T., KAWABE, K.** : Photoproduction of ions from chlorophyll *a* in solution studied by flash photolysis and photoconductivity. - Photochem. Photobiol. *22* : 129 - 134, 1975.

22872 - **INABA, T., KAJIWARA, T.** : [Physiological studies of cucumber downy mildew disease. With special reference to the relationship between lesion development, sporulability in lesions, and photosynthesis of host cucumber leaves.] - Bull. nat. Inst. agr. Sci., Ser. C *29* : 65 - 139, 1975. [In Jap., ab : E.]

*22873 - **INADA, K.** : [Selective utilization of light quality in the protected horticulture.] - Chem. Regulat. Plants *6* (1) : 9 - 18, 1971. [Ps; in Jap.]

22874 - **INADA, Y., YAMAZAKI, S., MIYAKE, S., HIROSE, S., OKADA, M., MIHAMA, H.** : Fibrin membrane endowed with biological function. II. Chloroplast adenosine triphosphatase embedded in fibrin membrane. - Biochem. biophys. Res. Commun. *67* : 1275 - 1280, 1975.

22875 - **INCROPERA, F.P.** : Leaf photosynthesis : The influence of environmental variables. - J. environm. Quality *4* : 440 - 447, 1975.

22876 - **INGLE, R.K., COLMAN, B.** : Carbonic anhydrase levels in blue-green algae. - Can. J. Bot. *53* : 2385 - 2387, 1975.

22877 - **INNIS, G.S.** : Role of total systems models in the grassland biome study. - In.

PATTEN, B.C. (ed.) : Systems Analysis and Simulation in Ecology. Vol. III. Pp. 13 - 47. Academic Press, New York - San Francisco - London 1975.

22878 - **INOSAKA, M., ITO, K., NUMAGUCHI, H., ARATANI, M., MAEDA, K.** : [Studies on the productivity of some tropical grasses. I. Relationship between heading rate and dry matter yield under different cutting treatment.] - Jap. J. trop. Agr. *18* : 87 - 92, 1975. [Dry-matter distribution; in Jap., ab : E.]

22879 - **INOSAKA, M., ITO, K., NUMAGUCHI, H., HIGASHI, S.** :[Studies on the productivity of some tropical grasses. II. Effect of N fertilizer applied at various growth stages on the yield of three varieties of Panic grass.] - Jap. J. trop. Agr. *18* : 93 - 98, 1975. [Dry-matter distribution; in Jap., ab : E.]

22880 - **INOUE, K., UCHIJIMA, Z., HORIE, T., IWAKIRI, S.** : Studies of energy and gas exchange within crop canopies (10). Structure of turbulence in rice crop. - J. agr. Meteorol. *31* : 71 - 82, 1975.

22881 - **INOUE, Y.** : [Multiple flash activation of the latent oxygen evolving system accumulated in the leaves greened under intermittent illumination.]-Tampakusitsu Kakusan Koso [Protein, nucl. Acid, Enzyme] *20* : 1301 - 1308, 1975. [In Jap.]

22882 - **INOUE, Y.** : Multiple-flash activation of the water-photolysis system in wheat leaves as observed by delayed emission. - Biochim. biophys. Acta *396* : 402 - 413, 1975.

22883 - **INOUE, Y., ICHIKAWA, T., KOBAYASHI, Y., SHIBATA, K.** : Multiple flash activation of the water-splitting system in wheat leaves grown under intermittent illumination. - In : **AVRON, M.** (ed.) : Proceedings of the Third International Congress on Photosynthesis. Vol. III. Pp. 1833 - 1840. Elsevier, Amsterdam - Oxford - New York 1975.

22884 - **INOUE, Y., KOBAYASHI, Y., SAKAMOTO, E., SHIBATA, K.** : Multiple flash activation of the water-photolysis system in intermittently illuminated wheat leaves. - Plant Cell Physiol. *16* : 327 - 335, 1975.

22885 - **IRVINE, J.E.** : Relations of photosynthetic rates and leaf and canopy characters to sugarcane yield. - Crop Sci. *15* : 671 - 676, 1975.

22886 - **ISAAKIDOU, J., PAPAGEORGIOU, G.** : A fluorimetric study of Mg^{2+}-induced structural changes in thylakoid membrane protein. - Arch. Biochem. Biophys. *168* : 266 - 272, 1975.

22887 - **ISANGALIN, F.Sh., SIBEL'DINA, L.A., KAYUSHIN, L.P., PROKHORENKO, B.R., KUTYURIN, V.M.** : Metod yadernogo magnitnogo rezonansa i primenenie ego v issledovanii khlorofillov. [Method of nuclear magnetic resonance and its use in chlorophyll studies.] - In : Metody Issledovaniya Fotokhimicheskikh Reaktsiĭ Fotosinteza *in Vitro* i *in Vivo*. Pp. 76 - 92. Pushchino 1975. [In R.]

22888 - **ISENRING, H.-P., ZASS, E., SMITH, K., FALK, H., LUISIER, J.-L., ESCHENMOSER, A.** : Über enolisierte Derivate der Chlorophyllreihe. 13^2-desmethoxycarbonyl-17^3-desoxy-13^2,17^3-cyclochlorophyllid a-enol und eine Methode zur Einführung von Magnesium in porphinoide Ligandsysteme unter milden Bedingungen. - Helv. chim. Acta *58* : 2357 - 2367, 1975.

22889 - **ITOH, S., MURATA, N.** : Studies on induction and decay kinetics of delayed light emission in spinach chloroplasts. - In : **AVRON, M.** (ed.) : Proceedings of the Third International Congress on Photosynthesis. Vol. I. Pp. 115 - 126. Elsevier, Amsterdam - Oxford - New York 1975.

*B22890 - **IVANCHENKO, V.M.** : Fotosintez i Strukturnoe Sostoyanie Khloroplastov. [Photosynthesis and Structural State of Chloroplasts.] - Nauka i Tekhnika, Minsk. 1974. [In R.]

22891 - **IVANCHENKO, V.M., URBANOVICH, T.A., MARSHAKOVA, M.I., KRUCHININA, S.S.** : O nekotorykh svoĭstvakh izolirovannykh khloroplastov, poteryavshikh i sokhranivshikh vneshnyuyu membranu pri vydelenii. [Some properties of isolated chloroplasts which lost or retained the external membrane in the course of isolation.] - In : Fiziologo-biokhimicheskie Aspekty Rosta i Razvitiya Rasteniĭ. Pp. 13 - 20. Nauka i Tekhnika, Minsk 1975. [In R.]

22892 - **IVANKINA, N.G., NOVAK, V.A.** : Zavisimost' svetoindutsirovannykh izmeneniĭ vnutrikletochnogo ėlektricheskogo potentsiala lista ėlodei ot fotofosforilirovaniya. [Dependence of the light-induced changes of intracellular electric potential of *Elodea* leaf on photophosphorylation.] - Élektronnaya Obrabotka Materialov *1975* (1) : 66 - 67, 1975. [In R.]

22893 - **IVANOV, A.F., RAKHTEENKO, L.I., SAVEL'EV, V.V., MOISEENKO, E.I., KLIMENKOVA, L.K.** : Rost seyantsev drevesnykh rasteniĭ pod vliyaniem razlichnykh vidov kaliĭnykh udobreniĭ v sochetanii s azotno-fosfornymi. [Growth of seedlings of woody plants under various potassium fertilization combined with nitrogen-phosphorus fertilizers.] - In : Ėkologo-biologicheskie Issledovaniya Rastitel'nykh Soobshchestv. Pp. 91 - 101. Nauka i Tekhnika, Minsk 1975. [In R.]

22894 - **IVANOV, B.N., AKULOVA, E.A., MUZAFAROV, E.N., RUZIEVA, R.Kh., SHMELEVA, V.L.** : Vliyanie kvertsetin-glyukozil-kumarata na reaktsii izolirovannykh khloroplastov v prisutstvii valinomitsina. [Effect of quercetin-glucosyl-coumarate on the reactions of isolated chloroplasts in the presence of valinomycin.] - Dokl. Akad. Nauk SSSR *222* : 1232 - 1235, 1975. [In R.]

22895 - **IWAKI, H.** : Computer simulation of vegetative growth of rice plants. - In : **MURATA, Y.** (ed.) : JIBP Synthesis. Vol. 11. Pp. 105 - 121. University of Tokyo Press, Tokyo 1975. [Ps.]

22896 - **IWAKI, H., HIROSAKI, S.** : A model for plant-growth under nongrazing condition. - In : **NUMATA, M.** (ed.) : JIBP Synthesis. (Ecological Studies in Japanese Grasslands - Productivity of Terrestrial Communities.) Vol. 13. Pp. 263 - 268. University of Tokyo Press, Tokyo 1975.

22897 - **IWAKIRI, S., INAYAMA, M.** : [Studies on the canopy photosynthesis of the horticultural crops in controlled environment. (4) Photosynthetic characteristics of single cucumber leaves.] - Nogyo Kisho [J. agr. Meteorol.] *30* : 161 - 166, 1975. [In Jap., ab : E.]

22898 - **IWAMURA, T., NAGAI, H., YAMAGUCHI, Y.** : Seasonal variation of the planktonic population in Lake Yunoko, as followed by the assay of chlorophyll, protein, RNA and DNA in the total harvested planktonic samples. - Int. Rev. ges. Hydrobiol. *60* : 97 - 113, 1975.

22899 - **IWANIJ, V., CHUA, N.-H., SIEKEVITZ, P.** : Synthesis and turnover of ribulose biphosphate carboxylase and of its subunits during the cell cycle of *Chlamydomonas reinhardtii*. - J. Cell Biol. *64* : 572 - 585, 1975.

22900 - **IZAWA, S., ORT, D.R., GOULD, J.M., GOOD, N.E.** : Electron transport reactions, energy conservation reactions and phosphorylation in chloroplasts. - In : **AVRON, M.** (ed.) : Proceedings of the Third International Congress on Photosynthesis. Vol. I. Pp. 449 - 461. Elsevier, Amsterdam - Oxford - New York 1975.

22901 - **IZVOSHCHIKOV, V.P., KONOVALENKO, V.V.** : Uchet fotosinteza list'ev pri ispol'zovanii odnominutnykh ėkspozitsiĭ. [Recording of photosynthesis of leaves with one minute exposures.] - Bot. Zh. *60* : 351 - 355, 1975. [In R.]

22902 - **JABBEN, M., MOHR, H.** : Stimulation of the Shibata shift by phytochrome in the cotyledons of the mustard seedling *Sinapis alba* L. - Photochem. Photobiol. *22* : 55 - 58, 1975. [Chl.]

22903 - **JACKSON, J.B.** : Electron transport pathways in chromatophores. - In : **AVRON, M.** (ed.) : Proceedings of the Third International Congress on Photosynthesis. Vol. I. Pp. 757 - 767. Elsevier, Amsterdam - Oxford - New York 1975.

22904 - **JACKSON, J.B., SAPHON, S., WITT, H.T.** : The extent of the stimulated electrical potential decay under phosphorylating conditions and the H^+/ATP ratio in *Rhodopseudomonas sphaeroides* chromatophores following short flash excitation. - Biochim. biophys. Acta *408* : 83 - 92, 1975.

22905 - **JACOBI, G., KLEMME, B., KRAPF, G., POSTIUS, C.** : Dark starvation and plant metabolism. IV. The alteration of enzyme activities. - Biochem. Physiol. Pflanzen *168* : 247 - 256, 1975.

22906 - **JACOBSON, B.S., FONG, F., HEATH, R.L.** : Carbonic anhydrase of spinach. Studies on its location, inhibition, and physiological function. - Plant Physiol. *55* : 468 - 474, 1975.

22907 - **JACQUARD, P.** : Concurrence intraspécifique et potentialités de rendement. - Ann. Amélior. Plantes *25* : 3 - 24, 1975. [Growth model.]

22908 - **JACQUES, G.L., VANDERLIP, R.L., ELLIS, R. Jr.** : Growth and nutrient accumulation and distribution in grain sorghum. II. Zn, Cu, Fe, and Mn uptake and distribution. - Agron. J. *67* : 611 - 616, 1975. [Dry-matter accumulation.]

22909 - **JACQUES, G.L., VANDERLIP, R.L., WHITNEY, D.A.** : Growth and nutrient accumulation and distribution in grain sorghum. I. Dry matter production and Ca and Mg uptake and distribution. - Agron. J. *67* : 607 - 611, 1975.

22910 - **JACQUINOT, L., POUZET, D.** : Modèles d'architectures de plantes, densité et rendement. I. - Utilization de l'énergie lumineuse. Aspects théoriques appliqués au Mil Pennisetum en zone sahélienne. - Oecol. Plant. *10* : 369 - 387, 1975.

22911 - **JADHAV, S.J., SALUNKHE, D.K.** : Formation and control of chlorophyll and glycoalkaloids in tubers of *Solanum tuberosum* L. and evaluation of glykoalkaloid toxicity. - Adv. Food Res. *21* : 308 - 354, 1975.

22912 - **JAGENDORF, A.T.** : Chloroplast membranes and coupling factor conformations. - Fed. Proc. *34* : 1718 - 1722, 1975.

22913 - **JAGENDORF, A.T.** : Mechanism of photophosphorylation. - In : **GOVINDJEE** (ed.) : Bioenergetics of Photosynthesis. Pp. 413 - 492. Academic Press, New York - San Francisco - London 1975.

22914 - **JAHN, O.L.** : Effect of washing "Hamlin" orange on chlorophyll and carotenoid changes during degreening. - J.amer. Soc. hort. Sci. *100* : 586 - 588, 1975.

22915 - **JAHNKE, L.S., FRENKEL, A.W.** : Evidence for the photochemical production of superoxide mediated by saponified chlorophyll. - Biochem. biophys. Res. Commun. *66* : 144 - 150, 1975.

22916 - **JAKRLOVÁ, J.** : Primary production and plant chemical composition in food-plain meadows. - Přírodověd. Práce Ústavů ČSAV Brně *9* (9) : 1 - 52, 1975.

22917 - **JAKUCS, P., VIRÁGH, K.** : Changes in the area and weight of light- and shade--adapted leaves and shoots of *Quercus petraea* and *Quercus cerris* in a Hungarian oak forest ecosystem. - Acta bot. Acad. Sci. hung. *21* : 25 - 36, 1975.

22918 - **JANKOVIĆ, M.M., POPOVIĆ, R., DIMITRIJEVIĆ, J.** : Energetske vrednosti organske produktivnosti nadzemnih delova prizemnih biljaka u zajednici *Festuco-quercetum petreae* M. JANK. [The energy values of the organic productivity in the aboveground parts of the ground flora and litter in the community *Festuco--quercetum petreae* M. JANK.] - Arh. biol. Nauka (Beograd) *26* : 141 - 164, 1975. [In Serbocroat., ab : E.]

22919 - **JANKOWSKA, K.** : Ekologia i produkcja pierwotna łąki w Ojcowskim parku narodowym i murawy kserotermicznej w rezerwacie stepowym Skowronno koło Pińczowa. [Ecology and primary production of a meadow in the Ojców National Park and of the xerothermic grassland in the Skowronno steppe reserve near Pińczów.] - Stud. Nat. Ser. A - pol. Akad. Nauk, Zakł. Ochr. Przyr. *11* : 1 - 79, 1975. [Ps; in Pol., ab : E.]

22920 - **JARVIS, P.G.** : Water transfer in plants. - In : **De VRIES, D.A.** (ed.) : Heat and Mass Transfer in the Environment of Vegetation. Pp. 369 - 394. Scripta, Washington, D.C. 1975. [Resistances.]

22921 - **JASSBY, A.D.** : An evaluation of ATP estimations of bacterial biomass in the presence of phytoplankton. - Limnol. Oceanogr. *20* : 646 - 648, 1975.

22922 - **JAYNES, J.M., VERNON, L.P.** : Photophosphorylation activity of membrane vesicles reconstituted from spinach photosystem 1 particles. - Plant. Physiol. *56* (Suppl.) : 31, 1975.

22923 - **JAYNES, J.M., VERNON, L.P., KLEIN, S.M.** : Photophosphorylation and related properties of reaggregated vesicles from spinach Photosystem I particles. - Biochim. biophys. Acta *408* : 240 - 251, 1975.

*22924 - **JEANNEAU, Y.** : Evolution histologique et cytologique des tissus entourant la mine creusée par la larve de *Phytomyza illicis* CURT. dans la feuille d'*Ilex aquifolium* L. et des tissus pathologiques néoformés, au cours du stade larvaire de l'insecte. - Bull. Soc. bot. France *118* : 589 - 620, 1971. [Chloroplast.]

*22925 - **JEANNEAU, Y.** : Etude de déformations pathologiques de la feuille de *Saintpaulia ionantha* H. WENDL., cult. *grandiflora*. - Bull. Soc. bot. France *121* : 251 - 268, 1974. [Chloroplast.]

22926 - **JEFFREY, S.W.** : Chloroplast pigment patterns in dinoflagellates. - Annu. Rep. mar. Biochem. Unit *1974-75* : 6 - 11, 1975.

22927 - **JEFFREY, S.W.** : The occurrence of chlorophylls c_1 and c_2 in algae. - Annu. Rep. mar. Biochem. Unit *1974-75* : 12 - 15, 1975.

22928 - **JEFFREY, S.W.** : Green algal pigments in the Central North Pacific Ocean. - Annu. Rep. mar. Biochem. Unit *1974-75* : 23 - 25, 1975.

22929 - **JEFFREY, S.W., HUMPHREY, G.F.** : New spectrophotometric equations for determining chlorophylls a, b, c_1 and c_2 in higher plants, algae and natural phytoplankton. - Biochem. Physiol. Pflanzen *167* : 191 - 194, 1975.

22930 - **JEFFREY, S.W., SIELICKI, M., HAXO, F.T.** : Chloroplast pigment patterns in dinoflagellates. - J. Phycol. *11* : 374 - 384, 1975.

22931 - **JEFFRIES, T.W., BUTLER, R.G.** : Multiple variant design for the enrichment of photosynthetic bacterial populations. - Can. J. Microbiol. *21* : 1046 - 1054, 1975.

22932 - **JENKINS, P.A.** : Seasonal trends in translocation of ^{14}C photosynthate and their association with wood formation in radiata pine seedlings. - N. Zeal. J. Forest Sci. *5* : 62 - 73, 1975.

22933 - **JENNINGS, R.C., FORTI, G.** : Evidence for energy migration from Photosystem I to Photosystem II and the effect of magnesium. - Biochim. biophys. Acta *376* : 89 - 96, 1975.

22934 - **JENNINGS, R.C., FORTI, G.** : Fluorescence induction in intact spinach chloroplasts. - Biochim. biophys. Acta *396* : 63 - 71, 1975.

22935 - **JENNINGS, R.C., FORTI, G.** : Involvement of oxygen during photosynthetic induction. - In : **AVRON, M.** (ed.) : Proceedings of the Third International Congress on Photosynthesis. Vol. I. Pp. 735 - 743. Elsevier, Amsterdam - Oxford - New York 1975.

22936 - **JENNINGS, R.C., GARLASCHI, F.M., FORTI, G.** : Nature of the slow fluorescence decline in chloroplasts and the involvement of the coupling factor. - In : **QUAGLIARIELLO, E., PAPA, S., PALMIERI, F., SLATER, E.C., SILIPRANDI, N.** (ed.) : Electron Transfer Chains and Oxidative Phosphorylation. Pp. 277 - 288. North-Holland Publ. Comp., Amsterdam 1975.

22937 - **JENSEN, C.R.** : Effects of salinity in the root medium. I. Yield, photosynthesis and water relationships at moderate evaporative demands and various light intensities. - Acta Agr. scand. *25* : 3 - 10, 1975.

22938 - **JENSEN, C.R.** : Effects of salinity in the root medium. II. Photosynthesis and transpiration in relation to superimposed water stress from change of evaporative demands and of root temperature for short periods. - Acta Agr. scand. *25* : 72 - 80, 1975.

*22939 - **JENSEN, K.F., BENDER, F.W., MASTERS, R.G.** : A two-cell chamber for measuring gas exchange in tree seedlings. - USDA Forest. Serv. Res. Note northeastern Forest Exp. Sta. *NE - 178* : 1 - 3, 1973.

22940 - **JENSEN, R.G., BAHR, J.T.** : Properties of ribulose diphosphate carboxylase as observed upon lysis of spinach chloroplasts. - In : **AVRON, M.** (ed.) : Proceedings of the Third International Congress on Photosynthesis. Vol. II. Pp. 1411 - 1420. Elsevier, Amsterdam - Oxford - New York 1975.

22941 - **JENSEN, S.E.** : Strålings- og energibalance. [Radiation and energy balance.] - Ugeskr. Agron. Horton. *4* (6) : 101 - 103, 1975. [In Dan.]

22942 - **JEWESS, P.J., KERR, M.W., WHITAKER, D.P.** : Inhibition of glycollate oxidase from pea leaves. - FEBS Lett. *53* : 292 - 296, 1975.

22943 - **JEWSON, D.H.** : The relation of incident radiation to diurnal rates of photosynthesis in Lough Neagh. - Int. Rev. ges. Hydrobiol. *60* : 759 - 767, 1975.

22944 - **JEWSON, D.H., WOOD, R.B.** : II. Lakes. 4. Europe. Some effects on integral photosynthesis of artificial circulation of phytoplankton through light gradients. - Verh. int. Verein. Limnol. *19* : 1037 - 1044, 1975.

22945 - **JHAMB, S., ZALIK, S.** : Plastid development in a virescens barley mutant and chloroplast microtubules. - Can. J. Bot. *53* : 2014 - 2025, 1975.

22946 - **JOHANSEN, C., LÜTGE, U.** : A comparison of potassium and chloride uptake by *Tradescantia albiflora* leaf cells at different KCl concentrations. - Aust. J. Plant Physiol. *2* : 471 - 479, 1975.

22947 - **JOHANSSON, B.C., BALTSCHEFFSKY, M.** : On the subunit composition of the coupling factor (ATPase) from *Rhodospirillum rubrum*. - FEBS Lett. *53* : 221 - 224, 1975.

22948 - **JOHANSSON, L.-G., LINDER, S.** : The seasonal pattern of photosynthesis of some vascular plants on a subarctic mire. - In : **WIELGOLASKI, F.E.** (ed.) : Fennoscandian Tundra Ecosystems. Part I. Plants and Microorganisms. Pp. 194 - 200. Springer-Verlag, Berlin - Heidelberg - New York 1975.

22949 - **JOHNSON, A.W.** : Porphyrins and related ring systems. - Chem. Soc. Rev. *4* : 1 - 26, 1975.

22950 - **JOHNSON, D.A., CALDWELL, M.M.** : Gas exchange of four arctic and alpine tundra plant species in relation to atmospheric and soil moisture stress. - Oecologia *21* : 93 - 108, 1975.

22951 - **JOHNSON, H.B.** : Gas-exchange strategies in desert plants. - In : **GATES, D.M., SCHMERL, R.B.** (ed.) : Perspectives of Biophysical Ecology. Pp. 105 - 120. Springer-Verlag, Berlin - Heidelberg - New York 1975.

22952 - **JOHNSON, R.R., WILLMER, C.M., MOSS, D.N.** : Role of awns in photosynthesis, respiration, and transpiration of barley spikes. - Crop Sci. *15* : 217 - 221, 1975.

22953 - **JOHNSTON, R., TALIAFERRO, C.M.** : Effects of temperature and light intensity on the expression of a variegated leaf pattern in Bermudagrass. - Crop Sci. *15* : 445 - 447, 1975. [Chl.]

22954 - **JOLCHINE, G., REISS-HUSSON, F.** : Studies on pigments and lipids in *Rhodopseudomonas spheroides* Y reaction center. - FEBS Lett. *52* : 33 - 36, 1975.

22955 - **JOLIOT, A.** : Fluorescence rise from 36 μs on following a flash at low temperature (+2° . -60°). - In : **AVRON, M.** (ed.) : Proceedings of the Third International Congress on Photosynthesis. Vol. I. Pp. 315 - 322. Elsevier, Amsterdam - Oxford - New York 1975.

22956 - **JOLIOT, P., JOLIOT, A.** : Comparative study of the 520 nm absorption change and delayed luminescence in algae. - In : **AVRON, M.** (ed.) : Proceedings of the Third International Congress on Photosynthesis. Vol. I. Pp. 25 - 39. Elsevier, Amsterdam - Oxford - New York 1975.

22957 - **JOLIOT, P., KOK, B.** : Oxygen evolution in photosynthesis. - In : **GOVINDJEE** (ed.) : Bioenergetics of Photosynthesis. Pp. 387 - 412. Academic Press, New York - San Francisco - London 1975.

22958 - **JOLIVET, E.** : Les carboxylations photosynthétiques. - In : **COSTES, C.** (ed.) : Photosynthèse et Production Végétale. Pp. 89 - 126. Gauthier-Villars, Paris 1975.

22959 - **JONAS, H.** : Bleaching of chlorophyll by digitoxin. - Z. Pflanzenphysiol. *77* : 42 - 53, 1975.

22960 - **JONES, H.G., FORD, M.A., PLUMLEY, R.** : The effect of vernalisation on photosynthesis in wheat. - Photosynthetica *9* : 24 - 30, 1975.

22961 - **JONES, L.W., BISHOP, N.I.** : Relation between hydrogen metabolism photosynthesis and nitrogen fixation in the blue green algae. - Plant Physiol. *56* (Suppl.) : 9, 1975.

22962 - **JONES, M.B.** : The effect of leaf age on leaf resistance and CO_2 exchange of the CAM plant *Bryophyllum fedtschenkoi*. - Planta *123* : 91 - 96, 1975.

22963 - **JONES, M.B., MANSFIELD, T.A.** : Circadian rhythms in plants. - Sci. Progr. *62* : 103 - 125, 1975. [Ps.]

22964 - **JORDAN, W.R., BROWN, K.W., THOMAS,J.C.** : Leaf age as a determinant in stomatal control of water loss from cotton during water stress. - Plant Physiol. *56* : 595 - 599, 1975. [Stomatal resistance.]

22965 - **JOSHI, G.V.** : Physiology of salt tolerance in plants. - Biovigyanam *1* : 21 - 39, 1975. [Ps.]

22966 - **JOSHI, M.M., IBRAHIM, I.K.A., HOLLIS, J.P.** : Hydrogen sulfide : Effects on the physiology of rice plants and relation to straighthead disease. - Phytopathology *65* : 1165 - 1170, 1975. [Ps.]

22967 - **JOYARD, J., DOUCE, R.** : Mn^{2+}-dependent ATPase of the envelope of spinach chloroplasts. - FEBS Lett. *51* : 335 - 340, 1975.

22968 - **JUDEL, G.K., LINSER, H., ZEID, F.A.** : Kupfer, Reinprotein und Phenoloxidase in der Blattfolge von *Helianthus annuus* im Verlaufe der Vegetationsperiode. - Z. Pflanzenern. Bodenk. *1975* (1) : 39 - 48, 1975.

22969 - **JUNGE, W.** : Physical aspects of the electron transport and photophosphorylation in green plants. - Ber. deut. bot. Ges. *88* : 283 - 301, 1975.

22970 - **JUNGE, W.** : Structural features of the primary charge separation across the functional membrane of photosynthesis in green plants. - In : **AVRON, M.** (ed.) : Proceedings of the Third International Congress on Photosynthesis. Vol. I. Pp. 273 - 286. Elsevier, Amsterdam - Oxford - New York 1975.

22971 - **JUNGE, W., AUSLÄNDER, W.** : Location of electron carriers in the functional membrane of photosynthesis in green plants. - In : **QUAGLIARIELLO, E., PAPA, S., PALMIERI, F., SLATER, E.C., SILIPRANDI, N.** (ed.) : Electron Transfer Chains and Oxidative Phosphorylation. Pp. 243 - 250. North-Holland Publishing Company, Amsterdam 1975.

22972 - **JUPIN, H.** : La théorie de Mitchell et la photosynthèse. - Rev. Quest. sci. *146* : 143 - 166, 1975.

22973 - **JUPIN, H., CATESSON, A.-M., GIRAUD, G., HAUSWIRTH, N.** : Chloroplastes à empilements granaires anormaux, appauvris en photosystème I, dans le phloème de *Robinia pseudoacacia* et de *Acer pseudoplatanus*. - Z. Pflanzenphysiol. *75* : 95 - 106, 1975.

22974 - **JURGENS, S., JOHNSON, R.R., BOYER, J.S.** : Relationship of crop metabolism and water status to irrigation need. - Water Resour. Cent. Res. Rep. *101* : 1 - 12, 1975. [Ps.]

22975 - **JYUNG, W.H., EHMANN, A., SCHLENDER, K.K., SCALA, J.** : Zinc nutrition and starch metabolism in *Phaseolus vulgaris* L. - Plant Physiol. *55* : 414 - 420, 1975. [Chloroplast.]

22976 - **KABANOVA, Yu.G., BORODKIN, S.O.** :Vliyanie mineral'nogo pitaniya na fotosinteticheskie i produktsionnye kharakteristiki fitoplanktona (Karibskoe more). [Influence of nutrients on photosynthetic and production characteristics of phytoplankton (Caribbean Sea).] - Okeanologiya *15* : 508 - 513, 1975. [In R, ab : E.]

22977 - **KACHRU, R.B., ANDERSON, L.E.** : Inactivation of pea leaf phosphofructokinase by light and dithiothreitol. - Plant Physiol. *55* : 199 - 202, 1975. [Ps.]

22978 - **KADOURI, A., ATSMON, D.** : Effect of ligh-dark regimes on elongation, number of chloroplasts and rate of DNA synthesis in cucumber hypocotyls. - Israel J. Bot. *24* : 46, 1975.

22979 - **KAGAMIYAMA, H., RAO, K.K., HALL, D.O., CAMMACK, R., MATSUBARA, H.** : *Equisetum* (horsetail) ferredoxin : characterization of the active centre and position of the four cysteine residues in this 2Fe-2S protein. - Biochem. J. *145* : 121 - 123, 1975.

22980 - **KAGAN-ZUR, V., LIPS, S.H.** : Studies on the intracellular location of enzymes of the photosynthetic carbon-reduction cycle. - Europe. J. Biochem. *59* : 17 - 23, 1975.

22981 - **KAGAN-ZUR, V., LIPS, S.H.** : Studies on the intracellular organization of enzymes of the photosynthetic CO_2 reduction cycle. - In : **AVRON, M.** (ed.) : Proceedings of the Third International Congress on Photosynthesis. Vol. II. Pp. 1469 - 1478. Elsevier, Amsterdam - Oxford - New York 1975.

22982 - **KAGAWA, T., BEEVERS, H.** : The development of microbodies (glyoxysomes and leaf peroxisomes) in cotyledons of germinating watermelon seedlings. - Plant Physiol. *55* : 258 - 264, 1975.

22983 - **KAGAWA, T., HATCH, M.D.** : Mitochondria as a site of C_4 acid decarboxylation in C_4-pathway photosynthesis. - Arch. Biochem. Biophys. *167* : 687 - 696, 1975.

22984 - **KAISER, H.W.** : Utilization of light energy by the maize canopy. - In : **DU PLESSIS, J.G., GROGAN, C.O., KÜHN, H.C., WALTERS, M.C.** (ed.) : Proceedings of the First South African Maize Breeding Symposium. Pp. 72 - 75. Div. Agr. Inf., Pretoria 1975.

22985 - **KAISER, H.W., DE JAGER, J.M.** : Effect of plant spacing on the vegetative development of a maize crop. - Gewasproduksie/Crop Production *4* : 13 - 17, 1975. [Ps.]

22986 - **KAISER, W., SCHULZ, S.** : On primary production in the Baltic. - Merentutkimuslaitok julk. / Havsforskningsinst. Skr. *239* : 29 - 33, 1975.

22987 - **KAK, S.N., KAUL, B.L.** : Mutagenic activity of hydrazine and its combinations with maleic hydrazide and X-rays in barley. - Cytobios *12* : 123 - 128, 1975. [Chl.]

22988 - **KAKHNOVICH, L.V., GRITS, M.G.** : Pigmentnyĭ fond khloroplastov v zavisimosti ot spektral'nogo sostava sveta. [Effect of spectral composition of light on pigment pool of chloroplasts.] - Fiziol. Rast. *22* : 461 - 465, 1975. [In R, ab : E.]

22989 - **KAKITANI, T.** : [Analysis of optical absorption curves and fluorescence curves of biomolecules. I.] - Seibutsu-Butsuri *15* : 13 - 22, 1975. [Car; in Jap., ab : E.]

22990 - **KALER, V.L.** : The metabolic and epigenetic control of chlorophyll biosynthesis. - In : **NASYROV, Yu.S., ŠESTÁK, Z.** (ed.) : Genetic Aspects of Photosynthesis. Pp. 295 - 301. Dr. W. Junk b.v. Publ., The Hague 1975.

22991 - **KALFF, J., KLING, H.J., HOLMGREN, S.H., WELCH, H.E.** : Phytoplankton, phytoplankton growth and biomass cycles in an unpolluted and in a polluted polar lake. - Verh. int. Verein. Limnol. *19* : 487 - 495, 1975.

22992 - **KALLIO, P., HEINONEN, S.** : CO_2 exchange and growth of *Rhacomitrium lanuginosum* and *Dicranum elongatum*. - In : **WIELGOLASKI, F.E.**(ed.) : Fennoscandian Tundra Ecosystems. Part 1. Plants and Microorganisms. Pp. 138 - 148. Springer-Verlag, Berlin - Heidelberg - New York 1975.

22993 - **KALLIO, P., KÄRENLAMPI, L.** : Photosynthesis in mosses and lichens. - In : **COOPER, J.P.** (ed.) : Photosynthesis and Productivity in Different Environments. Pp. 393 - 423. Cambridge University Press, Cambridge - London - New York - Melbourne 1975.

22994 - **KALLIO, P., VALANNE, N.** : On the effect of continuous light on photosynthesis in mosses. - In : **WIELGOLASKI, F.E.** (ed.) : Fennoscandian Tundra Ecosystems. Part 1. Plants and Microorganisms. Pp. 149 - 162. Springer-Verlag, Berlin - Heidelberg - New York 1975.

*22995 - **KAMEI, N., WAKAMATSU, K.** : [The effects on the photochemical activities of chloroplasts by addition of membrane permeable ions. II.] - Seikatsu Kagaku [Sci. hum. Life] *9*(3) : 116 - 120, 1973. [In Jap., ab : E.]

22996 - **KAMIENIETZKY, A., NELSON, N.** : Preparation and properties of chloroplast coupling factor 1 by sodium bromide treatment. - Plant Physiol. *55* : 282 - 287, 1975.

22997 - **KAMIYA, A., MIYACHI, S.** : Blue light-induced formation of phosphoenolpyruvate carboxylase in colorless *Chlorella* mutant cells. - Plant Cell Physiol. *16* : 729 - 736, 1975.

22998 - **KAMIYAMA, K., HORIE, M.** : Relative growth rate, net assimilation rate, and climate. - In : **MURATA, Y.** (ed.) : JIBP Synthesis. Vol. 11. Pp. 21 - 36. University of Tokyo Press, Tokyo 1975.

22999 - **KAMOTA, F., NAITO, Y.** : [Studies on photosynthesis and transpiration of vegetable crops. II. A linear electronic device for continuous measurement of stem and fruit enlargement in relation to water stress.] - Bull. veg. ornam. Crops Res. Sta., Ser. A *2* : 33 - 47, 1975. [In Jap., ab : E.]

23000 - **KAN, K.-S., THORNBER, J.P.** : The light-harvesting chlorophyll *a/b*-protein of *Chlamydomonas*. - Plant Physiol. *56* (Suppl.) : 11, 1975.

23001 - **KANAI, R., KASHIWAGI, M.** : *Panicum milioides*, a *Gramineae* plant having Kranz leaf anatomy without C_4-photosynthesis. - Plant Cell Physiol. *16* : 669 - 679, 1975.

23002 - **KANDA, M.** : Maximal growth rate and climate. - In : **MURATA, Y.** (ed.) : JIBP Synthesis. Vol. 11. Pp. 67 - 71. University of Tokyo Press, Tokyo 1975.

23003 - **KANDA, M.** : Efficiency for solar energy utilization. - In : **MURATA, Y.** (ed.) : JIBP Synthesis. Vol. 11. Pp. 187 - 198. University of Tokyo Press, Tokyo 1975.

23004 - **KANDA, M.** : The spectral composition of incident solar and sky radiation, and reflected and transmitted PAR in fied populations of rice plants. - Rep. Inst. agr. Res. Tohoku Univ. *26* : 1 - 14, 1975.

23005 - **KANDELER, R., HÜGEL, B., ROTTENBURG, T.** : Relations between photosynthesis and flowering in *Lemnaceae*. - In : **MARCELLE, R.** (ed.) : Environmental and Biological Control of Photosynthesis. Pp. 161 - 169. Dr. W. Junk b.v. Publ., The Hague 1975.

B23006 - **KANEMASU, E.T.** (ed.) : Measurement of Stomatal Aperture and Diffusive Resistance. - Coll. agr. Res. Cent., Washington State Univ. Bull. 809. 1975.

23007 - **KANEMASU, E.T., HIEBSCH, C.K.** : Net carbon dioxide exchange of wheat, sorghum, and soybean. - Can. J. Bot. *53* : 382 - 389, 1975.

23008 - **KANIUGA, Z., FRANCKOWIAK, B.** : Studies on the enzyme system involved in electron and energy transport in isolated chloroplasts : The relationship between non-cyclic electron transport and phosphorylation. - Pol. ekol. Stud. *1* : 27 - 32, 1975.

23009 - **KANNANGARA, C.G., JENSEN, C.J.** : Biotin carboxyl carrier protein in barley chloroplast membranes. - Europe. J. Biochem. *54* : 25 - 30, 1975.

23010 - **KAO, O.H.W., EDWARDS, M.R., BERNS, D.S.** : Physical-chemical properties of *C*-phycocyanin isolated from an acido-thermophilic eukaryote, *Cyanidium caldarium*. - Biochem. J. *147* : 63 - 70, 1975.

23011 - **KAPIL, R.N., PUGH, T.D., NEWCOMB, E.H.** : Microbodies and an anomalous "microcylinder" in the ultrastructure of plants with crassulacean acid metabolism. - Planta *124* : 231 - 244, 1975.

23012 - **KAPINUS, E.I., IVNITSKAYA, I.N., DILUNG, I.I.** : Nekotorye osobennosti tusheniya fluorestsentsii khlorofilla i ego analogov okislitelyami. [Some peculiarities of fluorescence quenching of chlorophyll and its derivatives by oxidants.] - Biofizika *20* : 411 - 413, 1975. [In R, ab : E.]

23013 - **KAPLAN, A., GALE, J.** : Separation of respiration from dark fixation of carbon dioxide in CAM plants. - Israel J. Bot. *24* : 59, 1975.

23014 - **KAPLAN, D., ROTH-BEJERANO, N., LIPS, S.H.** : Photosynthesis and the induction of nitrate reductase in plants. - In : **AVRON, M.** (ed.) : Proceedings of the Third International Congress on Photosynthesis. Vol. II. Pp. 1517 - 1524. Elsevier, Amsterdam - Oxford - New York 1975.

23015 - **KAPPEN, L., LANGE, O.L., SCHULZE, E.-D., EVENARI, M., BUSCHBOM, U.** : Primary production in lower plants (lichens) in the desert and its physiological basis. - In : **COOPER, J.P.** (ed.) : Photosynthesis and Productivity in Different Environments. Pp. 133 - 143. Cambridge University Press, Cambridge - London - New York - Melbourne 1975.

23016 - **KARABASHEV, G.S., BEKASOVA, O.D.** : O sootnoshenii mezhdu kontsentratsieĭ i lyuminestsentsieĭ pigmentov okeanicheskogo fitoplanktona. [Relationship between concentration and luminescence of pigments of the oceanic phytoplankton.]-In : Gidrofizicheskie i Opticheskie Issledovaniya v Indiĭskom Okeane. Pp. 95 - 99. Nauka, Moskva 1975. [In R.]

23017 - **KARABASHEV, G.S., SOLOV'EV, A.N.** : Zakonomernosti prostranstvenno-vremennoĭ izmenchivosti intensivnosti fluorestsentsii pigmentov v kletkakh zhivogo fitoplanktona. [Peculiarities of space-time variability of pigment fluorescence intensity in cells of living phytoplankton.] - Tr. Inst. Okeanol. Akad. Nauk SSSR *102* (Ėkosistemy Pelagiali Tikhogo Okeana) : 153 - 164, 1975. [In R, ab : E.]

23018 - **KARAKASHIAN, M.W.** : Phase shifting the photosynthetic rhythm in anucleate *Acetabularia* with periods of darkness. - Protoplasma *83* : 176, 1975.

23019 - **KARANOV, E., VASILEV, G., POGONCHEVA, E.** : Retarding effect of itaconic acid and of some of its esters on chlorophyll destruction in detached leaves. - Dokl. SKhA im. G. Dimitrova *8* (2) : 15 - 17, 1975.

23020 - **KARAPETYAN, N.V.** : Evolution of photosystems of photosynthetic organisms. - Origins Life *6* : 253 - 256, 1975.

23021 - **KARAPETYAN, N.V., KLIMOV, V.V., LANG, F., KRASNOVSKIĬ, A.A.** : Fluorescence induction of normal and mutant maize seedlings. - In : **NASYROV, Yu.S., ŠESTÁK, Z.** (ed.) : Genetic Aspects of Photosynthesis. Pp. 255 - 261. Dr. W. Junk b.v. Publ., The Hague 1975.

23022 - **KARAPETYAN, N.V., KONONENKO, A.A.** : Temperaturnaya zavisimost' fotoprevrashcheniĭ bakteriokhlorofillov purpurnykh serobakteriĭ i izolirovannykh khromatoforov. [Effect of temperature on photoinduced changes of bacteriochlorophylls in purple sulfur bacteria and isolated chromatophores.] - Mikrobiologiya *44* : 422 - 427, 1975. [In R , ab : E.]

23023 - **KARAVAEV, V.A., KUKUSHKIN, A.K.** : Adaptatsiya k temnote i dal'nemu krasnomu svetu list'ev vysshikh rasteniĭ v usloviyakh nedostatka kisloroda. [Adaptation of higher plant leaves to dark and far-red radiation under oxygen deficiency.] - Biofizika *20* : 88 - 92, 1975. [In R , ab : E.]

23024 - **KARAVAEV, V.A., KUKUSHKIN, A.K.** : Vliyanie predvaritel'nogo osveshcheniya svetom razlichnogo spektral'nogo sostava na bystruyu induktsiyu fluorestsentsii lista. [Effect of preliminary irradiation by different wavelengths on fast fluorescence induction in a leaf.] - Biofizika *20* : 739 - 740, 1975. [In R , ab : E.]

23025 - **KARAVAĬKO, N.N., OMANN, E.E., KULAEVA, O.N.** : Vliyanie tsitokinina na aktivnost' ryada fermentov v izolirovannykh semyadolyakh tykvy. [Effect of cytokinin on enzyme activity in isolated pumpkin cotyledons.] - Fiziol. Rast. *22* : 1031 - 1038, 1975. [Chl ; in R , ab : E.]

23026 - **KARAWYA, M.S., GHOURAB, M.G., EL-SHAMI, I.M.** : A study of β-carotene in certain Egyptian vegetable organs. - Egypt. J. pharm. Sci. *16* : 339 - 344, 1975.

23027 - **KÄRENLAMPI, L., TAMMISOLA, J., HURME, H.** : Weight increase of some lichens as related to carbon dioxide exchange and thallus moisture. - In : **WIELGOLASKI, F.E.** (ed.) : Fennoscandian Tundra Ecosystems. Part 1. Plants and Microorganisms. Pp. 135 - 137. Springer-Verlag, Berlin - Heidelberg - New York 1975.

23028 - **KARIMOVA, F.G., RYBKINA, G.V., SEDYKH, N.V., RATUSHNYAK, Yu.M., BEL'KOVICH, T.M., KHAMIDULLINA, N.G., BIGLOVA, S.G., VELIKANOVA, G.A.** : Vliyanie zasukhi na vodnyĭ rezhim khloroplastov. [Influence of drought on water regime of chloroplasts.] - In : Vodoobmen Rasteniĭ pri Neblagopriyatnykh Usloviyakh Sredy. Pp. 89 - 92. Shtiintsa, Kishinev 1975. [In R.]

23029 - **KARLANDER, E.P., SPEARING, A.M.** : Light-dependent chlorophyll degradation in *Chlorella*. - Plant Physiol. *56* (Suppl.) : 11, 1975.

23030 - **KARPILOV, Yu.S., AVDEEVA, T.A., PERSANOV, V.M.** : Localisation of carbon metabolism in two assimilation tissues of maize leaf. - In : **NASYROV, Yu.S., ŠESTÁK, Z.** (ed.) : Genetic Aspects of Photosynthesis. Pp. 177 - 185. Dr. W. Junk b.v. Publ., The Hague 1975.

23031 - **KARPILOV, Yu.S., BIL', K.Ya.** : Kolichestvennye, ul'trastrukturnye i funktsional'nye osobennosti khloroplastov assimilyatsionnykh tkaneĭ khlorofill'nogo mutanta kukuruzy. [Quantitative, ultrastructural and functional properties of chloroplasts of assimilatory tissues of the chlorophyll mutant of maize.] - In : Kolichestvennye Priznaki Mutantov. Pp. 96 - 99. Shtiintsa, Kishinev 1975. [In R.]

23032 - **KARPILOV, Yu.S., BIL', K.Ya., GUKASYAN, I.A.** : Uchastie ATFaz v transporte promezhutochnykh produktov fotosinteza mezhdu assimilyatsionnymi tkanyami C-4-rasteniĭ. [Participation of ATPases in transport of intermediate products of photosynthesis between assimilating tissues of C_4 plants.] - Fiziol. Rast. *22* : 1113 - 1120, 1975. [In R, ab : E.]

23033 - **KARPILOV, Yu.S., BIL', K.Ya., MALYSHEV, O.G.** : Zavisimost' skorosti perekhoda C^{14} iz dikarbonovykh kistot v pentozofosfatnyĭ tsikl ot sootnosheniya malata i aspartata i razmeshcheniya khloroplastov u C-4-rasteniĭ. [Effect of the ratio between malate and aspartate and the arrangement of chloroplasts in C_4 plants on the rate of ^{14}C-incorporation into the pentose phosphate cycle from dicarboxylic acids.] - Fiziol. Rast. *22* : 910 - 917, 1975. [In R, ab : E.]

23034 - **KASEMIR, H., BERGFELD, R., MOHR, H.** : Phytochrome-mediated control of prolamellar body reorganization and plastid size in mustard cotyledons. - Photochem. Photobiol. *21* : 111 - 120, 1975.

23035 - **KASEMIR, H., MASONER, M.** : Control of chlorophyll synthesis by phytochrome. II. The effect of phytochrome on aminolevulinate dehydratase in mustard seedling. - Planta *126* : 119 - 126, 1975.

23036 - **KASPRZYK, Z.** : Investigations on primary and secondary photosynthesis products and their metabolism in plants. - Pol. ecol. Stud. *1* : 97 - 106, 1975.

23037 - **KASSAM, A.H., ANDREWS, D.J.** : Effects of sowing date on growth, development and yield of photosensitive sorghum at Samaru, Northern Nigeria. - Exp. Agr. *11* : 227 - 240, 1975. [Growth analysis.]

23038 - **KASSAM, A.H., KOWAL, J.M.** : Water use, energy balance and growth of Gero millet at Samaru, northern Nigeria. - Agr. Meteorol. *15* : 333 - 342, 1975. [Ps.]

23039 - **KASSAM, A.H., KOWAL, J.M., HARKNESS, C.** : Water use and growth of groundnut at Samaru, Northern Nigeria. - Trop. Agr. *52* : 105 - 112, 1975. [Growth analysis.]

23040 - **KATADA, M., SATOMI, M.** : Ecology of marine algae. - In : **TOKIDA, J., HIROSE, H.** (ed.) : Advance of Phycology in Japan. Pp. 211 - 239. Dr. W. Junk b.v. Publ., The Hague 1975. [Ps.]

23041 - **KATOH, S.** : [Functional site of plastocyanin.] - Tampakushitsu, Kakusan, Koso [Protein, Nucleic Acid, Enzyme] *20* : 42 - 52, 1975. [In Jap.]

23042 - **KATOH, S., SATOH, K., YAMAGISHI, A., YAMAOKA, T.** : Fluorescence induction in chloroplasts isolated from the green alga, *Bryopsis maxima* I. Occurrence of the complete Kautsky effect in *Bryopsis* chloroplasts. - Plant Cell Physiol. *16* : 1093 - 1099, 1975.

23043 - **KATOH, T., OHKI, K.** : Loss of photosystem II induced by a nitrate deficiency in photoorganotrophically grown *Anabaena variabilis*. - Plant Cell Physiol. *16* : 815 - 828, 1975.

23044 - **KATYUZHANSKAYA, A.N., MOROZOVA, S.S., PEKHOV, A.V., MIKHAĬLOVA, N.S.** : Vliyanie usloviĭ khraneniya na kachestvo "lapki" pikhty sibirskoĭ i sostav poluchaemogo iz nee CO_2-ėkstrakta. [Effect of storage conditions on the quality of "lapka" of *Abies sibirica* and the composition of CO_2-extract obtained from it.] - Rastitel'nye Resursy *11* : 555 - 559, 1975. [Chl, Car; in R.]

23045 - **KATZFUSS, M.** : $^{14}CO_2$-Assimilation von Blättern zweier Apfelsorten im Verlaufe der natürlichen Alterung. - Arch. Gartenbau *23* : 257 - 264, 1975.

23046 - **KAUFMANN, K.J., DUTTON, P.L., NETZEL, T.L., LEIGH, J.S., RENTZEPIS, P.M.** : Picosecond kinetics of events leading to reaction center bacteriochlorophyll oxidation. - Science *188* : 1301 - 1304, 1975.

23047 - **KAUFMANN, K.J., NETZEL, T.L., DUTTON, P.L., LEIGH, J.S., RENTZEPIS, P.M.** : Picosecond absorption kinetics of *Rps sphaeroides*. - Biophys. J. *15* (2, Part 2) : 226 a, 1975.

23048 - **KAUL, K., SABHARWAL, P.S.** : Morphogenetic studies on *Haworthia* : Effects of inositol on growth and differentiation. - Am. J. Bot. *62* : 655 - 659, 1975. [Chl.]

23049 - **KAUL, R.** : Rapid method for assessing potential net photosynthesis in plant leaves. - Z. Pflanzenphysiol. *77* : 75 - 79, 1975.

23050 - **KAUR, B., MANJREKAR, S.P.** : Effect of dehydration on the stability of chlorophyll and β-carotene content of green leafy vegetables available in northern India. - J. Food Sci. Technol. *12* : 321 - 323, 1975.

*23051 - **KAWASHIMA, N., IMAI, A., TAMAKI, E.** : Studies on protein metabolism in higher plants. III. Changes in the soluble protein components with leaf growth. - Plant Cell Physiol. *8* : 447 - 458, 1967. [Fraction 1 protein.]

23052 - **KAZARYAN, V.O., DAVTYAN, V.A., GEVORKYAN, I.A.** : Reaktsiya rasteniĭ na isklyuchenie i dopolnitel'nuyu podkormku zhelezom korneĭ i list'ev. [Reaction of plants on the elimination and supplementary nutrition with iron of roots and leaves.] - Biol. Zh. Armenii *28* (10) : 3 - 10, 1975. [Chl ; in R, ab : Arm.]

*23053 - **KAZARYAN, V.O., MELKONYAN, A.S., DAVTYAN, V.A.** : Posledeĭstvie glubokogo rykhleniya pochvy na nekotorye fiziologicheskie pokazateli i urozhaĭnost' vinogradnoĭ lozy. [Aftereffect of deep soil ploughing on some physiological characteristics and yielding ability of grapevine.] - Biol. Zh. Armenii *24* (11) : 3 - 12, 1971. [Ps, Chl; in R, ab : Arm.]

23054 - **K"DREV, T., GEORGIEVA, M.** : Vliyanie na nedostiga na magnezii' v"rkhu s"d"rzhanieto na galaktolipidite i pigmentite v otdelni lista na tsarevichni rasteniya. [Magnesium deficiency effect on galactolipid and pigment content in individual maize leaves.] - Fiziol. Rast. (Sofia) *1* (3) : 10 - 16, 1975. [In Bulg., ab : E, R.]

23055 - **KE, B.** : Potentiometric titration of soluble and bound iron-sulfur proteins monitored by circular dichroism and electron paramagnetic resonance spectroscopy. - Bioelectrochem. Bioenerg. *2* : 93 - 105, 1975. [Ps.]

23056 - **KE, B.** : Some comments on the present status of the primary electron acceptor of photosystem I. - In : **AVRON, M.** (ed.) : Proceedings of the Third International Congress on Photosynthesis. Vol. I. Pp. 373 - 382. Elsevier, Amsterdam - Oxford - New York 1975.

23057 - **KE, B., SUGAHARA, K., SHAW, E.R.** : Further purification of "Triton subchloroplast fraction I" (TSF-I particles). Isolation of a cytochrome-free high-*P*-700 particle and a complex containing cytochromes *f* and b_6, plastocyanin and iron--sulfur protein(s). - Biochim. biophys. Acta *408* : 12 - 25, 1975.

23058 - **KEENAN, J.D.** : Bicarbonate utilization in *Anabaena*. - Physiol. Plant. *34* : 157 - 161, 1975.

23059 - **KEENER, M.E., McCREE, K.J.** : A test of the Duncan model of photosynthesis in plant communities. - Crop Sci. *15* : 214 - 216, 1975.

23060 - **KEFRBERG, O.F.** : Rol' sveta v dinamicheskoĭ regulyatsii fotosinteticheskogo metabolizma ugleroda. [Dynamic light regulation of carbon photosynthetic metabolism.] - In : Fotoregulyatsiya Metabolizma i Morfogeneza Rasteniĭ. Pp. 158 - 170, 252. Nauka, Moskva 1975. [In R.]

23061 - **KELLNER, E., VARGA, P., SLUŞANSCHI, H.** : Studiul comparativ al unor specii şi soiuri de graminee perene în cultură irigată. [A comparative study on several species and varieties of perennial grasses in irrigated cultures.] - Ann. Inst. Cercetări cereale Plante Tehnice-Fundulea *40* (Ser.C) : 335 - 345, 1975. [Growth analysis; in Roum., ab : E,R.]

23062 - **KELLY, G.J., LATZKO, E.** : Evidence for phosphofructokinase in chloroplasts. - Nature *256* : 429 - 430, 1975.

23063 - **KEMP, D.R.** : Subterranean clover as a winter forage crop in a subtropical environment. - Aust. J. exp. Agr. anim. Husb. *15* : 631 - 636, 1975. [Growth analysis.]

23064 - **KEMP, D.R.** : The growth of three tropical pasture grasses on the mid-north coast of New South Wales. - Aust. J. exp. Agr. anim. Husb. *15* : 637 - 644, 1975. [Growth analysis.]

23065 - **KEREKES, J.J.** : The relationship of primary production to basin morphometry in five small oligotrophic lakes in Terra Nova National Park in Newfoundland. - Symp. biol. hung. *15* : 35 - 48, 1975.

23066 - **KERKENI, A., KUKUSHKIN, A.K., SOLNTSEV, M.K.** : Issledovanie vliyaniya nekotorykh kofaktorov i razobshchiteleĭ fosforilirovaniya na termovysvechivanie zelenogo lista. [Effect of some cofactors and uncouplers of phosphorylation on thermoluminescence of green leaves.] - Fiziol. Rast. *22* : 776 - 781, 1975. [In R, ab : E.]

23067 - **KERR, J.P.** : The potential for maize production in New Zealand. - Proc. Agron. Soc. N. Zeal. *5* : 65 - 69, 1975.

23068 - **KERR, M.W.** : Studies on photorespiration enzymes from *Pisum sativum*. - In : **AVRON, M.** (ed.) : Proceedings of the Third International Congress on Photosynthesis. Vol. II. Pp. 1285 - 1289. Elsevier, Amsterdam - Oxford - New York 1975.

23069 - **KERSHAW, K.A.** : Studies on lichen-dominated systems. XIV. The comparative ecology of *Alectoria nitidula* and *Cladina alpestris*. - Can. J. Bot. *53* : 2608 - 2613, 1975. [Ps.]

23070 - **KERSHAW, K.A., FIELD, G.F.** : Studies on lichen-dominated systems. XV. The temperature and humidity profiles in a *Cladina alpestris* MAT. - Can. J. Bot. *53* : 2614 - 2620, 1975.

23071 - **KESSEL, R., HEIYOUNG, LEE, K., ROWE, P.R.** : Production of intraspecific aneuploids in the genus *Solanum*. III. Intraspecific aneuploids. - Euphytica *24* : 585 - 595, 1975. [Chloroplasts in guard cells.]

23072 - **KESTLER, D.P., MAYNE, B.C., RAY, T.B., GOLDSTEIN, L.D., BROWN, R.H., BLACK, C.C.** : Biochemical components of the photosynthetic CO_2 compensation point of higher plants. - Biochem. biophys. Res. Commun. *66* : 1439 - 1446, 1975.

23073 - **KEYS, A.J., BIRD, I.F., CORNELIUS, M.J., KUMARASINGHE, S., WHITTINGHAM, C.P.** : Use of isotopes to explore the physiology and biochemistry of photorespiration and its effects on crop yields. - In : Tracer Techniques for Plant Breeding. Pp. 13 - 18. International Atomic Energy Agency, Vienna 1975.

23074 - **KHAĬLOV, M.** : Populyatsionnaya regulyatsiya fotosinteza i organotrofii u morskikh makrofitov. [Population regulation of photosynthesis and organotrophy in marine macrophytes.] - Dokl. Akad. Nauk SSSR *221* : 1204 - 1206, 1975. [In R.]

23075 - **KHALIDOVA, G.B., KOSITSIN, A.V.** : Vliyanie nedostatka tsinka na karboangidraznuyu aktivnost' khloroplastov tomatov. [Effect of zinc deficiency on the carbonic anhydrase activity of tomato chloroplasts.] - Bot. Zh. *60* : 552 - 558, 1975. [In R.]

23076 - **KHANGULOV, S.V., GOL'DFEL'D, M.G.** : O kinetike temnovogo vosstanovleniya tsentrov *P*700 v fotosinteze. [Kinetics of dark reduction on the centres of *P*700 in photosynthesis.] - Biofizika *20* : 652 - 655, 1975. [In R, ab : E.]

23077 - **KHARANYAN, N.N., VIKHREVA, V.N.** : Aktivnost' nekotorykh fermentov v list'yakh rasteniĭ fasoli, obrabotannykh khlorkholinkhloridom (CCC) v usloviyakh pochvennoĭ zasukhi. [Activity of some enzymes in the leaves of *Phaseolus* plants treated with CCC under soil drought.] - Fiziol. Rast. *22* : 806 - 809, 1975. [Ps; in R, ab : E.]

*23078 - **KHEBER, U.** : V"treklet"chni vzaimodeĭstviya. [Intracellular interactions.] - Fiziol. Rast. (Sofia) *1* (1) : 23 - 35, 1974. [Chloroplast; in Bulg., ab : E,R.]

23079 - **KHISAMUTDINOVA, V.I., VASIL'EVA, I.M., KUZ'MINA, G.G.** : Vzaimosvyaz' sostoyaniya vody i ènergeticheskogo obmena list'ev ozimoĭ pshenitsy. [Interrelation of water state and energy metabolism of winter wheat leaves.] - Fiziol. Biokhim. kul't. Rast. *7* : 481 - 485, 1975. [Ps; in R, ab : E.]

23080 - **KHISAMUTDINOVA, V.I., VASIL'EVA, I.M., KUZ'MINA, G.G., VERSHININ, A.A.** : Vliyanie khlorkholinkhlorida na ènergeticheskiĭ obmen i sostoyanie vody ozimoĭ pshenitsy v protsesse zakalivaniya. [Effect of CCC on energy exchange and the state of water in winter wheat during hardening.] - Fiziol. Rast. *22* : 1048 - 1054, 1975. [Ps, Chl; in R, ab : E.]

23081 - **KHOANG T'YUNG** : Sezonnaya dinamika massy nadzemnykh i podzemnykh organov rasteniĭ lugovoĭ stepi. [Seasonal dynamics of the mass of aboveground and underground organs of meadow steppe plants.] - Vest. mosk. Univ., Biol. Pochvoved. *30* (3) : 58 - 63, 1975. [In R, ab : E.]

23082 - **KHODASEVICH, Ė.V., ARNAUTOVA, A.I.** : O sostoyanii pigmentov u nezhelteyushchikh khvoĭnykh v svyazi s vozrastom lista. [State of pigments in non-yellowing conifers in connection with the leaf age.] - In : **SHLYK, A.A.** (ed.) : Biosintez i Sostoyanie Khlorofillov v Rastenii. Pp. 216 - 226, 248. Nauka i Tekhnika, Minsk 1975. [In R, ab : E.]

23083 - **KHOJA, T.M., WHITTON, B.A.** : Heterotrophic growth of filamentous blue-green algae. - Brit. phycol. J. *10* : 139 - 148, 1975. [Chl, biliproteins.]

23084 - **KHOLMOGOROV, V.E.** : Obnaruzhenie paramagnitnykh chastits v pervichnykh reaktsiyakh fotosinteza. [Observation of paramagnetic particles in primary reactions of photosynthesis.] - In : Metody Issledovaniya Fotokhimicheskikh Reaktsiĭ Fotosinteza *in Vitro* i *in Vivo*. Pp. 64 - 75. Pushchino 1975. [In R.]

23085 - **KHOLUPENKO, I.P., KARPOV, E.A., KUZINA, N.V.** : Vliyanie kratkovremennogo izmeneniya soderzhaniya assimilyatov v list'yakh soi na ikh ottok k plodam. [Effect of short-term changes in the content of photosynthates in soybean leaves on their outflow to fruits.] - Fiziol. Rast. *22* : 723 - 728, 1975. [In R, ab : E.]

23086 - **KHOR'KOV, E.I., RESH, F.M., KRYLOVA, T.S.** : Izmenenie fotokhimicheskoĭ aktivnosti khloroplastov ozimoĭ pshenitsy v zavisimosti ot prodolzhitel'nosti yarovizatsii. [Change in photochemical activity in winter wheat chloroplasts in dependence on vernalization duration.] - Fiziol. Biokhim. kul't. Rast. *7* : 246 - 250, 1975. [In R, ab : E.]

23087 - **KIEFER, D.A., LASKER, R.** : Two blooms of *Gymnodinium splendens*, an unarmored dinoflagellate. - Fishery Bull. *73* : 675 - 678, 1975. [Ps, Chl.]

23088 - **KIM, E.-B.** : [The standing crop and its turnover rate of algae on the stone in the Cheonji stream, Cheju Do.] - Korean J. Limnol. *8* (3-4) : 1 - 5, 1975. [In Korean, ab : E.]

23089 - **KIM, V.A., VOZNYAK, V.M., EVSTIGNEEV, V.B.** : Fotokhimicheskaya generatsiya kation-radikala bakteriokhlorofilla *b* i anion-radikalov bakteriokhlorofilla *b* i bakteriofeofitina *b*, soderzhashchikhsya v fotosinteziruyushchikh bakteriyakh *Rhodopseudomonas viridis*. [Photochemical generation of bacteriochlorophyll *b* cation-radical and anion-radicals of bacteriochlorophyll *b* and bacteriopheophytin *b* of photosynthetic bacteria *Rhodopseudomonas viridis*.] - Biofizika *20* : 208 - 213, 1975. [In R, ab : E.]

23090 - **KIMENOV, G.P., MINKOV, I.N.** : On the behaviour of *Haberlea rhodopensis* FRIV. and *Ramonda serbica* PANČ. to the poikiloxerophytic type of plants. - Dokl. bolg. Akad. Nauk *28* : 829 - 831, 1975. [Ps.]

23091 - **KIRA, T.** : Primary production of forests. - In : **COOPER, J.P.** (ed.) : Photosynthesis and Productivity in Different Environments. Pp. 5 - 40. Cambridge University Press, Cambridge - London - New York - Melbourne 1975.

23092 - **KIRCHANSKI, S.J.** : The ultrastructural development of the dimorphic plastids of *Zea mays* L. - Amer. J. Bot. *62* : 695 - 705, 1975.

23093 - **KIRCHANSKI, S.J.** : Thiocarbamyl nitro blue tetrazolium (TC-NBT) : reagent for ultrastructural localization of photosynthetic electron transport. - Plant Physiol. *56* (Suppl.) : 31, 1975.

23094 - **KIRICHENKO, E.B., SMOLYGINA, L.D., SERDYUK, O.P.** : Izmenenie kompleksa polipeptidov lamell pri razvitii plastid mezofilla i obkladki *Zea mays* L. [Change in lamella polypeptide complex in developing plastids of mesophyll and bundle sheath of *Zea mays* L.] - Dokl. Akad. Nauk SSSR *222* : 979 - 982, 1975. [In R.]

*23095 - **KIRK, J.T.O.** : The relation of chlorophyll synthesis to protein synthesis in the growing thylakoid membrane. - Portugal. Acta biol., Sér. A *14* : 127 - 152, 1974.

23096 - **KIRK, J.T.O.** : A theoretical analysis of the contribution of algal cells to the attenuation of light within natural waters. I. General treatment of suspensions of pigmented cells. - New Phytol. *75* : 11 - 20, 1975.

23097 - **KIRK, J.T.O.** : A theoretical analysis of the contribution of algal cells to the attenuation of light within natural waters. II. Spherical cells. - New Phytol. *75* : 21 - 36, 1975. [Chl.]

23098 - **KIRK, J.T.O.** : The contribution of phytoplankton to the attenuation of light within natural waters : a theoretical analysis. - In : **AVRON, M.** (ed.) : Proceedings of the Third International Congress on Photosynthesis. Vol. I. Pp. 245 - 253. Elsevier, Amsterdam - Oxford - New York 1975. [Chl.]

23099 - **KIRST, G.O.** : Wirkung unterschiedlicher Konzentrationen von NaCl und anderen osmotisch wirksamen Substanzen auf die CO_2-Fixierung der einzelligen Alge *Platymonas subcordiformis*. - Oecologia *20* : 237 - 254, 1975.

23100 - **KISELEV, B.A., KOZLOV, Yu.N., EVSTIGNEEV, V.B.** : Ėlektrokhimicheskie metody v issledovanii fotokhimicheskikh reaktsiĭ. [Electrochemical methods in studies of photochemical reactions.] - In : Metody Issledovaniya Fotokhimicheskikh Reaktsiĭ Fotosinteza *in Vitro* i *in Vivo*. Pp. 116 - 124. Pushchino 1975. [In R.]

23101 - **KISS, A.S.** : Összefüggés a levéllemez különböző részeinek magnézium- és fehérjetartalma között. [Relationship between different Mg and protein contents in leaf-blade parts.] - Bot. Közlem. *62* : 117 - 120, 1975. [Bioenergetics; in Hung., ab : G.]

23102 - **KITAJIMA, M., BUTLER, W.L.** : Quenching of chlorophyll fluorescence and primary photochemistry in chloroplasts by dibromothymoquinone. - Biochim. biophys. Acta *376* : 105 - 115, 1975.

23103 - **KITAJIMA, M., BUTLER, W.L.** : Excitation spectra for Photosystem I and Photosystem II in chloroplasts and the spectral characteristics of the distribution of quanta between the two photosystems. - Biochim. biophys. Acta *408* : 297 - 305, 1975.

23104 - **KJELVIK, S.** : Plantenes primærproduksjon på landjorda, dens størrelse og fordeling på ulike vagetasjonstyper. [Primary plant production on the land, its magnitude and distribution for different types of vegetation.] - Blyttia *33* : 213 - 221, 1975. [In Norweg.]

23105 - **KJELVIK, S., WIELGOLASKI, F.E., JAHREN, A.** : Photosynthesis and respiration of plants studied by field technique at Hardangervidda, Norway. - In : **WIELGOLASKI, F.E.** (ed.) : Fennoscandian Tundra Ecosystems. Part 1. Plants and Microorganisms. Pp. 184 - 193. Springer-Verlag, Berlin - Heidelberg - New York 1975.

23106 - **KLAPWIJK, D., de LINT, P.J.A.L.** : Growth rates of tomato seedlings and seasonal radiation. - Neth. J. agr. Sci. *23* : 259 - 268, 1975. [Growth analysis.]

23107 - **KLARER, D.M., HICKMAN, M.** : The effect of thermal effluent upon the standing crop of an epiphytic algal community. - Int. Rev. ges. Hydrobiol. *60* : 17 - 62, 1975. [Chl.]

23108 - **KLEIN, S., HAREL, E., NE'EMAN, E., KATZ, E., MELLER, E.** : Accumulation of δ-aminolevulinic acid and its relation to chlorophyll synthesis and development of plastid structure in greening leaves. - Plant Physiol. *56* : 486 - 496, 1975.

23109 - **KLEIN, S., MANORI, I., NE'EMAN, E., KATZ, E.** : The effect of cycloheximide and δ-aminolevulinic acid on structural development of chloroplasts. - In : **AVRON, M.** (ed.) : Proceedings of the Third International Congress on Photosynthesis. Vol. III. Pp. 2089 - 2095. Elsevier, Amsterdam - Oxford - New York 1975.

23110 - **KLEIN, S.M., JAYNES, J.M., VERNON, L.P.** : Comparison of the photosynthetic membrane of vegetative cells and heterocysts of the blue-green alga *Anabaena flos-aquae*. - In : **AVRON, M.** (ed.) : Proceedings of the Third International Congress on Photosynthesis. Vol. I. Pp. 703 - 713. Elsevier, Amsterdam - Oxford - New York 1975.

23111 - **KLEIN, S.M., VERNON, L.P.** : The isolation of a chlorophyll *P*700-containing polypeptide from *Anabaena flos-aquae*. - Plant Physiol. *56* (Suppl.) : 31, 1975.

*23112 - **KLEINKOPF, G.E., WALLACE, A.** : Physiological basis for salt tolerance in *Tamarix ramosissima*. - Plant Sci. Lett. *3* : 157 - 163, 1974. [Ps.]

23113 - **KLEINKOPF, G.E., WALLACE, A., CHA, J.W.** : Sodium relationship in desert plants : 4. Some physiological responses of *Atriplex confertifolia* to different levels of sodium chloride. - Soil Sci. *120* : 45 - 48, 1975. [Ps.]

23114 - **KLEPPER, L.A.** : Inhibition of nitrite reduction by photosynthetic inhibitors. - Weed Sci. *23* : 188 - 190, 1975.

23115 - **KLEUDGEN, H.K., LICHTENTHALER, H.K.** : Die Wirkung von Phytochrom auf die Bildung von Einzelcarotinoiden in etiolierten *Hordeum*-Keimlingen. - Z. Naturforsch. *30 C* : 67 - 68, 1975.

23116 - **KLEUDGEN, H.K., LICHTENTHALER, H.K.** : Induction and reversion of prenyl-lipid synthesis in etiolated barley seedlings by red and far-red light. - In : **AVRON, M.** (ed.) : Proceedings of the Third International Congress on Photosynthesis. Vol. III. Pp. 2017 - 2020. Elsevier, Amsterdam - Oxford - New York 1975.

23117 - **KLIMOV, S.V., SHUL'GIN, I.A., NICHIPOROVICH, A.A.** : Ob ėnergeticheskom znachenii orientatsii list'ev u podsolnechnika. [The energetic significance of leaf orientation in sunflower.]-Vest. mosk. Univ., Ser. VI. Biol., Pochvoved. *30* (3) : 64 - 68, 1975. [In R, ab : E.]

23118 - **KLIMOV, V.V., KARAPETYAN, N.V., KRASNOVSKIĬ, A.A.** : Deĭstvie detergenta tritona X-100 na fotoindutsirovannve izmeneniya vykhoda fluorestsentsii khloroplastov. [Effect of detergent Triton X-100 on photoinduced changes in fluorescence yield of chloroplasts.] - Mol. Biol. *9* : 219 - 226, 1975. [In R, ab : E.]

*23119 - **KLIMOV, V.V., KRAKHMALEVA, I.N., KARAPETYAN, N.V., KRASNOVSKIĬ, A.A.** : Priroda ingibiruyushchego deĭstviya detergenta triton X-100 na fotosinteticheskuyu tsep' perenosa ėlektrona khloroplastov i khromatoforov. [Nature of inhibitory action of the detergent Triton X-100 on photosynthetic electron transport chain in chloroplasts and chromatophores.] - In : Itogi Issledovaniya Mekhanizma Fotosinteza. Pp. 70 - 79. Pushchino 1974. [In R.]

23120 - **KLIMOV, V.V., KRAKHMALEVA, I.N., SHUVALOV, V.A., KARAPETYAN, N.V., KRASNOVSKIĬ, A.A.** : Zavisimost' vykhoda fluorestsentsii khloroplastov i khromatoforov ot sostoyaniya reaktsionnykh tsentrov. [Dependence of fluorescence yield of chloroplasts and chromatophores on the state of reaction centres.] - Dokl. Akad. Nauk SSSR *221* : 1207 - 1210, 1975. [In R.]

23121 - **KLUGE, M., BLEY, L., SCHMID, R.** : Malate synthesis in crassulacean acid metabolism (CAM) via a double CO_2 dark fixation ? - In : **MARCELLE, R.** (ed.) : Environmental and Biological Control of Photosynthesis. Pp. 281 - 288. Dr. W. Junk b.v. Publ., The Hague 1975.

23122 - **KNAFF, D.B.** : The effect of *o*-phenanthroline on the midpoint potential of the primary electron acceptor of Photosystem II. - Biochim. biophys. Acta *376* : 583 - 587, 1975.

23123 - **KNAFF, D.B.** : The effect of pH on the midpoint oxidation-reduction potentials of components associated with plant photosystem II. - FEBS Lett. *60* : 331 - 335, 1975.

23124 - **KNAFF, D.B., BUCHANAN, B.B.** : Cytochrome *b* and photosynthetic sulfur bacteria. - Biochim. biophys. Acta *376* : 549 - 560, 1975.

23125 - **KNECHT, G.N.** : Response of radish to high CO_2. - HortScience *10* : 274 - 275, 1975.

23126 - **KNOBLOCH, K.** : Respiratory electron flow and ATPase system in photosynthetically grown *Rhodopseudomonas palustris*. - Z. Naturforsch. *30 C* : 342 - 348, 1975.

23127 - **KNOBLOCH, K.** : Energy-linked pyridine nucleotide transhydrogenase activity in photosynthetically grown *Rhodopseudomonas palustris*. - Z. Naturforsch. *30 C* : 771 - 776, 1975.

23128 - **KNOECHEL, R., KALFF, J.** : Algal sedimentation : the cause of a diatom - blue--green succession. - Verh. int. Verein. Limnol. *19* : 745 - 754, 1975. [Ps.]

23129 - **KNOTH, R.** : Struktur und Funktion der genetischen Information in den Plastiden. XIV. Die Auswirkung der Plastomutationen *en:alba-1* von *Antirrhinum majus* und *en : gilva-1* von *Pelargonium zonale* auf die Feinstruktur der Plastiden. - Biol. Zentralbl. *94* : 681 - 694, 1975.

23130 - **KNOX, R.S.** : Excitation energy transfer and migration : theoretical considerations. - In : **GOVINDJEE** (ed.) : Bioenergetics of Photosynthesis. Pp. 183 - 221. Academic Press, New York - San Francisco - London 1975.

23131 - **KNOX, R.S., GHOSH, V.J.** : Quenching of singlet excitations by triplet excitations. - Photochem. Photobiol. *22* : 149 - 150, 1975. [Chl.]

23132 - **KOBAYASHI, Y., INOUE, Y., SHIBATA, K.** : Light-dependent fluorescence variations in intact leaves. - Plant Cell Physiol. *16* : 767 - 776, 1975.

23133 - **KOBLENTS-MISHKE, O.I., PELEVIN, V.N., SEMENOVA, M.A.** : Pigmenty fitoplanktona i ispol'zovanie solnechnoĭ ėnergii v protsesse fotosinteza. [Phytoplankton pigments and the utilization of solar energy in photosynthesis.] - Tr. Inst. Okeanol. Akad. Nauk SSSR *102* : 104 - 152, 1975. [In R, ab : E.]

23134 - **KOCH, D.W., ESTES, G.O.** : Influence of potassium stress on growth, stomatal behavior, and CO_2 assimilation in corn. - Crop Sci. *15* : 697 - 699, 1975.

23135 - **KOCHEV, H., TRACZYK, T.**: Investigation on the primary production of the ground synusium and plant fall in the *Piceetum excelsae myrtillosum* association. - Fitologiya (Sofia) *1975* (2) : 3 - 11, 1975.

23136 - **KOCHEV, Kh., GORUNOVA, D.** : Izsledvane p"rvichnata produktsiya na trevostoya na asotsîatsiya *Agrostis capillaris* L. + *Festuca fallax* THUILL. na Vitosha planina. [Primary production of an *Agrostis capillaris* L. + *Festuca fallax* THUILL. association grass stand on the Vitosha Mountain.]- Fitologiya (Sofia) *1975* (1) : 7 - 18, 1975. [In Bulg., ab : E, R.]

23137 - **KOCHEV, Kh., GORUNOVA, D.** : Izsledvane kalorichnostta na biomasata na vidove rasteniya v s"stava na nyakoi asotsiatsii v Sofiĭsko. [Biomass caloric contents of plant species from some associations of the Sofia district.] - Fitologiya (Sofia) *1975* (3) : 3 - 12, 1975. [In Bulg., ab:E, R.]

23138 - **KOCHEV, Kh., GORUNOVA, D.** : P"rvichna produktsiya na trevnata pokrivka i rastitelen opad v lazarkinevo-bukova asotsiatsiya (*Fagetum sylvaticae asperulosum*) na Vitosha planina. [Primary production of the grass cover and plant fall off in the *Fagetum sylvaticae asperulosum* association in the Vitosha Mountains.] - In : V Chest' na Akad. Dakî Iordanov. Pp. 161 - 174. Bulg. Akad. Nauk, Sofia 1975. [In Bulg., ab : E.]

23139 - **KOCHUBEĬ, S.M., SHADCHINA, T.M., KONONENKO, A.A., LUKASHEV, E.P., TIMOFEEV, K.N., MATORIN, D.N.** : Effect of galactolipase on photoreactions in light chloroplast fragments. - Photosynthetica *9* : 255 - 260, 1975.

23140 - **KOCHUBEĬ, S.M., SHADCHINA, T.M., OSTROVSKAYA, L.K.** : Action of hydrolytic enzymes on the fluorescence spectra of pigment-lipoprotein complexes of Photosystem I. - Photosynthetica *9* : 391 - 394, 1975.

23141 - **KOCHUBEĬ, S.M., SHADCHINA, T.M., OSTROVSKAYA, L.K.** : O razlichii stroeniya molekulyarnykh agregatov s dlinnovolnovoĭ fluorestsentsieĭ v fotosistemakh I i II. [Differences in structure of molecular aggregates with long-wave fluorescence in photosystems I and II.] - Mol. Biol. (Moskva) *9* : 190 - 193, 1975. [In R, ab : E.]

23142 - **KODAMA, T., IGARASHI, Y., MINODA, Y.** : Isolation and culture conditions of a bacterium grown on hydrogen and carbon dioxide. - Agr. biol. Chem. *39* : 77 - 82, 1975.

23143 - **KODAMA, T., IGARASHI, Y., MINODA, Y.** : Material balance and efficiency of energy conversion for the autotrophic growth of a hydrogen bacterium. - Agr. biol. Chem. *39* : 83 - 87, 1975.

23144 - **KOESTER, V.J., GALLOWAY, L., FONG, F.K.** : Redox properties of the *in vitro* 700-nm absorbing chlorophyll *a*-water dimer. - Naturwissenschaften *62* : 530, 1975.

23145 - **KOH, S., KUMURA, A.** : [Studies on matter production in wheat plant. II. Carbon dioxide balance and efficiency of solar energy utilization in wheat stand.] - Proc. Crop Sci. Soc. Jap. *44* : 335 - 342, 1975. [In Jap., ab : E.]

23146 - **KOHEN, E., HIRSCHBERG, J.G., KOHEN, C., WOUTERS, A., PEARSON, A., SALMON, J.-M., THORELL, B.**: Multichannel microspectrofluorometry for topographic and spectral analysis. - Biochim. biophys. Acta *396* : 149 - 154, 1975. [NAD(P).]

23147 - **KOK, B.** : Prospects of photosynthetic energy production. - In : **MORGENTHALER, G.W., SILVER, A.N.** (ed.) : Energy Delta, Supply vs. Demand. Pp. 519 - 526. American Astronautical Society, Tarzana, Cal. 1975.

23148 - **KOK, B., RADMER, R., FOWLER, C.F.** : Electron transport in photosystem II. - In : **AVRON, M.** (ed.) : Proceedings of the Third International Congress on Photosynthesis. Vol. I. Pp. 485 - 496. Elsevier, Amsterdam - Oxford - New York 1975.

23149 - **KOLBASINA, Ė.I.** : Soderzhanie plastidnykh pigmentov v list'yakh ozimoĭ pshenitsy v osenne-zimne-vesenniĭ period. [Content of plastid pigments in winter wheat leaves in the autumn-winter-spring period.] - Fiziol. Biokhim. kul't. Rast. *7* : 465 - 469, 1975. [In R, ab : E.]

23150 - **KOLESNIKOV, P.A., PETROCHENKO, E.I., ZORE, S.V., PSHENOVA, K.V., MUTUSKIN, A.A., MAKOVKINA, L.E.** : Pogloshchenie kisloroda khloroplastami list'ev gorokha na svetu. [Oxygen uptake by chloroplasts of pea leaves in light.] - Fiziol. Rast. *22* : 34 - 39, 1975. [In R, ab : E.]

23151 - **KOLLER, D.** : Effects of environmental stress on photosynthesis - conclusions. - In : **COOPER, J.P.** (ed.) : Photosynthesis and Productivity in Different Environments. Pp. 587 - 589. Cambridge University Press, Cambridge - London - New York - Melbourne 1975.

23152 - **KOLLER, K.P., WEHRMEYER, W.** : *B*-phycoerythrin from *Rhodella violacea*. Characterization of two isoproteins. - Arch. Microbiol. *104* : 255 - 261, 1975.

23153 - **KOLLMAN, V.H., SHAPIRO, S.L., CAMPILLO, A.J.** : Photosynthetic studies with a 10-psec resolution streak camera. - Biochem. biophys. Res. Commun. *63* : 917 - 923, 1975.

*23154 - **KOLOMEĬCHENKO, V.** : Osobennosti fotosinteza zernobobovykh kul'tur. [Peculiarities of photosynthesis in leguminous crops.]- Zernov. maslich. Kul'tury *5* (1) : 41 - 42, 1970. [In R.]

*23155 - **KOLOMEĬCHENKO, V.V.** : Metodika opredeleniya pokazateleĭ fotosinteticheskoĭ deyatel'nosti v posevakh zernobobovykh kul'tur. [Methods of determining indices of photosynthetic activity in leguminous crops.] - In : Metody Issledovaniĭ s Zernobobovymi Kul'turami. Vol. 2. Pp. 33 - 39. VASKHNIL, Orel 1971. [In R.]

*23156 - **KOLOMEĬCHENKO, V.V.** : Formirovanie i rabota fotosinteticheskogo apparata bobovykh kul'tur pri oroshenii. [Formation and activity of photosynthetic apparatus of irrigated leguminous crops.] - Tr. tul'skoĭ gos. sel'skokhoz. opyt. Sta. *3* : 84 - 97, 1972. [In R.]

*23157 - **KOLOMEĬCHENKO, V.V.** : K metodike izucheniya fotosinteticheskoĭ deyatel'nosti v posevakh sel'skokhozyaĭstvennykh kul'tur. [Methods of studying photosynthetic activity in stands of agricultural crops.] - Tr. tul'skoĭ gos. sel'skokhoz. opyt. Sta. *4* : 198 - 209, 1972. [In R.]

23158 - **KOLOMEĬCHENKO, V.V.** : Uskorennye metody opredeleniya ploshchadi list'ev v polevykh usloviyakh. [Rapid methods of determining leaf area in field conditions.] - In : Intensifikatsiya Zemledeliya. Vol. 36. Pp. 106 - 109. VASKHNIL, Moskva 1975. [In R.]

23159 - **KOLOTENKOV, P.V.** : Vliyanie vysushivaniya i khraneniya semyan na povrezhdenie geneticheskogo apparata kletki. [Effect of drying and storage of seeds on the damage of cell genetic appartus.] - In : Khimicheskie Supermutageny v Selektsii. Pp. 98 - 102. Nauka, Moskva 1975. [Chl mutations; in R.]

23160 - **KOMISSAROV, G.G., SHUMOV, Yu.S.** : Ėksperimental'nye metody izucheniya fotoėlektrokhimicheskikh svoĭstv plenok fotosinteticheskikh pigmentov. [Experimental methods for studying photoelectrochemical properties of layers of photosynthetic pigments.] - In : Metody Issledovaniya Fotokhimicheskikh Reaktsiĭ Fotosinteza *in Vitro* i *in Vivo*. Pp. 137 - 147. Pushchino 1975. [In R.]

23161 - **KONDRAT'EVA, E.N., PETUSHKOVA, Yu.P., ZHUKOV, V.G.** : Rost i okislenie soedineniĭ sery *Thiocapsa roseopersicina* v temnote. [Growth and oxidation of sulphur compounds by *Thiocapsa roseopersicina* in the darkness.] - Mikrobiologiya *44* : 389 - 394, 1975. [Chemolithoautotrophy; in R, ab : E.]

*23162 - **KONOVALENKO, V.V.** : K voprosu izucheniya fotosinteza ozimykh pshenits. [On measuring photosynthesis of winter wheat.] - Sb. nauch. Tr. molod. Uchen. (Krasnodar) *4* : 145 - 150, 1974. [In R.]

23163 - **KONOVALOV, I.N., SAZYKINA, N.A., MUKHINA, V.A.** : O prochnosti khlorofill-belkovo-lipoidnogo kompleksa i aktivnosti fotofosforilirovaniya u rasteniĭ pri adaptatsii. [Stability of the chlorophyll-protein-lipid complex and photophosphorylation activity in plants during adaptation.] - In : Issledovaniya po Fiziologii Rasteniĭ v Zapolyar'e. Pp. 97 - 106. Polyarno-al'piĭskiĭ bot. Sad, Apatity 1975. [In R.]

23164 - **KONUSHEV, S.I.** : Pervichnaya produktsiya i rastvorennoe organicheskoe veshchestvo v zalive Petra Velikogo. [Primary production and dissolved organic matter in the Peter the Great Gulf.] - Tr. tikhookeansk. okeanogr. Inst. *9* (Gidrobiologicheskie Issledovaniya v Yaponskom More i Tikhom Okeane) : 9 - 14, 1975. [In R.]

23165 - **KOPECEK, K., FÜLLER, F., RATZMANN, W., SIMONIS, W.** : Lichtabhängige Insektizidwirkungen auf einzellige Algen. - Ber. deut. bot. Ges. *88* : 269 - 281, 1975. [Ps, Chl.]

*23166 - **KOPPEL, V.** : Radioaktiivse süsiniku assimileerimisest okasõuna lehtedes fotosünteesil. [Assimilation of radioactive carbon in *Datura* leaves during photosynthesis.] - Tartu Riikliku Ülikooli Toimetised [Trans. Tartu State Univ.] *270* : 53 - 62, 1971. [In Eston., ab : R.]

23167 - **KÖRNER, C.** : Wasserhaushalt und Spaltenverhalten alpiner Zwergsträucher. - - Verh. Ges. Ökol. *1975* : 23 - 30, 1975. [Stomatal resistance.]

*23168 - **KORNILOV, A.A.** : Fotosintez i urozhaĭnost' zernobobovykh kul'tur na severnom Kavkaze. [Photosynthesis and cropping capacity of some leguminous plants in North Caucasus.] - Fiziol. Rast. *21* : 1139 - 1144, 1974. [In R, ab : E.]

23169 - **KORNYUSHENKO, G.A.** : Issledovanie geterogennosti karotinoidov v khloroplastakh vysshikh rasteniĭ. [Carotenoids heterogeneity in chloroplasts of higher plants.] - Fiziol. Biokhim. kul't. Rast. *7* : 3 - 12, 1975. [In R, ab : E.]

23170 - **KORZH, B.V.** : Ispol'zovanie korotkikh seriĭ impul'snogo osveshcheniya dlya izucheniya protsessov fotosinteza i dykhaniya zelenykh rasteniĭ.[Use of short series of impulse illumination for studying processes of photosynthesis and respiration in green plants.] - Byull. vses. nauch.-issled. Inst. Rastenievod. N.I. Vavilova *56* : 42 - 53, 1975.[In R.]

23171 - **KORZHEVA, G.F., NILOVSKAYA, N.T.** : O nekotorykh osobennostyakh gazbobmena CO_2 i O_2 u rasteniĭ na svetu. [Some characteristics of CO_2 and O_2 gas exchange in plants in the light.] - Fiziol. Rast. *22* : 712 - 717, 1975. [In R, ab : E.]

23172 - **KOSHKIN, V.A., LIMAR', R.S., GALKIN, V.I.** : Potentsial'naya intensivnost' fotosinteza pshenitsy pri optimal'nykh sochetaniyakh osveshchennosti i temperatury lista. [Potential photosynthetic rate in wheat under optimum combination of illuminanco and leaf temperature.] - Byull. vses. Inst. Rastenievod. N.I. Vavilova *56* : 53 - 60, 1975. [n R.]

23173 - **KOSKE, T.J., SVEC, L.V.** : Some effects of maleic hydrazide on light reactions of photosynthesis in isolated chloroplasts from *Phaseolus vulgaris* plants. - - Can. J. Plant Sci. *55* : 145 - 149, 1975.

*23174 - **KOSMAKOVA, V.E., PROZUMENSHCHIKOVA, L.T., SKRIPCHENKO, A.F.** : Vliyanie molibdena na fiziologo-biokhimicheskie protsessy i urozhaĭ soi na lugovo-burykh--pochvakh Primorskogo kraya. [Effect of molybdenum on physiological and biochemical processes and yield of soybean on meadow brown soils of Primorskiĭ district.] - Uch. Zap. dal'nevost. gos. Univ. (Vladivostok) *17* (Mikroèlementy v Rastenievodstve Dal'nego Vostoka) : 3 - 31, 1968. [Ps; in R.]

23175 - **KOSMAKOVA, V.E., SKRIPCHENKO, A.F., PROZUMENSHCHIKOVA, L.T., NIKONOVA, N.S., BELYAVSKAYA, E.A., ZVEREVA, E.G.** : Reaktsiya rasteniĭ soi na pereuvlazhnenie pochvy. [Reaction of soybeans to overflooding of soil.] - Sel'skokhoz. Biol. *10* : 359 - 363, 1975. [Ps; in R, ab : E.]

*23176 - **KOSMAKOVA, V.E., YAKIMOVA, L.P.** : Ėffektivnost' sovmestnogo vneseniya molibdena i medi pod soyu. [Effectivity of simultaneous supply of molybdenum and copper to soybean.] - Uch. Zap. dal'nevost. gos. Univ. (Vladivostok) *17* (Mikroėlementy v Rastenievodstve Dal'nego Vostoka) : 32 - 39, 1968. [Ps, Chl, Car; in R.]

23177 - **KOSOVEL, V., TALARICO-BISIACCHI, L.** : Note preliminari sulle variazioni stagionali del contenuto in pigmenti idrosolubili e liposolubili nell'apparato fotosintetico di "*Gracillaria verrucosa*" (HUDS.) PAPENFUSS. [Preliminary notes on seasonal variations in photosynthetic pigments of *Gracilaria verrucosa* (HUDS.) PAPENFUSS.] - Inform. bot. ital. *7* (1) : 19 - 21, 1975. [In Ital., ab : E.]

23178 - **KÖST, H.-P., RÜDIGER, W., CHAPMAN, D.J.** : Über die Bindungen zwischen Chromophor und Protein in Biliproteiden, I. Abbauversuche und Spektraluntersuchungen an Biliproteiden. - Liebigs Ann. Chem. *1975* : 1582 - 1593, 1975.

23179 - **KÖST-REYES, E., KÖST, H.-P., RÜDIGER, W.** : Über die Bindungen zwischen Chromophor und Protein in Biliproteiden, II. Nachweis von Cystein als bindende Aminosäure in *B*-Phycoerythrin. - Liebigs Ann. Chem. *1975* : 1594 - 1600, 1975.

23180 - **KOSTYUK, V.S., GRIDNĖVA, N.V., POPOV, A.M., RĖZNYK, M.I.** : Vplyv dolomitovogo pylu na vmist khlorofilu ta intensyvnist' fotosyntezu ї dykhannya v lystkakh dekoratyvnykh roslyn. [Effect of dolomite dust on the chlorophyll content and rates of photosynthesis and respiration in leaves of ornamental plants.] - - Ukr. bot. Zh. *32* : 375 - 377, 1975. [In Ukr., ab : E.]

23181 - **KOTAŃSKA, M.** : Primary productivity in the meadow of the *Hieracio-Nardetum strictae* association in the Gorce mountains (Southern Poland). - Bull. Acad. pol. Sci., Sér. Sci. biol. Cl. II *23* : 623 - 627, 1975.

23182 - **KOTERA, A., SAITO, T., ISO, N., MIZUNO, H., TAKI, N.** : Solution properties of phycocyanin. II. Studies of the molecular shape and size by using the shell model. - Bull. chem. Soc. Jap. *48* : 1176 - 1179, 1975.

23183 - **KOUCHKOVSKY, Y. de** : Study of light-induced pH changes of chloroplast suspensions with non-covalent (fluorescein) and covalent (FITC) fluorescent probes. - In : **AVRON, M.** (ed.) : Proceedings of the Third International Congress on Photosynthesis. Vol. II. Pp. 1013 - 1020. Elsevier, Amsterdam - Oxford - New York 1975.

23184 - **KOUCHKOVSKY, Y. de** : Investigations on the photosynthetic electron and proton transfer reactions, and on chlorophyll fluorescence in chloroplasts and algae with the plastoquinone antagonist dibromothymoquinone (DBMIB). - In : **AVRON, M.** (ed.) : Proceedings of the Third International Congress on Photosynthesis. Vol. II. Pp. 1081 - 1093. Elsevier, Amsterdam - Oxford - New York 1975.

23185 - **KOUCHKOVSKY, Y. de** : Study of the chlorophyll fluorescence in chloroplasts and algae with the plastoquinone antagonist dibromothymoquinone. - Biochim. biophys. Acta *376* : 259 - 267, 1975.

23186 - **KOUCHKOVSKY, Y. de** : Mesure rapide du pH au niveau des membranes photosynthétiques avec une sonde fluorescente. - Physiol. vég. *13* : 182, 1975.

23187 - **KOUCHKOVSKY, Y. de, SIGALAT, C.** : Fractionnement et caractérisation de l'appareil photosynthétique de l'Algue bleue unicellulaire *Anacystis nidulans*. II. Analyse fonctionnelle. - Physiol. vég. *13* : 831 - 851, 1975.

23188 - **KOWALCZEWSKI, A.** : Algal primary production in the zone of submerged vegetation of a eutrophic lake. - Verh. int. Verein. Limnol. *19* : 1305 - 1308, 1975.

23189 - **KOWALCZEWSKI, A.** : Periphyton primary production in the zone of submerged vegetation of Mikołajskie Lake. - Ekol. pol. *23* : 509 - 543, 1975.

23190 - **KOZHOVA, O.M., ZAGORENKO, G.F., MAKSIMOV, V.N.** : Pervichnaya produktsiya pelagiali oz. Khubsugul (MNR). [Primary production of the Lake Khubsugul (Mongolian People's Republic) pelagic zone.] - Gidrobiol. Zh. *11* (3) : 5 - 9, 1975. [In R, ab : E.]

*23191 - **KOZLOV, Yu.N.** : Vliyanie prirody tsentral'nogo atoma metalla na okislitel'no-

-vosstanovitel'nye svoĭstva metallofeofitinov. [Effect of the nature of the central metal atom on the oxidation-reduction properties of metal-containing pheophytins.] - In : Biologicheskiĭ i Nauchno-Tekhnicheskiĭ Progress. Pp. 57-- 60. Pushchino 1974. [In R.]

23192 - **KRAAYENHOF, R., SLATER, E.C.** : Studies of chloroplast energy conservation with electrostatic and convalent fluorophores. - In : **AVRON, M.** (ed.) : Proceedings of the Third International Congress on Photosynthesis. Vol. II. Pp. 985 - 996. Elsevier, Amsterdam - Oxford - New York 1975.

23193 - **KRALJEVIĆ-BALALIĆ, M.** : Nasledivanje sadržaja hlorofila i karotinoida kod nekih genotipova vulgare pšenice. [Inheritance of chlorophyll and carotenoid content in some genotypes of vulgare wheat.] - Savrem. Poljoprivreda *23* (3-4): 29 - 46, 1975. [In Croat., ab : E.]

23194 - **KRÁLOVIČ, J.** : Effect of some insecticides on the photosynthesis of alfalfa. - In : **KOIVISTOINEN, P.** (ed.) : Pesticides. Environmental Quality and Safety. Suppl. Vol. III. Pp. 554 - 556. Thieme, Stuttgart 1975.

23195 - **KRAPF, G., JACOBI, G.** : CO_2-fixation and enzyme activities in isolated chloroplasts and in leaves after dark starvation of spinach plants. - In : **AVRON, M.** (ed.) : Proceedings of the Third International Congress on Photosynthesis. Vol. II. Pp. 1479 - 1487. Elsevier, Amsterdam - Oxford - New York 1975.

23196 - **KRAPF, G., JACOBI, G.** : Dark starvation and plant metabolism. II. CO_2 fixation in isolated chloroplasts. - Planta *124* : 135 - 143, 1975.

23197 - **KRAPF, G., JACOBI, G.** : Dark starvation and plant metabolism. III. CO_2 fixation and the distribution of radioactive intermediates in leaf discs from spinach plants. - Planta *124* : 145 - 152, 1975.

23198 - **KRAPF, G., JACOBI, G.** : The pH dependence of CO_2 fixation and carbon assimilation in isolated spinach chloroplasts. - Plant Sci. Lett. *5* : 67 - 71, 1975.

23199 - **KRASICHKOVA, G.V., LIPKIND, B.I., GILLER, Yu.E.** : O vliyanii ingibitorov transkriptsii i translyatsii na aktivnost' fotosistem khloroplastov gorokha. [Effect of inhibitors of transcription and translation on the activity of photosystems of pea chloroplasts.] - Dokl. Akad. Nauk tadzh. SSR *18* (10) : 53 - 56, 1975. [In R, ab : Tajik.]

23200 - **KRASNOVSKIĬ, A.A.** : Fotoretseptory rastitel'noĭ kletki i puti svetovogo regulirovaniya. [Plant cell photoreceptors and the pathways of light regulation.] - In : Fotoregulyatsiya Metabolizma i Morfogeneza Rasteniĭ. Pp. 5 - 15, 249. Nauka, Moskva 1975. [Ps; in R.]

23201 - **KRASNOVSKIĬ, A.A., BRIN, G.P., NIKANDROV, V.V.** : Vozbuzhdenie svetom vosstanovlennykh piridinnukleotidov : perenos ėlektrona na ferredoksin i metilviologen. [Light stimulation of reduced pyridinenucleotides : the transfer of electron on ferredoxin and methylviologen.] - Dokl. Akad. Nauk SSSR *220* : 1214 - - 1217, 1975. [In R.]

23202 - **KRASNOVSKIĬ, A.A., NIKANDROV, V.V., BRIN, G.P., GOGOTOV, I.N., OSHCHEPKOV, V. P.** : Fotoobrazovanie vodoroda v rastvorakh khlorofilla, NAD-H i khloroplastakh. [Photo-formation of hydrogen in chlorophyll solutions, NAD-H and chloroplasts.] - Dokl. Akad. Nauk SSSR *225* : 711 - 713, 1975. [In R.]

23203 - **KRASNOVSKIĬ, A.A., PUSHKINA, E.M., DROZDOVA, N.N., BUBLICHENKO, N.V., UMRIKHINA, A.V.** : Pervichnye stadii obratimogo fotookisleniya bakteriokhlorofilla *b*. [Primary stages of reversible photooxidation of bacteriochlorophyll *b*.] - Dokl. Akad. Nauk SSSR *221* : 1457 - 1460, 1975. [In R.]

23204 - **KRASNOVSKIĬ, A.A. (ml.)** : Fotokhemilyuminestsentsiya pigmentov v rastvorakh. [Photochemiluminescence of pigments in solutions.] - In : Metody Issledovaniya Fotokhimicheskikh Reaktsiĭ Fotosinteza *in Vitro* i *in Vivo*. Pp. 160 - 172. Pushchino 1975. [In R.]

23205 - **KRASNOVSKIĬ, A.A., ml., LEBEDEV, N.N., LITVIN, F.F.** : Obnaruzhenie tripletnykh sostoyaniĭ khlorofilla i ego predshestvennikov v list'yakh i khloroplastakh po fosforestsentsii i zamedlennoĭ fluorestsentsii pri -196°. [Triplet states of chlorophyll and its precursors in leaves and chloroplasts detected by phosphorescence and delayed fluorescence at -196 °C.] - Dokl. Akad. Nauk SSSR *225* : 207 - 210, 1975. [In R.]

23206 - **KRASNOVSKIĬ, A.A., ml., LITVIN, F.F.** : Mekhanizmy dlitel'nogo poslesvecheniya fotosinteticheskikh pigmentov. [Mechanisms of delayed light emission of photosynthetic pigments.] - Izv. Akad. Nauk SSSR, Ser. fiz. *39* : 1968 - 1971, 1975. [In R.]

23207 - **KRAUSE, C.** : Adenylatmengen und Phosphorylierungsaktivitäten in pflanzlichen Zellen und Geweben - eine Literaturübersicht unter Berücksichtigung methodischer Aspekte. - Photosynthetica *9* : 412 - 453, 1975.

23208 - **KRAUSE, G.H.** : Light-induced movement of metal cations in chloroplasts. Evidence from chlorophyll fluorescence and light scattering. - In : **AVRON, M.** (ed.) : Proceedings of the Third International Congress on Photosynthesis. Vol. II. Pp. 1021 - 1030. Elsevier, Amsterdam - Oxford - New York 1975.

23209 - **KRAUSE, G.H., SANTARIUS, K.A.** : Relative thermostability of the chloroplast envelope. - Planta *127* : 285 - 299, 1975. [Ps, Chl.]

23210 - **KREMER, B.P.** : $^{14}CO_2$-fixation by endosymbiotic alga *Platymonas convolutae* within the Turbellarian *Convoluta roscoffensis*. - Mar. Biol. *31* : 219 - 226, 1975.

23211 - **KREMER, B.P.** : Mannitmetabolismus in der marinen Braunalge *Fucus serratus*. - - Z. Pflanzenphysiol. *74* : 255 - 263, 1975.

23212 - **KREMER, B.P.** : Photosynthetische $^{14}CO_2$-Assimilation durch den Endosymbionten *Platymonas convolutae*. - Naturwissenschaften *62* : 97 - 98, 1975.

23213 - **KREMER, B.P.** : Physiologisch-chemische Charakteristik verschiedener Thallusbereiche von *Fucus serratus*. - Helgoländer wiss. Meeresunters. *27* : 115 - - 127, 1975. [Ps, Chl.]

23214 - **KREMER, B.P., VOGL, R.** : Zur chemotaxonomischen Bedeutung des [^{14}C]-Markierungsmusters bei *Rhodophyceen*. - Phytochemistry *14* : 1309 - 1314, 1975.

23215 - **KRENDELEVA, T.E., KUKARSKIKH, G.P., MATORIN, D.N., TIMOFEEV, K.N., RUBIN, A. B.** : Vliyanie vitamina K_3 na fotosinteticheskiĭ ėlektronnyĭ transport. [Effect of vitamin K_3 on the photosynthetic electron transport.] - Biokhimiya *40* : 57 - 62, 1975. [In R, ab : E.]

23216 - **KRENZER, E.G., Jr., MOSS, D.N.** : Carbon dioxide enrichment effects upon yield and yield components in wheat. - Crop Sci. *15* : 71 - 74, 1975.

23217 - **KRENZER, E.G., Jr., MOSS, D.N., CROOKSTON, R.K.** : Carbon dioxide compensation points of flowering plants. - Plant Physiol. *56* : 194 - 206, 1975.

23218 - **KRIEDEMANN, P.E., LOVEYS, B.R.** : Hormonal influences on stomatal physiology and photosynthesis. - In : **MARCELLE, R.** (ed.) : Environmental and Biological Control of Photosynthesis. Pp. 227 - 236. Dr. W. Junk b.v. Publ., The Hague 1975.

23219 - **KRIEDEMANN, P.E., LOVEYS, B.R., DOWNTON, W.J.S.** : Internal control of stomatal physiology and photosynthesis. II. Photosynthetic responses to phaseic acid. - Aust. J. Plant Physiol. *2* : 553 - 567, 1975.

23220 - **KRISHNAMURTHY, K., RAJASHEKARA, B.G., RAGHUNATHA, G., JAGANNATH, M.K., RAMACHANDRA PRASAD, T.V., VENUGOPAL, N., BOMMEGOWDA, A.** : Comparative growth and yield of sorghum hybrid and its parents. - Mysore J. agr. Sci. *9* : 596 - 601, 1975.[Growth analysis.]

23221 - **KRISTENSEN, K.J., JENSEN, S.E.** : A model for estimating actual evapotranspiration from potential evapotranspiration. - Nord. Hydrol. *6* : 170 - 188, 1975. [Growth analysis.]

23222 - **KRISTKALNE, S., VĪTOLA, Ā., GUBARE, G., KREICBERGS, O.** : Gurku augšana, attīstība un produktivitāte pēcapēnojuma periodā. [Growth, development and productivity of cucumbers in the period after low illuminance.] - In : Tautsaimniecība Derīgo Augu Agrotehnika. Pp. 133 - 141. Zinātne, Rīga 1975. [In Latv., ab : R.]

23223 - **KROGMAN, K.K., HOBBS, E.H.** : Yield and morphological response of rape (*Brassica campestris* L. cv. Span) to irrigation and fertilizer treatments. - Can. J. Plant Sci. *55* : 903 - 909, 1975. [Growth analysis.]

23224 - **KRONESTEDT, E., WALLES, B.** : On the presence of plastids and the eyespot apparatus in a porfiromycin-bleached strain of *Euglena gracilis*. - Protoplasma *84* : 75 - 82, 1975.

23225 - **KROOPNICK, P.M.** : Respiration, photosynthesis, and oxygen isotope fractionation in oceanic surface water. - Limnol. Oceanogr. *20* : 988 - 992, 1975.

23226 - **KRUCHININA, S.S., DASHKEVICH, E.M.**: Dinamika soderzhaniya zelenykh pigmentov v list'yakh kartofelya pod vliyaniem khlorofosa i dilora. [Dynamics of contents of green pigments in potato leaves affected by Chlorophos and Dilore.] - In : Pitanie i Obmen Veshchestv u Rasteniĭ. Pp. 69 - 72. Nauka i Tekhnika, Minsk 1975. [In R.]

23227 - **KRUPA, Z., BASZYNSKI, T.** : Requirement of galactolipids for Photosystem I activity in lyophilized spinach chloroplasts. - Biochim. biophys. Acta *408* : 26 - 34, 1975.

23228 - **KRUPA, Z., KRUPA, D.** : Galaktolipidy chloroplastów. [Chloroplast galactolipids.] - Wiad. bot. *19* : 43 - 58, 1975. [In Pol.]

23229 - **KRZYWICKA, A.M., WAGNER, G.H.** : Laboratory culture for ^{14}C-labeling of *Chlorella* and *Oedogonium*. - Int. J. appl. Rad. Isotop. *26* : 515 - 518, 1975.

23230 - **KSËNZHEK, O.S., APOSTOLOVA, R.D.** : Sravnenie ėnergeticheskikh kharakteristik membrany vodorosli *Acetabularia* na svetu i v temnote. [Energetic characteristics of *Acetabularia* membrane in light and dark.] - Biofizika *20* : 656 - 660, 1975. [In R, ab : E.]

23231 - **KU, S.B., EDWARDS, G.E.** : Photosynthesis in mesophyll protoplasts and bundle sheath cells of various types of C_4 plants. IV. Enzymes of respiratory metabolism and energy utilizing enzymes of photosynthetic pathways. - Z. Pflanzenphysiol. *77* : 16 - 32, 1975.

23232 - **KUBÍČEK, F.** : Primary production and phenology of the herb layer in the oak--hornbeam ecosystem. - In : **BISKUPSKÝ, V.** (ed.) : Research Project Báb IBP Progress Report II. Pp. 117 - 131. Veda, Bratislava 1975.

23233 - **KUBÍČEK, F.** : Príspevok k meraniu listovej plochy *Asperula odorata* L. [Contribution to measuring the leaf area of *Asperula odorata* L.] - Biológia (Bratislava) *30* : 791 - 794, 1975. [In Slovak, ab : E, R.]

23234 - **KUBÍČEK, F.** : The leaf area of the herb layer in the oak-hornbeam forest in Báb. - In : **BISKUPSKÝ, V.** (ed.) : Research Project Báb IBP Progress Report II. Pp. 147 - 156. Veda, Bratislava 1975.

23235 - **KUDO, K.** : Economic yield and climate. - In : **MURATA, Y.** (ed.) : JIBP Synthesis. Vol. 11. Pp. 199 - 220. Univ. Tokyo Press, Tokyo 1975. [Growth analysis.]

23236 - **KÜHBAUCH, W., SÜSS, A., LANG, V.** : Wanderung von ^{14}C-Assimilaten und ^{14}C-Herbiziden in Bärenklaupflanzen (*Heracleum sphondylium*). - Angew. Bot. *49* : 253-- 262, 1975.

23237 - **KÜHBAUCH, W., VOIGTLÄNDER, G.** : Morphologische Entwicklung und Kohlenhydratstoffwechsel von Lieschgrass (*Phleum pratense* L.). - Landwirt. Forsch. *28* : 303 - 309, 1975.

23238 - **KUKUSHKIN, A.K.** : O perenose neĭtral'nogo vozbuzhdeniya ot fotosistemy 2 k fotosisteme 1 v vysshikh rasteniyakh. [Transfer of neutral excitation from photosystem 2 to photosystem 1 in higher plants.] - Biofizika *20* : 159, 1975. [In R, ab : E.]

23239 - **KUKUSHKIN, A.K., TIKHONOV, A.N.** : O nekotorykh voprosakh teoreticheskogo opisaniya svetovykh stadiĭ fotosinteza. [Theoretical description of light-induced processes of photosynthesis.] - Biofizika *20* : 745 - 746, 1975. [In R, ab: E.]

23240 - **KUKUSHKIN, A.K., TIKHONOV, A.N.** : O nekotorykh osobennostyakh khromaticheskikh perekhodov fluorestsentsii vysshikh rasteniĭ. [Some properties of chromatic transitions of higher plants fluorescence.] - Biofizika *20* : 746 - 747, 1975. [In R, ab : E.]

23241 - **KUKUSHKIN, A.K., TIKHONOV, A.N., BLYUMENFEL'D, L.A., RUUGE, E.K.** : Teoreticheskie aspekty kinetiki pervichnykh protsessov fotosinteza vysshikh rasteniĭ i vodorosleĭ.[Theoretical aspects of the kinetics of primary photosynthetic processes in higher plants and algae.] - Fiziol. Rast. *22* : 241 - 250, 1975. [In R, ab : E.]

23242 - **KULANDAIVELU, G., GNANAM, A.** : Effect of growth regulators and herbicides on photosynthetic partial reactions in isolated leaf cells. - Physiol. Plant. *33* : 234 - 240, 1975.

23243 - **KULIKOV, G.V., YAROSLAVTSEVA, Z.P., CHEMARIN, N.G.** : Osobennosti fotosinteza i krakhmalonakopleniya introdutsirovannykh v Krymu vechnozelenykh i listopadnykh derev'ev i kustarnikov. [Peculiarities of photosynthesis and starch accumulation in evergreen and deciduous trees and bushes introduced in Crimea.] - - Bot. Zh. *60* : 482 - 489, 1975. [In R, ab : E.]

23244 - **KULL, U., HOFFMANN, F.** : Einfluß von Zeatin auf Atmung und Photosynthese der Mesophyll-Protoplasten von *Petunia*. - Biol. Plant. *17* : 31 - 37, 1975.

23245 - **KUMAR, A.** : Variety, standing crop and net community productivity of the vegetation on a hard ground and stabilized dunes near Pilani, Rajasthan. - Ann. Arid Zone *14* : 124 - 134, 1975.

23246 - **KUMAR, N.C.** : Eco-phsiological studies of *Dendrophthoe falcata* infection. - - Bull. bot. Soc. Bengal *29* (1) : 33 - 38, 1975.[Chl.]

23247 - **KUMBAR, M., McCOLL, R.** : Effect of aromatic molecules on the aggregation of C-phycocyanin. Quantum chemical calculations on phycocyanobilin and phycoerythrobilin. - Res. Commun. chem. Pathol. Pharmacol. *11* : 627 - 637, 1975.

23248 - **KÜMMEL, H.W., GRIMME, L.H.** : The inhibition of carotenoid biosynthesis in green algae by SANDOZ H 6706 : accumulation of phytoene and phytofluene in *Chlorella fusca*. - Z. Naturforsch. *30c* : 333 - 336, 1975.

23249 - **KÜMMEL, H.W., GRIMME, L.H.** : Zum multiplen Wirkungsspektrum des Versuchsherbizids SAN H 6706. - Mitteil. biol. Bundesanstalt *165* : 178, 1975. [Ps, Chl, Car.]

23250 - **KUMURA, A.** : Dry matter partition and climatic factors. - In : **MURATA, Y.** (ed.) : JIBP Synthesis. Vol. 11. Pp. 49 - 59. Univ. Tokyo Press, Tokyo 1975.

23251 - **KUMURA, A.** : Leaf area development and climate. - In : **MURATA, Y.** (ed.) : JIBP Synthesis. Vol. 11. Pp. 60 - 66. Univ. Tokyo Press, Tokyo 1975.

23252 - **KUMURA, A.** : Comparison of growth characteristics between species. - In : **MURATA, Y.** (ed.) : JIBP Synthesis. Vol. 11. Pp. 221 - 233.Univ. Tokyo Press, Tokyo 1975.

23253 - **KUNERT, K.-J., BÖGER, P.** : Absence of plastocyanin in the alga *Bumilleriopsis* and its replacement by cytochrome 553. - Z. Naturforsch. *30c*: 190 - 200, 1975.

23254 - **KUNG, S.D., GRAY, J.C., WILDMAN, S.G., CARLSON, P.S.** : Polypeptide composition of Fraction 1 protein from parasexual hybrid plants in the genus *Nicotiana*. - Science *187* : 353 - 355, 1975.

23255 - **KUNG, S.D., MARSHO, T.V.** : Oxygenase activity in crystallized fraction 1 protein from tobacco. - Plant Physiol. *56* (Suppl.) : 26, 1975.

23256 - **KUNG, S.D., SAKANO, K., GRAY, J.C., WILDMAN, S.G.** : The evolution of Fraction I protein during the origin of a new species of *Nicotiana*. - J. mol. Evol. *7*: 59 - 64, 1975.

23257 - **KÜNSTLE, E., MITSCHERLICH, G.** : Photosynthese, Transpiration und Atmung in einem Mischbestand im Schwarzwald. I. Teil : Photosynthese. - Allg. Forst-Jagdzeit. *146* (3/4): 45 - 63, 1975.

23258 - **KUPKA, J., TRUONG QUANG TAN** : Fotosyntéza a obsah chlorofylu v ontogenezi kukuřice. [Photosynthesis and chlorophyll content in the ontogenesis of maize.] - Rostlinná Výroba (Praha) *21* : 403 - 408, 1975. [In Czech, ab : E, R.]

23259 - **KURINNYĬ, F.I.** : Sostoyanie pigmentov v list'yakh sakharnoĭ svekly pod vliyaniem bora i margantsa. [State of pigments in sugar beet leaves affected by boron and manganese.] - In : Udobrenie Sakharnoĭ Svekly. Pp. 119 - 122. VNIS, Kiev 1975. [In R.]

23260 - **KURKOVA, E.B.** : Strukturnye izmeneniya khloroplastov v svyazi s izmeneniyami intensivnosti fotosinteza kak rezul'tat obezvozhivaniya lista. [Structural modifications of chloroplasts caused by changes in photosynthetic rate as a result of dehydration of the leaf.] - Fiziol. Rast. *22* : 1121 - 1126, 1975. [In R, ab : E.]

*23261 - **KURUSHIMA, M., TSUKAMOTO, A.** : Oxidation of reduced pyridine nucleotides by molecular oxygen in spinach chloroplasts. - Nat. Sci. Rep. Ochanomizu Univ. *23* (1) : 43 - 47, 1972.

B23262 - **KUSHNIRENKO, M.D.** : Fiziologiya Vodoobmena i Zasukhoustoĭchivosti Plodovykh Rasteniĭ. [Physiology of Water Balance and Drought Resistance of Fruit Trees.] - Shtiintsa, Kishinev 1975. [Chl; in R.]

23263 - **KUSHNIRENKO, M.D., KRYUKOVA, E.V., PECHERSKAYA, S.N., KANASH, E.V., MEDVEDEVA, T.N.** : Rol' khloroplastov v vodnom obmene i ustoĭchivosti rasteniĭ k zasukhe. [Role of of chloroplasts in water balance and drought resistance of plants.] - In : Vodoobmen Rasteniĭ pri Neblagopriyatnykh Usloviyakh Sredy. Pp. 43 - 50. Shtiintsa, Kishinev 1975. [In R.]

B23264 - **KUSHNIRENKO, M.D., KURCHATOVA, G.P., KRYUKOVA, E.V.** : Metody Otsenki Zasukhoustoĭchivosti Plodovykh Rasteniĭ. [Methods of Determining Drought Resistance in Fruit Trees.] - Shtiintsa, Kishinev 1975. [Chl; in R.]

23265 - **KUTZELNIGG, H., MEYER, B., SCHÖTZ, F.** : Untersuchungen an Plastom-Mutanten von *Oenothera*. II. Überblick über die Ultrastruktur der mutierten Plastiden. - - Biol. Zentralbl. *94* : 513 - 526, 1975.

23266 - **KUTZELNIGG, H., MEYER, B., SCHÖTZ, F.** : Untersuchungen an Plastom-Mutanten von *Oenothera*. III. Vergleichende ultrastrukturelle Charakterisierung der Mutanten. - Biol. Zentralbl. *94* : 527 - 538, 1975.

23267 - **KUZ'MENKO, L.V.** : Fitoplankton i pervichnaya produktsiya Araviĭskogo morya v period zimnego mussona. [Phytoplankton and primary production of Arabian Sea during the winter monsoon.] - In : Biologicheskie Issledovaniya v Tropicheskoĭ Zone Okeana. Pp. 23 - 30. Naukova Dumka, Kiev 1975. [In R.]

23268 - **KUZ'MENKO, L.V.** : Sootnoshenie mezhdu pervichnoĭ produktsieĭ i biomassoĭ fitoplanktona v Araviĭskom more. [Relation of primary production and biomass of phytoplankton in the Arabian Sea.] - Ėkologiya *1975* (5) : 43 - 48, 1975. [In R.]

23269 - **KUZ'MINA, A.I.** : Soderzhanie pigmentov v planktone tropikov zapadnoĭ Patsifiki. [Pigment content in plankton in the tropics of the Western Pacific Ocean.] - Tr. tikhookean. okeanolog. Inst. *9* (Gidrobiologicheskie Issledovaniya v Yaponskom More i Tikhom Okeane): 28 - 46, 1975. [In R.]

23270 - **KUZ'MINA, A.I.** : Soderzhanie khlorofilla "*a*" v planktone pribrezhnykh vod Avstralii. [Chlorophyll *a* level in plankton of the coastal waters of Australia.] - Tr. tikhookean. okeanolog. Inst. *9* (Gidrobiologicheskie Issledovaniya v Yaponskom More i Tikhom Okeane): 47 - 55, 1975. [In R.]

23271 - **KVĚT, J.** : Growth and mineral nutrients in shoots of *Typha latifolia* L. - In : Limnology of Shallow Waters. (Symp. Biol. Hung.). Pp. 113 - 123. Akad. Kiadó, Budapest 1975. [Biomass.]

23272 - **KVITKO, K.V., TUGARINOV, V.V., HO, P.T., CHUNAEV, A.S., TEMPER, E.E., MUKHAMADIEV, B.T.** : Mutation analysis as a method of studying genotype structure of green algae. - In : **NASYROV, Yu.S., ŠESTÁK, Z.** (ed.) : Genetic Aspects of Photosynthesis. Pp. 225 - 236. Dr. W. Junk B.V. Publ., The Hague 1975.

23273 - **KYLIN, A., OKKEH, A., SUNDBERG, I., TILLBERG, J.-E.** : Number of photophosphorylative sites, regulation of their activity, and differences in their performance. - In : **AVRON, M.** (ed.) : Proceedings of the Third International Congress on Photosynthesis. Vol. II. Pp. 1047 - 1054. Elsevier, Amsterdam - Oxford - New York 1975.

23274 - **KYLIN, A., QUATRANO, R.S.** : Metabolic and biochemical aspects of salt tolerance. - In : **POLJAKOFF-MAYBER, A., GALE, J.** (ed.) : Plants in Saline Environments. Pp. 147 - 167. Springer-Verlag, Berlin - Heidelberg - New York 1975. [Ps.]

23275 - **KYVASK, V., MILIUS, A., PORK, M.** : Sezonnye izmeneniya biomassy i soderzhaniya khlorofilla "*a*" fitoplanktona v ėvtrofnykh ozerakh Ėstonii. [Seasonal changes in biomass and chlorophyll *a* content of phytoplankton in eutrophic lakes of Estonia.] - In : Osnovy Bioproduktivnosti Vnutrennikh Vodoemov Pribaltiki. Pp. 132 - 133. Vil'nyus 1975. [In R.]

23276 - **LACH, H.-J., BÖGER, P.** : Solubilisierung von "high potential" Cytochrom *b*-559 aus Spinat-Chloroplasten. - Z. Naturforsch. *30c* : 628 - 633, 1975.

23277 - **LaCROIX, L.J., LIER, J.B.** : Carotenoids in *durum* wheat: developmental patterns during two growing seasons. - Can. J. Plant Sci. *55* : 679 - 684, 1975.

23278 - **LADYGIN, V.G.** : Ob otsenke produktivnosti pigmentnykh mutantov odnokletochnykh zelenykh vodorosleĭ. [Assessment of productivity of pigment mutants in unicellular green algae.] - Fiziol. Rast. *22* : 317 - 323, 1975. [In R, ab : E.]

23279 - **LADYGIN, V.G., SEMENOVA, G.A., TAGEEVA, S.V.** : Nepreryvnost' khloroplasta *Chlamydomonas reinhardi* v techenie zhiznennogo tsikla. II. Formirovanie khloroplasta zigoty i plastid zoospor v protsesse polovogo razmnozheniya. [Continuity of chloroplast of *Chlamydomonas reinhardi* during life cycle. II. Formation of zygote chloroplast and zoospore plastids during sexual reproduction.] - Tsitologiya *7* : 115 - 121, 1975. [In R, ab : E.]

23280 - **LADYGIN, V.G., SEMYONOVA, G.A., TAGEEVA, S.V.** : Electron-microscopic study of chloroplasts in zygotes of *Chlamydomonas reinhardi*. - In : **NASYROV, Yu.S., ŠESTAK, Z.** (ed.) : Genetic Aspects of Photosynthesis. Pp. 43 - 48. Dr. W. Junk B.V. Publ., The Hague 1975.

23281 - **LAGOUTTE, B., DURANTON, J.** : A manganese protein complex within the chloroplast structures. - FEBS Lett. *51* : 21 - 24, 1975.

23282 - **LAING, W.A., OGREN, W.L., HAGEMAN, R.H.** : Bicarbonate stabilization and activation of soybean RuDP carboxylase. - In : **AVRON, M.** (ed.) : Proceedings of the Third International Congress on Photosynthesis. Vol. II. Pp. 1407 - 1410. Elsevier, Amsterdam - Oxford - New York 1975.

23283 - **LAING, W.A., OGREN, W.L., HAGEMAN, R.H.** : Bicarbonate stabilization of ribulose 1,5-diphosphate carboxylase. - Biochemistry *14* : 2269 - 2275, 1975.

23284 - **LAIR, N.** : Sur la production des Copépodes dans deux lacs du Massif Central français. - Verh. int. Verein. Limnol. *19* : 3204 - 3211, 1975. [Chl.]

23285 - **LAIR, N., CHARUAU, M.P., BERTHON, J.L.** : Repartition spatio-temporelle diurne des Crustaces planctoniques dans les lacs de Tazenat et Pavin (Massif Central francais).- Ann. Sta. biol. Besse-en-Chandesse *1974-1975* (9) : 25 - 58, 1974-1975. [Chl.]

23286 - **LAǏSK, A., OYA, V.** : Aktivnost' ribulozodifosfatkarboksilazy v fotosinteziruyushchikh list'yakh osiny pri razlichnoǐ intensivnosti sveta. [Ribulose bisphosphate carboxylase activity in photosynthesizing aspen leaves at varying illuminances.] - Izv. Akad. Nauk ėst.SSR, Biol. *24* (1): 10 - 17, 1975. [In R, ab : E, Est.]

23287 - **LAKHANOV, A.P.** : Rol' temperaturnogo faktora v ontogeneze rasteniǐ fasoli. [Role of temperature in ontogenesis of *Phaseolus* plants.] - Fiziol. Rast. *22* : 1001 - 1006, 1975. [Chl; in R, ab : E.]

23288 - **LAKHANOV, A.P.** : Vozmozhnost' diagnostirovaniya ustoǐchivosti zernobobovykh kul'tur k nizkim polozhitel'nym temperaturam po fiziologicheskim pokazatelyam. [Possibility of diagnosing resistance of pulses to low positive temperatures from physiological indexes.] - Fiziol. Biokhim. kul't. Rast. *7* : 517 - 521, 1975. [Chl; in R, ab : E.]

23289 - **LAKSO, A.N., KLIEWER, W.M.**: The influence of temperature on malic acid metabolism in grape berries. I. Enzyme responses. - Plant Physiol. *56* : 370 - 372, 1975.

23290 - **LAMOLA, A.A., MANION, M.L., ROTH, H.D., TOLLIN, G.** : Photooxidation of chlorins by quinones studied by nuclear magnetic resonance techniques. - Proc. nat. Acad. Sci. USA *72* : 3265 - 3269, 1975.

23291 - **LANCER, H.A., COHEN, C.E., SCHIFF, J.A.** : Changing ratios of phototransformable protochlorophyll (Pchlide) in leaves of bean seedlings developing in the dark. - Plant Physiol. *56* (Suppl.) : 33, 1975.

23292 - **LAND, E.J.** : Molecular aspects of biological energy and electron transport : photosynthesis. - In : **ADAMS, G.E., FELDER, E.M., MICHAEL, B.D.** (ed.) : Fast Processes in Radiation Chemistry and Biology. Pp. 135 - 146. John Wiley & Sons, London 1975.

23293 - **LAND, E.J.** : Photochemistry of polyenes. - Photochem. Photobiol. *22* : 286 - - 288, 1975. [Car.]

23294 - **LAND, L.S., LANG, J.C., SMITH, B.N.** : Preliminary observations on the carbon isotopic composition of some reef coral tissues and symbiotic zooxanthellae. - Limnol. Oceanogr. *20* : 283 - 287, 1975. [Ps.]

23295 - **LANDSBERG, J.J., BEADLE, C.L., BISCOE, P.V., BUTLER, D.R., DAVIDSON, B., INCOLL, L.D., JAMES, G.B., JARVIS, P.G., MARTIN, P.J., NEILSON, R.E., POWELL, D.B.B., SLACK, E.M., THORPE, M.R., TURNER, N.C., WARRIT, B., WATTS, W.R.** : Diurnal energy, water and CO_2 exchanges in an apple (*Malus pumila*) orchard. - - J. appl. Ecol. *12* : 659 - 684, 1975.

23296 - **LANGE, O.L.** : Plant water relations. - Progress in Botany *37* : 78 - 97, 1975. [Ps.]

23297 - **LANGE, O.L., SCHULZE, E.-D., EVENARI, M., KAPPEN, L., BUSCHBOM, U.** : The temperature-related photosynthetic capacity of plants under desert conditions II. Possible controlling mechanisms for the seasonal changes of the photosynthetic response to temperature. - Oecologia *18* : 45 - 53, 1975.

23298 - **LANGE, O.L., SCHULZE, E.-D., KAPPEN, L., BUSCHBOM, U., EVENARI, M.** : Adaptations of desert lichens to drought and extreme temperatures. - In : **HADLEY, N.F.** (ed.) : Environmental Physiology of Desert Organisms. Pp. 20 - 37. Dowden, Hutchinson and Ross, Inc., Stroudsburg 1975. [Ps.]

23299 - **LANGE, O.L., SCHULZE, E.-D., KAPPEN, L., BUSCHBOM, U., EVENARI, M.** : Photosynthesis of desert plants as influenced by internal and external factors. - - In : **GATES, D.M., SCHMERL, R.B.** (ed.) : Perspectives of Biophysical Ecology. Pp. 121 - 143. Springer-Verlag, Berlin - Heidelberg - New York 1975.

23300 - **LANGE, O.L., SCHULZE, E.-D., KAPPEN, L., EVENARI, M., BUSCHBOM, U.** : CO_2 exchange pattern under natural conditions of *Caralluma negevensis*, a CAM plant of the Negev desert. - Photosynthetica *9* : 318 - 326, 1975.

23301 - **LÄNNERGREN, C., SKJOLDAL, H.R.** : The spring phytoplankton bloom in Lindåspollene, a land-locked Norwegian fjord. Autotrophic and heterotrophic activities in relation to nutrients. - In : 10th European Symposium on Marine Biology. Vol. 2. Pp. 363 - 391. Ostend 1975. [Chl and productivity.]

B23302 - **LARCHER, W.** : Physiological Plant Ecology. - Springer-Verlag, Berlin - Heidelberg - New York 1975. [Ps, dry-matter production, Chl, Car.]

23303 - **LARCHER, W.** : Produktionsökologie alpiner Zwergstrauchbestände auf dem Patscherkofel bei Innsbruck. - Verhandl. Ges. Ökol. (Wien) *1975* : 3 - 7, 1975. [Dry-matter production.]

23304 - **LARCHER, W., CERNUSCA, A., SCHMIDT, L., GRABHERR, G., NÖTZEL, E., SMEETS, N.** : Mt. Patscherkofel, Austria. - Ecol. Bull. (Stockholm) *20* (**ROSSWALL, T., HEAL, O.W.** (ed.) : Structure and Function of Tundra Ecosystems) : 125 - 139, 1975. [Ps, production.]

23305 - **LARSON, D.W., KERSHAW, K.A.** : Acclimation in arctic lichens. - Nature *254* : 421 - 423, 1975. [Growth analysis.]

23306 - **LARSON, D.W., KERSHAW, K.A.** : Measurement of CO_2 exchange in lichens : a new method. - Can. J. Bot. *53* : 1535 - 1541, 1975.

23307 - **LARSON, D.W., KERSHAW, K.A.** : Studies on lichen-dominated systems. XIII. Seasonal and geographical variation of net CO_2 exchange of *Alectoria ochroleuca*. - - Can. J. Bot. *53* : 2598 - 2607, 1975.

23308 - **LARSON, D.W., KERSHAW, K.A.** : Studies on lichen-dominated systems. XVI. Comparative patterns of net CO_2 exchange in *Cetraria nivalis* and *Alectoria ochroleuca* collected from a raised-beach ridge. - Can. J. Bot. *53* : 2884 - 2892, 1975.

23309 - **LARSSON, C.** : Photosynthetic glycolate formation *via* phosphoglycolate in isolated spinach chloroplasts. - In : **AVRON, M.** (ed.) : Proceedings of the Third International Congress on Photosynthesis. Vol. II. Pp. 1321 - 1328. Elsevier, Amsterdam - Oxford - New York 1975.

23310 - **LARSSON, C., ALBERTSSON, P.Å.** : Photosynthetic $^{14}CO_2$ fixation by chloroplasts and chloroplast-containing subcellular particles. - In : **AVRON, M.** (ed.) : Proceedings of the Third International Congress on Photosynthesis. Vol. II. Pp. 1489 - 1498. Elsevier, Amsterdam - Oxford - New York 1975.

23311 - **LARSSON, C.-M., TILLBERG, J.-E.** : Effects of the commercial polychlorinated biphenyl mixture Aroclor 1242 on growth, viability, phosphate uptake, respiration and oxygen evolution in *Scenedesmus*. - Physiol. Plant. *33* : 256 - 260, 1975.

23312 - **LASCELLES, J.** : The regulation of heme and chlorophyll synthesis in bacteria. - Ann. New York Acad. Sci. *244* : 334 - 347, 1975.

23313 - **LASHCHINSKIĬ, N.N.** : O vliyanii derev'ev na strukturu travostoya v travyanykh borakh nizhnego Priangar'ya. [Effect of trees on the structure of grass stand in grass pine forests in lower reaches of Angara.] - Bot. Zh. *60* : 1721 - - 1727, 1975. [In R.]

23314 - **LASKER, R.** : Field criteria for survival of anchovy larvae : The relation between inshore chlorophyll maximum layers and successfull first feeding. - - Fish. Bull. *73* : 453 - 462, 1975.

23315 - **LASSIG, J., NIEMI, Å.** : Parameters of production in the Baltic measured during cruises with R/V Aranda in June and July 1970 and 1971. - Merentutkimuslait. Julk./Havsforskningsinst. Skr. *239* : 34 - 40, 1975. [Chl.]

23316 - **LASSITER, R.R.** : The effect of higher trophic level components in an aquatic ecosystem model. - In : Proceedings : Biostimulation - Nutrient Assessment Workshop. EPA-660/3-75-034. Pp. 87 - 112. Nat. environm. Res. Center, Corvallis, Oregon 1975. [Chl.]

23317 - **LATAŁA, A.** : Investigations of the changes in the trans-membrane potential of the leaf cell of the moss *Funaria hygrometrica*, under the influence of light. - Bull. Acad. pol. Sci., Sér. Sci. biol. *23* : 717 - 723, 1975. [Ps.]

23318 - **LAUDI, G., MANZINI, M.L.** : Chlorophyll content and plastid ultrastructure in leaflets of *Metasequoia glyptostroboides*. - Protoplasma *84* : 185 - 190, 1975.

23319 - **LAUDI, G., MEDEGHINI-BONATTI, P.** : Ultrastructures of plastids in embryos of *Picea excelsa* L. and *Larix decidua* L. during germination in darkness. - Caryologia *28* : 133 - 147, 1975.

*23320 - **LÄUGER, P., POHL, G.W., STEINEMANN, A., TRISSL, H.-W.** : Artificial lipid membranes as possible tools for the study of elementary photosynthetic reactions. - In : **ESTRADA-O., S., GITLER, C.** (ed.) : Perspectives in Membrane Biology. Pp. 645 - 659. Academic Press, New York - London 1974.

23321 - **LAURIÈRE, C., SKAKOUN, A., DAUSSANT, J.** : Adaptation de méthodes immunochimiques à l'étude de l'évolution ontogénique de protéines : fraction I protéique et malate déshydrogénase des feuilles de Blé en voie de sénescence. - Physiol. vég. *13* : 467 - 478, 1975.

23322 - **LAURITIS, J.A., VIGIL, E.L., SHERMAN, L., SWIFT, H.** : Photosynthetically-linked oxidation of diaminobenzidine in blue-green algae. - J. Ultrastruct. Res. *53* : 331 - 344, 1975.

23323 - **LAVAL-MARTIN, D.** : Chloroplasts of cherry tomato fruits and leaves : Physiological differences and lamellar characteristics. - Plant Physiol. *56* (Suppl.): 87, 1975.

23324 - **LAVAL-MARTIN, D., QUENNEMET, J., MONEGER, R.** : Pigment evolution in *Lycopersicon esculentum* fruits during growth and ripening. - Phytochemistry *14* : 2357 - - 2362, 1975.

23325 - **LAVAL-MARTIN, D., QUENNEMET, J., MONEGER, R.** : Remarques sur l'évolution lipochromique et ultrastructurale des plastes durant la maturation du fruit de tomate "cerise". - Colloq. int. CNRS *238* (Facteurs et Régulation de la Maturation des Fruits) : 347 - 354, 1975.

23326 - **LAVOREL, J.** : Fast and slow phases of luminescence in *Chlorella*. - Photochem. Photobiol. *21* : 331 - 343, 1975.

23327 - **LAVOREL, J.** : Luminescence. - In : **GOVINDJEE** (ed.) : Bioenergetics of Photosynthesis. Pp. 223 - 317. Academic Press, New York - San Francisco - London 1975.

23328 - **LAVOREL, J.** : On the system II recombination reaction. - In : **AVRON, M.** (ed.): Proceedings of the Third International Congress on Photosynthesis. Vol. I. Pp. 145 - 148. Elsevier, Amsterdam - Oxford - New York 1975.

23329 - **LAWLOR, D.W., FOCK, H.** : Photosynthesis and photorespiratory CO_2-evolution of water-stressed sunflower leaves. - Planta *126* : 247 - 258, 1975.

23330 - **LAWLOR, D.W., MILFORD, G.F.J.** : The control of water and carbon dioxide flux in water-stressed sugar beet. - J. exp. Bot. *26* : 657 - 665, 1975.

23331 - **LAWRENCE, T.** : Inheritance if a temperature-sensitive yellow foliage character in Russian wild ryegrass. - Can. J. Plant Sci. *55* : 709 - 710, 1975.

23332 - **LAWRIE, A.C., WHEELER, C.T.** : Nitrogen fixation in the root nodules of *Vicia faba* L. in relation to the assimilation of carbon. I. Plant growth and metabolism of photosynthetic assimilates. - New Phytol. *74* : 429 - 436, 1975.

23333 - **LAWRIE, A.C., WHEELER, C.T.** : Nitrogen fixation in the root nodules of *Vicia faba* L. in relation to the assimilation of carbon. II. The dark fixation of carbon dioxide. - New Phytol. *74* : 437 - 445, 1975.

23334 - **LAWS, E.A.** : The importance of respiration losses in controlling the size distribution of marine phytoplankton. - Ecology *56* : 419 - 426, 1975. [Ps.]

23335 - **LAYCOCK, M.V.** : The amino acid sequence of cytochrome *f* from the brown alga *Alaria esculenta* (L.) GREV. - Biochem. J. *149* : 271 - 279, 1975.

23336 - **LAZARO, J.J., CHUECA, A., LOPEZ GORGE, J., MAYOR, F.** : Fructose-1,6-diphosphatase from spinach leaf chloroplasts : molecular weight transitions of the purified enzyme. - Plant Sci. Lett. *5* : 49 - 55, 1975.

23337 - **LÁZARO, J.J., CHUECA, A., LÓPEZ GORGÉ, J., MAYOR, F.** : Properties of spinach chloroplast fructose-1,6-diphosphatase. - Phytochemistry *14* : 2579 - 2583, 1975.

23338 - **LAZENBY, A., LOVETT, J.V.** : Growth of pasture species on the Northern Tablelands of New South Wales. - Aust. J. agr. Res. *26* : 269 - 280, 1975.[Modelling growth.]

23339 - **LEBEDEV, S.I., BARANSKIĬ, P.I., LITVINENKO, L.G., SHIYAN, L.T.** : Fiziologo--biokhimicheskie osobennosti rasteniĭ posle predposevnogo vozdeĭstviya postoyannym magnithym polem. [Physiological and biochemical characteristics of plants after presowing treatment with stationary magnetic field.] - Fiziol. Rast. *22* : 103 - 109, 1975. [Ps, Chl; In R, ab : E.]

23340 - **LEBEDEV, S.I., CHEPELEV, V.V., ALEĬNIKOV, I.M.** : Izmenenie fotokhimicheskikh svoĭstv khlorofilla *a* pri vzaimodeĭstvii s karotinoidami. [Change in photochemical properties of chlorophyll *a* under interaction with carotenoids.] - - Fiziol. Biokhim. kul't. Rast. *7* : 356 - 359, 444, 1975. [In R, ab : E.]

*23341 - **LEBEDEV, V.M.** : Vliyanie urovnya fosfornogo pitaniya na intensivnost' i ėnergeticheskuyu ėffektivnost' fotosinteza yabloni. [Effect of phosphorus supply on the rate and energetic efficiency of apple photosynthesis.] - Sb. nauch. Rabot vsesoyuz. nauchno-issled. Inst. Sadovod. (Michurinsk) *14* : 171 - 178, 1970. [In R.]

23342 - **LeBLANC, F., RAO, D.N.** : Effects of air pollutants on lichens and bryophytes. - In : **MUDD, J.B., KOZLOWSKI, T.T.** (ed.) : Responses of Plants to Air Pollution. Pp. 237 - 272. Academic Press, New York - San Francisco - London 1975. [Chl.]

23343 - **LEBLOVÁ, S., MAREŠ, J.** : Thermally stable phosphoenolpyruvate carboxylase from pea, tobacco and maize green leaves. - Photosynthetica *9* : 177 - 184, 1975.

23344 - **LECLERC, J.-C., HOARAU, J.** : A comparative study of *Porphyridium* Chl *a* bands and Chl *a* bands *in vitro*. - In : **AVRON, M.** (ed.) : Proceedings of the Third International Congress on Photosynthesis. Vol. I. Pp. 235 - 243. Elsevier, Amsterdam - Oxford - New York 1975.

23345 - **LECLERC, J.C., HOARAU, J., GUÉRIN-DUMARTRAIT, E.** : An analysis of *Porphyridium* absorption bands with a digital spectrophotometer. - Photochem. Photobiol. *22* : 41 - 48, 1975.

23346 - **LEDENT, J.F.** : Leaf azimuth in wheat canopies. - Cereal Res. Commun. *3* : 279-- 287, 1975.

23347 - **LEE, A.G.** : A photosynthetic structure. - Nature *258* : 568, 1975.

23348 - **LEE, A.G.** : Segregation of chlorophyll *a* incorporated into lipid bilayers. -- Biochemistry *14* : 4397 - 4403, 1975.

23349 - **LEE, A.G.** : Fluorescence studies of chlorophyll *a* incorporated into lipid mixtures, and their interpretation of "phase" diagrams. - Biochim. biophys. Acta *413* : 11 - 23, 1975.

23350 - **LEE, L.P., HECHT, A.** : Chloroplasts of monoploid and diploid *Oenothera hookeri*. - Amer. J. Bot. *62* : 268 - 272, 1975.

23351 - **LEE, S.S.** : Effect of light intensity on photoreaction in mesophyll and bundle sheath chloroplasts isolated from corn leaves (*Zea may* L.). - Plant Physiol. *56* (Suppl.) : 46, 1975.

23352 - **LEE, S.Y., AMEMIYA, A., TANAKA, I.** : [The influence of low temperature on the photosynthesis of *Japonica-Indica* rice hybrid and its parents.] - Proc. Crop Sci. Soc. Jap. *44* : 370 - 371, 1975. [In Jap.]

23353 - **LEECH, R.M.** : Plastid development in isolated etiochloroplasts and isolated etioplasts. - In : **SUNDERLAND, N.** (ed.) : Perspectives in Experimental Biology. Vol. 2. Botany. Pp. 145 - 162. Pergamon Press, Oxford - New York - Toronto - Sydney - Paris - Braunschweig 1975.

23354 - **LEFORT-TRAN, M.** : Mitochondries et chloroplastes chez *Euglena* en culture synchrone. - In : Les Cycles Cellulaires et Leur Blocage chez Plusieurs Protistes. Coll. int. CNRS 240. Pp. 297 - 308. Édit. CNRS, Paris 1975.

23355 - **LEGENCHENKO, B.I., MIKUL'SKAYA, S.A., TALANOVA, K.S., LAGUN, L.P.** : O vliyanii impul'snogo dozhdevaniya na fiziologiyu i produktivnost' ovsa. [Effect of impulse irrigation on physiology and productivity of oats.] - In : Fiziologo--Biokhimicheskie Aspekty Rosta i Razvitiya Rasteniĭ. Pp. 156 - 163, 175. Nauka i Tekhnika, Minsk 1975. [In R.]

23356 - **LEGOCKA, J., SZWEYKOWSKA, A.** : Effect of kinetin on nucleic acid synthesis in senescing and young leaves in *Brassica oleracea* L. var. *gongylodes* L. - Acta Soc. Bot. Pol. *44* : 553 - 565, 1975. [Chl.]

23357 - **LEGOCKA, J., WOŹNY, A., SZWEYKOWSKA, A.** : Ultrastructural study of chloroplasts isolated by various fractionation methods. - Acta Soc. Bot. Pol. *44* : 443 -- 448, 1975.

23358 - **LEHOCZKI, E.** : New chlorophyll-*b* forms in a chlorophyll-detergent photosynthetic model system. - Biochim. biophys. Acta *408* : 223 - 227, 1975.

23359 - **LEHOCZKI, E., CSATORDAY, K.** : Energy transfer by chlorophyll *b* in detergent micelles. - Biochim. biophys. Acta *396* : 86 - 92, 1975.

23360 - **LEICKNAM, J.P., HENRY, M., PLUS, R., GILET, R., KLEO, J.** : Infra-red spectrometry of isolated chlorophylls : Association constants and isotopic effects. - In : **AVRON, M.** (ed.) : Proceedings of the Third International Congress on Photosynthesis. Vol. I. Pp. 205 - 222. Elsevier, Amsterdam - Oxford - New York 1975.

*23361 - **LEICKNAM, J.-P., HENRY, M., ROUX, E.** : Examen par spectrométrie infrarouge de complexes chlorophylle "*a*"-phosphates. Influence de la lumière. - Compt. rend. Acad. Sci. Paris, Sér. B *278* : 467 - 470, 1974.

23362 - **LEISER, M., GROMET-ELHANAN, Z.** : Effect of an artificially induced diffusion potential on postillumination and acid-base phosphorylation in *Rhodospirillum rubrum* chromatophores. - In : **AVRON, M.** (ed.) : Proceedings of the Third International Congress on Photosynthesis. Vol. II. Pp. 941 - 949. Elsevier, Amsterdam - Oxford - New York 1975.

23363 - **LEISER, M., GROMET-ELHANAN, Z.** : Postillumination adenosine triphosphate synthesis in *Rhodospirillum rubrum* chromatophores. I. Conditions for maximal yields. - J. biol. Chem. *250* : 84 - 89, 1975.

23364 - **LEKHOTSKI, É., CHATORDAI, K., SALAI, L., SABAD, Ya.** : Peredacha énergii vozbuzhdeniya mezhdu molekulami khlorofilla *a* v rastvorakh detergenta. [Transfer of excitation energy between molecules of chlorophyll *a* in detergent solutions.] - Biofizika *20* : 44 - 45, 1975. [In R, ab : E.]

23365 - **LEMASSON, C., ETIENNE, A.L.** : Photo-inactivation of System II centers by carbonyl cyanide *m*-chlorophenylhydrazone in *Chlorella pyrenoidosa*. - Biochim. biophys. Acta *408* : 135 - 142, 1975.

23366 - **LEMBI, C.A.** : The fine structure of the flagellar apparatus of *Carteria*. - J. Phycol. *11* : 1 - 9, 1975. [Chloroplast.]

23367 - **LEMEUR, R., ROSENBERG, N.J.** : Reflectant induced modification of soybean canopy radiation balance. II. A quantitative and qualitative analysis of radiation reflected from a green soybean canopy. - Agron. J. *67* : 301 - 306, 1975.

23368 - **LEMEUR, R., ROSENBERG, N.J.** : Reflectant induced modification of the radiation balance for increased crop water use efficiency. - In : **de VRIES, D.A., AFGAN, N.H.** (ed.) : Heat and Mass Transfer in the Biosphere. Part I. Transfer Processes in the Plant Environment. Pp. 479 - 488. Scripta Book Comp., J. Wiley and Sons, Washington, D.C. 1975.

*23369 - **LEMOINE, Y.** : Étude de l'ultrastructure des chloroplastes en fonction de leur éclairement chez un mutant chlorophyllien de tabac. Analyse des relations avec la teneur en pigments et l'activité photosynthétique. - J. Microscopie *20* : 193 - 214, 1974.

23370 - **LENDZIAN, K., BASSHAM, J.A.** : Regulation of glucose-6-phosphate dehydrogenase in spinach chloroplasts by ribulose 1,5-diphosphate and NADPH/$NADP^+$ ratios. - Biochim. biophys. Acta *296* : 260 - 275,1975.

23371 - **LENZ, F., DAUNICHT, H.J.** : Photosynthese bei Erdbeeren. - Erwerbsobstbau *17* : 148 - 152, 1975.

23372 - **LENZ, F., DÖRING, H.W.** : Fruit effects on growth and water consumption in *Citrus*. - Gartenbauwissenschaft *40* : 257 - 260, 1975. [Photosynthates.]

23373 - **LERBS, S., WOLLGIEHN, R.** : Association of cytoplasmic ribosomes with chloroplast membranes as determined by discrimination between newly synthesized cytoplasmic rRNA and chloroplast ribosomal precursor RNA. - Biochem. Physiol. Pflanzen *168* : 167 - 174, 1975.

23374 - **LERMAN, J.C.** : How to interpret variations in the carbon isotope ratio of plants : biologic and environmental effects. - In : **MARCELLE, R.** (ed.) : Environmental and Biological Control of Photosynthesis. Pp. 323 - 335. Dr. W. Junk b.v. Publ., The Hague 1975.

23375 - **LESNIKOV, M.F., SAVITSKAITE, D.V.** : Dinamika obrazovaniya karotina v rasteniyakh v zavisimosti ot urovnya azotnogo pitaniya i orosheniya. [Dynamic of carotene formation in plants as a function of nitrogen nutrient level and irrigation.] - Agrokhimiya *1975* (8) : 69 - 73, 1975. [In R.]

23376 - **LESPINAT, P.-A., ANDRÉ, M., BOUREAU, M.** : Étude quantitative du carbone libéré par les racines de maïs cultivé en conditions contrôlées sous $^{14}CO_2$. - - Physiol. vég. *13* : 137 - 151, 1975.

23377 - **LEVADNAYA, G.D.** : Izuchenie produktivnosti perifitona v Novosibirskom vodokhranilishche. [Periphyton productivity in the Novosibirsk Reservoir.] - Izv. sib. Otd. Akad. Nauk SSSR, Ser. biol. Nauk *1975* (3) : 32 - 38, 1975. [Ps; in R, ab : E.]

23378 - **LEVANON, H., SCHERZ, A.** : EPR study of electron spin polarization in the photoexcited triplet state of chlorophyll *a* and *b*. - Chem. Phys. Lett. *31* : 119- - 124, 1975.

23379 - **LEVI, C., GIBBS, M.** : Carbon dioxide fixation in isolated *Kalanchoe* chloroplasts. - Plant Physiol. *56* : 164 - 166, 1975.

23380 - **LEVIN, E.S.** : Diskretnaya model' gazoobmena vnutri lista s uchetom diffuzii. [Discrete model of gas exchange inside the leaf with respect to diffusion.] - - Byull. vses. nauch.-issled. Inst. Rastenievod. N.I. Vavilova *56* : 60 - 64, 1975. [In R.]

23381 - **LEVSHIN, L.V., SLAVNOVA, T.D., YUZHAKOV, V.I.** : Priroda kontsentratsionnogo tusheniya lyuminestsentsii v rastvorakh khlorofilla. [Nature of concentration quenching of luminescence in chlorophyll solutions.] - Biofizika *20* : 150 - - 152, 1975. [In R, ab : E.]

23382 - **LEWAK, S., KACPERSKA-PALACZ, A.** : Regulation of metabolic processes and photosynthesis. - Pol. ecol. Stud. *1* : 107 - 116, 1975.

23383 - **LEWIN, R.A.** : A marine *Synechocystis (Cyanophyta, Chroococcales)* epizoic on ascidians. - Phycologia *14* : 153 - 160, 1975.

23384 - **LEWIN, R.A., WITHERS, N.W.** : Extraordinary pigment composition of a procaryotic alga. - Nature *256* : 735 - 737, 1975.

23385 - **LEWIS, D.H.** : Comparative aspects of the carbon nutrition of mycorrhizas. - - In : **SANDERS, F.E., MOSSE, B., TINKER, P.B.** (ed.) : Endomycorrhizas. Pp. 119 - 148. Academic Press, London - New York - San Francisco 1975. [Photosynthates.]

23386 - **LEWIS, O.A.M.** : An ^{15}N-^{14}C study of the role of the leaf in the nitrogen nutrition of the seed of *Datura stramonium* L. - J. exp. Bot. *26* : 361 - 366, 1975.

23387 - **LEWIS, W.M., Jr.** : A theoretical comparison of the attenuation of light energy and quanta in waters of divergent optical properties. - Arch. Hydrobiol. *75* : 285 - 296, 1975.

*23388 - **LEXA, D., REIX, M.** : Étude des propriétés d'oxydoréduction de complexes métalliques de divers cycles tétrapyrroliques, I. - Potentiels de demi-palier et nature du complexe. - J. Chim. phys. *71* : 511 - 516, 1974. [Chl.]

*23389 - **LEXA, D., REIX, M.** : Étude des propriétés d'oxydoréduction de complexes métalliques de divers cycles tétrapyrroliques, II. - Identifications et stabilités des monoradicaux. - J. Chim. phys. *71* : 517 - 524, 1974. [Chl.]

23390 - **LEY, A.C., BABCOCK, G.T., SAUER, K.** : Flash kinetics and light intensity dependence of oxygen evolution in the blue-green alga *Anacystis nidulans*. - Biochim. biophys. Acta *387* : 379 - 387, 1975.

23391 - **LEY, J. DE** : The recognition of bioenergetic processes. - Proc. roy. Soc. London, B *189* : 235 - 248, 1975. [Ps.]

23392 - **LHOSTE, A.-M., GARREC, J.-P.** : Étude biométrique des effets du fluor sur la structure des chloroplastes de Mais (*Zea mays* L. var. INRA 260). - J. Microscop. Biol. cell. *24* : 351 - 364, 1975.

23393 - **LI, B.D.** : O soderzhanii fotosinteticheskikh pigmentov u morskikh bentosnykh vodorosleǐ iz razlichnykh ėkologicheskikh nish. [Contents of photosynthetic pigments in marine benthic algae from different ecological sites.] - In : Biologiya Shel'fa. P. 101. Vladivostok 1975. [In R.]

23394 - **LI, E.H., MILES, C.D.** : Effects of cadmium on photoreaction II of chloroplasts. - Plant Sci. Lett. *5* : 33 - 40, 1975.

23395 - **LI, Y.-S.** : Salts and chloroplast fluorescence. - Biochim. biophys. Acta *376*: 180 - 188, 1975.

23396 - **LI, Y.-S.** : The sites of induced limiting steps of the electron transport reactions in chloroplasts isolated with low-salt medium. - In : **AVRON, M.** (ed.): Proceedings of the Third International Congress on Photosynthesis. Vol. I. Pp. 665 - 670. Elsevier, Amsterdam - Oxford - New York 1975.

23397 - **LIBERA, W., ZIEGLER, I., ZIEGLER, H.** : The action of sulfite on the HCO_3^--fixation and the fixation pattern of isolated chloroplasts and leaf tissue slices. - Z. Pflanzenphysiol. *74* : 420 - 433, 1975.

23398 - **LICHTENTHALER, H.K.** : Regulation of prenyl chain synthesis in etiolated *Hordeum* seedlings by far-red and white light. - Physiol. Plant. *33* : 241 - 244, 1975.

23399 - **LICHTENTHALER, H.K.** : Control of light-induced carotenoid synthesis in *Raphanus* seedlings by phytochrome. - Physiol. Plant. *34* : 357 - 358, 1975.

23400 - **LICHTENTHALER, H.K., BECKER, K.** : Kinetic of lipoquinone synthesis in etiolated *Raphanus* seedlings in continuous far-red and white light. - Z. Pflanzenphysiol. *75* : 296 - 302, 1975.

23401 - **LICHTENTHALER, H.K., BECKER, K.** : The influence of continuous far-red and white light on prenyl chain synthesis in plastids of *Raphanus* seedlings. - - Planta *122* : 255 - 258, 1975.

23402 - **LICHTENTHALER, H.K., GRUMBACH, K.H.** : Observations on the turnover of thylakoids and their prenyl lipids in *Hordeum vulgare* L. - In : **AVRON, M.** (ed.) : Proceedings of the Third International Congress on Photosynthesis. Vol. III. Pp. 2007 - 2015. Elsevier, Amsterdam - Oxford - New York 1975.

23403 - **LICHTENTHALER, H.K., KLEUDGEN, H.K.** : Phytochromsystem und Lipochinonsynthese in den Plastiden etiolierter *Hordeum*-Keimlinge. - Z. Naturforsch. *30C* : 64 - - 68, 1975.

23404 - **LICHTENTHALER, H.K., STRAUB, V., GRUMBACH, K.H.** : Unequal formation of prenyl- -lipids in a plant tissue culture and in leaves of *Nicotiana tabacum* L. - - Plant Sci. Lett. *4* : 61 - 65, 1975. [Chl, Car.]

23405 - **LIDDLE, M.J.** : A theoretical relationship between the primary productivity of vegetation and its ability to tolerate trampling. - Biol. Conserv. *8* : 251 - 255, 1975.

23406 - **LIDSTER, P.D., PORRITT, S.W., EATON, G.W., MASON, J.** : Spartan apple breakdown as affected by orchard factors, nutrient content and fruit quality. - - Can. J. Plant Sci. *55* : 443 - 446, 1975. [Chl.]

B23407 - **LIEN, S., SAN PIETRO, A.** : An Inquiry into Biophotolysis of Water to Produce Hydrogen. - Dept. Plant Sci., Indiana Univ., Bloomington 1975.

23408 - **LIETH, H.** : Measurement of caloric values. - In : **LIETH, H., WHITTAKER, R.H.** (ed.) : Primary Productivity of the Biosphere. Pp. 119 - 129. Springer-Verlag, Berlin - Heidelberg - New York 1975.

23409 - **LIETH, H.** : Historical survey of primary productivity research. - In : **LIETH, H., WHITTAKER, R.H.** (ed.) : Primary Productivity of the Biosphere. Pp. 7 - 16. Springer-Verlag, Berlin - Heidelberg - New York 1975.

23410 - **LIETH, H.** : Primary productivity of the major vegetation units of the world. - In : **LIETH, H., WHITTAKER, R.H.** (ed.) : Primary Productivity of the Biosphere. Pp. 203 - 215. Springer-Verlag, Berlin - Heidelberg - New York 1975.

23411 - **LIETH, H.** : Modeling the primary productivity of the world. - In : **LIETH, H., WHITTAKER, R.H.** (ed.) : Primary Productivity of the Biosphere. Pp. 237 - 263. Springer-Verlag, Berlin - Heidelberg - New York 1975.

23412 - **LIETH, H.** : Some prospects beyond production measurement. - In : **LIETH, H., WHITTAKER, R.H.** (ed.) : Primary Productivity of the Biosphere. Pp. 285 - 304. Springer-Verlag, Berlin - Heidelberg - New York 1975.

B23413 - **LIETH, H., WHITTAKER, R.H.** (ed.) : Primary Productivity of the Biosphere. - - Springer-Verlag, Berlin - Heidelberg - New York 1975.

23414 - **LIKENS, G.E.** : Primary productivity of inland aquatic ecosystems. - In : **LIETH, H., WHITTAKER, R.H.** (ed.) : Primary Productivity of the Biosphere. Pp. 185 - - 202. Springer-Verlag, Berlin - Heidelberg - New York 1975.

23415 - **LILLEY, R. McC., FITZGERALD, M.P., RIENITS, K.G., WALKER, D.A.** : Criteria of inactness and the photosynthetic activity of spinach chloroplast preparations. - New Phytol. *75* : 1 - 10, 1975.

23416 - **LILLEY, R. McC., WALKER, D.A.** : Carbon dioxide assimilation by leaves, isolated chloroplasts, and ribulose bisphosphate carboxylase from spinach. - Plant Physiol. *55* : 1087 - 1092, 1975.

23417 - **LIMAR', R.S.** : Fiziologicheskie osobennosti yaponskikh pshenits gruppy Norin. [Physiological peculiarities of Japanese wheats of the group Norin.] - Byull. vses. nauch.-issled. Inst. Rastenievod. N.I. Vavilova *56* : 64 - 67, 1975. [Chl, Car; in R.]

23418 - **LIMPÁROVÁ, M.** : Suitability of the gravimetric method for the study of the intensity of photosynthesis of maize leaves (*Zea mays* L.). - Acta Fac. Rerum natur. Univ. Comenianae, Physiol. Plant. *10* : 37 - 43, 1975.

23419 - **LIN, C.H., STOCKING, C.R.** : Glyceraldehyde 3-phosphate dehydrogenase in greening maize leaves. - Plant Physiol. *56* (Suppl.) : 48, 1975.

23420 - **LIN, L., THORNBER, P.** : Dark recovery kinetics of reaction center preparation from native and UQ-depleted chromatophores of *Chromatium vinosum* (strain D). - Plant Physiol. *56* (Suppl.) : 47, 1975.

23421 - **LIN, L., THORNBER, J.P.** : Isolation and partial characterization of the photochemical reaction center of *Chromatium vinosum* (strain D). - Photochem. Photobiol. *22* : 37 - 40, 1975.

23422 - **LINDEN, J.C., SCHILLING, N., BRACKENHOFER, H., KANDLER, O.** : Asymmetric labelling of maltose during photosynthesis in $^{14}CO_2$. - Z. Pflanzenphysiol. *76* : 176-- 181, 1975.

23423 - **LINDSAY, D.C.** : Growth rates of *Cladonia rangiferina* (L.) WEB. on South Georgia. - Brit. Antarct. Surv. Bull. *40* : 49 - 53, 1975. [Dry-matter production.]

23424 - **LINDSEY, D.W., GUDAUSKAS, R.T.** : Effects of maize dwarf mosaic virus on water relations of corn. - Phytopathology *65* : 434 - 440, 1975. [Diffusion resistances.]

23425 - **LINSER, H., ZEID, F.A.** : Reinprotein, Chlorophyll, Carotin und Kohlenhydrate bei *Daucus carota* im Verlauf der Vegetationsperiode des ersten Jahres unter dem Einfluss von Wachstumsregulatoren. - Z. Pflanzenern. Bodenkunde *1975* : 181 - 196, 1975.

23426 - **LINVILL, D.E., DALE, R.F.** : Population density and sampling location effects on net radiation measurements over corn. - Agron. J. *67* : 463 - 468, 1975.

23427 - **LIPS, S.H.** : Enzyme content of plant microbodies as affected by experimental procedures. - Plant Physiol. *55* : 598 - 601, 1975. [Photorespiration.]

23428 - **LIPSKAYA, G.A.** : Geterogennost' khlorofilla i aktivnost' reaktsii Khilla pri posledeĭstvii kobal'tovykh mikroudobreniĭ. [Heterogeneity of chlorophyll and activity of Hill reaction as aftereffect of cobalt micronutrient fertilizers.] - Fiziol. Biokhim. kul't. Rast. *7* : 522 - 525, 1975. [In R, ab : E.]

23429 - **LIPSKAYA, G.A., ZELENAYA, L.A.** : Vliyanie kobal'ta na temnoustoĭchivost' pigmentnogo apparata yachmenya. [Effect of cobalt on dark resistance of the pigment apparatus of barley.] - Fiziol. Rast. *22* : 277 - 281, 1975. [In R, ab : E.]

23430 - **LITTLE, C.H.A.** : Inhibition of cambial activity in *Abies balsamea* by internal water stress : role of abscisic acid. - Can. J. Bot. *53* : 3041 - 3050, 1975. [Ps.]

23431 - **LITTLE, C.H.A., LOACH, K.** : Effect of gibberellic acid on growth and photosynthesis in *Abies balsamea*. - Can. J. Bot. *53* : 1805 - 1810, 1975.

23432 - **LITVIN, F.F., BALASHOV, S.P., SINESHCHEKOV, V.A.** : Issledovanie pervichnykh fotokhimicheskikh prevrashcheniĭ bakteriorodopsina v purpurnykh membranakh i kletkakh *Halobacterium halobium* metodom nizkotemperaturnoĭ spektrofotometrii. [Investigation of primary photochemical conversions of bacteriorhodopsin in purple membranes and cells of *Halobacterium halobium* by low temperature spectrophotometry.] - Bioorg. Khim. *1* : 1767 - 1777, 1975. [In R, ab : E.]

23433 - **LITVIN, F.F., SHUBIN, V.V., SINESHCHEKOV, V.A.** : Issledovanie razlichnykh tipov agregatov khlorofilla *a* v rastvorakh i plenkakh metodom absorbtsionnoĭ i lyuminestsentnoĭ proizvodnoĭ spektroskopii. [Two types of chlorophyll *a* aggregates in solutions and films studied by the derivative absorption and luminescence spectroscopy.] - Biofizika *20* : 202 - 207, 1975. [In R, ab : E.]

23434 - **LITVIN, F.F., SINESHCHEKOV, V.A.** : Molecular organization of chlorophyll and energetics of the initial stages in photosynthesis. - In : **GOVINDJEE** (ed.) : Bioenergetics of Photosynthesis. Pp. 619 - 661. Academic Press, New York - San Francisco - London 1975.

23435 - **LITVINENKO, L.G.** : Ob uchastii riboflavina v kislorodnom zvene fotosinteza. [Participation of riboflavin in oxygen link of photosynthesis.] - Fiziol. Biokhim. kul't. Rast. *7* : 283 - 285, 1975. [In R, ab : E.]

23436 - **LOACH, P.A.** : A proposal for the primary photochemical events of bacterial photosynthesis. - Biophys. J. *15* : 225a, 1975.

23437 - **LOACH, P.A., KUNG, M. (CHU), HALES, B.J.** : Characterization of the phototrap in photosynthetic bacteria. - Ann. New York Acad. Sci. *244* : 297 - 319, 1975

*23438 - **LOBODA, N.I., NEKRASOV, L.I., SHCHEGOLEVA, N.A.** : Reaktsiya vosstanovleniya metilovogo krasnogo, fotosensibilizirovannaya adsorbirovannym na kapronovykh plenkakh khlorofillom. [Reaction of methyl red reduction photosensibilized by chlorophyll adsorbed on capron layers.] - Zh. fiz. Khim. *48* : 338 - 341, 1974. [In R.]

23439 - **LOEBLICH, A.R. III** : A seawater medium for dinoflagellates and the nutrition of *Cachonina niei*. - J. Phycol. *11* : 80 - 86, 1975. [Ps.]

*23440 - **LOEBLICH, L.A.** : Action spectra and effect of light intensity on growth, pigments and photosynthesis in *Dunaliella salina*. - J. Protozool. *21* : 420, 1974.

23441 - **LOFTUS, M.E., SELIGER, H.H.** : Some limitations of the *in vivo* fluorescence technique. - Chesapeake Sci. *16* : 79 - 92, 1975. [Chl.]

23442 - **LOHONYAI, N.** : A fotoszintézis nyomon követése polarográfiás módszerrel. [Photosynthesis followed by polarography.] - Kertészeti Egyetem Közleményeiből [Publ. Univ. Horticult.] *39* : 231 - 237, 1975. [In Hung., ab : E, R.]

23443 - **LÖHR, E.** : Analyse der Stoffproduktion in Bewachsungen von Gerste. - Pflanzenphysiol. Lab. Univ. Kopenhagen 1975. [Ps, Chl.]

23444 - **LOMMEN, P.W., SMITH, S.K., YOCUM, C.S., GATES, D.M.** : Photosynthetic model. - - In : **GATES, D.M., SCHMERL, R.B.** (ed.) : Perspectives of Biophysical Ecology. Pp. 33 - 43. Springer-Verlag, Berlin - Heidelberg - New York 1975.

23445 - **LONDON, R.E., KOLLMAN, V.H., MATWIYOFF, N.A., MUELLER, D.D.** : Biosynthetic and biophysical information from ^{13}C-^{13}C multiplets by ^{13}C nuclear magnetic resonance. - Report *LA-UR-75-1985* : 470 - 484, 1975. [Ps.]

23446 - **LONERGAN, T.A., SARGENT, M.L.** : The effects of acetazolamide on carbonic anhydrase, photosystem activity and oxygen evolution in *Euglena gracilis*. - - Plant Physiol. *56* (Suppl.) : 26, 1975.

23447 - **LONG, S.P., INCOLL, L.D., WOOLHOUSE, H.W.** : C_4 photosynthesis in plants from cool temperate regions, with particular reference to *Spartina townsendii*. - - Nature *257* : 622 - 624, 1975.

23448 - **LOOMIS, R.S., GERAKIS, P.A.** : Productivity of agricultural ecosystems. - In : **COOPER, J.P.** (ed.) : Photosynthesis and Productivity in Different Environments. Pp. 145 - 172. Cambridge Univ. Press, Cambridge - London - New York - Melbourne 1975.

23449 - **LOOS, E.** : Action spectra for light-induced pH changes in chloroplast suspensions from yellow-green alga *Bumilleriopsis filiformis*. - In : **AVRON, M.** (ed.): Proceedings of the Third International Congress on Photosynthesis. Vol. II. Pp. 997 - 1000. Elsevier, Amsterdam - Oxford - New York 1975.

23450 - **LOPATA, W.-D., ULLRICH, H.** : Untersuchungen zu stofflichen und strukturellen Veränderungen an Pflanzen unter NO_2 Einfluß. - Staub-Reinhalt. Luft *35* (5) : 196 - 200, 1975. [Ps, Chl.]

23451 - **LORD, J.M., ARMITAGE, T.L., MERRETT, M.J.** : Ribulose 1,5-diphosphate carboxylase synthesis in *Euglena*. II. Effect of inhibitors on enzyme synthesis during regreening and subsequent transfer to darkness. - Plant Physiol. *56* : 600 - - 604, 1975.

23452 - **LORD, J.M., BROWN, R.H.** : Purification and some properties of *Chlorella fusca* ribulose 1,5-diphosphate carboxylase. - Plant Physiol. *55* : 360 - 364, 1975.

23453 - **LORD, J.M., CODD, G.A., STEWART, W.D.P.** : Serological comparison of ribulose-1,5-diphosphate carboxylase from *Euglena gracilis*, *Chlorella fusca* and several blue-green algae. - Plant Sci. Lett. *4* : 377 - 383, 1975.

23454 - **LORD, J.M., MERRETT, M.J.** : Ribulose diphosphate carboxylase synthesis in *Euglena*. Increased enzyme activity after transferring regreening cells to darkness. - Plant Physiol. *55* : 890 - 892, 1975.

23455 - **LORIMER, G.H., ANDREWS, T.J., TOLBERT, N.E.** : *In vivo* and *in vitro* studies with oxygen-18 on the mechanism of photorespiratory glycolate synthesis. - In: C.R. Colloq. Int. Isot. Oxygène 1972. Pp. 169 - 174. Serv. Radioagron., St--Paul-lez-Durance 1975.

23456 - **LOSSOW, K., SIKOROWA, A., DROZD, H., MUCHOWA, A., NEJRANOWSKA, H., SOBIERAJSKA, M., WIDUTO, J., ZMYSŁOWSKA, I.** : Results of research on the influence of aeration on the physico-chemical systems and biological complexes in the Starodworskie Lake obtained hitherto. - Pol. Arch. Hydrobiol. *22* : 195 - 216, 1975.

23457 - **LOUWERSE, W., EIKHOUDT, J.W.** : A mobile laboratory for measuring photosynthesis, respiration and transpiration of field crops. - Photosynthetica *9* : 31 - - 34, 1975.

23458 - **LOZIER, R.H., BOGOMOLNI, R.A., STOECKENIUS, W.** : Bacteriorhodopsin : a light-driven proton pump in *Halobacterium halobium*. - Biophys. J. *15* : 955 - 962, 1975.

23459 - **LOZOVA, G.I.** : Lipidy khloroplastiv ta ĭkh funktsional'na rol'. [Lipids of chloroplasts and their functional role.] - Ukr. bot. Zh. *32* : 681 - 696, 810, 1975. [In Ukr., ab : E, R.]

23460 - **LOZOVA, G.I., TRUTNĖVA, I.A.** : Osoblyvosti pigmentvmisnykh kompleksiv z deyakykh predstavnykiv mokhiv i paporoteĭ. [Peculiarities of pigment-containing complexes of certain representatives of mosses and ferns.] - Ukr. bot. Zh. *32* : 199 - 204, 265, 1975. [In Ukr., ab : E, R.]

23461 - **LOZOVA, G.I., TRUTNĖVA, I.A.** : Izofermenty ėlektroforetychnykh fraktsiĭ pigmentvmisnykh kompleksiv. [Isoenzymes of electrophoretical fractions in pigment-containing complexes.] - Ukr. bot. Zh. *32* : 651 - 655, 675, 1975. [In Ukr., ab : E, R.]

23462 - **LUCAS, W.J.** : Photosynthetic fixation of 14 carbon by internodal cells of *Chara corallina*. - J. exp. Bot. *26* : 331 - 346, 1975.

23463 - **LUCAS, W.J.** : The influence of light intensity on the activation and operation of the hydroxyl efflux system of *Chara corallina*. - J. exp. Bot. *26* : 347 - - 360, 1975. [Ps.]

23464 - **LUDLOW, C.J., WOLF, F.T.** : Photosynthesis and respiration rates of ferns. - - Amer. Fern J. *65* : 43 - 48, 1975.

23465 - **LUDLOW, M.M.** : Effect of water stress on the decline of leaf net photosynthesis with age. - In : **MARCELLE, R.** (ed.) : Environmental and Biological Control of Photosynthesis. Pp. 123 - 134. Dr. W. Junk b.v. Publ., The Hague 1975.

23466 - **LUDWIG, J.A., REYNOLDS, J.F., WHITSON, P.D.** : Size-biomass relationships of several Chihuahuan desert shrubs. - Amer. Midland Naturalist *94* : 451 - 461, 1975.

23467 - **LUDWIG, L.J., CHARLES-EDWARDS, D.A., WITHERS, A.C.** : Tomato leaf photosynthesis and respiration in various light and carbon dioxide environments. - In : **MARCELLE, R.** (ed.) : Environmental and Biological Control of Photosynthesis. Pp. 29 - 36. Dr. W. Junk b.v. Publ., The Hague 1975.

23468 - **LUEBS, R.E., LAAG, A.E., NASH, P.A.** : Evapotranspiration of dryland barley with different plant spacing patterns. - Agron. J. *67* : 339 - 342, 1975. [Growth analysis.]

23469 - **LUGANSKAYA, A.N., KRASNOVSKIĬ, A.A.** : Uchastie kislôroda v fotosensibilizirovannom khlorofillom vosstanovlenii metilviologena v vodnom rastvore detergenta. [Participation of oxygen in chlorophyll photosensitized reduction of methylviologen in aqueous solution of detergent.] - Biofizika *20* : 999 - 1003, 1975. [In R, ab : E.]

23470 - **LUKASHEV, E.P., NOKS, P.P., KONONENKO, A.A., VENEDIKTOV, P.S., RUBIN, A.B.** : Vliyanie temperatury na temnovoe vosstanovlenie fotookislennogo bakteriokhlorofilla P_{870} u fotosinteziruyushchikh bakteriĭ *Rhodospirillum rubrum*. [Effect of temperature on dark reduction of photooxidized bacteriochlorophyll P_{870} in photosynthesizing bacteria of *Rhodospirillum rubrum*.] - Nauch. Dokl. vyssh. Shkoly, biol. Nauki *18* (7) : 48 - 55, 1975. [In R.]

23471 - **LUK'YANOVA, L.M., MARKOVSKAYA, E.F.** : O sutochnykh izmeneniyakh soderzhaniya ksantofillov v list'yakh manzhetki v Zapolyâr'e. [Daily changes in the content of xanthophylls in the leaves of *Alchemilla vulgaris* growing beyond the polar circle.] - Fiziol. Rast. *22* : 490 - 499, 1975. [In R, ab : E.]

23472 - **LUMPKIN, O.** : ^{25}Mg and ^{14}N nuclear quadrupole resonances in chlorophyll-*a* and magnesium phthalocyanine. - J. chem. Phys. *62* : 3281 - 3283, 1975.

23473 - **LUMSDEN, J., HALL, D.O.** : Chloroplast manganese and superoxide. - Biochem. biophys. Res. Commun. *64* : 595 - 602, 1975.

23474 - **LUMSDEN, J., HALL, D.O.** : Superoxide dismutase in photosynthetic organisms provides an evolutionary hypothesis. - Nature *257* : 670 - 672, 1975.

23475 - **LÜNING, K., DRING, M.J.** : Reproduction, growth and photosynthesis of gametophytes of *Laminaria saccharina* grown in blue and red light. - Mar. Biol. *29* : 195 - 200, 1975.

23476 - **LUNNEY, C.A., DAVIS, G.J., JONES, M.N.** : Unusual structure associated with peripheral reticulum in chloroplasts of *Myriophyllum spicatum* L. - J. Ultrastruct. Res. *50* : 293 - 296, 1975.

23477 - **LURIE, S., MALKIN, S.** : The effect on stimulated luminescence of treatments inhibiting photosystem II. - In : **AVRON, M.** (ed.) : Proceedings of the Third International Congress on Photosynthesis. Vol. I. Pp. 83 - 91. Elsevier, Amsterdam - Oxford - New York 1975.

*23478 - **LÜTTGE, U.** : Localized ion transport in complex systems of higher plants as related to respiration and photosynthesis. - In : **BRODA, E., LOCKER, A., SPRINGER-LEDERER, H.** (ed.) : Proceedings of the First European Biophysics Congress. Vol. 3. Pp. 119 - 123. Verlag Wiener med. Akad., Wien 1971.

23479 - **LÜTTGE, U.** : Salt glands. - In : **BAKER, D.A., HALL, J.L.** (ed.) : Ion Transport in Plant Cells and Tissues. Pp. 335 - 376. North-Holland Publ. Comp., Amsterdam - Oxford, Amer. Elsevier Publ. Comp., New York 1975. [Ps.]

23480 - **LÜTTGE, U., BALL, E., TROMBALLA, H.-W.** : Potassium independence of osmoregulated oscillations of malate^{2-} levels in the cells of CAM-leaves. - Biochem. Physiol. Pflanzen *167* : 267 - 283, 1975.

23481 - **LÜTTGE, U., KLUGE, M., BALL, E.** : Effects of osmotic gradients on vacuolar malic acid storage. A basic principle in oscillatory behavior of Crassulacean acid metabolism. - Plant Physiol. *56* : 613 - 616, 1975.

23482 - **LÜTZ, C.** : Biochemische und cytologische Untersuchungen zur Chloroplastenentwicklung. I. Die chemische Charakterisierung der Prolamellarkörper aus Etioplasten von *Avena sativa* L. - Z. Pflanzenphysiol. *75* : 346 - 359, 1975.

23483 - **LÜTZ, C.** : Biochemische und cytologische Untersuchungen zur Chloroplastenentwicklung. II. Isolierung und Charakterisierung eines Glykoproteins aus den Prolamellarkörpern der Etioplasten von *Avena sativa* L. - Z. Pflanzenphysiol. *76* : 130 - 142, 1975.

23484 - **LUTZ, H.U., BEYELER, W., PFLUGSHAUPT, C., BACHOFEN, R.** : Possible role of firmly bound ATP in the energy transduction of photosynthetic membranes. - J. supramol. Struct. *3* : 498 - 509, 1975.

23485 - **LUTZ, M.** : Resonance Raman spectroscopy of the chlorophylls in photosynthetic structures at low temperature. - In : **JOUSSOT-DUBIEN, J.** (ed.) : Lasers in Physical Chemistry and Biophysics. Pp. 451 - 463. Elsevier sci. Publ. Comp., Amsterdam 1975.

23486 - **LUZHNOVA, M.I., SHEKHTMAN, L.M.** : Mode of action of meturine. - Pesticide Biochem. Physiol. *5* :205 - 210, 1975. [Ps.]

23487 - **LUZZANA, M.R.,PENNISTON, J.T.** : Electrode measurement of oxygen tension with 1-ms time resolution. - Biochim. biophys. Acta *396* : 157 - 164, 1975.

23488 - **LYAKHNOVICH, Ya.P.** : Fraktsionirovanie pigmentnogo apparata vyrashchennykh na svetu i zatemnennykh kletok khlorelly metodom gel'-elektroforeza. [Gel--electrophoresis fractionation of pigment apparatus of light and dark grown *Chlorella* cells.] - In : **SHLYK, A.A.** (ed.) : Biosintez i Sostoyanie Khlorofillov v Rastenii. Pp. 207 - 215, 247 - 248. Nauka i Tekhnika, Minsk 1975. [In R, ab : E.]

23489 - **LYAKHNOVICH, Ya.P., GONTAREVA, T.V.** : O sostoyanii pigmentov v avtosporakh khlorelly pri polnom zatemnenii. [Pigment state in *Chlorella* autospores in full darkness.] - In : Fiziologo-Biokhimicheskie Aspekty Rosta i Razvitiya Rasteniĭ. Pp. 44 - 49. Nauka i Tekhnika, Minsk 1975. [In R.]

23490 - **LYMAN, H., JUPP, A.S., LARRINUA, I.** : Action of nalidixic acid on chloroplast replication in *Euglena gracilis*. - Plant Physiol. *55* : 390 - 392, 1975.

23491 - **LYSENKO, G.G., BEKINA, R.M., RUBIN, B.A.** : Osobennosti funktsionirovaniya sistem fotopogloshcheniya kisloroda v khloroplastakh gorokha. [Peculiarities of functioning of oxygen photoabsorption systems in pea chloroplasts.] - Nauch. Dokl. vyssh. Shkoly, biol. Nauki *18* (3) : 64 - 70, 1975. [In R.]

23492 - **LYTTLETON, J.W.** : Activation and stabilisation of low K_m carbon dioxide fixation by chloroplast extracts. - Plant Sci. Lett. *4* : 385 - 389, 1975.

23493 - **MA, P., HUNT, L.A.** : Photosynthesis of newly matured leaves during the ontogeny of barley grown at different nutrient levels. - Can. J. Bot. *53* : 2389 - - 2398, 1975.

23494 - **MACDONALD, I.R.** : Effect of vacuum infiltration on photosynthetic gas exchange in leaf tissue. - Plant Physiol. *56* : 109 - 112, 1975.

23495 - **MACDONALD, R.E., LANYI, J.K.** : Light-driven leucine transport in *Halobacterium halobium* cell envelope vesicles. - Biophys. J. *15* (2, Part 2) : 65a, 1975.

23496 - **MACDONALD, R.E., LANYI, J.K.** : Light-induced leucine transport in *Halobacterium halobium* envelope vesicles : A chemiosmotic system. - Biochemistry *14* : 2882 - 2889, 1975.

*23497 - **MACFADYEN, A.** : Soil metabolism in relation to ecosystem energy flow and to primary and secondary production. - In : **PHILLIPSON, J.** (ed.) : Methods of Study in Soil Ecology. Pp. 167 - 172. UNESCO, Paris 1970.

23498 - **MACHOLD, O.** : On the molecular nature of chloroplast thylakoid membranes. - - Biochim. biophys. Acta *382* : 494 - 505, 1975.

23499 - **MADGWICK, H.A.I.** : Effects of partial defoliation on the growth of *Liriodendron tulipifera* L. seedlings. - Ann. Bot. *39* : 1111 - 1115, 1975. [Growth analysis.]

23500 - **MADGWICK, H.A.I., SATOO, T.** : On estimating the aboveground weights of tree stands. - Ecology *56* : 1446 - 1450, 1975.

23501 - **MADIGAN, M.T., BROCK, T.D.** : Photosynthetic sulfide oxidation by *Chloroflexus aurantiacus*, a filamentous, photosynthetic, gliding bacterium. - J. Bacteriol. *122* : 782 - 784, 1975.

23502 - **MADSEN, E.** : Effect of CO_2-enrichment on growth, development, fruit production and fruit quality of tomato - from a physiological view point. - In : **CHOUARD, P., de BILDERLING, N.** (ed.) : Phytotronics in Agricultural and Horticultural Research. Phytotronics III. Pp. 318 - 330. Gauthier-Villars, Paris 1975. [Ps.]

*23503 - **MAE, T., VONK, C.R.** : Effect of light and growth substances on flowering of *Iris* x *hollandica* cv. Wedgwood. - Acta bot. neerl. *23* : 321 - 331, 1974. [Ps.]

23504 - **MAGALHÃES, A.C., PETERS, D.B., HAGEMAN, R.H.** : Temperature effects on nitrate reductase activity and net photosynthesis of soybean leaves. - Plant Physiol. *56* (Suppl.) : 36, 1975.

23505 - **MAGAZZU', G., ANDREOLI, C., MUNAO', F.** : Ciclo annuale del fitoplancton e della produzione primaria del Basso Tirreno (1969 - 1970). [Phytoplankton and primary production cycle in the South Tyrrhenian Sea (1969 - 1970).] - Mem. Biol. mar. Oceanograf. N.S. *5* (2) : 25 - 48, 1975. [In Ital., ab : E.]

23506 - **MAGNUSSON, R.P., McCARTY, R.E.** : Influence of adenine nucleotides on the inhibition of photophosphorylation in spinach chloroplasts by N-ethylmaleimide. - J. biol. Chem. *250* : 2593 - 2598, 1975.

23507 - **MAGNUSSON, R.P., McCARTY, R.E.** : Multiple nucleotide binding sites on membrane bound spinach chloroplasts coupling factor 1. - Fed. Proc. *34* : 595, 1975.

23508 - **MAGYAROSY, A.C., BUCHANAN, B.B.** : Effect of bacterial infiltration on photosynthesis of bean leaves. - Phytopathology *65* : 777 - 780, 1975.

23509 - **MAHON, J.D., EGLE, K., FOCK, H.A.** : A radio-gas chromatographic method for determining the specific radioactivity of glycolic acid in ^{14}C-labeled leaf tissue. - Can. J. Biochem. *53* : 609 - 614, 1975.

23510 - **MAIRANOVSKY, V.G., ENGOVATOV, A.A., IOFFE, N.T., SAMOKHVALOV, G.I.** : Electron--donor and electron-acceptor properties of carotenoids. Electrochemical study of carotenes. - J. electroanal. Chem. *66* : 123 - 137, 1975.

23511 - **MAKSIMOVA, I.V., DAL', E.S.** : Vydelenie glikolevoĭ kisloty kletkami *Chlorella pyrenoidosa.* [Production of glycolic acid by the cells of *Chlorella pyrenoidosa.*] - Mikrobiologiya *44* : 1057 - 1063, 1975. [In R, ab : E.]

23512 - **MALIGA, P., SZ.-BREZNOVITS, A., MARTON, L., JOO, F.** : Non-Mendelian streptomycin-resistant tobacco mutant with altered chloroplasts and mitochondria. - - Nature *255* : 401 - 402, 1975.

23513 - **MALIK, C.P., SETHI, R.S.** : Histochemical studies in stomatal apparatus of *Phaseolus mungo* LINN. I. Localization of enzymes and structural material. - - Acta histochem. *52* : 303 - 323, 1975. [ATPase.]

23514 - **MALIK, C.P., SETHI, R.S.** : Histochemical studies in stomatal apparatus of *Phaseolus mungo* LINN. IV. Mechanism of stomatal action. - Acta histochem. *53*: 1 - 11, 1975. [Ps enzymes.]

23515 - **MALKIN, R.** : Photochemical properties of a photosystem I subchloroplast fragment. - Arch. Biochem. Biophys. *169* : 77 - 83, 1975.

23516 - **MALKIN, R.** : Primary reactions in photosynthesis. - Photochem. Photobiol. *22*: 292 - 294, 1975.

23517 - **MALKIN, R., APARICIO, P.J.** : Identification of a g = 1.90 high-potential iron--sulfur protein in chloroplasts. - Biochem. biophys. Res. Commun. *63* : 1157 - - 1160, 1975.

23518 - **MALKIN, R., BEARDEN, A.J.** : Laser-flash-activated electron paramagnetic resonance studies of primary photochemical reactions in chloroplasts. - Biochim. biophys. Acta *396* : 250 - 259, 1975.

*23519 - **MALKIN, R., KNAFF, D.B.** : The role of C-550 in photosystem II of photosynthesis. - Fed. Proc. *33* : 1329, 1974.

23520 - **MALKIN, S.** : On the cooperation between photosynthetic units of photosystem II. - In : **AVRON, M.** (ed.) : Proceedings of the Third International Congress on Photosynthesis. Vol. I. Pp. 199 - 204. Elsevier, Amsterdam - Oxford - New York 1975.

23521 - **MALKINA, I.S.** : Rastyazhenie kletok mezofilla sazhentsev razlichnoĭ stepeni tenevynoslivosti. [Stretching of mesophyll cells in seedlings of different shade-endurance.] - Lesovedenie *1975* (4) : 69 - 74, 1975. [Growth; in R, ab : E.]

23522 - **MALLOT, P.G., DAVY, A.J., JEFFERIES, R.L., HUTTON, M.J.** : Carbon dioxide exchange in leaves of *Spartina anglica* HUBBARD. - Oecologia *20* : 351 - 358, 1975.

23523 - **MALOFEEV, V.M.** : Chuvstvitel'nost' fotosinteticheskoĭ funktsii lista k narusheniyu vodnogo balansa u fasoli s ėksperimental'no poluchennoĭ kseromorfnost'yu. [Sensitivity of the photosynthetic function of leaves to the disturbance in the water balance in the bean with experimentally obtained xeromorphism.] - Sel'skokhoz. Biol. *10* : 745 - 749, 1975. [In R, ab : E.]

23524 - **MALOFEEV, V.M.** : Samoregulyatsiya fotosinteticheskoĭ funktsii rasteniĭ pri rezkoĭ smene pogody v polevykh usloviyakh. [Self-regulation of photosynthetic function in plants after a rapid change in weather in field conditions.] - - Dokl. TSKhA (Moskva) *209* : 5 - 9, 1975. [In R.]

23525 - **MALOFEEVA, I.V., KONDRATIEVA, E.N., RUBIN, A.B.** : Ferredoxin-linked nitrate reductase from the phototrophic bacterium *Ectothiorhodospira shaposhnikovii.* - FEBS Lett. *53* : 188 - 189, 1975.

23526 - **MANANKOV, M.K.** : Vliyanie gibberellina na nekotorye fiziologicheskie protsessy vinograda. [Effect of gibberellin on some physiological processes in grapevine.] - Fiziol. Biokhim. kul't. Rast. *7* : 301 - 305, 1975. [Chl, Car; in R, ab : E.]

23527 - **MANCINELLI, A.L., LINDQUIST, P., ANDERSON, O.R., EISENSTADT, F.A.** : Photocontrol of seed germination. VII. Preliminary observation on the development of the photosynthetic apparatus in light-inhibited seeds of cucumber (*Cucumis sativus*). - Bull. Torrey bot. Club *102* (3) : 93 - 99, 1975.

23528 - **MANCINELLI, A.L., YANG, C.-P.H., LINDQUIST, P., ANDERSON, O.R., RABINO, I.** : Photocontrol of anthocyanin synthesis. III. The action of streptomycin on the synthesis of chlorophyll and anthocyanin. - Plant Physiol. *55* : 251 - 257, 1975.

23529 - **MANDAHAR, C.L., GARG, I.D.** : Effect of ear removal on sugars and chlorophylls of barley leaves. - Photosynthetica *9* : 407 - 409, 1975.

23530 - **MANENTI, G.** : The structure of variegated leaves of *Acer negundo* L. A light and electron microscope study. - Israel J. Bot. *24* : 61 - 70, 1975.

23531 - **MANETAS, Y., AKOYUNOGLOU, G.** : The stoichiometric relationship between δ-amino-levulinic acid and protochlorophyllide. - Plant Sci. Lett. *5* : 375 - 378, 1975.

23532 - **MANGEL, M.** : The enhancement of photocurrents in bilayer lipid membranes by phycocyanin : pH and surface charge dependence. - Biochem. biophys. Res. Commun. *66* : 393 - 396, 1975.

23533 - **MANGEL, M., BERNS, D.S., ILANI, A.** : Dependence of photosensitivity of bileaflet lipid membranes upon the chlorophyll and carotenoid content. - J. Membrane Biol. *20* : 171 - 180, 1975.

23534 - **MANN, K.H., CHAPMAN, A.R.O.** : Primary production of marine macrophytes. - In : **COOPER, J.P.** (ed.) : Photosynthesis and Productivity in Different Environments. Pp. 207 - 223. Cambridge Univ. Press, Cambridge - London - New York - Melbourne 1975.

23535 - **MANOS, P.J., GOLDTHWAITE, J.** : A kinetic analysis of the effects of gibberellic acid, zeatin, and abscisic acid on leaf tissue senescence in *Rumex*. - Plant Physiol. *55* : 192 - 198, 1975. [Chl.]

23536 - **MANOS, P.J., GOLDTHWAITE, J.** : Leaf tissue senescence. Constant responsiveness to hormones despite a seasonal cycle in senescence rate. - Plant Physiol. *55* : 951 - 953, 1975. [Chl.]

23537 - **MANTAI, K.E.** : The physiology of *Cladophora glomerata* in Lake Erie. - Plant Physiol. *56* (Suppl.) : 12, 1975. [Ps.]

23538 - **MAR, T., BREBNER, J., ROY, G.** : Induction kinetics of delayed light emission in spinach chloroplasts. - Biochim. biophys. Acta *376* : 345 - 353, 1975.

23539 - **MARCELLE, R.** : Effect of photoperiod on the CO_2 and O_2 exchanges in leaves of *Bryophyllum daigremontianum* (BERGER). - In : **MARCELLE, R.** (ed.) : Environmental and Biological Control of Photosynthesis. Pp. 349 - 356. Dr. W. Junk b.v. Publ., The Hague 1975.

B23540 - **MARCELLE, R.** (ed.) : Environmental and Biological Control of Photosynthesis. - Dr. W. Junk b.v. Publ., The Hague 1975.

23541 - **MARCHANT, R.H.** : An analysis of the effect of added thiols on the rate of oxygen uptake during Mehler reactions which involve superoxide production. - In: **AVRON, M.** (ed.) : Proceedings of the Third International Congress on Photosynthesis. Vol. I. Pp. 637 - 643. Elsevier, Amsterdam - Oxford - New York 1975.

23542 - **MARGALEF, R.** : Fitoplancton invernal de la laguna costera de Alvarado (Mexico). [Winter phytoplankton in the coastal lagoon of Alvarado (Mexico).] - An. Inst. bot. A.J. Cavanilles *32* : 381 - 387, 1975. [Chl; in Span.]

23543 - **MARGULIES, M.M., TIFFANY, H.L., MICHAELS, A.** : Vectorial discharge of nascent polypeptides attached to chloroplast thylakoid membranes. - Biochem. biophys. Res. Commun. *64* : 735 - 739, 1975.

23544 - **MARINUS, J., BODLAENDER, K.B.A.** : Response of some potato varieties to temperature. - Potato Res. *18* : 189 - 204, 1975. [Growth analysis.]

23545 - **MARK, A.F.** : Photosynthesis and dark respiration in three alpine snow tussocks (*Chionochloa* spp.) under controlled environments. - New Zeal. J. Bot. *13* : 93 - 122, 1975.

23546 - **MARKLEY, J.L., ULRICH, E.L., BERG, S.P., KROGMANN, D.W.** : Nuclear magnetic resonance studies of the copper binding sites of blue copper proteins : Oxidized, reduced, and apoplastocyanin. - Biochemistry *14* : 4428 - 4433, 1975.

23547 - **MARKOWSKI, A., GRZESIAK, S., SCHRAMEL, M.** : Indexes of the susceptibility of various species of cultivated plants to sulphur dioxide action. - Bull. Acad. pol. Sci., Sér. Sci. biol. *23* : 637 - 646, 1975. [Chl.]

23548 - **MAROC, J., GARNIER, J.** : La photooxydation du cytochrome *b*-559, en présence de carbonylcyanure-*p*-trifluorométhoxyphenylhydrazone et de 2,5-dibromo-3--méthyl-6-isopropyl-*p*-benzoquinone, ou de *p*-benzoquinone, chez trois mutants non-photosynthétiques de *Chlamydomonas reinhardti*. - Biochim. biophys. Acta *387* : 52 - 68, 1975.

23549 - **MARSCHNER, H., POSSINGHAM, J.V.** : Effect of K^+ and Na^+ on growth of leaf discs of sugar beet and spinach. - Z. Pflanzenphysiol. *75* : 6 - 16, 1975. [Chl.]

23550 - **MARSHAKOVA, M.I., SHERSTENIKINA, A.V.** : Osobennosti fiziologicheskikh protsessov v rasteniyakh kartofelya pri razlichnykh rezhimakh mineral'nogo pitaniya. [Properties of physiological processes in potato plants under various mineral nutrition.] - In : Fiziologo-Biokhimicheskie Aspekty Rosta i Razvitiya Rasteniĭ. Pp. 148 - 155, 175. Nauka i Tekhnika, Minsk 1975. [Chl, Car; in R.]

23551 - **MARSHALL, P.E., KOZLOWSKI, T.T.** : Changes in mineral contents of cotyledons and young seedlings of woody angiosperms. - Can. J. Bot. *53* : 2026 - 2031, 1975. [Ps.]

23552 - **MARSHO, T.V.** : Slow redox changes involving photosystem II acceptor pools in *Ulva*. - Plant Physiol. *56* (Suppl.) : 10, 1975.

23553 - **MARSHO, T.V., HOMMERSAND, M.H.** : Slow 514 nm absorption phases and oxygen exchange transients in *Ulva*. - Biochim. biophys. Acta *376* : 354 - 365, 1975.

23554 - **MARTIN, D.B., NOVOTNY, J.F.** : Nutrient limitation of summer phytoplankton growth in two Missouri River reservoirs. - Ecology *56* : 199 - 205, 1975.

23555 - **MARTIN, M., SABATER, B.** : Fosforilacion por cambio de pH en cloroplastos de cebada. [Phosphorylation in barley chloroplasts after change in pH.] - An. Inst. bot. Cavanilles *32* : 279 - 285, 1975. [In Span.]

23556 - **MARTINČIČ, A., GAMS, M., VOGELNIK, K., BATIČ, F., VRHOVŠEK, D.** : Zimska fotosintetska aktivnost vrste *Ilex aquifolium* L. [Net photosynthetic activity of the holey, *Ilex aquifolium* L., in winter conditions.] - Biol. Vestn. (Ljubljana) *23* : 45 - 52, 1975. [In Slovenian, ab : E.]

23557 - **MARUTA, H.** : [Studies on matter production in hop plants. I. Seasonal changes in matter economy.] - Proc. Crop Sci. Soc. Jap. *44* : 22 - 28, 1975. [Growth analysis; in Jap., ab : E.]

23558 - **MARX, J.L.** : Laser spectroscopy : Probing biomolecular functions. - Science *188* : 1002, 1004, 1975. [Ps, Chl.]

23559 - **MARYNICK, D.S., MARYNICK, M.C.** : A mathematical treatment of rate data obtained in biological flow systems under nonsteady state conditions. - Plant Physiol. *56* : 680 - 683, 1975. [CO_2 measurement.]

23560 - **MASAROVIČOVÁ, E., DUDA, M.** : Correlations among phenological phases, climatic factors and quantitative changes of chlorophylls in selected forest ecosystem plant species. - Biológia (Bratislava) *30* : 781 - 789, 1975.

23561 - **MASHARIPOV, P.M., KUCHKAROVA, M.A., AMANOV, M.A.** : Vliyanie nekotorykh azotfiksiruyushchikh sinezelenykh i protokokkovykh vodorosleĭ na fiziologicheskie protsessy pshenitsy. [Effect of some nitrogen fixing blue-green and protococcous algae on physiological processes of wheat.] - In : Vodorosli i Griby Sredneĭ Azii. Pp. 129 - 134. Fan, Tashkent 1975. [Ps, Chl; in R.]

23562 - **MASHCHUK, P.A.** : Uplyŭ trykhloratsėtatu natryyu na kol'kasts' khlarafilu ŭ raslinakh lubinu roznykh sartoŭ. [Effect of sodium trichloracetate on chlorophyll amount in plants of different cultivars of *Lupinus*.] - Vestsi Akad. Navuk belarus. SSR, Ser. biyal. Navuk *1975* (2) : 33 - 36, 137 - 138, 1975. [In Belorus., ab : R.]

3563 - **MASKIEWICZ, R., BRUICE, T.C., BARTSCH, R.G.** : The acid-base properties and kinetics of dissolution of the Fe_4S_4 cores of *Chromatium* ferredoxin and high potential iron protein. - Biochem. biophys. Res. Commun. *65* : 407 - 412, 1975.

23564 - **MASONER, M., KASEMIR, H.** : Control of chlorophyll synthesis by phytochrome. I. The effect of phytochrome on the formation of 5-aminolevulinate in mustard seedlings. - Planta *126* : 111 - 117, 1975.

23565 - **MASSIE, D.R., NORRIS, K.H.** : High-intensity spectrophotometer interfaced with a computer for food quality measurement. - Trans. ASAE *18* : 173 - 176, 1975. [Pigments *in vivo*.]

23566 - **MASTERS, B.R., MAUZERALL, D.** : The effects of plastoquinones on the photoconductivity and photovoltage of chlorophyll *a* containing lipid bilayers : coupling of photons, protons and electrons. - Biophys. J. *15* (2, Pt. 2): 303a, 1975.

23567 - **MASYUK, N.P., RADCHENKO, M.I.** : O pigmentakh zelenykh vodorosleĭ v svyazi s nekotorymi voprosami taksonomii *Chlorophycophyta*. [Green algae pigments in relation to some problems of *Chlorophycophyta* taxonomy.] - In : Flora, Sistematika i Filogeniya Rasteniĭ. Pp. 92 - 101. Naukova Dumka, Kiev 1975. [In R.]

23568 - **MATAR, Y., DÖRING, H.-W., MARSCHNER, H.** : Auswirkungen von NaCl und Na_2SO_4 auf Substanzbildung, Mineralstoffgehalt und Inhaltsstoffe bei Spinat und Salat. - Z. Pflanzenernähr. Bodenkunde *3* : 295 - 307, 1975. [Chl, Car.]

*23569 - **MATHEKE, G.E.M., HORNER, R.** : Primary productivity of the benthic microalgae in the Chukchi Sea near Barrow, Alaska. - J. Fish. Res. Board Can. *31* : 1779- - 1786, 1974. [Chl.]

*23570 - **MATHEW, C., RAMADASAN, A.** : Chlorophyll content in certain cultivars and hybrids of coconut. - J. Plant Crops *1* (Suppl. - The Proceedings of the National Symposium on Plantation Crops): 96 - 98, 1973.

23571 - **MATHIESON, A.C., NORALL, T.L.** : Physiological studies of subtidal red algae. - J. exp. mar. Biol. Ecol. *20* : 237 - 247, 1975. [Ps.]

23572 - **MATHIESON, A.C., NORALL, T.L.** : Photosynthetic studies of *Chondrus crispus*. - - Mar. Ecol. *33* : 207 - 213, 1975.

23573 - **MATHIS, P., VERMEGLIO, A.** : Chlorophyll radical cation in Photosystem II of chloroplasts. Millisecond decay at low temperature. - Biochim. biophys. Acta *396* : 371 - 381, 1975.

23574 - **MATHIS, P., VERMEGLIO, A., HAVEMAN, J.** : Primary reactions of photosynthesis in green plants. A study of photosystem-2 at low temperature. - In : **JOUSSOT-DUBIEN, J.** (ed.) : Lasers in Physical Chemistry and Biophysics. Pp. 465 - - 475. Elsevier sci. Publ. Comp., Amsterdam 1975.

23575 - **MATIENKO, E.B.** : Grebenchatye tilakoidy v karotinoidoplastakh subépidermal'nykh kletok ploda tykvy. [Crested thylakoids in carotenoidoplasts of subepidermal cells of the pumpkin *Cucurbita pepo* L. var. *aurantiformis* HORT.] - Tsitologiya *17* : 1323 - 1325, 1975. [In R, ab : E.]

23576 - **MATORIN, D.N., VENEDIKTOV, P.S., MAKEVNINA, M.G.** : Primenenie metoda registratsii poslesvecheniya zelenykh vodorosleĭ dlya opredeleniya zagryaznennosti fitotoksicheskimi veshchestvami pochvy i vody. [Use of the method of recording of the green algae delayed light emission for the determination of soil and water polution by phytotoxic substances.] - Nauch. Dokl. vyssh. Shkoly, biol. Nauki *18* (12) : 122 - 125, 1975. [In R.]

23577 - **MATSON, R.S., KIMURA, T.** : Immunological quantitation of chloroplast ferredoxin. - Biochim. biophys. Acta *396* : 293 - 300, 1975.

23578 - **MATSUDA, M., BAUMGARTNER, A.** : Ökosystematische Simulation des Nutzeffektes der Sonnenenergie für Wälder. - Forstwiss. Centralbl. *94* : 89 - 104, 1975.

23579 - **MATSUI, T., EGUCHI, H., SOEJIMA, Y., HAMAKOGA, M.** : Control of artificial light for plants. II. Automatic control of light intensity and spectral composition. - Environ. Control Biol. *13* : 109 - 116, 1975. [PhAR.]

23580 - **MATSUZAKI, E., KAMIMURA, Y., YAMASAKI, T., YAKUSHIJI, E.** : Purification and properties of cytochrome *f* from *Brassica komatsuna* leaves. - Plant Cell Physiol. *16* : 237 - 246, 1975.

23581 - **MATTHEIS, J.R., REBEIZ, C.A.** : The net synthesis of protochlorophyllide from tetrapyrroles by isolated plastids. - Plant Physiol. *56* (Suppl.) : 31, 1975.

23582 - **MATTSON, W.J., ADDY, N.D.** : Phytophagous insects as regulators of forest primary production. - Science *190* : 515 - 522, 1975.

*23583 - **MATWIYOFF, N.A., BURNHAM, B.F.** : Carbon-13 NMR spectroscopy of tetrapyrroles. - Ann. New York Acad. Sci. *206* : 365 - 382, 1973. [Chl.]

23584 - **MAUZERALL, D., HONG, F.T.** : Photochemistry of porphyrins in membranes and photosynthesis. - In : **SMITH, K.M.** (ed.) : Porphyrins and Metalloporphyrins. Pp. 701 - 725. Elsevier sci. Publ. Comp., Amsterdam - Oxford - New York 1975.

23585 - **MAY, D.S.** : Genetic and physiological adaptation of the Hill reaction in altitudinally-diverse populations of *Taraxacum*. - Photosynthetica *9* : 293 - 298, 1975.

23586 - **MAYNE, B.C., DEE, A.M., EDWARDS, G.E.** : Photosynthesis in mesophyll protoplasts and bundle sheath cells of various types of C_4 plants. III. Fluorescence emission spectra, delayed light emission, and *P*700 content. - Z. Pflanzenphysiol. *74* : 275 - 291, 1975.

23587 - **McARTHUR, J.A., HESKETH, J.D., BAKER, D.N.** : Cotton. - In : **EVANS, L.T.** (ed.): Crop Physiology. Pp. 297 - 325. Cambridge Univ. Press, London - New York 1975. [Ps.]

23588 - **McARTHUR, J.M., HIKICHI, M.** : Fraction 1 protein concentration in plants by analytical ultracentrifuge. - Anal. Biochem. *66* : 12 - 17, 1975.

23589 - **McCANN, D.J., SADDLER, H.D.W.** : Photobiological energy conversion - a practical proposition ? - In : Feasibility of Alternative Renewable Resources. (Symp. Solar Energy Resources.) Pp. 15 - 40. Soc. social Responsibility Sci., Canberra 1975. [Ps.]

*23590 - **McCARTHY, J.J., TAYLOR, W.R., LOFTUS, M.E.** : Significance of nanoplankton in the Chesapeake Bay estuary and problems associated with the measurement of nanoplankton productivity. - Mar. Biol. *24* : 7 - 16, 1974.

23591 - **McCARTY, R.E., PORTIS, A.R., Jr., MAGNUSSON, R.P.** : Effects of adenine nucleotides on the hydrogen ion concentration gradient in illuminated chloroplasts and on the reactivity of coupling factor 1 to N-ethylmaleimide. - In : **AVRON, M.** (ed.) : Proceedings of the Third International Congress on Photosynthesis. Vol. II. Pp. 975 - 984. Elsevier, Amsterdam - Oxford - New York 1975.

23592 - **McCAUGHEY, J.H., DAVIES, J.A.** : Energy exchange in a corn canopy. - Can. J. Plant Sci. *55* : 691 - 704, 1975.

23593 - **McCOLL, R.H.S.** : Availability of soil and sediment phosphorus to a planktonic alga. - New Zeal. J. mar. Freshwater Res. *9* : 169 - 182, 1975. [Chl.]

23594 - **McCONNELL, R.L., TOWNSEND, C.E.** : Inheritance of a chlorophyll deficiency in diploid alsike clover. - Crop Sci. *15* : 583 - 584, 1975.

23595 - **McCORMICK, J.M., QUINN, P.T.** : Phytoplankton diversity and chlorophyll-*A* in a polluted estuary. - Mar. Pollut. Bull. *6* (7) : 105 - 106, 1975.

23596 - **McCRACKEN, M.D., ADAMS, M.S., TITUS, J., STONE, W.** : Diurnal course of photosynthesis in *Myriophyllum spicatum* and *Oedogonium*. - Oikos *26* : 355 - 361, 1975.

23597 - **McFADDEN, B.A., LORD, J.M., ROWE, A., DILKS, S.** : Composition, quaternary structure, and catalytic properties of D-ribulose-1,5-bisphosphate carboxylase from *Euglena gracilis*. - Europ. J. Biochem. *54* : 195 - 206, 1975.

23598 - **McFADDEN, B.A., TABITA, F.R., KUEHN, G.D.** : Ribulose-diphosphate carboxylase from the hydrogen bacteria and *Rhodospirillum rubrum*. - In : **COLOWICK, S.P., KAPLAN, N.O.** (ed.) : Methods in Enzymology. Vol. 42. Pp. 461 - 472. Academic Press, New York - San Francisco - London 1975.

23599 - **McFEETERS, R.F.** : Substrate specificity of chlorophyllase. - Plant Physiol. *55* : 377 - 381, 1975.

23600 - **McINTOSH, A.R., CHU, M., BOLTON, J.R.** : Flash photolysis electron spin resonance studies of the light reversible counterpart to signal 1 at low temperatures in spinach subchloroplast particles - the true primary acceptor of photosystem I? - In : **AVRON, M.** (ed.) : Proceedings of the Third International Congress on Photosynthesis. Vol. I. Pp. 389 - 398. Elsevier, Amsterdam - Oxford - New York 1975.

23601 - **McINTOSH, A.R., CHU, M., BOLTON, J.R.** : Flash photolysis electron spin resonance studies of the electron acceptor species at low temperatures in Photosystem I of spinach subchloroplast particles. - Biochim. biophys. Acta *376* : 308 - 314, 1975.

23602 - **McKINION, J.M., BAKER, D.N., HESKETH, J.D., JONES, J.W.** : Part 4. - SIMCOT II: A simulation of cotton growth and yield. - In : Computer Simulation of a Cotton Production System. Users Manual. Pp. 27 - 82. Agr. Res. Serv., US Dept. Agr. 1975.

23603 - **MCKINION, J.M., JONES, J.W., HESKETH, J.D.** : A system of growth equations for the continuous simulation of plant growth. - Trans. ASAE *18* : 975 - 979, 984, 1975. [Ps.]

23604 - **McKINION, J.M., JONES, J.W., HESKETH, J.D., LANE, H.C.** : Simulation of plant growth : Morphogenetic control of leaf area expansion. - In : Proceedings 1975 Beltwide Cotton Production Research Conference. Pp. 56 - 59. Nat. Cotton Council, Mississippi 1975.

23605 - **McLAUGHLIN, S.B., Jr., BARNES, R.L.** : Effects of fluoride on photosynthesis and respiration of some South-East American forest trees. - Environ. Pollut. *8* : 91 - 96, 1975.

23606 - **McMURRAY, G., OLIVE, J.H.** : Summer phytoplankton photosynthesis in a northeastern Ohio glacial lake. - Ohio J. Sci. *75* : 238 - 250, 1975.

*23607 - **MEDEGHINI-BONATTI, P., FORNASIERO-BARONI, R.** : Plastidi in foglioline di gemme di *Larix decidua*. [Plastids in bud leaflets of *Larix decidua*.] - Atti Soc. Nat. Mat. Modena *105* : 119 - 125, 1974. [In Ital.]

23608 - **MEDERSKI, H.J., CHEN, L.H., CURRY, R.B.** : Effect of leaf water deficit on stomatal and nonstomatal regulation of net carbon dioxide assimilation. - - Plant Physiol. *55* : 589 - 593, 1975.

23609 - **MEDERSKI, H.J., CURRY, R.B., CHEN, L.H.** : Effects of irradiance and leaf water deficit on net carbon dioxide assimilation and mesophyll and transport resistances. - Plant Physiol. *55* : 594 - 597, 1975.

23610 - **MEDINA, E.** : Dark CO_2 fixation, habitat preference and evolution within the *Bromeliaceae*. - Evolution *28* : 677 - 686, 1975.

23611 - **MEHARD, C.W., PRÉZELIN, B.L., HAXO, F.T.** : Isolation and characterization of dinoflagellate and chrysophyte cytochrome-*f* (553-4). - Phytochemistry *14* : 2379 - 2382, 1975.

23612 - **MEIDNER, H.** : Water supply, evaporation, and vapour diffusion in leaves. - J. exp. Bot. *26* : 666 - 673, 1975. [Fluxes of water vapour and CO_2.]

*23613 - **MEIER, P.G., GANNON, J.J., BENDER, M.E.** : Oxygen production in experimental channels. - In : **JENKINS, S.H.** (ed.) : Advances in Water Pollution Research. Vol. I. Pp. I-9/1 - I-9/17. Pergamon Press, Oxford - New York 1971. [Chl.]

23614 - **MEISCH, H.-U., BIELIG, H.-J.** : Effect of vanadium on growth, chlorophyll formation and iron metabolism in unicellular green algae. - Arch. Microbiol. *105*: 77 - 82, 1975.

23615 - **MELANDRI, B.A., ZANNONI, D., CASADIO, R., BACCARINI-MELANDRI, A.** : Energy conservation and transduction in photosynthesis and respiration of facultative photosynthetic bacteria. - In : **AVRON, M.** (ed.) : Proceedings of the Third International Congress on Photosynthesis. Vol. II. Pp. 1147 - 1162. Elsevier, Amsterdam - Oxford - New York 1975.

23616 - **MELCAREK, P.K., BROWN, G.N.** : *In vivo* chlorophyll fluorescence monitoring with solid state photosensors. - Physiol. Plant. *35* : 147 - 151, 1975.

23617 - **MELCAREK, P.K., BROWN, G.N.** : Fluorescence properties of hemlock and black locust under various temperature regimes. - Plant Physiol. *56* (Suppl.) : 9, 1975.

23618 - **MELIS, A., HOMANN, P.H.** : Kinetic analysis of the fluorescence induction in 3-(3,4-dichlorophenyl)-1,1-dimethylurea poisoned chloroplasts. - Photochem. Photobiol. *21* : 431 - 437, 1975.

23619 - **MELLER, E., BELKIN, S., HAREL, E.** : The biosynthesis of δ-aminolevulinic acid in greening maize leaves. - Phytochemistry *14* : 2399 - 2402, 1975.

23620 - **MELLINGER, M.V., McNAUGHTON, S.J.** : Structure and function of successional vascular plant communities in central New York. - Ecol. Monogr. *45* : 161 - 182, 1975. [Biomass.]

23621 - **MEL'NIKOV, S.S.** : Ėndosimbioz khloroplastov s zhivotnymi kletkami. [Endosymbiosis of chloroplasts with animal cells.] - In : **SHLYK, A.A.** (ed.) : Biosintez i Sostoyanie Khlorofillov v Rastenii. Pp. 227 - 239, 248. Nauka i Tekhnika, Minsk 1975. [In R, ab : E.]

23622 - **MENDE, D., WIESSNER, W.** : Indications for the existence of a special electron acceptor of photosystem II in photoheterotrophically cultivated *Chlamydobotrys stellata* not belonging to the normal non-cyclic electron transport. - In: **AVRON, M.** (ed.) : Proceedings of the Third International Congress on Photosynthesis. Vol. I. Pp. 505 - 513. Elsevier, Amsterdam - Oxford - New York 1975.

23623 - **MENDIOLA-MORGENTHALER, L., EIKENBERRY, E.F., PRICE, C.A.** : Proteins of the chloroplast and cytoplasmic ribosomes of *Euglena*. - Plant Cell Physiol. *16*: 981 - 994, 1975.

23624 - **MENKE, W., KOENIG, F., RADUNZ, A., SCHMID, G.H.** : The isolation of some polypeptides from the thylakoid membrane, their localization and function. - FEBS Lett. *49* : 372 - 375, 1975.

23625 - **MEREZHINSKIĬ, Yu.G., MITROFANOV, B.A., IVANISHCHEV, V.N.** : Intensivnost' fotosinteza v zavisimosti ot urovnya postupleniya i nakopleniya gerbitsidov v rasteniyakh. [Photosynthetic rate in dependence on the level of herbicides uptake and accumulation in plants.] - Fiziol. Biokhim. kul't. Rast. *7* : 345 - - 350, 1975. [In R, ab : E.]

23626 - **MERGENHAGEN, D., SCHWEIGER, H.G.** : Circadian rhythm of oxygen evolution in cell fragments of *Acetabularia mediterranea*. - Exp. Cell Res. *92* : 127 - 130, 1975.

23627 - **MERGENHAGEN, D., SCHWEIGER, H.G.** : The effect of different inhibitors of transcription and translation on the expression and control of circadian rhythm in individual cells of *Acetabularia*. - Exp. Cell Res. *94* : 321 - 326, 1975. [Ps.]

23628 - **MÉRIAUX, S., WEBER, M., ROLLIN, H., RUTTEN, P.** : Effets des modalités d'application d'une contrainte hydrique sur quelques aspects de la morphologie et de l'activité de la feuille de *Vitis vinifera* L., variété Cabernet-Sauvignon. - - Compt. rend. Acad. Sci. Paris, Sér. D *281* : 1235 - 1237, 1975.

23629 - **MERLYN, M.B., ZELITCH, I.** : Photosynthetic CO_2 uptake by tobacco callus cells. - Plant Physiol. *56* (Suppl.) : 72, 1975.

23630 - **MERZLYAK, M.N., YUFEROVA, S.G.** : Okislenie lipidnykh komponentov v izolirovannykh khloroplastakh pod deĭstviem sveta. [Oxidation of lipid components in isolated chloroplasts induced by light.] - Fiziol. Rast. *22* : 896 - 902, 1975. [In R, ab : E.]

23631 - **METZNER, H.** : Photosynthese - Umwandlung der Sonnenenergie. - Umschau Wiss. Tech. *75* : 435 - 441, 1975.

23632 - **METZNER, H.** : Water decomposition in photosynthesis ? A critical reconsideration. - J. theor. Biol. *51* : 201 - 231, 1975.

23633 - **METZNER, H., FISCHER, K., LUPP, G.** : Energy conservation in photosynthesis models. II. Salt effects on oxygen evolution. - Photosynthetica *9* : 327 - 330, 1975.

23634 - **MEYER, C.P., CANNY, M.J.** : CO_2 storage in *Eucalyptus* oil glands : A hypothesis disproved. - Aust. J. Plant Physiol. *2* : 647 - 658, 1975.

23635 - **MEYER, T.E., AMBLER, R.P., BARTSCH, R.G., KAMEN, M.D.** : Amino acid sequence of cytochrome-*c*' from the purple photosynthetic bacterium *Rhodospirillum rubrum* S1. - J. biol. Chem. *250* : 8416 - 8421, 1975.

23636 - **MICHAEL, S.D., SPURR, A.R.** : An albino mutant in *Plantago insularis* requiring thiamine pyrophosphate - effects of TPP on chloroplast ultrastructure. - New Phytol. *74* : 227 - 234, 1975.

23637 - **MICHAELS, A., MARGULIES, M.M.** : Amino acid incorporation into protein by ribosomes bound to chloroplast thylakoid membranes : Formation of discrete products. - Biochim. biophys. Acta *390* : 352 - 362, 1975.

23638 - **MICHEL, J.M.** : Effects of CCC on photosynthesis in *Euglena*. - In : **MARCELLE, R.** (ed.) : Environmental and Biological Control of Photosynthesis. Pp. 217 - - 225. Dr. W. Junk b.v. Publ., The Hague 1975.

23639 - **MICHEL, J.M., SIRONVAL, C.** : 77 °K fluorescence spectra of dark-grown *Euglena gracilis* subjected to short light flashes. - Plant Sci. Lett. *4* : 419 - 425, 1975.

23640 - **MIDGLEY, D.** : Investigations into the use of gas-sensing membrane electrodes for the determination of carbon dioxide in power station waters. - Analyst *100* : 386 - 399, 1975.

23641 - **MIGINIAC-MASLOW, M., HOARAU, A.** : The reducing power and the CO_2 fixation activity of isolated chloroplasts. - Physiol. Plant. *35* : 186 - 190, 1975.

23642 - **MIKHAĬLOVA, S.A., CHAĬKA, M.T., MANANKINA, E.E.** : Fraktsionirovanie pigmentnogo fonda membran pri razvitii khloroplastov, obrazuyushchikh i ne obrazuyushchikh grany. [Membrane pigment fractionation of developing chloroplasts forming and not forming grana.] - In : **SHLYK, A.A.** (ed.) : Biosintez i Sostoyanie Khlorofillov v Rastenii. Pp. 197 - 206, 247. Nauka i Tekhnika, Minsk 1975. [In R, ab : E.]

23643 - **MIKHAĬLOVA, T.P., YAKOVUK, A.S., ZOZULYA, L.V.** : Nekotorye fiziologo-biokhimicheskie osobennosti raznokachestvennykh semyan tabaka. [Certain physiological and biochemical peculiarities of tobacco seeds of different quality.] - Fiziol. Biokhim. kul't. Rast. *7* : 311 - 316, 1975. [Ps; in R, ab : E.]

23644 - **MIKHEEVA, T.M.** : Otsenka velichin biomassy fitoplanktona v ozerakh mira. [Evaluation of phytoplankton biomass in world lakes.] - Gidrobiol. Zh. *11* (3) : 90 - 104, 1975. [In R.]

23645 - **MIKOSHIBA, K., HORIUCHI, J., HORIE, M.** : Growth pattern and climate. 3.5.4 Developmental analysis in soybean. - In : **MURATA, Y.** (ed.) : JIBP Synthesis. Vol. 11. Pp. 149 - 159. Univ. Tokyo Press, Tokyo 1975. [Growth analysis.]

23646 - **MIKULSKI, J.S., ADAMCZAK, B., BITTEL, L., BOHR, R., BRONISZ, D., DONDERSKI, W., GIZIŃSKI, A., ŁUŚCIŃSKA, M., REJEWSKI, M., STRZELCZYK, E., WOLNOMIEJSKI, N., ZAWIŚLAK, W., ŻYTKOWICZ, R.** : Basic regularities of productive processes in the Iława lakes and in the Gopło lake from the point of view of utility values of the water. - Pol. Arch. Hydrobiol. *22* : 101 - 122, 1975.

23647 - **MILES, D.** : Genetic analysis of photosynthesis (photosynthesis mutants, electron transport, phosphorylation). - Stadler Symp. *7* : 135 - 154, 1975.

23648 - **MILFORD, G.F.J.** : Effects of mist irrigation on the physiology of sugar beet. - Ann. appl. Biol. *80* : 247 - 250, 1975. [Ps.]

23649 - **MILFORD, G.F.J., PEARMAN, I.** : The relationship between photosynthesis and the concentrations of carbohydrates in the leaves of sugar beet. - Photosynthetica *9* : 78 - 83, 1975.

23650 - **MILICĂ, C.I., PĂCURAR, I.** : Particularităţi fiziologice ale biotipurilor extrase din soiul Bezostaia 1. [Physiological peculiarities of the biotypes derived from the variety Bezostaya 1.] - Ann. Inst. Cercet. Cereale Plante teh. - Fundulea *40* (Ser. C) : 47 - 65, 1975. [Growth analysis; in Roum., ab : E, R.]

23651 - **MILLER, D.E.** : Simple flow plugs for regulating gas composition. - Proc. Soil Sci. Soc. Amer. *39* : 587, 1975. [For Ps measurement.]

23652 - **MILLER, J.R., TOCHER, R.D.** : Photosynthesis and respiration of *Arceuthobium tsugense (Loranthaceae)*. - Amer. J. Bot. *62* : 765 - 769, 1975.

23653 - **MILLER, P.C.** : A comparison in short-term effects of thermal addition on photosynthesis and plant-water stress in three ecosystems. - In : Environmental

Effects of Cooling Systems at Nuclear Power Plants. Pp. 623 - 636. Int. at. Energy Agency, Vienna 1975.

23654 - **MILLER, P.C., COLLIER, B.D., BUNNELL, F.L.** : Development of ecosystem modeling in the tundra biome. - In : **PATTEN, B.C.** (ed.) : Systems Analysis and Simulation in Ecology. Vol. III. Pp. 95 - 115. Academic Press, New York - San Francisco - London 1975.

23655 - **MILLER, R.W., MACDOWALL, F.D.H.** : The Tiron free radical as a sensitive indicator of chloroplastic photoautoxidation. - Biochim. biophys. Acta *387* : 176-- 187, 1975.

23656 - **MILLS, J., BARBER, J.** : Energy-dependent cation-induced control of chlorophyll *a* fluorescence in isolated intact chloroplasts. - Arch. Biochem. Biophys. *170*: 306 - 314, 1975.

23657 - **MINE, A., MATSUNAKA, S.** : Mode of action of bentazon : effect on photosynthesis. - Pestic. Biochem. Physiol. *5* : 444 - 450, 1975.

23658 - **MIROSHNICHENKO, Yu.M.** : Osobennosti sezonnoĭ dinamiki produktivnosti v fitotsenozakh afro-aziatskoĭ aridnoĭ oblasti. [Peculiarities of seasonal dynamics of productivity in phytocoenoses of afro-asiatic arid area.] - Bot. Zh. *60* : 1164 - 1178, 1975. [In R.]

23659 - **MITAMURA, O., SAIJO, Y.** : Decomposition of urea associated with photosynthesis of phytoplankton in coastal waters. - Mar. Biol. *30* : 67 - 72, 1975.

23660 - **MITCHELL, C.A., STOCKING, C.R.** : Kinetics and energetics of light-driven chloroplast glutamine synthesis. - Plant Physiol. *55* : 59 - 63, 1975.

23661 - **MITCHELL, D.S., TUR, N.M.** : The rate of growth of *Salvinia molesta (S. auriculata* AUCT.) in laboratory and natural conditions. - J. appl. Ecol. *12* : 213 - 225, 1975. [Growth analysis.]

23662 - **MITCHELL, P.** : Proton translocation mechanisms and energy transduction by adenosine triphosphatases : an answer to criticism. - FEBS Lett. *50* : 95 - 97, 1975.

23663 - **MITCHELL, S.F.** : Phosphate, nitrate, and chloride in a eutrophic coastal lake in New Zealand. - New Zeal. J. mar. Freshwater Res. *9* : 183 - 198, 1975. [Chl.]

23664 - **MITCHELL, S.F.** : Some effects of agricultural development and fluctuations in water level on the phytoplankton productivity and zooplankton of a New Zealand reservoir. - Freshwater Biol. *5* : 547 - 562, 1975.

23665 - **MITROFANOV, B.A., GOĬTSA, N.I., ROGACHENKO, A.D., OKANENKO, A.S.** : Vodnyĭ defitsit lista i vlagoobespechennost' rasteniĭ kak vazhnye faktory intensivnosti i produktivnosti fotosinteza. [Leaf water deficit and water supply of plants as important factors influencing photosynthetic rate and productivity.] - In: Vodoobmen Rasteniĭ pri Neblagopriyatnykh Usloviyakh Sredy. Pp. 183 - 187. Shtiintsa, Kishinev 1975. [In R.]

23666 - **MITSCHERLICH, G., KÜNSTLE, E.** : Photosynthese, Transpiration und Atmung in einem Mischbestand im Schwarzwald. II. Teil : Transpiration. - Allgem. Forst-Jagdzeit. *146* (3/4) : 88 - 100, 1975.

23667 - **MITSUI, A.** : Multiple utilization of tropical and subtropical marine photosynthetic organisms. - In : The 3rd International Ocean Development Conference. Vol. III. Pp. 13 - 29. Seino Print. Co., Tokyo 1975.

23668 - **MIYACHI, S.** : Photosynthesis in *Chlorella*. - In: **TOKIDA, J., HIROSE, H.** (ed.): Advance of Phycology in Japan. Pp. 193 - 201. Dr. W. Junk b.v. Publ., The Hague 1975.

23669 - **MIYASAKA, A., MURATA, Y., IWATA, T.** : Leaf area development and leaf senescence in relation to climatic and other factors. - In : **MURATA, Y.** (ed.) : JIBP Synthesis. Vol. 11. Pp. 72 - 85. Univ. Tokyo Press, Tokyo 1975. [Growth analysis.]

23670 - **MŁODZIANOWSKI, F., SIWECKI, R.** : Ultrastructural changes in chloroplasts of *Populus tremula* L. leaves affected by the fungus *Melampsora pinitorqua* BRAUN. ROSTR. - Physiol. Plant Pathol. *6* : 1 - 3, 1975.

23671 - **MOCK, J.J., PEARCE, R.B.** : An ideotype of maize. - Euphytica *24* : 613 - 623, 1975. [Ps.]

23672 - **MOGILEVA, G.A., ZELENSKIĬ, M.I., SHITOVA, I.P.** : Izmeneniya fotokhimicheskoĭ aktivnosti khloroplastov v ontogeneze pshenits.[Changes in the photochemical activity of chloroplasts in the course of wheat ontogenesis.] - Byull. vses. nauch.-issled. Inst. Rastenievod. N.I. Vavilova *56* : 68 - 73, 1975. [In R.]

23673 - **MOHANTY, P., BOYER, J.S.** : Quantum yield of photosynthesis in leaves and chloroplasts when the leaves have low water potentials. - Plant Physiol. *56* (Suppl.) : 61, 1975.

23674 - **MOHR, H., KASEMIR, H.** : Control of plastid development and chlorophyll synthesis by phytochrome. - Proc. ind. nat. Sci. Acad. *41 B* : 503 - 525, 1975.

23675 - **MOK, M.C., GABELMAN, W.H., SKOOG, F.** : Carotenoid synthesis in carrot tissue cultures. - Plant Physiol. *56* (Suppl.) : 29, 1975.

23676 - **MOLCHANOV, M.I., TRUSOVA, V.M.** : Vklyuchenie ^{32}P v fosfolipidy differentsiruyushchikhsya khloroplastov. [Incorporation of ^{32}P in phospholipids of differentiating chloroplasts.] - Dokl. Akad. Nauk SSSR *220* : 1222 - 1225, 1975. [In R.]

23677 - **MOLCHANOV, M.I., TRUSOVA, V.M., OPARIN, A.I.** : Aminoatsilfosfatidilglitseriny differentsiruyushchikhsya khloroplastov. [Aminoacylphosphatidylglycerines of the differentiating chloroplasts.] - Dokl. Akad. Nauk SSSR *220* : 975 - 977, 1975. [In R.]

23678 - **MOLDAU, Kh.** : Optimal'noe raspredelenie assimilyatov pri defitsite vody. (Matematicheskaya model'.) [Optimal distribution of assimilates at limited water supply. (A mathematical model.)] - ENSV Tead. Akad. Toimetised, Biol. [Izv. Akad. Nauk est. SSR, Biol.] *24* : 3 - 9, 1975. [In R, ab : E, Est.]

23679 - **MOLDAU, Kh.A.** : Zavisimost' soprotivleniya ust'its ot meteorologicheskikh faktorov pri vodnom defitsite. [Influence of meteorological factors on stomatal resistance during water stress.] - In : **GAZALIĬ, G.I., BEĬDEMAN, I.N.** (ed.) : Vodnyĭ Obmen v Osnovnykh Tipakh Rastitel'nosti SSSR kak Ėlement Krugovorota Veshchestva i Ėnergii. Pp. 42 - 49. Nauka, sibir. Otd., Novosibirsk 1975. [Model; in R.]

*23680 - **MOLINE, H.E.** : Ultrastructure of *Datura stramonium* leaves infected with the physalis mottle strain of belladonna mottle virus. - Virology *56* : 123 - 133, 1973. [Chloroplasts.]

23681 - **MOLINE, H.E., JENSEN, S.G.** : Histochemical evidence for glycogen-like deposits in barley yellow dwarf virus-infected barley leaf chloroplasts. - J. Ultrastruct. Res. *53* : 217 - 221, 1975.

*23682 - **MOLYAKA, O.N., DUBYNA, D.V.** : Biologiya, tsenologiya ta zapasy sukhotsvitu bagnovogo (*Gnaphalium uliginosum* L.) na pidtoplenykh ploshchakh Kremenchuts'kogo vodoĭmyshcha. [Biology, cenology and resources of *Gnaphalium uliginosum* L. on water-covered areas of the Kremenchug reservoir.] - In : Roslynnî Resursy Ukraïny, Ïkh Vyvchennya ta Ratsional'ne Vykorystannya. Pp. 93 - 98, 206. Naukova Dumka, Kyïv 1973. [Chl, Car; in Ukr., ab : R.]

*23683 - **MOLYAKA, O.N., DUBYNA, D.V., LYUBENKO, M.O.** : Urozhaĭnist' i khimichnyĭ sklad deyakykh priberezhno-vodnykh roslyn Kremenchuts'kogo vodoĭmyshcha. [Productivity and chemical composition of some littoral aquatic plants of the Kremenchug reservoir.] - In : Introduktsiya ta Aklimatyzatsiya Roslyn na Ukraïni. Resp. Mizhvid. Zbyr. Vol. 6. Pp. 183 - 189, 219. Naukova Dumka, Kyïv 1973. [Car; in Ukr., ab : R.]

*23684 - **MOLYAKA, O.N., MARUNCHENKO, Yu.M., ROMODAN, V.N.** : Biologiya, produktyvnist' ta khimichni osoblyvosti vydiv rodu *Typha* L. Kremenchuts'kogo vodoĭmyshcha. [Biology, productivity and chemical properties of *Typha* species from the Kremenchug reservoir.] - In : Roslynni Resursy Ukraïny, Ïkh Vyvchennya ta Ratsional'na Vykorystannya. Pp. 78 - 83, 205. Naukova Dumka, Kyïv 1973. [Chl, Car; in Ukr., ab : R.]

*23685 - **MONHEIMER, R.H.** : Sulfate uptake as a measure of planktonic microbial production in freshwater ecosystems. - Can. J. Microbiol. *20* : 825 - 831, 1974.[Ps.]

23686 - **MONHEIMER, R.H.** : Planktonic microbial heterotrophy : its significance to community biomass production. - Verh. int. Verein. Limnol. *19* : 2658 - 2663, 1975. [Ps.]

23687 - **MONTALBINI, P., CAPPELLI, C.** : Variazioni del contenuto di clorofilla, plastochinoni ed ubichinone nelle foglie di pesco (cv. Elberta) attaccate da *Taphrina deformans* (BERK.) TUL. [Changes in the chlorophyll, plastoquinones and ubiquinone content of peach leaves (cv. Elberta) infected by *Taphrina deformans* (BERK.) TUL.] - Phytopathol. mediterr. *14* : 46 - 47, 1975. [In Ital.]

B23688 - **MONTEITH, J.L.** (ed.) : Vegetation and the Atmosphere. Vol. 1. Principles. - - Academic Press, London 1975. [Ps.]

23689 - **MONTES, G., BRADBEER, J.W.** : The biogenesis of the photosynthetic membranes and the development of photosynthesis in greening maize leaves. - In : **AVRON, M.** (ed.) : Proceedings of the Third International Congress on Photosynthesis. Vol. III. Pp. 1867 - 1876. Elsevier, Amsterdam - Oxford - New York 1975.

23690 - **MONTIES, B.** : Compartimentation des polyphénols présents dans les feuilles et les chloroplastes d'angiospermes : mise en évidence par extraction progressive. - Compt. rend. Acad. Sci. Paris, Sér. C *280* : 1331 - 1334, 1975.

23691 - **MONTIES, B.** : Rayonnement ultraviolet et photosynthèse. - In : **COSTES, C.** (ed.) : Photosynthèse et Production Végétale. Pp. 193 - 215. Gauthier-Villars, Paris 1975.

23692 - **MOONEY, H.A., BJÖRKMAN, O., BERRY, J.** : Photosynthetic adaptations to high temperature. - In : **HADLEY, N.F.** (ed.) : Environmental Physiology of Desert Organisms. Pp. 138 - 151. Dowden, Hutchinson and Ross, Inc., Stroudsburg, Pa. 1975.

23693 - **MOONEY, H.A., HARRISON, A.T., MORROW, P.A.** : Environmental limitations of photosynthesis on a California evergreen shrub. - Oecologia *19* : 293 - 301, 1975.

23694 - **MOORBY, J., JARMAN, P.D.** : The use of compartmental analysis in the study of the movement of carbon through leaves. - Planta *122* : 155 - 168, 1975. [Photosynthates.]

23695 - **MOORBY, J., MILTHORPE, F.L.** : Potato. - In : **EVANS, L.T.** (ed.) : Crop Physiology. Pp. 225 - 257. Cambridge Univ. Press, London - New York 1975. [Ps.]

23696 - **MOORBY, J., MUNNS, R., WALCOTT, J.** : Effect of water deficit on photosynthesis and tuber metabolism in potatoes. - Aust. J. Plant Physiol. *2* : 323 - 333, 1975.

*23697 - **MOORE, F.D.** : Pigment synthesis in nucleate and enucleate *Acetabularia mediterranea*. - Protoplasma *75* : 482, 1972.

23698 - **MOORE, K.G.** : Changes in leaf composition in *Parthenocissus tricuspidata* PLANCH. during growth and senescence of short shoots. - Ann. Bot. *39* : 631 - - 637, 1975. [Chl.]

23699 - **MOORE, P.D.** : Carbon dioxide fluctuations. - Nature *255* : 108, 1975. [Ps.]

23700 - **MOORE, P.D.** : Carbon dioxide flux in arctic ecosystems. - Nature *257* : 90, 1975.

23701 - **MOORE, P.D.** : Hill reaction in the mountains. - Nature *257* : 537 - 538, 1975.

23702 - **MORDASOVA, N.V.** : Nekotorye rezul'taty opredeleniya khlorofilla "*a*" i feofitina v vode i osadkakh Chernogo morya fluorestsentnym metodom. [Chlorophyll *a* and pheophytin determinations with the fluorescence method in the water and sediments of the Black Sea.] - Okeanologiya *15* : 1035 - 1039, 1975. [In R, ab : E.]

23703 - **MORESHET, S.** : Effects of phenyl-mercuric acetate on stomatal and cuticular resistance to transpiration. - New Phytol. *75* : 47 - 52, 1975.

23704 - **MORGAN, N.L., GRIFFITHS, W.T.** : The development of photosystem 1 in greening barley. - Biochem. Soc. Trans. *3* : 391 - 392, 1975.

23705 - **MORGENTHALER, J.-J., MARSDEN, M.P.F., PRICE, C.A.** : Factors affecting the separation of photosynthetically competent chloroplasts in gradients of silica sols. - Arch. Biochem. Biophys. *168* : 289 - 301, 1975.

23706 - **MORITA, K., KONO, M.** : Relationship between the change in the chloroplastic nitrogen fractions and the rate of oxygen evolution in rice plants. - Soil Sci. Plant Nutr. *21* : 263 - 271, 1975.

23707 - **MÖRSCHEL, E., WEHRMEYER, W.** : Cryptomonas biliprotein : phycocyanin-645 from a *Chroomonas* species. - Arch. Microbiol. *105* : 153 - 158, 1975.

*23708 - **MORTENSON, L.E., NAKOS, G.** : Bacterial ferredoxins and/or iron-sulfur proteins as electron carriers. - In : **LOVENBERG, W.** (ed.) : Iron-Sulfur Proteins. Vol. 1. Pp. 37 - 64. Academic Press, New York - London 1973.

23709 - **MOSHKOV, D.A., MUZAFAROV, E.N., MOSHKOVA, Z.G., AKULOVA, E.A.** : Issledovanie struktury khloroplastov metodom zamorazhivaniya-zameshcheniya. [The structure of chloroplasts studied by technique of freezing-replacement.] - Fiziol. Rast. *22* : 891 - 895, 1975. [In R, ab : E.]

*23710 - **MOSS, G.P.** : Carotenoids and polyterpenoids. - In : **OVERTON, K.H.** (ed.) : Terpenoids and Steroids. Vol. 1. Pp. 198 - 220. Chem. Soc., London 1971.

23711 - **MOUDRIANAKIS, E.N., TIEFERT, M.A.** : Transformations of CF1-bound nucleotides and non-participation of conventional adenylate kinase in these reactions. - - Biophys. J. *15* : 277a, 1975.

23712 - **MOUSSEAU, M.** : The effect of daylength on daily CO_2 balances of *Sinapis alba* L. - In : **MARCELLE, R.** (ed.) : Environmental and Biological Control of Photosynthesis. Pp. 135 - 147. Dr. W. Junk b.v. Publ., The Hague 1975.

23713 - **MOWER, R.L., HANCOCK, J.G.** : Mechanism of honeydew formation by *Claviceps* species. - Can. J. Bot. *53* : 2826 - 2834, 1975. [Photosynthates.]

23714 - **MOYSE, A.** : La photorespiration. - In : **COSTES, C.** (ed.) : Photosynthèse et Production Végétale. Pp. 127 - 146. Gauthier-Villars, Paris 1975.

23715 - **MOYSE, A.** : Internal control of chloroplasts activity. - In : **AVRON, M.** (ed.): Proceedings of the Third International Congress on Photosynthesis. Vol. II. Pp. 1463 - 1467. Elsevier, Amsterdam - Oxford - New York 1975.

23716 - **MUDD, J.B.** : Sulfur dioxide. - In : **MUDD, J.B., KOZLOWSKI, T.T.** (ed.) : Responses of Plants to Air Pollution. Pp. 9 - 22. Academic Press, New York - San Francisco - London 1975. [Ps, Chl.]

23717 - **MUDD, J.B.** : Peroxyacyl nitrates. - In : **MUDD, J.B., KOZLOWSKI, T.T.** (ed.) : Responses of Plants to Air Pollution. Pp. 97 - 119. Academic Press, New York - San Francisco - London 1975. [Ps.]

23718 - **MUKHAMADIEV, B.T., KVITKO, K.V.** : The relation of pigment composition in algae mutant cells to their resistance to inhibitors of photophosphorylation. - In : **NASYROV, Yu.S., ŠESTÁK, Z.** (ed.) : Genetic Aspects of Photosynthesis. Pp. 203 - 208. Dr. W. Junk B.V. Publ., The Hague 1975.

23719 - **MUKHIN, E.N., NEZNAĬKO, N.F., CHUGUNOV, V.A., EROKHIN, Yu.E.** : O dvukh formakh rastvorimogo ferredoksina iz list'ev *Pisum sativum* L. [Two forms of soluble ferredoxin from the leaves of *Pisum sativum* L.] - Dokl. Akad. Nauk SSSR *221* : 228 - 231, 1975. [In R.]

23720 - **MUKHIN, E.N., NEZNAĬKO, N.F., SHKUROPATOV, A.Ya., STOLOVITSKIĬ, Yu.M., EVSTIGNEEV, V.B.** : O fotosensibilizirovannom khlorofillom *a* vosstanovlenii ferredoksina v vodnom bufernom rastvore detergenta. [Chlorophyll *a*-photosensibilized reduction of ferredoxin in aqueous buffer solution of detergent.] - Dokl. Akad. Nauk SSSR *220* : 978 - 981, 1975. [In R.]

23721 - **MUKOHATA, Y., YAGI, T.** : Electron transport coupled to quasi-arsenylation in isolated chloroplasts. - J. Bioenerg. *7* : 111 - 121, 1975.

23722 - **MUKOHATA, Y., YAGI, T., SUGIYAMA, Y.** : Inhibition and uncoupling of the ADP--regulated electron transport in isolated chloroplasts. - J. Bioenerg. *7* : 103 - 109, 1975.

23723 - **MUKOHATA, Y., YAGI, T., SUGIYAMA, Y., MATSUNO, A., HIGASHIDA, M.** : Regulation of electron transport in isolated chloroplasts by sequential binding of adenine nucleotides to the coupling factor protein. - J. Bioenerg. *7* : 91 - 102, 1975.

23724 - **MURAI, T., KATOH, T.** : Photosystem I-dependent oxidation of organic acids in blue-green alga, *Anabaena variabilis*. - Plant Cell Physiol. *16* : 789 - 797, 1975.

23725 - **MURAKAMI, S.** : Structure of ribulose-1,5-diphosphate carboxylase crystal formed *in vivo* and *in vitro*. - In: **AVRON, M.** (ed.) : Proceedings of the Third International Congress on Photosynthesis. Vol. III. Pp. 2073 - 2080. Elsevier, Amsterdam - Oxford - New York 1975.

23726 - **MURAKAMI, S., TORRES-PEREIRA, J., PACKER, L.** : Structure of the chloroplast membrane - relation to energy coupling and ion transport. - In : **GOVINDJEE** (ed.) : Bioenergetics of Photosynthesis. Pp. 555 - 618. Academic Press, New York - San Francisco - London 1975.

23727 - **MURAKAMI, T.** : [An approach to the measurement of photosynthetic rate in single leaves by the aeration method. II. The effects of aeration velocity, CO_2--concentration and air stirring on the photosynthetic rate of mulberry leaf.] - Bull. sericult. exp. Sta. (Tokyo) *26* : 279 - 307, 1975. [In Jap., ab : E.]

23728 - **MURATA, N., FORK, D.C.** : Temperature dependence of chlorophyll *a* fluorescence in relation to the physical phase of membrane lipids in algae and higher plants. - Plant Physiol. *56* : 791 - 796, 1975.

23729 - **MURATA, N., FORK, D.C., TROUGHTON, J.** : Control of photosynthesis by temperature. - Carnegie Inst. Year Book *74* : 766 - 776, 1975.

23730 - **MURATA, N., TROUGHTON, J.H., FORK, D.C.** : Relationships between the transition of the physical phase of membrane lipids and photosynthetic parameters in *Anacystis nidulans* and lettuce and spinach chloroplasts. - Plant Physiol. *56* : 508 - 517, 1975.

B23731 - **MURATA, Y.** (ed.) : JIBP Synthesis. Vol. 11. Crop Productivity and Solar Energy Utilization in Various Climates in Japan. - Univ. Tokyo Press, Tokyo 1975.

23732 - **MURATA, Y.** : The effect of climatic factors and ageing on net assimilation rate of crop stands. - In : **MURATA, Y.** (ed.) : JIBP Synthesis. Vol. 11. Pp. 172 - 186. Univ. Tokyo Press, Tokyo 1975.

23733 - **MURATA, Y.** : Estimation and simulation of rice yield from climatic factors. - Agr. Meteorol. *15* : 117 - 131, 1975.

23734 - **MURATA, Y.** : The effect of solar radiation, temperature, and aging on net assimilation rate of crop stands - from the analysis of the "Maximal growth rate experiment" of IBP/PP. I. The case of rice plants. - Proc. Crop Sci. Soc. Jap. *44* : 153 - 159, 1975.

23735 - **MURATA, Y.** : The effect of solar radiation, temperature, and aging on net assimilation rate of crop stands - from the analysis of the "Maximal growth rate experiment" of IBP/PP. II. The case of maize and soybean plants. - Proc. Crop Sci. Soc. Jap. *44* : 160 - 165, 1975.

23736 - **MURATA, Y., MATSUSHIMA, S.** : Rice. - In : **EVANS, L.T.** (ed.) : Crop Physiology. Pp. 73 - 99. Cambridge Univ. Press, London - New York 1975. [Productivity.]

23737 - **MURATA, Y., TOGARI, Y.** : Summary of data. - In : **MURATA, Y.** (ed.) : JIBP Synthesis. Vol. 11. Crop Productivity and Solar Energy Utilization in Various Climates in Japan. Pp. 9 - 19. Univ. Tokyo Press, Tokyo 1975. [Ps.]

23738 - **MUREĬ, I.A.** : Vliyanie zagushcheniya posevov tomata na rost i vodnyĭ rezhim stebleĭ i korneĭ. [Effect of thickness of tomato sowing on growth and water regime of roots and stems.] - Fiziol. Rast. *22* : 814 - 825, 1975. [Biomass; in R, ab : E.]

23739 - **MUREŞAN, T., HURDUC, N., ŢERBEA, M., COSMIN, O., SARCA, T.** : Dependenţa dintre intensitatea fenomenului heterozis, productivitatea fotosintezei şi unii indici fiziologi la porumb. III. Dinamica conţinutului de pigmenţi foliari (clorofile, carotinoizi). [Relationship between heterosis intensity, photosynthetic productivity and some physiological indices in maize. III. Dynamics of leaf pigments content (chlorophylls, carotenoids).] - An. Inst. Cercet. Cereale Plante teh. - Fundulea *40* (Ser. C): 165 - 176, 1975. [In Roum., ab : E, R.]

23740 - **MUREŞAN, T., HURDUC, N., ŢERBEA, M., COSMIN, O., SARCA, T.** : Dependenţa dintre intensitatea fenomenolui heterozis, productivitatea fotosintezei şi unii indici fiziologici la porumb IV. Dinamica acumulării substanţei uscate în perioada umplerii bobului la unele linii consangvinizate. [The relationship between heterosis intensity, photosynthesis productivity and some physiological indices in maize IV. Dynamics of dry matter accumulation during the kernel filling stage in some inbred lines.] - An. Inst. Cercet. Cereale Plante teh. - Fundulea *40* (Ser. C) : 177 - 186, 1975. [In Roum., ab : E, R.]

23741 - **MURPHY, P.G.** : Net primary productivity in tropical terrestral ecosystems. - - In : **LIETH, H., WHITTAKER, R.H.** (ed.) : Primary Productivity of the Biosphere. Pp. 217 - 231. Springer-Verlag, Berlin - Heidelberg - New York 1975.

23742 - **MURZAEVA, S.V., AKULOVA, E.A.** : O lokalizatsii katalazy i peroksidazy v izolirovannykh khloroplastakh gorokha i ikh fragmentakh. [Catalase and peroxidase localization in isolated pea chloroplasts and their fragments.] - Biokhimiya *40* : 166 - 168, 1975. [In R, ab : E.]

23743 - **MURZAEVA, S.V., AKULOVA, E.A., TAUKELEVA, Sh.N., SHUBIN, L.M.** : Fluorestsentnyĭ metod opredeleniya perekisi vodoroda v izolirovannykh khloroplastakh. [Fluorescence method of hydrogen peroxide determination in isolated chloroplasts.] - Biokhimiya *40* : 973 - 977, 1975. [In R, ab : E.]

23744 - **MUSCATINE, L., POOL, R.R., TRENCH, R.K.** : Symbiosis of algae and invertebrates : aspects of the symbiont surface and the host-symbiont interface. - Trans. amer. microscop. Soc. *94* : 450 - 469, 1975. [Symbiotic chloroplasts.]

23745 - **MUSSELMAN, R.C., LESTER, D.T., ADAMS, M.S.** : Localized ecotypes of *Thuja occidentalis* L. in Wisconsin. - Ecology *56* : 647 - 655, 1975. [Ps.]

23746 - **MUSTÁRDY, L.A., MACHOWICZ, É., FALUDI-DÁNIEL, Á.** : Light-induced conformational changes of thylakoid membrane in normal and carotenoid deficient maize leaves. - Plant Physiol. *56* (Suppl.) : 47, 1975.

23747 - **MUSZYŃSKI, S.** : The ultrastructure of chloroplasts in variegata irregulare mutants of garden petunias (*Petunia hybrida* hort. *superbissima*). - Acta Soc. Bot. Pol. *44* : 25 - 28, 1975.

23748 - **MUTUSKIN, A.A.** : Plastotsianin - med'soderzhashchiĭ belok tsepi perenosa ėlektronov khloroplastov. [Plastocyanin - a copper-containing protein of the electron transport chain in chloroplasts.] - In : Rastitel'nye Belki i Ikh Biosintez. Pp. 190 - 194. Nauka, Moskva 1975. [In R.]

23749 - **MUZAFAROV, E.N.** : Bestrahlungsansatz zum Spektralphotometer SPECORD UV VIS des VEB Carl Zeiss JENA. - Jenaer Rundschau *20* : 143, 1975. [Ps.]

23750 - **MUZAFAROV, E.N., AKULOVA, E.A.** : Plastotsianin - perenoschik ėlektrona v khloroplastakh. [Plastocyanin - electron transmitter in chloroplasts.] - Izv. Akad. Nauk SSSR, Ser. biol. *1975* : 253 - 261, 1975. [In R, ab : E.]

23751 - **MYERS, J., GRAHAM, J.-R.** : Photosynthetic unit size during the synchronous life cycle of *Scenedesmus*. - Plant Physiol. *55* : 686 - 688, 1975.

23752 - **NAABER, L.Kh.** : O pigmentakh rasteniĭ zharkikh mestoobitaniĭ. [Pigments of plants from hot habitats.] - In : Fiziologiya i Biokhimiya Dikorastushchikh Kormovykh Rasteniĭ Uzbekistana. Pp. 3 - 14. FAN, Tashkent 1975. [In R.]

23753 - **NAABER, L.Kh.** : Vliyanie vlazhnosti pochvy na pigmentnyĭ kompleks pustynnykh rasteniĭ. [Effect of soil humidity on the pigment complex of desert plants.] - In : Fiziologiya i Biokhimiya Dikorastushchikh Kormovykh Rasteniĭ Uzbekistana. Pp. 84 - 98. FAN, Tashkent 1975. [In R.]

23754 - **NAD', A., BOKANI, A., ILLIK, M., BACH, B., DOMAN, N.G.** : O kharaktere nasledovaniya nekotorykh pokazateleĭ fotosinteticheskoĭ sposobnosti rasteniĭ. [Nature of inheritance of some properties connected with the photosynthetic ability of plants.] - Sel'skokhoz. Biol. *10* : 736 - 739, 1975. [In R, ab : E.]

23755 - **NADAKAVUKAREN, M.J., McCRACKEN, D.A., BERTAGNOLLI, B.L.** : Scanning electron microscopy of isolated chloroplasts. - Plant Physiol. *56* (Suppl.) : 32, 1975.

23756 - **NAGARAJAH, S.** : Effect of debudding on photosynthesis in leaves of cotton. - - Physiol. Plant. *33* : 28 - 31, 1975.

23757 - **NAGARAJAH, S.** : The relation between photosynthesis and stomatal resistance of each leaf surface in cotton leaves. - Physiol. Plant. *34* : 62 - 66, 1975.

23758 - **NAGATO, K., SUZUKI, S., SADO, T.** : [Process of dry matter accumulation into the kernel and grain characteristics of japonica and indica rice.] - Proc. Crop Sci. Soc. Jap. *44* : 431 - 437, 1975. [In Jap., ab : E.]

23759 - **NAIMAN, R.J., GERKING, S.D.** : Interrelationships of light, chlorophyll, and primary production in a thermal stream. - Verh. int. Limnol. *19* : 1659 - 1664, 1975.

23760 - **NAKAJIMA, T., YABUSHITA, Y., TABUSHI, I.** : Amino acid synthesis through biogenetic-type CO_2 fixation. - Nature *256* : 60 - 61, 1975.

23761 - **NAKAMURA, Y., YAMADA, M.** : Fatty acid synthesis by spinach chloroplasts I. Property of fatty acid synthesis from acetate. - Plant Cell Physiol. *16* : 139 - 149, 1975. [Ps, Chl.]

23762 - **NAKAMURA, Y., YAMADA, M.** : Fatty acid synthesis by spinach chloroplasts. III. Relationship between fatty acid synthesis and photophosphorylation. - Plant Cell Physiol. *16* : 163 - 174, 1975.

23763 - **NALBORCZYK, E., GEJ, B.** : Comparative investigations on the photosynthetic productivity of different varieties of spring wheat. - Pol. ecol. Stud. *1* : 71 - 80, 1975.

23764 - **NALBORCZYK, E., LaCROIX, L.J., HILL, R.D.** : Environmental influences on light and dark CO_2 fixation by *Kalanchoe daigremontiana*. - Can. J. Bot. *53* : 1132- - 1138, 1975.

23765 - **NASAR, S.A.K., MUNSHI, J.D.** : Studies on primary production in a freshwater pond. - Jap. J. Ecol. *25* : 21 - 23, 1975.

23766 - **NASYROV, Yu.S.** : Avtonomiya khloroplastov ? [Chloroplast autonomy?] - Priroda *1975* (11) : 42 - 48, 1975. [In R.]

B23767 - **NASYROV, Yu.S.** : Fotosintez i Genetika Khloroplastov. [Photosynthesis and Genetics of Chloroplasts.] - Nauka, Moskva 1975. [In R.]

23768 - **NASYROV, Yu.S., GILLER, Yu.E., USMANOV, P.D.** : Genetic control of chlorophyll biosynthesis and formation of its forms *in vivo*. - In : **NASYROV, Yu.S., ŠESTÁK, Z.** (ed.) : Genetic Aspects of Photosynthesis. Pp. 133 - 145. Dr. W. Junk B.V. Publ., The Hague 1975.

23769 - **NASYROV, Yu.S., KVITKO, K.V., ŠESTÁK, Z.** : Actual problems of genetics of photosynthesis. - In : **NASYROV, Yu.S., ŠESTÁK, Z.** (ed.) : Genetic Aspects of Photosynthesis. Pp. 369 - 374. Dr. W. Junk B.V. Publ., The Hague 1975.

B23770 - **NASYROV, Yu.S., ŠESTÁK, Z.** (ed.) : Genetic Aspects of Photosynthesis. - Dr. W. Junk B.V. Publ., The Hague 1975.

23771 - **NATO, A., DELEENS, E.** : Chlorophyll accumulation, CO_2 fixation enzymes, and Hill activity, in greening roots of *Lens culinaris*. - Physiol. Plant. *34* : 121 - 124, 1975.

23772 - **NATO, A., DELEENS, E.** : Relative contribution of ribulose 1,5-diphosphate carboxylase and phosphoenolpyruvate carboxylase to CO_2 fixation activity in roots of dark-grown or light-grown *Lens culinaris* seedlings. - Physiol. Plant. *34* : 309 - 313, 1975.

23773 - **NÁTR, L.** : Influence of mineral nutrition on photosynthesis and the use of assimilates. - In : **COOPER, J.P.** (ed.) : Photosynthesis and Productivity in Different Environments. Pp. 537 - 555. Cambridge Univ. Press, Cambridge - London - New York - Melbourne 1975.

23774 - **NÁTR, L.** : Některé problémy tvorby a distribuce asimilátů z hlediska celistvosti rostliny. [Some problems of the formation and distribution of photosynthates from the point of view of the plant as an entirety.] - Acta Univ. Agr. (Brno), Fac. Agron. A *23* : 845 - 853, 1975. [In Czech, ab : E, G, R.]

23775 - **NÁTR, L., APEL, P.** : Příspěvek k fotosyntetické charakteristice osin u jarního ječmene. [Contribution to photosynthetic characteristics of awns in spring barley. - Rostlinná Výroba (Praha) *21* : 75 - 82, 1975. [In Czech, ab : E, R.]

23776 - **NÁTR, L., KOUSALOVÁ, I.** : Tvorba a distribuce asimilátů ve vztahu k výnosu zrna u obilnin. [Formation and distribution of photosynthates in relation to grain yields in cereals.] - Stud. Inform. ÚVTI, zákl. Vědy Zemědělství *1975* (2) : 1 - 51, 1975. [In Czech, ab : E, R.]

23777 - **NÁTR, L., KOUSALOVÁ, I., KOPECKÝ, M., VU VAN VU** : Produkční potenciál a akumulační kapacita jarního ječmene. [Production potential and accumulation capacity of spring barley.] - Rostlinná Výroba (Praha) *21* : 419 - 427, 1975. [In Czech, ab : E, R.]

23778 - **NÁTR, L., LORENC, M.** : Měření listové plochy na základě analýzy obrazu přístrojem QTM (Quantimet). [Leaf area measurement by an equipment for quantitative image analysis (Quantimet).] - Rostlinná Výroba (Praha) *21* : 329 - 334, 1975. [In Czech, ab : E, R.]

23779 - **NÁTR, L., VU VAN VU** : Citlivost odrůd jarního ječmene na deficit minerálních živin. [Sensitiveness of summer barley cultivars to a deficit of mineral nutrients.] - Rostlinná Výroba (Praha) *21* : 1271 - 1275, 1975. [Ps; in Czech, ab : E, R.]

23780 - **NÁTR, L., VU VAN VU, KOUSALOVÁ, I.** : Vliv stupňovaných dávek dusíku na fotosyntetickou charakteristiku jarního ječmene. [The effect of gradated nitrogen rations on the photosynthetic characteristics of spring barley.] - Rostlinná Výroba (Praha) *21* : 393 - 402, 1975. [In Czech, ab : E, R.]

*23781 - **NAVARRO, S., COSTA, F., CARPENA, O.** : Influencia del hierro y manganeso en la formacion y evolucion de clorofilas y carotenoides en citrus. [Iron and manganese effects on the formation and evolution of chlorophyll and carotenoids in *Citrus*.] - In : Atti VI Simposio Internazionale di Agrochimica sul "Trasporto delle Molecole Organiche nelle Piante". Pp. 397 - 409. 1966. [In Span.]

23782 - **NAYLOR, D.G., TEARE, I.D.** : An improved, rapid, field method to measure photosynthesis with $^{14}CO_2$. - Agron. J. *67* : 404 - 406, 1975.

23783 - **NAYLOR, D.G., TEARE, I.D., KANEMASU, E.T.** : Photosynthesis in field-grown sorghum. - Fyton *33* : 97 - 102, 1975.

23784 - **NAYLOR, D.G., TEARE, I.D., NICKELL, C.D.** : Variations in photosynthesis of soybean under nonstressed and stressed conditions. - Fyton *33* : 103 - 109, 1975.

23785 - **NAZAROV, S.K.** : Fotosinteticheskiĭ metabolizm ugleroda u nekotorykh rasteniĭ o. Vaĭgach. [Photosynthetic carbon metabolism in certain plants of Vaigach Island.] - Bot. Zh. *60* : 1626 - 1631, 1975. [In R.]

23786 - **NEALES, T.F.** : The gas exchange patterns of CAM plants. - In : **MARCELLE, R.** (ed.) : Environmental and Biological Control of Photosynthesis. Pp. 299 - 310. Dr. W. Junk b.v. Publ., The Hague 1975.

23787 - **NEALES, T.F., HEW, C.S.** : Two types of carbon fixation in tropical orchids. - - Planta *123* : 303 - 306, 1975.

23788 - **NEDUKHA, E.M.** : Sovremennye predstavleniya o proiskhozhdenii plastid. [Modern ideas of plastids origin.] - Tsitol. Genet. *9* : 266 - 274, 1975. [In R, ab : E.]

23789 - **NEDUKHA, E.M.** : Ėlektronnotsitokhimicheskoe izuchenie lokalizatsii aktivnosti glyukozo-6-fosfatazy v tsitoplazme kletok funarii vlagomernoĭ. [Electron-cytochemical study of localization of glucose-6-phosphatase activity of cytoplasm of *Funaria hygrometrica* cells.] - Tsitol. Genet. *9* : 523 - 528, 1975. [In R, ab : E.]

23790 - **NEGHASSI, H.M., HEERMANN, D.F., SMIKA, D.E.** : Wheat yield models with limited soil water. - Trans. ASAE *18* : 549 - 557, 1975. [Primary production.]

23791 - **NEILSON, R.E., JARVIS, P.G.** : Photosynthesis in Sitka spruce (*Picea sitchensis* (BONG.) CARR.). VI. Response of stomata to temperature. - J. appl. Ecol. *12* : 879 - 891, 1975.

23792 - **NEKRASOV, L.I., LOBODA, N.I.** : Izuchenie adsorbtsii i fotokhimicheskikh svoĭstv khlorofilla "*a*" na monosloe al'bumina. [Adsorption and photochemical properties of chlorophyll *a* on albumin monolayer.] - Vestn. mosk. Univ., Khimiya *16* (1) : 19 - 22, 1975. [In R, ab : E.]

23793 - **NEKRASOV, L.I., NOVIKOVA, N.A.** : Kataliticheskoe posledeĭstvie ionov zheleza na intensivnost' fotosinteza. [The catalytic residual effect of Fe^{2+} ions on the rate of photosynthesis.] - Vestn. mosk. Univ., Khimiya *1975* (3) : 272 - - 275, 1975. [In R, ab : E.]

23794 - **NELLE, R., TISCHNER, R., HARNISCHFEGER, G., LORENZEN, H.** : Correlation between pigment systems and photosynthetic activity during the developmental cycle of *Chlorella*. - Biochem. Physiol. Pflanzen *167* : 463 - 472, 1975.

23795 - **NELSON, C.J., ASAY, K.H., HORST, G.L.** : Relationship of leaf photosynthesis to forage yield of tall fescue. - Crop Sci. *15* : 476 - 478, 1975.

23796 - **NELSON, C.J., ASAY, K.H., PATTON, L.D.** : Photosynthetic responses of tall fescue to selection for longevity below the CO_2 compensation point. - Crop Sci. *15* : 629 - 633, 1975.

23797 - **NELSON, C.J., TREHARNE, K.J., LLOYD, E.J.** : Genetic variation in enzyme activity of tall fescue leaf blades. - Crop Sci. *15* : 771 - 774, 1975. [Ps.]

23798 - **NELSON, N., BENGIS, C.** : Reaction center *P*700 from chloroplasts. - In : **AVRON, M.** (ed.) : Proceedings of the Third International Congress on Photosynthesis. Vol. I. Pp. 609 - 620. Elsevier, Amsterdam - Oxford - New York 1975.

23799 - **NELSON, N.,BENGIS, C., SILVER, B.L., GETZ, D., EVANS, M.C.W.** : Electron spin resonance studies of bound ferredoxin in chloroplast photosystem I reaction center. - FEBS Lett. *58* : 363 - 365, 1975.

23800 - **NELSON, N., KAMIENIETZKY, A., DETERS, D.W., NELSON, H.** : Subunit structure and function of CF_1. - In : **QUAGLIARIELLO, E., PAPA, S., PALMIERI, F., SLATER, E. C., SILIPRANDI, N.** (ed.) : Electron Transfer Chains and Oxidative Phosphorylation. Pp. 149 - 154. North-Holland Publishing Co., Amsterdam - Oxford, American Elsevier Publ. Co., Inc., New York 1975.

23801 - **NELSON, S.D., MAYO, J.M.** : The occurrence of functional non-chlorophyllous guard cells in *Paphiopedilum* spp. - Can. J. Bot. *53* : 1 - 7, 1975. [Leaf resistances.]

23802 - **NELSON, S.H., HWANG, K.E.** : Water usage by potato plants at different stages of growth. - Amer. Potato J. *52* : 331 - 339, 1975. [Dry-matter accumulation.]

23803 - **NEMEC, S.** : Vessel blockage by myelin forms in citrus with and without rough-lemon decline symptoms. - Can. J. Bot. *53* : 102 - 108, 1975. [Photosynthates.]

23804 - **NESTEROVICH, N.D., NOVIKOVA, A.A.** : Posledeĭstvie razlichnoĭ prodolzhitel'nosti osveshcheniya na rost i soderzhanie khlorofilla v list'yakh seyantsev drevesnykh rasteniĭ. [Aftereffect of irradiance of various duration on the growth and chlorophyll content in leaves of seedlings of woody plants.] - In : Pitanie i Obmen Veshchestv u Rasteniĭ. Pp. 73 - 76. Nauka i Tekhnika, Minsk 1975. [In R.]

23805 - **NESTOROVA, S.** : Vliyanie na narastvashchite normi na azota i fosfora v"rkhu listnata pov"rkhnost i chistata produktivnost na fotosintezata. [Effect of increasing doses of nitrogen and phosphorus fertilizers on leaf area and net photosynthetic productivity.] - Pochvoznanie Agrokhim. *10* (6) : 78 - 88, 1975. [Dry matter accumulation; in Bulg., ab : E, R.]

23806 - **NEUHOLD, J.M.** : Introduction to modeling in the biomes. - In : **PATTEN, B.C.** (ed.) : Systems Analysis and Simulation in Ecology. Vol. III. Pp. 7 - 12. Academic Press, New York - San Francisco - London 1975.

23807 - **NEUMANN, K.-H., SCHWAB, B.** : Untersuchungen über den Einfluß von Gibberellinsäurespritzungen auf den Ertrag, die Anatomie der Wurzel und die Karotinverteilung bei Karotten. - Z. Pflanzenernähr. Bodenkunde *1975* : 19 - 23, 1975.

23808 - **NEUVILLE, D., DASTE, P.** : Nouvelles observations concernant la production d'un pigment bleu par la Diatomée *Navicula ostrearia* (GAILLON)BORY en culture *in vitro*. - Compt. rend. Acad. Sci. Paris, Sér. D *280* : 2889 - 2891, 1975. [Ps.]

23809 - **NEVEUX, J., FIALA, M., JACQUES, G., PANOUSE, M.** : Phytoplancton et matériel particulaire a Banyuls-sur-mer (Golfe du Lion), 1973. - Vie Milieu, Sér. B, Oceanogr. *25* : 85 - 97, 1975. [Chl.]

23810 - **NEVRYANSKAYA, A.D.** : Pigmenty kolosa novykh mutantov i iskhodnykh sortov ozimoĭ pshenitsy. [Spike pigments of new mutants and initial cultivars of winter wheat.] - Izv. Akad. Nauk mold. SSR, Ser. biol. khim. Nauk *1975* (3) : 17 - 22, 91, 1975. [In R.]

23811 - **NEVRYANSKAYA, A.D.** : Pigmenty list'ev novykh mutantov i iskhodnykh sortov ozimoĭ pshenitsy. [Leaf pigments of new mutants and initial cultivars of winter wheat.] - Izv. Akad. Nauk mold. SSR, Ser. biol. khim. Nauk *1975* (5) : 7 - 10, 90, 1975. [In R.]

23812 - **NEWELL, E., RIENITS, K.G.** : Chlorophyll synthesis in a yellow mutant of wheat. - Aust. J. Plant Physiol. *2* : 543 - 552, 1975.

23813 - **NG, T.T., WILSON, J.R., LUDLOW, M.M.** : Influence of water stress on water relations and growth of a tropical (C_4) grass, *Panicum maximum* var. *trichoglume*. - Aust. J. Plant Physiol. *2* : 581 - 595, 1975. [Growth analysis.]

23814 - **NICHIPOROVICH, A.A.** : The genetics of photosynthesis and rational means of breeding highly productive plants. - In : **NASYROV, Yu.S., ŠESTÁK, Z.** (ed.) : Genetic Aspects of Photosynthesis. Pp. 315 - 341. Dr. W. Junk B.V. Publ., The Hague 1975.

23815 - **NICHIPOROVICH, A.A.** : Realizatsiya regulyatornoĭ funktsii sveta v zhiznedeyatel'nosti rasteniya kak tselogo i v ego produktivnosti. [Regulatory function of light in the plant as a whole and plant productivity.] - In : Fotoregulyatsiya Metabolizma i Morfogeneza Rasteniĭ. Pp. 228 - 244, 254. Nauka, Moskva 1975. [In R.]

B23816 - **NICHIPOROVICH, A.A., SHUL'GIN, I.A.** (ed.) : Fotosintez i Ispol'zovanie Ėnergii Solnechnoĭ Radiatsii. Bibliograficheskiĭ Ukazatel' Otechestvennykh Rabot, Izdannykh v 1967 - 1972 gg. [Photosynthesis and Utilization of Energy of Solar Radiation. Bibliographic Guide to Soviet Papers Published in 1967 to 1972.] - Biblioteka Akad. Nauk SSSR, Leningrad 1975. [In R.]

23817 - **NICHOLS, R., HO, L.C.** : An effect of ethylene on the distribution of ^{14}C-sucrose from the petals to other flower parts in the senescent cut inflorescence of *Dianthus caryophyllus*. - Ann. Bot. *39* : 433 - 438, 1975.

23818 - **NICHOLSON, S.A.** : Foliage biomass in temperate tree species. - Amer. Midland Naturalist *93* : 44 - 52, 1975.

*23819 - **NICHOLSON, S.A., BEST, D.G.** : Root:shoot and leaf area relationships of macrophyte communities in Chautauqua Lake, New York. - Bull. Torrey bot. Club *101* : 96 - 100, 1974.

23820 - **NICHOLSON-GUTHRIE, C.S., TURNER, F.R., HUDOCK, G.A.** : Abnormal chloroplast structures in a mutant of *Chlamydomonas reinhardi*. - Exp. Cell Res. *93* : 240- - 244, 1975.

23821 - **NIELSEN, N.C.** : Electrophoretic characterization of membrane proteins during chloroplast development in barley. - Europe. J. Biochem. *50* : 611 - 623, 1975.

23822 - **NIELSEN, N.C., HENNINGSEN, K.W., SMILLIE, R.M.** : Chloroplast membrane proteins in wild-type and mutant barley. - In : **AVRON, M.** (ed.) : Proceedings of the Third International Congress on Photosynthesis. Vol. III. Pp. 1603 - 1614. Elsevier, Amsterdam - Oxford - New York 1975.

23823 - **NIELSEN, O.F.** : Macromolecular physiology of plastids. XIII. The effect of photoinactive protochlorophyllide on the function of protochlorophyllide holochrome. - Biochem. Physiol. Pflanzen *167* : 195 - 206, 1975.

23824 - **NIEMI, Å.** : Ecology of phytoplankton in the Tvärminne area, SW coast of Finland. II. Primary production and environmental conditions in the archipelago and the sea zone. - Acta bot. fenn. *105* : 3 - 73, 1975.

23825 - **NIETH, K.-F., DREWS, G.** : Formation of reaction centers and light-harvesting bacteriochlorophyll-protein complexes in *Rhodopseudomonas capsulata*. - Arch. Microbiol. *104* : 77 - 82, 1975.

23826 - **NIETH, K.F., DREWS, G., FEICK, R.** : Photochemical reaction centers from *Rhodopseudomonas capsulata*. - Arch. Microbiol. *105* : 43 - 45, 1975.

23827 - **NIIHARA, Y.** : [Physiological studies of *Laminaria japonica* var. *ochotensis*. The effect of temperature, light intensity and salinity upon photosynthesis and respiration of young sporophytes.] - Sci. Reports Hokkaido Fish. exp. Sta. *17* : 11 - 17, 1975. [In Jap., ab : E.]

23828 - **NIKITSENKA, U.P., RAKHTSEENKA, L.I.** : Uplyŭ mineral'nykh ugnaennyaŭ na rost i zhytstsyadzeĭnasts' seyantsaŭ sasny i elki na vypratsavanykh tarfyanikakh Belarusi. [Effect of mineral fertilizers on growth and physiological activity of pine and spruce seedlings on worked-out peat bogs of Belorussia.] - Vestsi Akad. Navuk belarus. SSR, Ser. biyal. Navuk *1975* (1) : 40 - 44, 1975. [Ps; in Belorus., ab : R.]

23829 - **NIKOLAEVA, E.S.** : O pozelenenii zarodysheĭ nekotorykh *Angiospermae*. [Greening of embryos of *Angiospermae*.] - Bot. Zh. *60* : 517 - 522, 1975. [In R.]

23830 - **NIKOLAEVA, L.F., FLOROVA, N.B., KLEVKOV, A.P.** : Obrazovanie dlinnovolnovykh form khlorofilla u khvoĭnykh v otsutstvie sveta. [Formation of long-wave forms of chlorophyll in conifers in the absence of light.] - Nauch. Dokl. vyssh. Shkoly, biol. Nauki *18* (7) : 76 - 79, 1975. [In R.]

23831 - **NIKOLAEVA, L.F., RUBIN, B.A., FLOROVA, N.B.** : Perspektivy ispol'zovaniya metoda fluorestsentnoĭ spektroskopii dlya vyyavleniya potentsial'noĭ produktivnosti sortov pshenitsy. [Prospects of the use of the method of fluorescence spectroscopy to evaluate the potential productivity of wheat cultivars.] - - Sel'skokhoz. Biol. *10* : 343 - 347, 1975. [Chl; in R, ab : E.]

*23832 - **NIKOLAEVSKIĬ, V.S., MIROSHNIKOVA, A.T.** : Biofizicheskie metody opredeleniya dopustimykh norm zagryazneniya vozdukha dlya rasteniĭ. [Biophysical methods for determining the acceptable quantities of air pollution for plants.] - Uch. Zap. perm. gos. Univ. *256* : 211 - 225, 1971. [Ps; in R.]

23833 - **NIKOLOV, Kh., GEORGIEVA, S.** : Khlorofil'nye khimery v M_1 i ikh znachenie pri uchete indutsirovannykh khlorofil'nykh mutatsiĭ u yachmenya. [M_1 chlorophyll chimeras and their relation to estimation of induced chlorophyll mutations in barley.] - Genetika *11* (6) : 47 - 50, 1975. [In R, ab : E.]

23834 - **NILSEN, S., KRISTIANSEN, E., HALLDAL, P.** : A system for continuous measurement of photosynthetic rate and dark respiration at constant CO_2-level. - Physiol. Plant. *35* : 59 - 61, 1975.

23835 - **NIMBALKAR, J.D., JOSHI, G.V.** : Physiological studies in senescent leaves of sugarcane var. Co.740. - Indian J. exp. Biol. *13* : 384 - 386, 1975.

23836 - **NINNEMANN, H., STRASSER, R.J.** : Electrophoretic differences in the chromoprotein patters of flashed and flashed-induced bean leaves. - In : **AVRON, M.** (ed.) : Proceedings of the Third International Congress on Photosynthesis. Vol. III. Pp. 1639 - 1646. Elsevier, Amsterdam - Oxford - New York 1975.

23837 - **NISHIMURA, K., KAWATA, T., ASADA, K., NAKAJIMA, M.** : Effect of Hill reaction inhibitors on the photoreduction of ferricyanide and cyclic photophosphorylation by spinach chloroplasts. - Agr. biol. Chem. *39* : 867 - 872, 1975.

B23838 - **NISHIMURA, M.** : Kôgôsei no Shoki-katei. [Molecular Mechanisms of the Primary Processes of Photosynthesis.] - Kyoritsu Shuppan, Tokyo 1975. [In Jap.]

23839 - **NISHIMURA, M.** : [Primary processes in energy transfer of photosynthesis.] - - Tampakushitsu Kakusan Koso [Protein, Nucleic Acid, Enzyme] *20* : 301 - 317, 1975. [In Jap.]

23840 - **NISHIMURA, M., AKAZAWA, T.** : Photosynthetic activities of spinach leaf protoplasts. - Plant Physiol. *55* : 712 - 716, 1975.

23841 - **NISHIMURA, M., GRAHAM, D., AKAZAWA, T.** : Effect of oxygen on photosynthesis by spinach leaf protoplasts. - Plant Physiol. *56* : 718 - 722, 1975.

23842 - **NISHIZAKI, Y.** : Sigmoidal stimulation by solutes of light-induced proton translocation in thylakoid membranes. - Plant Cell Physiol. *16* : 589 - 594, 1975.

23843 - **NISIZAWA, K., IKAWA, T.** : Photosynthesis in algae other than *Chlorella*. - In : **TOKIDA, J., HIROSE, H.** (ed.) : Advance of Phycology in Japan. Pp. 201 - 209. Dr. W. Junk b.v. Publ., The Hague 1975.

23844 - **NIVAL, P., NIVAL, S., THIRIOT, A.** : Influence des conditions hivernales sur les productions phyto- et zooplanctoniques en Méditerranée Nord-Occidentale. V. Biomasse et production zooplanctonique-relations phyto-zooplancton. - Mar. Biol. *31* : 249 - 270, 1975.

23845 - **NIYAZMUKHAMEDOVA, M.B.** : O fotosinteze nekotorykh pigmentnykh mutantov *Arabidopsis thaliana* (L.). [Photosynthesis of some pigment mutants of *Arabidopsis thaliana* (L.).] - Dokl. Akad. Nauk tadzh.SSR *18* (12) : 47 - 50, 1975. [In R, ab : Tajik.]

23846 - **NOBEL, P.S.** : Effective thickness and resistance of the air boundary layer adjacent to spherical plant parts. - J. exp. Bot. *26* : 120 - 130, 1975.

23847 - **NOBEL, P.S.** : Chloroplasts. - In : **BAKER, D.A., HALL, J.L.** (ed.) : Ion Transport in Plant Cells and Tissues. Pp. 101 - 124. North-Holland Publ. Comp., Amsterdam - Oxford, Amer. Elsevier Publ. Comp., New York 1975.

23848 - **NOBEL, P.S., ZARAGOZA, L.J., SMITH, W.K.** : Relation between mesophyll surface area, photosynthetic rate, and illumination level during development for leaves of *Plectranthus parviflorus* HENCKEL. - Plant Physiol. *55* : 1067 - 1070, 1975.

23849 - **NOBLE, R.D.** : Effects of light intensity on photosynthesis and chloroplast ultrastructure in lethal-yellow soybean. - Plant Physiol. *56* (Suppl.) : 12, 1975.

23850 - **NOKES, M.A., SIMIC, M.** : X-ray sensitivity of photosynthetic and reproductive systems in *Chlorella*. - Photochem. Photobiol. *21* : 265 - 268, 1975.

23851 - **NOLAN, W.G., BISHOP, D.G.** : The site of inhibition of photosynthetic electron transfer by amphotericin *B*. - Arch. Biochem. Biophys. *166* : 323 - 329, 1975.

23852 - **NOLAN, W.G., PARK, R.B.** : Comparative studies on the polypeptide composition of chloroplast lamellae and lamellar fractions. - Biochim. biophys. Acta *375*: 406 - 421, 1975.

*23853 - **NOLLENDORF, V.F.** : Optimizatsiya mineral'nogo pitaniya rasteniĭ (tomaty, ogurtsy) mikroėlementami na osnove tkanevoĭ diagnostiki. [Optimization of the mineral nutrition of plants (tomatoes, cucumbers) with trace nutrients based on tissue diagnosis.] - In : Mikroėlementy - Regulyatory Zhiznedeyatel'nosti i Produktivnosti Rasteniĭ. Pp. 93 - 103. Zinatne, Riga 1971. [Chl, Car; in R.]

23854 - **NORDLIE, F.G., KELSO, D.P.** : Trophic relationships in a tropical estuary. - - Rev. Biol. trop. *23* : 77 - 99, 1975.

*23855 - **NORGÅRD, S., LIAAEN-JENSEN, S.** : Chemical reactions of peridinin. - In : Comm. Abstr. 3rd International Carotenoid Symposium. Pp. 13 - 14. Cluj 1972.

23856 - **NORMAN, J.M., JARVIS, P.G.** : Photosynthesis in Sitka spruce (*Picea sitchensis* (BONG.) CARR.). V. Radiation penetration theory and a test case. - J. appl. Ecol. *12* : 839 - 878, 1975.

23857 - **NORRIS, J.R., SCHEER, H., KATZ, J.J.** : Models for antenna and reaction center chlorophylls. - Ann. New York Acad. Sci. *244* : 260 - 280, 1975.

23858 - **NORTON, I.L., WELCH, M.H., HARTMAN, F.C.** : Evidence for essential lysyl residues in ribulosebisphosphatecarboxylase by use of the affinity label 3-bromo--1,4-dihydroxy-2-butanone 1,4-bisphosphate. - J. biol. Chem. *250* : 8062 - 8068, 1975.

23859 - **NOVAK, V.A., IVANKINA, N.G.** : Sravnitel'noe izuchenie svetoindutsirovannykh izmeneniĭ ėlektricheskikh potentsialov rasteniĭ. [Comparative study of light--induced changes in electric potentials of plants.] - Fiziol. Rast. *22* : 49 - 54, 1975. [In R, ab : E.]

23860 - **NOVAK-HOFER, I., SIEGENTHALER, P.-A.** : Isoelectric focusing of membrane proteins from spinach chloroplasts. - FEBS Lett. *60* : 47 - 50, 1975.

23861 - **NOVIKAVA, A.A.** : Uplyŭ roznaĭ pratsyaglastsi asvyatlennya na rost seyantsaŭ sasny zvychaĭnaĭ i nakaplenne khlarafilu ŭ ikh iglitsy. [Effect of various duration of irradiance on the growth of pine seedlings and accumulation of chlorophyll in their needles.] - Vestsi Akad. Navuk belarus.SSR, Ser. biyal. Navuk *1975* (1) : 10 - 13, 135, 1975. [In Belorus., ab : R.]

23862 - **NOWAKOWSKI, W., KONOPKO, E., LUBAŃSKA, G.** : Utilisation de l'énergie solaire dans la productivité photosynthétique au cours de quatre phases vegetatives de differentes variétés du *Triticum durum*. - Acta agrobot. (Warszawa) *28* : 197 - 204, 1975.

23863 - **NOWAKOWSKI, W., LUBAŃSKA, G.** : Wpływ IAA i GA_3 na natężenie fotosyntezy, oddychania i transpiracji siewek pszenicy jarej i kukurydzy w warunkach niedostatku wody w glebie. [The effect of IAA and GA_3 on photosynthesis, respiration and transpiration of spring wheat and corn seedlings under conditions of insufficient water in the soil.] - Acta agrobot.(Warszawa) *28* : 79 - 87, 1975. [In Pol., ab : E.]

23864 - **NYE, P.H., BREWSTER, J.L., BHAT, K.K.S.** : The possibility of predicting solute uptake and plant growth response from independently measured soil and plant characteristics. I. The theoretical basis of the experiments. - Plant Soil *42* : 161 - 170, 1975. [Growth analysis.]

23865 - **NYGAARD, R.T.** : Acclimatization effect in photosynthesis and respiration. - - In : **WIELGOLASKI, F.E.** (ed.) : Fennoscandian Tundra Ecosystems. Part I. Plants and Microorganisms. Pp. 163 - 167. Springer-Verlag, Berlin - Heidelberg - - New York 1975.

23866 - **OBEN, G., MARCELLE, R.** : The effects of CCC and GA on some biochemical and photochemical activities of primary leaves of bean plants. - In : **MARCELLE, R.** (ed.) : Environmental and Biological Control of Photosynthesis. Pp. 211 - - 216. Dr. W. Junk b.v. Publ., The Hague 1975. [Ps.]

23867 - **OBER, K.** : Effects of diazepam on photosynthesis, respiration, rubidium uptake, and finestructure of *Scenedesmus obliquus* in synchronous cultures. - Arch. Microbiol. *102* : 129 - 137, 1975.

23868 - **O'BRIEN, M.J., POWLS, R.** : Glyceraldehyde 3-phosphate dehydrogenases of *Scenedesmus obliquus* : Two enzymes with different pyridine nucleotide dependence. - In : **AVRON, M.** (ed.) : Proceedings of the Third International Congress on Photosynthesis. Vol. II. Pp. 1431 - 1440. Elsevier, Amsterdam - Oxford - New York 1975.

23869 - **O'CARRA, P.** : Heme-cleavage : biological systems and chemical analogs. - In : **SMITH, K.M.** (ed.) : Porphyrins and Metalloporphyrins. Pp. 123 - 153. Elsevier sci. Publ. Comp., Amsterdam - Oxford - New York 1975. [Phycobilins.]

23870 - **ODINTSOVA, M.S., YURINA, N.P.** : Ribosomal proteins and the origin of plastids. - In : **NASYROV, Yu.S., ŠESTÁK, Z.** (ed.) : Genetic Aspects of Photosynthesis. Pp. 37 - 41. Dr. W. Junk B.V. Publ., The Hague 1975.

23871 - **ODUM, E.P.** : Diversity as a function of energy flow. - In : **van DOBBEN, W.H., LOWE-McCONNEL, R.H.** (ed.) : Unifying Concepts in Ecology. Pp. 11 - 14. Dr. W. Junk b.v. Publ., The Hague; Pudoc, Wageningen 1975.

23872 - **OECHEL, W.C., HICKLENTON, P., SVEINBJORNSSON, B., MILLER, P.C., STONER, W.** : Temperature acclimation of photosynthesis in *D. fuscescens* growing *in situ* in the Arctic and Subarctic. - In : Proceedings of the Circumpolar Conference on Northern Ecology. Pp. 1131 - 1144. Nat. Res. Counc. Canada, Ottawa 1975.

23873 - **OGAWA, T.** : Two steps of gas exchange in leaf photosynthesis. - Physiol. Plant. *35* : 91 - 95, 1975.

23874 - **OGAWA, T., BOVEY, F., INOUE, Y., SHIBATA, K.** : Early stages of greening in etiolated bean leaves. - In : **AVRON, M.** (ed.) : Proceedings of the Third International Congress on Photosynthesis. Vol. III. Pp. 1829 - 1832. Elsevier, Amsterdam - Oxford - New York 1975.

23875 - **OGAWA, T., BOVEY, F., SHIBATA, K.** : An intermediate in the phytylation of chlorophyllide *a in vivo*. - Plant Cell Physiol. *16* : 199 - 202, 1975.

23876 - **OGAWA, T., NAKAMURA, K., SHIBATA, K.** : Chlorophyll composition in the two photosystems of marine green algae. - Arch. Hydrobiol. *49* (Suppl. Algol. Stud. *14*) : 37 - 48, 1975.

23877 - **OGREN, W.L.** : Control of photorespiration in soybean and maize. - In : **MARCELLE, R.** (ed.) : Environmental and Biological Control of Photosynthesis. Pp. 45- - 52. Dr. W. Junk b.v. Publ., The Hague 1975.

23878 - **OGURA, N., NAKAGAWA, H., TAKEHANA, H.** : [Effect of high temperature-short term storage of mature green tomato fruits on changes of their chemical composition after ripening at room temperature. (Studies on the storage temperature of tomato fruits. Part I.)] - J. agr. chem. Soc. Jap. [Nippon Nogei Kagaku Kaishi] *49* : 189 - 196, 1975. [In Jap., ab : E.]

23879 - **OHAD, I.** : Control by light of synthesis of membrane proteins of cytoplasmic and chloroplastic origin and their role in the formation and function of the active center of PSII and PSI in *Chlamydomonas reinhardi Y-1*. - In : Les Cycles Cellulaires et Leur Blocage chez Plusieurs Protistes. Coll. int. CNRS 240. Pp. 267 - 268. Édit. CNRS, Paris 1975.

23880 - **OHAD, I., DREWS, G.** : Action spectrum for the synthesis of chlorophyll and chloroplast membrane proteins of cytoplasmic origin in *Chlamydomonas reinhardi y-1*. - In : **AVRON, M.** (ed.) : Proceedings of the Third International Congress on Photosynthesis. Vol. III. Pp. 1907 - 1912. Elsevier, Amsterdam - Oxford - New York 1975.

23881 - **OH-HAMA, T., HASE, E.** : Syntheses of δ-aminolevulinic acid and chlorophyll during chloroplast formation in *Chlorella protothecoides*. - Plant Cell Physiol. *16* : 297 - 303, 1975.

23882 - **OH-HAMA, T., SENGER, H.** : The development of structure and function in chloroplasts of greening mutants of *Scenedesmus* III. Biosynthesis of δ-aminolevulinic acid. - Plant Cell Physiol. *16* : 395 - 405, 1975.

23883 - **OHIRA, K., YAMAYA, T., OJIMA, K.** : Studies on the greening of cultured soybean and *Ruta* cells I. Pigmentation as influenced by the composition of the medium. - Tohoku J. agr. Res. *26* : 136 - 148, 1975.

23884 - **OHKI, K., KATOH, T.** : Photoorganotrophic growth of a blue-green alga, *Anabaena variabilis*. - Plant Cell Physiol. *16* : 53 - 64, 1975. [Ps, Chl.]

23885 - **OHTA, N., SAGER, R., INOUYE, M.** : Identification of a chloroplast ribosomal protein altered by a chloroplast mutation in *Chlamydomonas*. - J. biol. Chem. *250* : 3655 - 3659, 1975.

23886 - **OKAMURA, M.Y., ISAACSON, R.A., FEHER, G.** : Primary acceptor in bacterial photosynthesis : Obligatory role of ubiquinone in photoactive reaction centers of *Rhodopseudomonas spheroides*. - Proc. nat. Acad. Sci. USA *72* : 3491 - 3495, 1975.

23887 - **OKANENKO, A.S., GULYAEV, B.I., MANUIL'SKIĬ, V.D.** : Ustoĭchivost' fotosinteza k obezvozhivaniyu lista u razlichnykh vidov sel'skokhozyaĭstvennykh rasteniĭ. [Resistance of photosynthesis to leaf water stress in various species of crop plants.] - In : Vodoobmen Rasteniĭ pri Neblagopriyatnykh Usloviyakh Sredy. Pp. 108 - 114. Shtiintsa, Kishinev 1975. [In R.]

23888 - **OKKEH, A., KYLIN, A.** : The ATP levels obtained in photophosphorylation, glycolysis and oxidative phosphorylation in *Scenedesmus* titrated with desaspidin. - Physiol. Plant. *33* : 118 - 123, 1975.

23889 - **OKKEH, A., TILLBERG, J.-E., KYLIN, A.** : The effect of light intensity on total photophosphorylation and the ATP level in *Scenedesmus*. - Physiol. Plant. *33* : 124 - 127, 1975.

23890 - **OKOMBI, G., BILLOT, J., HARTMANN, C.** : Variations des teneurs en chlorophylles et en caroténoïdes chez la Cerise au cours de la conservation du fruit cueilli à différents stades de la croissance et de la maturation. - Physiol. vég. *13*: 417 - 426, 1975.

23891 - **OKU, T., HAYASHI, H., TOMITA, G.** : Oxygen evolution in dark-developed spruce chloroplasts. - Plant Cell Physiol. *16* : 101 - 108, 1975.

23892 - **OKU, T., TOMITA, G.** : The reversible photoconversion of *Chenopodium* chlorophyll protein and its control by the apoprotein structure. - Plant Cell Physiol. *16* : 1009 - 1016, 1975.

23893 - **OKUBO, T.** : Developmental analysis in maize. - In : **MURATA, Y.** (ed.) : JIBP Synthesis. Vol. 11. Pp. 159 - 164. Univ. Tokyo Press, Tokyo 1975. [Growth analysis.]

23894 - **OKUBO, T., HIROSAKI, S., OKUNO, T.** : A model for plant-growth under grazing condition. - In : **NUMATA, M.** (ed.) : JIBP Synthesis. (Ecological Studies in Japanese Grasslands - Productivity of Terrestrial Communities.) Vol. 13. Pp. 268 - 275. Univ. Tokyo Press, Tokyo 1975.

23895 - **OKUBO, T., KAWANABE, S., HOSHINO, M.** : [Chlorophyll amount for analysis of matter production in forage crops. II. Seasonal variations in maximum crop growth rate and leaf photosynthesis, and their correlations with chlorophyll content in alfalfa and ladino clover.] - J. jap. Soc. Grassland Sci. *21* : 124 - 135, 1975. [In Jap., ab : E.]

23896 - **OKUBO, T., KAWANABE, S., HOSHINO, M.** : [Chlorophyll amount for analysis of matter production in forage crops. III. Leaf photosynthesis under dim light, maximum crop growth rate under different plant density, and their dependence on chlorophyll content in alfalfa.] - J. jap. Soc. Grassland Sci. *21* : 136 - - 145, 1975. [In Jap., ab : E.]

23897 - **OKUBO, T., TAKAHASHI, S., AKIYAMA, T.** : [Chlorophyll amount for analysis of matter production in forage crops. IV. Maximum crop growth rate in the swards of four Italian ryegrass clones different in chlorophyll content.] - J. jap. Soc. Grassland Sci. *21* : 280 - 290, 1975. [In Jap., ab : E.]

23898 - **OKUNTSOV, M.M., VERKHOTUROVA, G.S.** : Deĭstvie sveta na metabolizm malonovoĭ kisloty v list'yakh fasoli. [Effect of light on the maleic acid metabolism in bean leaves.] - Nauch. Dokl. vyssh. Shkoly, biol. Nauki *18* (5) : 66 - 70, 1975. [In R.]

*23899 - **OKUNTSOV, M.M., VRUBLEVSKAYA, K.G., ZAĬTSEVA, T.A.** : Vliyanie krasnogo i sinego sveta na soderzhanie adenozinfosfatov v ėtiolirovannykh rasteniyakh pshenitsy. [Effect of red and blue light on the content of adenosine phosphates in etiolated wheat plants.] - Nauch. Dokl. vyssh. Shkoly, biol. Nauki *17* (3) : 76 - 81, 1974. [In R.]

23900 - **O'LEARY, J.W.** : Environmental influence of total water consumption by whole plants. - In : **GATES, D.M., SCHMERL, R.B.** (ed.) : Perspectives of Biophysical Ecology. Pp. 203 - 212. Springer-Verlag, Berlin - Heidelberg - New York 1975. [Growth analysis, resistances.]

23901 - **OLECH, K.** : Influence of the absence of light on the ensuing photosynthetic activity of the leaves of some higher plants. - Pol. ekol. Stud. *1* : 65 - 70, 1975.

23902 - **OLESEN, P.** : Plasmodesmata between mesophyll and bundle sheath cells in relation to the exchange of C_4-acids. - Planta *123* : 199 - 202, 1975.

23903 - **OLIVEIRA, L.** : On the occurrence of helical polysomes in developing chloroplasts of mesophyll leaf cells of a *Triticale*. - Caryologia *28* : 467 - 476, 1975.

23904 - **OLIVEIRA, L.** : Ribosomes associated with developmental changes occurring in proplastids of *Triticale* root cells during experimental greening. - J. submicroscop. Cytol. *7* : 97 - 105, 1975.

23905 - **OLIVEIRA, L.** : On the morphology and nature of the plastid inclusions of leaf cells of a *Triticale*. - J. submicroscop. Cytol. *7* : 271 - 280, 1975.

23906 - **OLIVER, D.J., JAGENDORF, A.T.** : Inhibition of the coupling factor from spinach chloroplasts by trinitrobenzenesulfonic acid. - Fed. Proc. *34* : 596, 1975.

23907 - **OLOFINBOBA, M.O.** : Studies on seedlings of *Theobroma cacao* L., variety F_3 Amazon. I. Role of cotyledons in seedling development. - Turrialba *25* : 121 - - 127, 1975. [Ps.]

23908 - **OLOFINBOBA, M.O.** : Studies on seedlings of *Theobroma cacao L.*, variety F_3 Amazon. II. Distribution of radio-carbon photoassimilated by 43-day-old seedlings under low soil water content. - Turrialba *25* : 420 - 424, 1975.

23909 - **OLSON, J.M., GIDDINGS, T.H., Jr., SHAW, E.K.** : Dissociation of the bacteriochlorophyll reaction-center complex from green bacteria. - Biophys. J. *15*: 225a, 1975.

23910 - **ONDOK, J.P.** : Photosynthetically active radiation in a stand of *Phragmites* *communis* TRIN. IV. Stochastic model. - Photosynthetica *9* : 201 - 210, 1975.

23911 - **O'NEILL, R.V.** : Modeling in the Eastern deciduous forest biome. - In : **PATTEN, B.C.** (ed.) : Systems Analysis and Simulation in Ecology. Vol. III. Pp. 49 - 72. Academic Press, New York 1975. [Production.]

23912 - **ONG, C.K., MARSHALL, C.** : Assimilate distribution in *Poa annua* L. - Ann. Bot. *39* : 413 - 421, 1975.

23913 - **ONWUEME, I.C., LAWANSON, A.O.** : Chlorophyll accumulation in cowpea (*Vigna*) leaves and melon (*Colocynthis*) cotyledons as influenced by prior heat stress and seedling age. - Fyton *33* : 69 - 73, 1975.

23914 - **OPARIN, A.I., ODINTSOVA, M.S., YURINA, N.P.** : Chloroplast ribosomes as ribosomes of the prokaryotic type. - Biochem. Physiol. Pflanzen *168* : 175 - 183, 1975.

23915 - **OPHIR, I., TALMON, A., POLAK-CHARCON, S., BEN-SHAUL, Y.** : Aspects of structure and photosynthetic competence of *Euglena* plastids under conditions of greening and degreening. - Protoplasma *84* : 283 - 295, 1975.

23916 - **ÖQUIST, G.** : The spectral and photochemical properties of subchloroplast membrane particles from *Pinus silvestris*. - Physiol. Plant. *34* : 300 - 305, 1975.

23917 - **ÖQUIST, G., MARTIN, B.** : Isolated photoreactions in chloroplasts prepared from *Pinus silvestris*. - In : **AVRON, M.** (ed.) : Proceedings of the Third International Congress on Photosynthesis. Vol. I. Pp. 729 - 734. Elsevier, Amsterdam - Oxford - New York 1975.

23918 - **ORLOVA, I.F.** : Posledeĭstvie molibdena na urozhaĭ i kachestvo gorokha. [After-effect of molybdenum on the yield and quality of pea plants.] - Nauch. Tr. orl. obl. sel'sko-khoz. opyt. Sta. *7* : 132 - 142, 1975. [Chl, Car; in R.]

23919 - **ORSENIGO, M., RASCIO, N.** : Chloroplast fine structure and morphogenesis in the japonica-2 maize mutant. 1 Mesophyll plastids. - Cytobios *14* : 161 - 170, 1975.

23920 - **ORT, D.R.** : Quantitative relationship between photosystem I electron transport and ATP formation. - Arch. Biochem. Biophys. *166* : 629 - 638, 1975.

23921 - **ORT, D.R., DILLEY, R.A.** : The dependence of photophosphorylation on time of illumination. - Plant Physiol. *56* (Suppl.) : 45, 1975.

23922 - **OSADA, A.** : Some characteristics in photosynthetic activity of leaves of indica rice varieties. - In : Ass. Jap. Agr. Sci. Soc. Symposium on Rice in Asia. Pp. 210 - 222. Univ. Tokyo Press, Tokyo 1975.

23923 - **OSAFO. D.M., MILBOURN, G.M.** : The growth of maize III. The effect of date of sowing and bitumen mulch on dry-matter yields. - J. agr. Sci. *85* : 271 - 279, 1975. [Growth analysis.]

23924 - **OSAFUNE, T., HASE, E.** : Some structural characteristics of chloroplasts in the "glucose-bleaching" and re-greening cells of *Chlorella protothecoides*. - Biochem. Physiol. Pflanzen *168* : 533 - 542, 1975.

23925 - **OSHIMA, E.** : Developmental analysis in sugar beet. - In : **MURATA, Y.** (ed.) : JIBP Synthesis. Vol. 11. Pp. 164 - 171. Univ. Tokyo Press, Tokyo 1975. [Growth analysis.]

23926 - **OSMOND, C.B.** : Environmental control of photosynthetic options in crassulacean plants. - In : **MARCELLE, R.** (ed.) : Environmental and Biological Control of Photosynthesis. Pp. 311 - 321. Dr. W. Junk b.v. Publ., The Hague 1975.

23927 - **OSMOND, C.B., AKAZAWA, T., BEEVERS, H.** : Localization and properties of ribulose diphosphate carboxylase from castor bean endosperm. - Plant Physiol. *55* : 226 - 230, 1975.

23928 - **OSMOND, C.B., BJÖRKMAN, O.** : Pathways of CO_2 fixation in the CAM plant *Kalanchoë daigremontiana*. II. Effects of O_2 and CO_2 concentration on light and dark CO_2 fixation. - Aust. J. Plant Physiol. *2* : 155 - 162, 1975.

23929 - **OSMOND, C.B., ZIEGLER, H.** : Schwere Pflanzen und leichte Pflanzen : Stabile Isotope im Photosynthesestoffwechsel und in der biochemischen Ökologie. - Naturwiss. Rundschau *28* : 323 - 328 1975.

23930 - **OSMOND, C.B., ZIEGLER, H., STICHLER, W., TRIMBORN, P.** : Carbon isotope discrimination in alpine succulent plants supposed to be capable of Crassulacean Acid Metabolism (CAM). - Oecologia *18* : 209 - 217, 1975.

23931 - **OSTROFSKY, M.L., DUTHIE, H.C.** : Primary productivity and phytoplankton of lakes on the Eastern Canadian Shield. - Verh. int. Verein. Limnol. *19* : 732 - - 738, 1975.

23932 - **OSTROFSKY, M.L., DUTHIE, H.C.** : Primary productivity, phytoplankton and limiting nutrient factors in Labrador lakes. - Int. Rev. ges. Hydrobiol. *60* : 145 - - 158, 1975.

B23933 - **OSTROVSKAYA, L.K.** (ed.) : Fotokhimicheskie Sistemy Khloroplastov. [Photochemical Systems of Chloroplasts.] - Naukova Dumka, Kiev 1975. [In R.]

23934 - **OSTROVSKAYA, L.K., KOCHUBEĬ, S.M., SHADCHINA, T.M.** : Deĭstvie fosfolipazy *a* na spektral'nye i fotokhimicheskie svoĭstva khloroplastov i ikh fragmentov. [Effect of phospholipase *a* on spectral and photochemical characteristics of chloroplasts and their fragments.] - Biokhimiya *40* : 169 - 174, 1975. [In R, ab : E.]

23935 - **OSTROVSKAYA, L.K., YAKOVENKO, G.M., GAMAYUNOVA, M.S., MANUIL'SKAYA, S.V., GRIGORA, M.Yu., MIKHNO, A.I.** : Lipidnyĭ sostav soderzhashchikh fotosistemu I fragmentov mezhgrannykh tilakoidov i tilakoidov gran khloroplastov gorokha. [Lipid composition of Photosystem I-containing fragments of intergrana and grana thylakoids of pea chloroplasts.] - Fiziol. Biokhim. kul't. Rast. *7* : 451 - 455, 1975. [In R, ab : E.]

23936 - **OUDSHOORN, H.P., THOMAS, J.B.** : Curve analysis of the red absorption band of chlorophyll *b* in *Ulva lactuca*. - Acta bot. neerl. *24* : 49 - 53, 1975.

23937 - **OUTLAW, W.H., Jr., FISHER, D.B.** : Compartmentation in *Vicia faba* leaves. I. Kinetics of ^{14}C in the tissues following pulse labeling. - Plant Physiol. *55*: 699 - 703, 1975.

23938 - **OUTLAW, W.H., Jr., FISHER, D.B.** : Compartmentation in *Vicia faba* leaves. III. Photosynthesis in the spongy and palisade parenchyma. - Aust. J. Plant Physiol. *2* : 435 - 439, 1975.

23939 - **OUTLAW, W.H., Jr., TOLBERT, N.E.** : Carbon metabolism in the mesophyll tissues of *Vicia faba*. - Plant Physiol. *56* (Suppl.) : 72, 1975.

23940 - **OVERNELL, J.** : Potassium and photosynthesis in the marine diatom *Phaeodactylum tricornutum* as related to washes with sodium chloride. - Physiol. Plant. *35* : 217 - 224, 1975.

23941 - **OVERNELL, J.** : The effect of heavy metals on photosynthesis and loss of cell potassium in two species of marine algae, *Dunaliella tertiolecta* and *Phaeodactylum tricornutum*. - Mar. Biol. *29* : 99 - 103, 1975.

23942 - **OVERNELL, J.** : The effect of some heavy metal ions on photosynthesis in a freshwater alga. - Pestic. Biochem. Physiol. *5* : 19 - 26, 1975.

23943 - **OVERTON, W.S.** : The ecosystem modeling approach in the coniferous forest biome. - In : **PATTEN, B.C.** (ed.) : Systems Analysis and Simulation Ecology. Vol. III. Pp. 117 - 138. Academic Press, New York 1975. [Production.]

23944 - **OWEN, W.J., ROGERS, L.J., HAYES, J.D.** : Inhibition of Hill reaction and photophosphorylation by 1,1,1-trichloro-2,2-bis(*p*-chlorophenyl)ethane. - J. exp. Bot. *26* : 692 - 701, 1975.

23945 - **OWONUBI, J.J., KANEMASU, E.T., POWERS, W.L.** : The micro-climate of narrow- and wide-row sorghum with equal plant densities. - Agr. Meteorol. *15* : 61 - - 69, 1975.

23946 - **OZOLINA, I.A., MOCHALKIN, A.I.** : O zashchitnoĭ roli karotinoidnykh pigmentov v rastenii. [Protective role of carotenoid pigments in the plant.] - Izv. Akad. Nauk SSSR, Ser. Biol. *1975* : 387 - 392, 1975. [In R, ab : E.]

23947 - **PACKARD, T.T., DORTCH, Q.** : Particulate protein-nitrogen in North Atlantic surface waters. - Mar. Biol. *33* : 347 - 354, 1975. [Chl.]

23948 - **PACKER, L., TORRES-PEREIRA, J., CHANG, P., HANSEN, S.** : Stabilization of chloroplast membranes as measured by light induced quenching of acridine dyes. - - In : **AVRON, M.** (ed.) : Proceedings of the Third International Congress on Photosynthesis. Vol. II. Pp. 867 - 872. Elsevier, Amsterdam - Oxford - New York 1975.

23949 - **PACOLD, I., ANDERSON, L.E.** : Chloroplast and cytoplasmic enzymes. VI. Pea leaf 3-phosphoglycerate kinases. - Plant Physiol. *55* : 168 - 171, 1975.

23950 - **PALIWAL, K.V., MALIWAL, G.L., NANAWATI, G.C.** : Effect of bicarbonate-rich irrigation waters on the growth, nutrient uptake and synthesis of proteins and carbohydrates in wheat. - Plant Soil *43* : 523 - 536, 1975. [Dry-matter production.]

23951 - **PAMPLIN, E.J., CHAPMAN, J.M.** : Sucrose suppression of chlorophyll synthesis in tissue culture : changes in the activity of the enzymes of the chlorophyll biosynthetic pathway. - J. exp. Bot. *26* : 212 - 220, 1975.

23952 - **PANDYA, R.B., KHAN, M.I., GUPTA, S.K., DHINDSA, K.S.** : Germination, seedling growth and sugar metabolism of two species of *Brassica* under polyethylene glycol (PEG) induced water stress. - Biochem. Physiol. Pflanzen *167* : 439 - 445, 1975.

*23953 - **PAOLETTI, C., MATERASSI, R.** : Sulla estrazione e determinazione dei carotenoidi nelle microalghe. [Extraction and determination of carotenoids in microalgae.] - Riv. ital. Sostanze grasse *50* (5) : 128 - 130, 1973. [TLC, PC; in Ital., ab : E, F, G, Span.]

23954 - **PAPAGEORGIOU, G.** : Chlorophyll fluorescence : an intrinsic probe of photosynthesis. - In : **GOVINDJEE** (ed.) : Bioenergetics of Photosynthesis. Pp. 319 - - 371. Academic Press, New York - San Francisco - London 1975.

23955 - **PAPAGEORGIOU, G.** : On the mechanism of the PMS-effected quenching of chloroplast fluorescence. - Arch. Biochem. Biophys. *166* : 390 - 399, 1975.

23956 - **PAPAGEORGIOU, G.** : An analysis of the PMS-effected quenching of chloroplast fluorescence. - In : **AVRON, M.** (ed.) : Proceedings of the Third International Congress on Photosynthesis. Vol. I. Pp. 41 - 52. Elsevier, Amsterdam - Oxford- - New York 1975.

23957 - **PAPAGEORGIOU, G., TSIMILLI-MICHAEL, M., ISAAKIDOU, J.** : Quenching of excited chlorophyll *a in vivo* by nitrobenzene. - Biophys. J. *15* : 83 - 93, 1975.

23958 - **PAPPAS, P.N., HEBERLEIN, G.T.** : Applied nutrients and the mobilization of metabolites by laminar crown gall tumors in the excised primary leaf of *Phaseolus vulgaris*, var. Pinto L. - Plant Physiol. *56* (Suppl.) : 18, 1975. [Chl.]

23959 - **PAPPAS, P.N., HEBERLEIN, G.T.** : Factors affecting senescence of the excised primary leaf of *Phaseolus vulgaris*, var. Pinto L. - Plant Physiol. *56* (Suppl.): 64, 1975. [Chl.]

*23960 - **PARAMONOVA, T.K., KAMYSHENKO, L.K., GROZOVSKAYA, M.S.** : O deĭstvii ingibitorov sul'fgidril'nykh grupp belkov na biosintez pigmentov v zelenykh i postėtiolirovannykh prorostkakh. [Effect of inhibitors of sulfhydryl groups of proteins on pigment biosynthesis in green and postetiolated seedlings.] - In : Teoreticheskie i Prakticheskie Voprosy Ratsional'nogo Ispol'zovaniya Zhivotnykh i Rasteniĭ. Pp. 23 - 24. Zinatne, Riga 1973. [In R.]

23961 - **PÂRJOL, L.** : Cercetări privind rezistenţa la secetă a fasolei în diferite faze de vegetaţie. [Drought resistance of beans in various vegetational stages.] - Ann. Inst. Cercet. Cereale Plante teh. - Fundulea *40* (Ser. C) : 243 - 266, 1975. [Chl; in Roum., ab : E, R.]

23962 - **PARKER, J.** : Photosynthesis and respiration in black oak twig bark as affected by season and defoliation. - Plant Physiol. *56* (Suppl.) : 14, 1975.

23963 - **PARKER, R.R., SIBERT, J., BROWN, T.J.** : Inhibition of primary productivity through heterotrophic competition for nitrate in a stratified estuary. - J. Fish. Res. Board Can. *32* : 72 - 77, 1975.

23964 - **PAROMENSKAYA, L.N., MIKHAĬLOVA, L.D., BAGIYAN, L.G.** : Vliyanie gerbitsidov 2M-4X i 2M-4XM na fotosinteticheskie pigmenty v rasteniyakh gorokha. [Effect of herbicides 2M-4X and 2M-4XM on photosynthetic pigments in pea plants.] - - Fiziol. Rast. *22* : 421 - 423, 1975. [In R.]

23965 - **PARSON, W.W., CLAYTON, R.K., COGDELL, R.J.** : Excited states of photosynthetic reaction centers at low redox potentials. - Biochim. biophys. Acta *387* : 265- - 278, 1975.

23966 - **PARSON, W.W., COGDELL, R.J.** : The primary photochemical reaction of bacterial photosynthesis. - Biochim. biophys. Acta *416* : 105 - 149, 1975.

23967 - **PARTHASARATHY, S., PURUSHOTHAMAN, D.** : Changes in spectral properties of chlorophyll due to herbicide treatment. - Indian J. agr. Chem. *8* : 247 - 251, 1975.

23968 - **PARTHIER, B., KRAUSPE, R.** : Specificity and synthesis of plastid-specific aminoacyl-tRNA synthetases in *Euglena gracilis*. - In : Les Cycles Cellulaires et Leur Blocage chez Plusieurs Protistes. - Coll. int. CNRS 240. Pp. 233 - 239. Édit. CNRS, Paris 1975. [Chl.]

23969 - **PARTHIER, B., KRAUSPE, R., MUNSCHE, D., WOLLGIEHN, R.** : The biogenesis of chloroplasts. - In : **HARBORNE, J.B., VAN SUMERE, C.F.** (ed.) : The Chemistry and Biochemistry of Plant Proteins. Pp. 167 - 210. Academic Press, London - New York - San Francisco 1975.

23970 - **PASCHENKO, V.Z., PROTASOV, S.P., RUBIN, A.B., TIMOFEEV, K.N., ZAMAZOVA, L.M., RUBIN, L.B.** : Probing the kinetics of Photosystem I and Photosystem II fluorescence in pea chloroplasts on a picosecond pulse fluorometer. - Biochim. biophys. Acta *408* : 143 - 153, 1975.

23971 - **PASHCHENKO, V.Z., RUBIN, A.B., RUBIN, L.B.** : Izmerenie dlitel'nosti fluorestsentsii khlorofilla na impul'snom fluorometre s vozbuzhdeniem ot lazera s sinkhronizatsieĭ mod. [Measurement of duration of chlorophyll fluorescence using an impulse fluorometer with synchrone mode laser excitation.] - Kvant. Ėlektronika *2* : 1336 - 1340, 1975. [In R.]

23972 - **PASICHNYĬ, A.P., MAKARENKO, K.I.** : Ėlektricheskie potentsialy list'ev i assimilyatsiya uglekisloty u rasteniĭ. [Electrical potentials of leaves and assimilation of carbon dioxide in plants.] - Fiziol. Biokhim. kul't. Rast. *7* : 30 - 34, 1975. [In R, ab : E.]

23973 - **PASSERA, C.** : Meccanismo di assimilazione della CO_2 e livello di fotorespirazione in piante di orzo di differente età. [Photosynthetic CO_2 fixation mechanisms and photorespiration level in barley plants at different age.] - Riv. Agron. *9* : 56 - 60, 1975. [In Ital., ab : E.]

23974 - **PASSERA, C., ALBUZIO, A.** : Effetto della benzilamminopurina sulle reazioni di fissazione della CO_2 in foglie di orzo. [Effect of benzylaminopurine on CO_2 fixation reactions in barley leaves.] - Agrochimica *19* : 480 - 490, 1975.[In Ital., ab : E, F, G, Span.]

23975 - **PATE, J.S.** : Pea. - In : **EVANS, L.T.** (ed.) : Crop Physiology. Pp. 191 - 224. Cambridge Univ. Press, London - New York 1975. [Ps.]

23976 - **PATTEN, B.C., EGLOFF, D.A., RICHARDSON, T.H.** : Total ecosystem model for a cove in Lake Texoma. - In : **PATTEN, B.C.** (ed.) : Systems Analysis and Simulation in Ecology. Vol. III. Pp. 205 - 421. Academic Press, New York - San Francisco - - London 1975. [Production.]

*23977 - **PATTERSON, D.T.** : Investigations of the photosynthetic characteristics of smooth pigweed, *Amaranthus hybridus* L. - J. Elisha Mitchell sci. Soc. *87* : 155 - 156, 1971.

23978 - **PATTERSON, D.T.** : Photosynthetic acclimation to irradiance in *Celastrus orbiculatus* THUNB. - Photosynthetica *9* : 140 - 144, 1975.

23979 - **PATTERSON, D.T., HILE, J.L.** : A CO_2 monitoring and control system for plant growth chambers. - Ohio J. Sci. *75* : 190 - 193, 1975.

23980 - **PATTERSON, D.T., KRAMER, P.J.** : *In situ* and laboratory measurements of photosynthesis in cotton and soybeans grown in controlled environments and out-of doors. - Plant Physiol. *56* (Suppl.) : 12, 1975.

23981 - **PAUL, J.S., SULLIVAN, C.W., VOLCANI, B.E.** : Photorespiration in diatoms. Mitochondrial glycolate dehydrogenase in *Cylindrotheca fusiformis* and *Nitzschia alba*. - Arch. Biochem. Biophys. *169* : 152 - 159, 1975.

23982 - **PAUL, J.S., VOLCANI, B.E.** : Photorespiration in diatoms. III. Glycolate:cytochrome *c* reductase in the diatom *Cylindrotheca fusiformis*. - Plant Sci. Lett. *5* : 281 - 285, 1975.

23983 - **PEACOCK, J.M.** : Temperature and leaf growth in *Lolium perenne*. I. The thermal microclimate : its measurement and relation to crop growth. - J. appl. Ecol. *12* : 99 - 114, 1975. [Growth analysis.]

23984 - **PEACOCK, J.M.** : Temperature and leaf growth in *Lolium perenne*. III. Factors affecting seasonal differences. - J. appl. Ecol. *12* : 685 - 697, 1975. [Growth analysis.]

23985 - **PEARCE, R.B., MOCK, J.J., BAILEY, T.B.** : Rapid method for estimating leaf area per plant in maize. - Crop Sci. *15* : 691 - 694, 1975.

23986 - **PEARCY, R.W., TROUGHTON, J.** : C_4 photosynthesis in tree form *Euphorbia* species from Hawaiian rainforest sites. - Plant Physiol. *55* : 1054 - 1056, 1975.

23987 - **PEARSON, B.R., NORRIS, R.E.** : Fine structure of cell division in *Pyramimonas parkeae* NORRIS and PEARSON (*Chlorophyta, Prasinophyceae*). - J. Phycol. *11* : 113 - 124, 1975. [Chloroplast.]

23988 - **PEARSON, L.C.** : Daily and seasonal patterns of photosynthesis in *Artemisia tridentata*. - J. Idaho Acad. Sci. *1975* : 11 - 19, 1975.

23989 - **PEAVEY, D.G., GIBBS, M.** : Photosynthetic enhancement studied in intact spinach chloroplasts. - Plant Physiol. *55* : 799 - 802, 1975.

23990 - **PECHENOV, V.A.** : Peredvizhenie ^{14}C-assimilyatov u sakharnoĭ svekly pri razlichnykh usloviyakh pitaniya i uvlazhneniya. [Transport of ^{14}C-assimilates in sugar beet under different conditions of nutrition and moistening.] - Fiziol. Biokhim. kul't. Rast. *7* : 286 - 290, 1975. [In R, ab : E.]

23991 - **PEDERSEN, A.** : Growth measurements of five *Sphagnum* species in South Norway. - Norw. J. Bot. *22* : 277 - 284, 1975. [Dry-matter accumulation.]

23992 - **PEET, M., ANDERSON, R., ADAMS, M.S.** : Effect of fire on big bluestem production. - Amer. Midland Naturalist *94* : 15 - 26, 1975. [Ps.]

23993 - **PEISER, G., YANG, S.F.** : Chlorophyll destruction in the presence of bisulfite. - Plant Physiol. *56* (Suppl.) : 13, 1975.

23994 - **PEISKER, M., APEL, P.** : Influence of oxygen on photosynthesis and photorespiration in leaves of *Triticum aestivum* L. 1. Relationship between oxygen concentration, CO_2 compensation point, and intracellular resistance to CO_2 uptake. - Photosynthetica *9* : 16 - 23, 1975.

*23995 - **PEKH, S.M., KRAĬNOVA, N.N., CHERNOV, I.A.** : Zavisimost' funktsional'noĭ aktivnosti khloroplastov ot osnovnykh komponentov smeseĭ dlya izvlecheniya i inkubatsii. [Relation between the functional activity of chloroplasts and the main components of mixtures for extraction and incubation.] - Nauch. Dokl. vyssh. Shkoly, biol. Nauki *17* (2) : 134 - 144, 1974. [In R.]

23996 - **PELEVIN, V.N., BYALKO, A.V., BEKASOVA, O.D., TSVETKOVA, A.M.** : Opredelenie kontsentratsii khlorofilla po spektru izlucheniya, vykhodyashchego iz morya. [Determination of chlorophyll concentration from the spectrum of radiation emitted by the sea.] - In : Gidrofizicheskie i Opticheskie Issledovaniya v Indiĭskom Okeane. Pp. 144 - 148. Nauka, Moskva 1975. [In R.]

*23997 - **PELL, E.J.** : The impact of ozone on the bioenergetics of plant systems. - In : **DUGGER, M.** (ed.) : Air Pollution Effects on Plant Growth. ACS Symposium Series No. 3. Pp. 106 - 114. Amer. chem. Soc., Washington 1974. [Ps.]

23998 - **PENNA, F.J., REED, D.W., KE, B.** : The orientation of photosynthetic pigments in chromatophore membranes and in purified reaction centers of *Rhodopseudomonas spheroides*. - In : **AVRON, M.** (ed.) : Proceedings of the Third International Congress on Photosynthesis. Vol. I. Pp. 421 - 425. Elsevier, Amsterdam - - Oxford - New York 1975.

23999 - **PENNER, D.** : Bentazone selectivity between soybean and Canada thistle. - Weed Res. *15* : 259 - 262, 1975. [Ps.]

24000 - **PENNING DE VRIES, F.W.T.** : Use of assimilates in higher plants. - In : **COOPER, J.P.** (ed.) : Photosynthesis and Productivity in Different Environments. Pp. 459 - 480. Cambridge Univ. Press, Cambridge - London - New York - Melbourne 1975.

24001 - **PENNING DE VRIES, F.W.T.** : The cost of maintenance processes in plant cells. - Ann. Bot. *39* : 77 - 92, 1975. [Carbon balance.]

24002 - **PERDON, A.A., del ROSARIO, E.J., JULIANO, B.O.** : Solubilization of starch synthetase bound to *Oryza sativa* starch granules. - Phytochemistry *14* : 949 - - 951, 1975.

24003 - **PERRIER, A.** : Assimilation nette, utilisation de l'eau et microclimat d'un champ de maîs. VI. Étude de la température de la surface du sol sous culture. - Ann. agron. *26* : 139 - 157, 1975.

24004 - **PERRIER, A.** : Methods of observation of heat and mass transfer in the lower atmosphere and in plant canopies. - In : **De VRIES, D.A., AFGAN, N.H.** (ed.) : Heat and Mass Transfer in the Biosphere. Part I. Transfer Processes in the Plant Environment. Pp. 229 - 249. Scripta Book Company, John Wiley and Sons, Washington 1975. [Ps.]

24005 - **PERRIER, A., ITIER, B., BERTOLINI, J.M., BLANCO de PABLOS, A.** : Mesure automatique du bilan d'énergie d'une culture. Exemples d'application. - Ann. agron. *26* : 19 - 40, 1975. [Resistances.]

24006 - **PERRY, L.J., Jr., CHAPMAN, S.R.** : Effects of clipping on dry matter yields of basin wildrye. - J. Range Management *28* : 271 - 274, 1975. [Dry-matter accumulation.]

24007 - **PERRY, M.W.** : Field environment studies on lupins. II The effects of time of planting on dry matter partition and yield components of *Lupinus angustifolius* L. - Aust. J. agr. Res. *26* : 809 - 818, 1975. [Production.]

24008 - **PERSANOV, V.M., KUZNETSOVA, L.G., KARPILOV, Yu.S.** : Lokalizatsiya reaktsiĭ tsikla C-4-dikarbonovykh kislot v kletochnykh strukturakh assimilyatsionnykh tkaneĭ lista kukuruzy. [Reactions of the cycle of C-4-dicarboxylic acids localized in cell structures of assimilation tissues of maize leaf.] - Fiziol. Rast. *22* : 479 - 483, 1975. [In R, ab : E.]

24009 - **PESANDO, J.M.** : Proton magnetic resonance studies of carbonic anhydrase. II. Group controlling catalytic activity. - Biochemistry *14* : 681 - 688, 1975.

24010 - **PESANDO, J.M., GROLLMAN, A.P.** : Proton magnetic resonance studies of carbonic anhydrase. III. Binding of sulfonamides. - Biochemistry *14* : 689 - 693, 1975.

24011 - **PESCHEK, G.A.** : Light inhibition of respiration and fermentation and its reversal by inhibitors and uncouplers of photophosphorylation in whole cells of a prokaryotic alga, *Anacystis nidulans*. - In : **AVRON, M.** (ed.) : Proceedings of the Third International Congress on Photosynthesis. Vol. II. Pp. 921 - 928. Elsevier, Amsterdam - Oxford - New York 1975.

24012 - **PETERBURGSKIĬ, A.V., KUDRYASHOV, V.S., TORMASOVA, E.E.** : Sravnitel'noe izuchenie éffektivnosti primeneniya vanadiya i molibdena na urozhaĭ, kachestvo i nekotorye fiziologicheskie protsessy bobovykh kul'tur. [Vanadium and molybdenum effects on yield, quality and certain physiological processes of leguminous plants.] - Fiziol. Biokhim. kul't. Rast. *7* : 234 - 240, 1975. [Ps, Chl; in R, ab : E.]

24013 - **PETERS, G.A.** : The *Azolla* - *Anabaena azollae* relationship III. Studies on metabolic capabilities and a further characterization of the symbiont. - Arch. Microbiol. *103* : 113 - 122, 1975. [Ps.]

24014 - **PETERSON, J.R., SMART, R.E.** : Foliage removal effects on 'Shiraz' grapevines. - Amer. J. Enol. Viticult. *26* : 119 - 124, 1975. [Growth analysis.]

24015 - **PETERSON, K.M., BILLINGS, W.D.** : Carbon dioxide flux from tundra and vegetation as related to temperature at Barrow, Alaska. - Amer. Midland Naturalist *94* : 88 - 98, 1975. [Source of CO_2 for vegetation.]

24016 - **PETERSON, L.W., HUFFAKER, R.C.** : Loss of ribulose 1,5-diphosphate carboxylase and increase in proteolytic activity during senescence of detached primary barley leaves. - Plant Physiol. *55* : 1009 - 1015, 1975.

24017 - **PETERSON, W.L., MAYO, J.M.** : Moisture stress and its effect on photosynthesis in *Dicranum polysetum*. - Can. J. Bot. *53* : 2897 - 2900, 1975.

24018 - **PETINOV, N.S., SINITSYNA, Z.A., KIRILLINA, V.I.** : Izmenenie sostoyaniya vody i fraktsionnyĭ sostav belkov v list'yakh fasoli i kukuruzy pri razlichnoĭ vodoobespechennosti. [Changes in the state of water and fractional composition of proteins in bean and maize leaves under different water supply.] - In : Vodoobmen Rasteniĭ pri Neblagopriyatnykh Usloviyakh Sredy. Pp. 86 - 88. Shtiintsa, Kishinev 1975. [Chloroplast; in R.]

24019 - **PETKOV, P.S.** : Vliyanie na usloviyata na otglezhdane v"rkhu biologichnite osobenosti na nyakoi sortove fasul. [Growing conditions as affecting the biological peculiarities of some bean varieties.] - Rasteniev"d. Nauki *12* (10) : 45 - 55, 1975. [Chl; in Bulg., ab : E, R.]

24020 - **PETR, J.** : Photosynthetic productivity of tall and dwarf varieties of pea (*Pisum sativum* L.). - Rostlinná Výroba (Praha) *21* : 885 - 895, 1975. [Growth analysis.]

24021 - **PETREA, V.** : Wirkung des Schwefelschwarzes auf einige physiologische Prozesse der Alge *Chlorella vulgaris*. - In : **NECHAS, I.** (ed.) : Izuchenie Intensivnoĭ Kul'tury Vodorosleĭ. Vol. II. Pp. 115 - 120. Třeboň 1975. [Ps, Chl.]

*24022 - **PETROSYAN, A.S.** : Opredelenie khlorofilla u nekroticheskikh gibridov pshenitsy. [Determination of chlorophyll in necrotic wheat hybrids.] - Biol. Zh. Arm. *26* : 108 - 109, 1973. [In R.]

*24023 - **PETROV, A., MANOLOV, P.** : Vliyanie na vegetativniya prirast v"rkhu razvitieto na plodovete pri krushoviya sort Dobra Luiza v nachaloto na vegetatsiyata. [Effect of vegetative growth on fruit development in the Dobra Luiza variety of pear at the beginning of vegetation.] - Fiziol. Rast. (Sofiya) *1* (2) : 81- - 92, 1974. [Photosynthates; in Bulg., ab : E, R.]

B24024 - **PETROV, V.E.** : Ėnergetika Assimiliruyushcheĭ Kletki i Fotosintez. [Energetics of Assimilating Cell and Photosynthesis.] - Izd. kazan. Univ., Kazan' 1975. [In R.]

24025 - **PETROV, V.E., SEĬFULLINA, N.Kh., SORVIN, S.V., KARABAEV, M.K.** : Ėnergetika assimiliruyushchikh kletok i fotosintez. III. Rol' sveta v razvitii povrezhdeniĭ, obuslovlennykh ėkstremal'noĭ temperaturoĭ. [The energetics of assimilating cells and photosynthesis. III. The role of light in development of injuries, induced by extremal temperature.] - Bot. Zh. *60* : 16 - 25, 1975. [In R, ab : E.]

24026 - **PETRUSHENKO, V.V., SLAVINA, N.G.** : Fotoėlektricheskaya aktivnost' prorostkov pshenitsy v svyazi s ikh ontogenezom i usloviyami pitaniya. [Photoelectrical activity of wheat seedlings in connection with their ontogenesis and conditions of nutrition.] - Fiziol. Biokhim. kul't. Rast. *7* : 251 - 255, 1975. [Ps; in R, ab : E.]

24027 - **PETTIGREW, G.W., AVIRAM, I., SCHEJTER, A.** : Physicochemical properties of two atypical cytochromes *c*, *Crithidia* cytochrome *c*-557 and *Euglena* cytochrome *c*-558. - Biochem. J. *149* : 155 - 167, 1975.

24028 - **PETTY, K.M., DUTTON, P.L.** : Physical aspects of membrane proton uptake and ubiquinone reduction. - Biophys. J. *15* : 280a, 1975.

24029 - **PFEIFER, R.F., McDIFFETT, W.F.** : Some factors affecting primary productivity of stream riffle communities. - Arch. Hydrobiol. *75* : 306 - 317, 1975.

24030 - **PFEIFHOFER, A.O., BELTON, J.C.** : Ultrastructural changes in chloroplasts resulting from fluctuations in NaCl concentration : Freeze-fracture of thylakoid membranes in *Dunaliella salina*. - J. Cell Sci. *18* : 287 - 299, 1975.

24031 - **PFENNIG, N.** : The phototrophic bacteria and their role in the sulfur cycle. - - Plant Soil *43* : 1 - 16, 1975.

24032 - **PFISTER, K., BUSCHMANN, C., LICHTENTHALER, H.K.** : Inhibition of the photosynthetic electron transport by bentazon. - In : **AVRON, M.** (ed.) : Proceedings of the Third International Congress on Photosynthesis. Vol. I. Pp. 675 - 681. Elsevier, Amsterdam - Oxford - New York 1975.

24033 - **PFLEGER, I., BERGMANN, H., ROTH, D.** : Untersuchungen zur Eignung des Refraktometerwertes für die Beregnungssteuerung. - Arch. Acker- Pflanzenbau Bodenkunde *19* : 121 - 131, 1975. [Ps.]

24034 - **PFLUGER, U.N., DAHL, J.S., LUTZ, H.U., BACHOFEN, R.** : Interaction of a coupling factor from *Rhodospirillum rubrum* with coupling factor deficient chromatophores. - Arch. Microbiol. *104* : 179 - 184, 1975.

24035 - **PFLUGSHAUPT, C., BACHOFEN, R.** : Bound ATP in chloroplast membranes : formation and effect of different inhibitors on the labelling. - J. Bioenerg. *7* : 49 - 60, 1975.

24036 - **PHAM THI, A.T., VIEIRA da SILVA, J.** : Action d'un traitement osmotique sur l'ultrastructure de feuilles de Cotonniers (*Gossypium hirsutum* L. et *G. anomalum* WAW et PEYR.). - Compt. rend. Acad. Sci. Paris, Sér. D *280* : 2857 - 2860, 1975. [Chloroplast.]

24037 - **PHAN, C.-T., HSU, H.** : L'éthylène et le déverdissage des tissus végétaux. - - Physiol. vég. *13* : 427 - 434, 1975.

24038 - **PHILLIPSON, J.** : Rainfall, primary production and 'carrying capacity' of Tsavo National Park (East), Kenya. - East Afr. Wildlife J. *13* : 171 - 201, 1975.

24039 - **PHIPPS, R.H., FULFORD, R.J., CROFTS, F.C.** : Relationships between the production of forage maize and accumulated temperature, Ontario heat units and solar radiation. - Agr. Meteorol. *14* : 385 - 397, 1975.

24040 - **PHUL, P.S., GUPTA, S.K., GILL, K.S.** : The relative importance of chlorophyll concentrations and greenness of leaves influencing grain yield in pearl millet. - Curr. Sci. *44* : 480 - 481, 1975.

24041 - **PICAUD, A., ACKER, S.** : Etude de la structure de membranes chloroplastiques isolées de la souche sauvage et d'un mutant sans activité du photosystème I de *Chlamydomonas reinhardtii*. - FEBS Lett. *54* : 13 - 17, 1975.

24042 - **PICK, U., ROTTENBERG, H., AVRON, M.** : Proton gradients, proton concentrations and photophosphorylation. - In : **AVRON, M.** (ed.) : Proceedings of the Third International Congress on Photosynthesis. Vol. II. Pp. 967 - 974. Elsevier, Amsterdam - Oxford - New York 1975.

24043 - **PICKETT, J.M.** : Growth of *Chlorella* in a nitrate-limited chemostat. - Plant Physiol. *55* : 223 - 225, 1975. [Ps.]

24044 - **PIEPER, R.D., DWYER, D.D., BANNER, R.E.** : Primary shoot production of blue grama grassland in south-central New Mexico under two soil nitroger levels. - - Southwest. Nat. *20* : 293 - 302, 1975.

24045 - **PIETERS, G.A.** : Thermography and plant physiology. - In : Thermography. (Proc. 1st Europe. Congr.; Bibl. Radiol. No. 6.) Pp. 210 - 217. Karger, Basel - New York 1975. [Leaf temperature in Ps measurements.]

24046 - **PIETERS, G.A., ZIMA, M.** : Photosynthesis of desiccating leaves of poplar. - - Physiol. Plant. *34* : 56 - 61, 1975.

24047 - **PIKE, C., ROSEN, J., GOLDEN, M.** : Iron deficiency in corn induced by excess zinc. - Plant Physiol. *56* (Suppl.) : 4, 1975. [Chl.]

24048 - **PILLAI, C.G.P., DAVIS, D.E.** : Mode of action of CGA-18762, CGA-17020, and CGA-24705. - Proc. South Weed sci. Soc. *28* : 308 - 314, 1975. [Ps.]

24049 - **PINEAU, B., DOUCE, R.** : Analysis of the protein composition of spinach chloroplast envelopes. - In : **AVRON, M.** (ed.) : Proceedings of the Third International Congress on Photosynthesis. Vol. III. Pp. 1667 - 1673. Elsevier, Amsterdam - - Oxford - New York 1975.

24050 - **PINEVICH, V.V., IVANOVA, S.B., LIPSKAYA, A.A.** : Characteristics of chloroplast DNA-protein complex. - In : **NASYROV, Yu.S., ŠESTÁK, Z.** (ed.) : Genetic Aspects of Photosynthesis. Pp. 31 - 36. Dr. W. Junk B.V. Publ., The Hague 1975.

24051 - **PINGREE, R.D., PUGH, P.R., HOLLIGAN, P.M., FORSTER, G.R.** : Summer phytoplankton blooms and red tides along tidal fronts in the approaches to the English Channel. - Nature *258* : 672 - 677, 1975. [Chl.]

24052 - **PIPINIS, I.A., SMALYUKAS, D.Yu.** : Biologiya prirosta korneĭ, nadzemnoĭ massy i produktivnost' tarana dubil'nogo v Litovskoĭ SSR. [Biology of growth of roots, shoot biomass and productivity of *Polygonum tinctorium* in Lithuanian SSR.] - In : Dubil'nye Rasteniya i Ikh Ispol'zovanie. Pp. 76 - 89. Inst. Bot. Akad. Nauk litov.SSR, Vil'nyus 1975. [Growth analysis; in R.]

24053 - **PIRSON, A.** : Synchrony, synchronisation and cell cycles. - In : Les Cycles Cellulaires et Leur Blocage chez Plusieurs Protistes. Coll. int. CNRS 240. Pp. 19 - 24. Édit. CNRS, Paris 1975. [Ps.]

24054 - **PISTORIUS, E.K., GEWITZ, H.-S., VOSS, H., VENNESLAND, B.** : Cyanide generation from carbonylcyanide *m*-chlorophenylhydrazone and illuminated grana or algae. - FEBS Lett. *59* : 162 - 166, 1975.

24055 - **PITMAN, M.G., LÜTTGE, U., LÄUCHLI, A., BALL, E.** : Diurnal changes in photosynthetic capacity of barley leaves. - Aust. J. Plant Physiol. *2* : 101 - 103, 1975.

24056 - **PLAKUNOVA, V.G.** : Obratimoe termicheskoe vytsvetanie membrannogo fioletovogo kompleksa bakteriorodopsina. [Reversible thermal decoloration of bacteriorhodopsin membrane complex.] - Nauch. Dokl. vyssh. Shkoly, biol. Nauki *18* (8) : 56 - 59, 1975. [In R.]

24057 - **PLAMONDON, A.P., GRANDTNER, M.M.** : Microclimat estival d'une sapinière à *Hylocomium* de la forêt Montmorency. - Natural. can. *102* : 73 - 87, 1975. [Energy balance.]

24058 - **PLANTE-CUNY, M.-R.** : Distribution selon la profondeur de la chlorophylle *a* fonctionnelle et des phéopigments sur les sédiments de la lagune Ebrié (Abidjan). - Compt. rend. Acad. Sci. Paris, Sér. D *281* : 1325 - 1328, 1975.

24059 - **PLATT, S.G., PLAUT, Z., BASSHAM, J.A.** : Steady state photosynthesis in whole leaves. - Plant Physiol. *56* (Suppl.) : 25, 1975.

24060 - **PLATT, T.** : Analysis of the importance of spatial and temporal heterogeneity in the estimation of annual production by phytoplankton in a small, enriched, marine basin. - J. exp. mar. Biol. Ecol. *18* : 99 - 109, 1975.

24061 - **PLATT, T., DENMAN, K.L.** : Spectral analysis in ecology. - Annu.Rev. Ecol. Systematics *6* : 189 - 210, 1975. [Chl.]

24062 - **PLATT, T., SUBBA RAO, D.V.** : Primary production of marine microphytes. - In : **COOPER, J.P.** (ed.) : Photosynthesis and Productivity in Different Environments. Pp. 249 - 280. Cambridge Univ. Press, Cambridge - London - New York - Melbourne 1975.

24063 - **PLAUT, Z., HALEVY, A.H., DISKIN, Y.** : Diurnal pattern of plant water status and CO_2 fixation of roses as affected by irrigation regimes. - J. amer. Soc. hort. Sci. *100* : 191 - 194, 1975.

24064 - **PLAUT, Z., LITTAN, A.** : Interaction between photosynthetic CO_2 fixation products and nitrate reduction in spinach and wheat leaves. - In : **AVRON, M.** (ed.) : Proceedings of the Third International Congress on Photosynthesis. Vol. II. Pp. 1507 - 1516. Elsevier, Amsterdam - Oxford - New York 1975.

24065 - **PLUS, R., LUTZ, M.** : Resonance Raman scattering of mesoporphyrin IX dimethylester in solution. - Spectroscop. Lett. *8* : 119 - 139, 1975.

24066 - **POINCELOT, R.P.** : Transport of metabolites across isolated envelope membranes of spinach chloroplasts. - Plant Physiol. *55* : 849 - 852, 1975.

24067 - **POLAK-CHARCON, S., PORAT, N., SHNEYOUR, A., BEN-SHAUL, Y.** : Ultrastructural organization of thylakoids in two photosynthetic mutants of *Euglena gracilis*. - In : **AVRON, M.** (ed.) : Proceedings of the Third International Congress on Photosynthesis. Vol. III. Pp. 1787 - 1800. Elsevier, Amsterdam - Oxford - New York 1975.

24068 - **POLEVAYA, V.S., SMOLOV, A.P., IGNAT'EV, A.R.** : Fotosinteticheskaya aktivnost' avtotrofnykh kul'tur tkani ruty. [Photosynthetic activity of the autotrophic tissue cultures of rue.] - Dokl. Akad. Nauk SSSR *225* : 230 - 231, 1975. [In R.]

24069 - **POLING, S.M., HSU, W.-J., YOKOYAMA, H.** : Structure-activity relationships of chemical inducers of carotenoid biosynthesis. - Phytochemistry *14* : 1933 - - 1938, 1975.

B24070 - **POLJAKOFF-MAYBER, A., GALE, J.** (ed.) : Plants in Saline Environments. - Springer-Verlag, Berlin - Heidelberg - New York 1975. [Ps.]

24071 - **POOVAIAH, B.W., MIZRAHI, Y., DOSTAL, H.C., CHERRY, J.H., LEOPOLD, A.C.** : Water permeability during tomato fruit development in normal and *rin* nonripening mutant. - Plant Physiol. *56* : 813 - 815, 1975. [Chl.]

24072 - **POPE, D.H.** : Effects of light intensity, oxygen concentration, and carbon dioxide concentration on photosynthesis in algae. - Microbial Ecol. *2* : 1 - 16, 1975.

*24073 - **POPE, D.H., BERGER, L.R.** : Algal photosynthesis at constant pO_2 and increased hydrostatic pressure. - In : **COLWELL, R.R., MORITA, R.Y.** (ed.) : Effect of the Ocean Environment on Microbial Activities. Pp. 203 - 207. Univ. Park Press, Baltimore - London 1974.

24074 - **POPOVA, L.P., VAKLINOVA, S.G.** : Effect of ferredoxine on incorporation and distribution of $^{14}CO_2$ upon carboxylation of phosphoenolpyruvic acid in maize. - Dokl. bolg.Akad. Nauk *28* : 817 - 820, 1975.

24075 - **POROKHNEVICH, N.V.** : Strukturnaya organizatsiya i pigmentnyĭ fond plastidnogo apparata edinitsy ploshchadi list'ev i produktivnost' l'na pri razlichnom snabzhenii tsinkom i med'yu. [Structural organization and the pigment fund of the plastid apparatus of the unit of leaf surface area and the productivity of flax under different zinc and copper nutrition.] - Sel'skokhoz. Biol. *10*: 511 - 517, 1975. [In R, ab : E.]

24076 - **POROKHNEVICH, N.V., IVANOV, N.P.** : Plastidnyĭ apparat lista i urozhaya yachmenya v svyazi s primeneniem tsinkovogo mikroudobreniya na torfyano-bolotnoĭ pochve. [Effect of zinc microfertilization on moor-peat soil on the plastid apparatus of the leaf and barley yield.] - Fiziol. Rast. *22* : 306 - 311, 1975. [In R, ab : E.]

24077 - **POROKHNEVICH, N.V., KHATUNTSOVA, V.V.** : Issledovanie geterogennosti khlorofilla i aktivnosti reaktsii Khilla izolirovannykh khloroplastov list'ev l'na v pervom semennom pokolenii v svyazi s udobreniem materinskikh rasteniĭ tsinkom i med'yu. [Chlorophyll heterogeneity and Hill reaction activity of isolated flax leaf chloroplasts in the first seed generation in relation to fertilization of present plants with zinc and copper.] - In : Fiziologo-biokhimicheskie Aspekty Rosta i Razvitiya Rasteniĭ. Pp. 37 - 43. Nauka i Tekhnika, Minsk 1975. [In R.]

24078 - **POROKHNEVICH, N.V., KONDRATENYA, L.K., IVANOVICH, V.A.** : Osobennosti nakopleniya khlorofillov *a* i *b* v list'yakh l'na pod vliyaniem tsinka pri razlichnom soderzhanii medi v pitatel'nom substrate. [Peculiarities of chlorophyll *a* and *b* accumulation in flax leaves affected by zinc under various content of copper in nutrient medium.] - In : Botanika. Vol. 17. Pp. 130 - 136. Nauka i Tekhnika, Minsk 1975. [In R.]

24079 - **POROKHNEVICH, N.V., VAKUL'CHIK, M.N.** : Izmeneniya plastidnogo apparata edinitsy ploshchadi list'ev i urozhaĭ yachmenya pri razlichnom snabzhenii tsinkom. [Changes in the plastid apparatus of area unit of leaves and yield of barley under various zinc supply.] - Agrokhimiya *1975* (8) : 90 - 94, 1975. [Chl; in R.]

24080 - **PORTER, T.K., REYNOLDS, J.H.** : Relationship of alfalfa cultivar yields to specific leaf weight, plant density, and chemical composition. - Agron. J. *67* : 625 - 629, 1975.

24081 - **PORTIS, A.R., Jr., MAGNUSSON, R.P., McCARTY, R.E.** : Conformational changes in coupling factor 1 may control the rate of electron flow in spinach chloroplasts. - Biochem. biophys. Res. Commun. *64* : 877 - 884, 1975.

24082 - **POSKUTA, J.** : Photorespiration as factor in photosynthesis. The effect of light, oxygen and growth regulators. - Pol. ecol. Stud. *1* : 117 - 125, 1975.

24083 - **POSKUTA, J., FRANKIEWICZ-JOZKO, A.** : Enhanced dark CO_2 fixation by maize leaves in relation to previous illumination and oxygen concentration. - In : **MARCELLE, R.** (ed.) : Environmental and Biological Control of Photosynthesis. Pp. 89 - 105. Dr. W. Junk b.v. Publ., The Hague 1975.

24084 - **POSKUTA, J., PARYS, E., OSTROWSKA, E.** : Growth, CO_2 exchange rates and yield of pea (*Pisum sativum* L.) cv. "Bordi" in the field conditions after pretreatment of seeds with gibberellic acid (GA_3). - Biul. warzywniczy *18* : 197 - 206, 1975.

24085 - **POSKUTA, J., PARYS, E., OSTROWSKA, E., WOLKOWA, E.** : Photosynthesis, photorespiration, respiration and growth of pea seedlings treated with gibberellic acid (GA_3). - In : **MARCELLE, R.** (ed.) : Environmental and Biological Control of Photosynthesis. Pp. 201 - 209. Dr. W. Junk b.v. Publ., The Hague 1975.

24086 - **POSSINGHAM, J.V., CRAN, D.G., ROSE, R.J., LOVEYS, B.R.** : Effects of green light on the chloroplasts of spinach leaf discs. - J. exp. Bot. *26* : 33 - 42, 1975.

24087 - **POSTEL, J.R., RAO, V.N.R.** : A comparison of membrane and glass-fibre filters for productivity experiments. - Proc. Indian Acad. Sci. *B 82* : 221 - 229, 1975.

24088 - **POTTER, J.R.** : The effect of temperature on growth and partitioning of photosynthate. - Plant Physiol. *56* (Suppl.) : 79, 1975.

24089 - **POUGET, R., OTTENWAELTER, M.** : Les problèmes posés par la sélection de porte--greffes résistants à la chlorose calcaire. - Vitis *13* : 292 - 296, 1975. [Chl.]

24090 - **POWELL, T.M., RICHERSON, P.J., DILLON, T.M., AGEE, B.A., DOZIER, B.J., GODDEN, D.A., MYRUP, L.O.** : Spatial scales of current speed and phytoplankton biomass fluctuations in Lake Tahoe. - Science *189* : 1088 - 1090, 1975. [Chl.]

*24091 - **PRABHAKAR, C.S., RAO, K.R.** : Photosynthetic studies of some bryophytes. - In: Use of Radiations and Radioisotopes in Studies of Plant Productivity. Pp. 717--725. Bombay 1974.

24092 - **PREBBLE, J., WEST, S.** : Carotenoids of cauliflower bud tissue. - Ann. Bot. *39* : 1097 - 1102, 1975.

24093 - **PRESSLAND, A.J.** : Productivity and management of mulga in South-western Queensland in relation to tree structure and density. - Aust. J. Bot. *23* : 965 - 976, 1975. [Growth analysis.]

*24094 - **PRIEZZHEV, N.I., KOLOMEĬCHENKO, V.V., KRUKHMALEV, P.Ya.** : Vliyanie udobreniĭ na fotosinteticheskuyu deyatel'nost' i urozhaĭ tomatov. [Effect of fertilizers on photosynthetic activity and yield of tomato.] - Tr. volgograd. opyt. Sta. VIR *6* : 61 - 76, 1969. [In R.]

24095 - **PRINCE, R.C., BACCARINI-MELANDRI, A., HAUSKA, G.A., MELANDRI, B.A., CROFTS, A. R.** : Asymmetry of an energy transducing membrane. The location of cytochrome c_2 in *Rhodopseudomonas spheroides* and *Rhodopseudomonas capsulata*. - Biochim. biophys. Acta *387* : 212 - 227, 1975.

24096 - **PRINCE, R.C., DUTTON, P.L.** : A kinetic completion of the cyclic photosynthetic electron pathway of *Rhodopseudomonas sphaeroides* : Cytochrome *b*-cytochrome c_2 oxidation-reduction. - Biochim. biophys. Acta *387* : 609 - 613, 1975.

24097 - **PRINCE, R.C., HAUSKA, G.A., CROFTS, A.R., MELANDRI, A., MELANDRI, B.A.** : Asymmetry of an energy-coupling membrane. The immunological localisation of cytochrome c_2 in *Rhodopseudomonads*. - In : **AVRON, M.** (ed.) : Proceedings of the Third International Congress on Photosynthesis. Vol. I. Pp. 769 - 776. Elsevier, Amsterdam - Oxford - New York 1975.

24098 - **PRINCE, R.C., LINDSAY, J.G., DUTTON, P.L.** : The Rieske iron-sulfur center in mitochondrial and photosynthetic systems : E_m/pH relationships. - FEBS Lett. *51* : 108 - 111, 1975.

4099 - **PRINS, H.B.A.** : Photosynthesis and the uptake of Rb^+ and Cl^- in leaves of *Vallisneria spiralis*. - Plant Physiol. *56* (Suppl.) : 43, 1975.

24100 - **PRIOUL, J.L., REYSS, A., CHARTIER, P.** : Relationships between carbon dioxide transfer resistances and some physiological and anatomical features. - In : **MARCELLE, R.** (ed.) : Environmental and Biological Control of Photosynthesis. Pp. 17 - 28. Dr. W. Junk b.v. Publ., The Hague 1975.

24101 - **PRISTAVU, N.** : Action of *p*-benzoquinone on the radioactive carbon metabolism in *Chlorella pyrenoidosa*. - In : **AVRON, M.** (ed.) : Proceedings of the Third International Congress on Photosynthesis. Vol. II. Pp. 1541 - 1546. Elsevier, Amsterdam - Oxford - New York 1975.

24102 - **PROCHASKA, L.J., GROSS, E.L.** : The effects of carboxyl group modification upon divalent cation binding and fluorescence changes in Triton X-100 sub-chloroplast particles. - Biophys. J. *15* (2, Pt. 2): 223a, 1975.

24103 - **PROCHASKA, L.J., GROSS, E.L.** : The effect of 1-ethyl-3(3-dimethylaminopropyl) carbodiimide on calcium binding and associated changes in chloroplast structure and chlorophyll *a* fluorescence in spinach chloroplasts. - Biochim. biophys. Acta *376* : 126 - 135, 1975.

24104 - **PROKHORCHIK, R.A., BORODINA, S.M.** : Nekotorye osobennosti vzaimosvyazi stroeniya i fotokhimicheskoĭ aktivnosti prostykh fenol'nykh soedineniĭ. [Some peculiarities of relationship of structure and photochemical activity of simple phenolic compounds.] - In : Pitanie i Obmen Veshchestv u Rasteniĭ. Pp. 77 - - 84. Nauka i Tekhnika, Minsk 1975. [Ps; in R.]

24105 - **PROMNITZ, L.C.** : A photosynthate allocation model for tree growth. - Photosynthetica *9* : 1 - 15, 1975.

*24106 - **PRONINA, N.B.** : O nekotorykh svoĭstvakh adenilatkinazy iz khloroplastov gorokha. [Some properties of adenylate kinase from pea chloroplasts.] - Nauch. Dokl. vyssh. Shkoly, biol. Nauki *17* (4) : 89 - 94, 1974.[In R.]

24107 - **PRONINA, N.B., MAKAROV, A.D., KHOLODENKO, N.Ya.** : O dvukh osnovnykh putyakh sinteza ATF v khloroplastakh gorokha. [Two main ways of ATP synthesis in pea chloroplasts.] - Fiziol. Biokhim. kul't. Rast. *7* : 382 - 386, 445, 1975. [In R, ab : E.]

24108 - **PROTSENKO, D.F., KIRICHENKO, F.G., MUSIENKO, N.N., SLANVYĬ, P.S.** : Osobennosti vodoobmena sortov ozimoĭ pshenitsy v ontogeneze v svyazi s zasukhoustoĭchivost'yu i termoustoĭchivost'yu. [Peculiarities of water balance of winter wheat varieties during ontogenesis in connection with their drought and thermal resistance.] - In : Vodoobmen Rasteniĭ pri Neblagopriyatnykh Usloviyakh Sredy. Pp. 33 - 38. Shtiintsa, Kishinev 1975. [Chl; in R.]

24109 - **PUDEK, M.R., RICHARDS, W.R.** : A possible alternate pathway of bacteriochlorophyll biosynthesis in a mutant of *Rhodopseudomonas sphaeroides*. - Biochemistry *14* : 3132 - 3137, 1975.

24110 - **PUECH, J., LENCREROT, P., DECAU, J.** : Effet d'une réduction de l'intensité lumineuse sur la photosynthèse globale d'une culture de Tournesol (*Helianthus annuus* L.). Incidence sur la production oléoprotéique de la graine. - Compt. rend. Acad. Sci. Paris, Sér. D *281* : 387 - 390, 1975.

24111 - **PUGH, P.R.** : Variations in the biochemical composition of the diatom *Coscinodiscus eccentricus* with culture age and salinity. - Mar. Biol. *33* : 195 - 205, 1975. [Chl.]

24112 - **PUPILLO, P., GIULIANI PICCARI, G.** : The reversible depolymerization of spinach chloroplast glyceraldehyde-phosphate dehydrogenase. Interaction with nucleotides and dithiothreitol. - Europe. J. Biochem. *51* : 475 - 482, 1975.

24113 - **PUROHIT, A.N., TREGUNNA, E.B.** : Inhibition of tobacco mosaic virus multiplication by glycidate and its reversal by glycine and serine. - Plant Sci. Lett. *5* : 177 - 182, 1975. [Photorespiration.]

24114 - **PURVES, S., HADLEY, G.** : Movement of carbon compounds between the partners in orchid mycorrhiza. - In : **SANDERS, F.E., MOSSE, B., TINKER, P.B.** (ed.) : Endomycorrhizas. Pp. 175 - 194. Academic Press, London - New York - San Francisco 1975.

24115 - **PUTTASWAMY, S., KRISHNAMURTHY, K.** : Pattern of dry matter accumulation and distribution in finger millet genotypes in relation to levels of spacing and nitrogen. - Mysore J. agr. Sci. *9* : 372 - 378, 1975.

24116 - **PYLIOTIS, N.A., GOODCHILD, D.J.** : Induced and normal crystalline inclusions in plastids revealed by freeze-fracturing. - Protoplasma *85* : 277 - 283, 1975.

24117 - **PYLIOTIS, N.A., GOODCHILD, D.J., GRIMME, L.H.** : The regreening of nitrogen--deficient *Chlorella fusca*. II. Structural changes during synchronous regreening. - Arch. Microbiol. *103* : 259 - 270, 1975.

24118 - **PYRINA, I.L., ELIZAROVA, V.A.** : Intensivnost' fotosinteza i soderzhanie khlorofilla u massovykh form fitoplanktona v vodoemakh razlichnogo tipa. [Photosynthetic rate and chlorophyll content in mass forms of phytoplankton in water reservoirs of various type.] - In : Izuchenie i Okhrana Vodnykh Resursov. P. 60. Nauka, Moskva 1975. [In R.]

24119 - **PYROZHENKO, S.U.** : Biokhimichna kharakterystyka *Spirulina platensis* (GOM.) GEITL. II. Vmist vuglevodiv u protsesi rostu kul'tury. [Biochemical characteristics of *Spirulina platensis* (GOM.)GEITL. II. Content of carbohydrates in the process of crop growth.] - Ukr. bot. Zh. *32* : 229 - 233, 268, 1975. [In Ukr., ab : E, R.]

24120 - **QUADER, H., KAUSS, H.** : Die Rolle einiger Zwischenstoffe des Galaktosylglyzerinstoffwechsels bei der Osmoregulation in *Ochromonas malhamensis*. - Planta *124* : 61 - 66, 1975. [Ps.]

24121 - **QUAYLE, J.R., PFENNIG, N.** : Utilization of methanol by *Rhodospirillaceae*. - - Arch. Microbiol. *102* : 193 - 198, 1975. [Ps.]

24122 - **QUEBEDEAUX, B., CHOLLET, R.** : Growth and development of soybean (*Glycine max* [L.]MERR.) pods. CO_2 exchange and enzyme studies. - Plant Physiol. *55* : 745 - - 748, 1975.

24123 - **QUEBEDEAUX, B., HARDY, R.W.F.** : Reproductive growth and dry matter production of *Glycine max* (L.) MERR. in response to oxygen concentration. - Plant Physiol. *55* : 102 - 107, 1975.

24124 - **QUEBEDEAUX, B., HARDY, R.W.F.** : O_2 effects on transport and accumulation of photosynthate from leaves to reproductive structures in the light. - Plant Physiol. *56* (Suppl.) : 17, 1975.

24125 - **QUEBEDEAUX, B., HAVELKA, U.D., LIVAK, K.L., HARDY, R.W.F.** : Effect of altered pO_2 in the aerial part of soybean on symbiotic N_2 fixation. - Plant Physiol. *56* : 761 - 764, 1975.

24126 - **QUEIROZ, O.** : Rhythmical characteristics at different levels of CAM regulation: physiological and adaptive significance. - In : **MARCELLE, R.** (ed.) : Environmental and Biological Control of Photosynthesis. Pp. 357 - 368. Dr. W. Junk b.v. Publ., The Hague 1975.

24127 - **RAAFAT, A., HERWIG, K.** : Accumulation of metabolites and retardation of leaf senescence by 6-benzylaminopurine or disbudding treatment in intact bean plants. - Bodenkultur *26* : 355 - 368, 1975. [Chl.]

24128 - **RABINOVITZ, H., REISFELD, A., EDELMAN, M.** : Isolation and characterization of ribulose diphosphate carboxylase (RuDPCase) from autotrophic *Euglena gracilis*. - Israel J. Bot. *24* : 57, 1975.

24129 - **RABINOWITCH, H.D., BUDOWSKI, P., KEDAR, N.** : Carotenoids and epoxide cycles in mature-green tomatoes. - Planta *122* : 91 - 97, 1975.

24130 - **RABINOWITZ, H., REISFELD, A., SAGHER, D., EDELMAN, M.** : Ribulose diphosphate carboxylase from autotrophic *Euglena gracilis*. - Plant Physiol. *56* : 345 - - 350, 1975.

24131 - **RACHKULIK, V.I., SITNIKOVA, M.V.** : Nekotorye voprosy identifikatsii rastitel'nosti. [Some problems of vegetation identification.] - Meteorol. Gidrol. *1975* (1) : 85 - 88, 1975. [Reflectance of vegetation; in R.]

B24132 - Radiation Techniques for Water-Use Efficiency Studies.-Tech. Rep. Ser. No.168. Int. at. Energy Agency, Vienna 1975. [Growth analysis.]

24133 - **RADMER, R., KOK, B.** : Energy capture in photosynthesis : Photosystem II. - Annu. Rev. Biochem. *44* : 409 - 433, 1975.

24134 - **RADMER, R.J.** : The mass spectrometric measurement of gas exchange in illuminated algae. - Plant Physiol. *56* (Suppl.) : 72, 1975.

24135 - **RADUNZ, A.** : Binding of antibodies onto the thylakoid membrane. I. Maximal antibody binding and adsorption of antibodies to lipids. - Z. Naturforsch. *30C* : 484 - 488, 1975.

24136 - **RADUNZ, A., SCHMID, G.H.** : The effect of antibodies to neoxanthin on electron transport on the oxygen evolving side of photosystem II and the reactions of this antiserum with various chloroplast preparations. - Z. Naturforsch. *30C* : 622 - 627, 1975.

24137 - **RADWAN, S.S.** : The lipids in plant tissue cultures compared to the lipids of photosynthetic and non-photosynthetic plants and plant organs. - Fette-Seifen -Anstrichmittel *77* (5) : 181 - 184, 1975.

24138 - **RAGGI, V.** : Incorporation of $^{14}CO_2$ in amides and amino acids in rusted bean leaves. - Can. J. Bot. *53* : 1244 - 1251, 1975.

24139 - **RAGHAVENDRA, A.S., DAS, V.S.R.** : Malonate-inhibition of allosteric phosphoenolpyruvate carboxylase from *Setaria italica*. - Biochem. biophys. Res. Commun. *66* : 160 - 165, 1975.

24140 - **RAI, H.** : Pigment cycle in a north German lake. - Verh. int. Verein. Limnol. *19* : 1220 - 1228, 1975.

24141 - **RAI, M., DAS, K.** : *Gamma* ray induced chlorophyll mutations in linseed. - Indian J. Genet. *35* : 462 - 466, 1975.

24142 - **RAI, R.S.V., MURTY, K.S., SAHU, G.** : Efficiency in ^{14}C photosynthesis and translocation in photosensitive rice varieties. - J. nucl. agr. Biol. *4* (4) : 81 - 82, 1975.

24143 - **RAJAN, A.K., BLACKMAN, G.E.** : Interacting effects of light and day and night temperatures on the growth of four species in the vegetative phase. - Ann. Bot. *39* : 733 - 743, 1975.

24144 - **RAJYALAKSHMI, T., PREMSWARUP, T.V.** : Primary productivity in River Godawari. - Indian J. Fish. *22* : 205 - 214, 1975.

24145 - **RAKHIMOV, A., RAKHIMOV, S.** : Izuchenie biokhimicheskikh osobennosteĭ ryaski maloĭ (soderzhanie karotina i askorbinovoĭ kisloty v biomasse ryaski maloĭ v kul'ture). [Biochemical characteristics of *Lemna minor* (carotene and ascorbic acid contents in the biomass of *Lemna minor* in culture).] - In : Vodorosli i Griby Sredneĭ Azii. Pp. 208 - 213. Fan, Tashkent 1975. [In R.]

24146 - **RAKHIMOV, A., YUNUSOV, I.I.** : Nekotorye biokhimicheskie osobennosti vol'fii beskoreshkovoĭ (*Wolffia arrhiza* WIMMER), vyrashchennoĭ na razlichnykh pitatel'nykh sredakh. [Some biochemical characteristics of *Wolffia arrhiza* WIMMER grown in different nutrient media.] - In : Vodorosli i Griby Sredneĭ Azii. Pp. 200 - 207. Fan, Tashkent 1975. [Car; in R.]

24147 - **RAKHIMOV, G.T.** : Aktivnost' khloroplastov izenya (*Kochia prostrata* (L.) SCHRAD.) v zavisimosti ot obezvozhivaniya list'ev. [Activity of chloroplasts of the summer-cypress (*Kochia prostrata* (L.) SCHRAD.) depending on leaf dehydration.] - Uzb. biol. Zh. *1975* (4) : 18 - 19, 1975. [In R, ab : Uz.]

24148 - **RAKHIMOV, G.T.** : Ob aktivnosti fotosinteticheskogo apparata ėspartsetov, vyrashchennykh v razlichnykh usloviyakh vlazhnosti pochvy. [Activity of the photosynthetic apparatus of *Onobrychis* species grown under various soil moisture.] - In : Fiziologiya i Biokhimiya Dikorastushchikh Kormovykh Rasteniĭ Uzbekistana. Pp. 99 - 106. Fan, Tashkent 1975. [In R.]

24149 - **RAKHMANKULOVA, M.E., KHODZHAEV, A.S.** : Otsenka metodov opredeleniya aktivnosti fotofosforilirovaniya v khloroplastakh list'ev khlopchatnika. [Evaluation of methods for determination of photophosphorylation activity in chloroplasts of cotton leaves.] - Fiziol. Biokhim. kul't. Rast. *7* : 526 - 530, 1975. [In R, ab : E.]

24150 - **RAKHTEENKO, I.N., MARTINOVICH, B.S.** : Produktivnost' smeshannykh lesnykh nasazhdeniĭ i ee zavisimost' ot kolichestvennykh sootnosheniĭ sostavlyayushchikh fitotsenoz komponentov. [Productivity of mixed forest stands and its dependence on quantitative relations between components of the phytocenoses.] -

In : Ėkologo-Biologicheskie Issledovaniya Rastitel'nykh Soobshchestv. Pp. 28 - - 37. Nauka i Tekhnika, Minsk 1975. [In R.]

24151 - **RAKITIN, L.Yu., RAKITIN, V.Yu.** : Laboratornaya ustanovka dlya provedeniya opytov po issledovaniyu rastitel'nykh ob"ektov v reguliruemoĭ gazovoĭ srede. [A laboratory device for studying fruits and other plant objects in physiological experiments under regulated conditions of gaseous medium.] - Fiziol. Rast. *22*: 654 - 659, 1975. [In R, ab : E.]

*24152 - **RAMAGE, R.** : Carotenoid chemistry. - In : **NEWMAN, A.A.** (ed.) : Chemistry of Terpenes and Terpenoids. Pp. 288 - 336. Academic Press, London - New York 1972.

24153 - **RAMAM, S.S.** : Primary production and nutrient cycling in tropical deciduous forest ecosystems. - Trop. Ecol. *16* : 140 - 146, 1975.

24154 - **RAPS, S., GREGORY, R.P.F.** : Specific organization of bulk chloroplast pigments. - In : **AVRON, M.** (ed.) : Proceedings of the Third International Congress on Photosynthesis. Vol. III. Pp. 1983 - 1990. Elsevier, Amsterdam - Oxford - New York 1975.

24155 - **RASCHKE, K.** : Stomatal action. - Annu. Rev. Plant Physiol. *26* : 309 - 340, 1975.

24156 - **RASCHKE, K.** : Simultaneous requirement of carbon dioxide and abscisic acid for stomatal closing in *Xanthium strumarium* L. - Planta *125* : 243 - 259, 1975. [Stomatal resistance.]

24157 - **RASCIO, N., ORSENIGO, M.** : Chloroplast fine structure and morphogenesis in the japonica-2 maize mutant. 2 Vascular bundle sheath plastids. - Cytobios *14* : 171 - 181, 1975.

24158 - **RASKIN, V.I., GUREVICH, G.M., YUNEVICH, V.I.** : Perenos ėnergii vozbuzhdeniya mezhdu molekulami protokhlorofillida i khlorofillida v ėtiolirovannykh list'yakh i rastvore protokhlorofillid-golokhroma do i posle osveshcheniya. [Excitation energy transfer between protochlorophyllide and chlorophyllide molecules in etiolated leaves and in protochlorophyllide-holochrome solution before and after illumination.] - In : **SHLYK, A.A.** (ed.) : Biosintez i Sostoyanie Khlorofillov v Rastenii. Pp. 19 - 41, 243 - 244. Nauka i Tekhnika, Minsk 1975. [In R, ab : E.]

24159 - **RATHNAM, C.K.M., DAS, V.S.R.** : Biophysical characterization of mesophyll and bundle sheath chloroplasts isolated from the leaves of *coracana*, an aspartate-type C-4 plant. IV. High-amplitude swelling and low-amplitude shrinkage of chloroplasts. - Biochem. Physiol. Pflanzen *167* : 327 - 341, 1975.

24160 - **RATHNAM, C.K.M., DAS, V.S.R.** : Energetic basis of the phloem transport of ^{14}C--assimilate in leaves of *Eleusine coracana*. - Biochem. Physiol. Pflanzen *167* : 565 - 575, 1975.

24161 - **RATHNAM, C.K.M., DAS, V.S.R.** : Aspartate-type C-4 photosynthetic carbon metabolism in leaves of *Eleusine coracana* GAERTN. - Z. Pflanzenphysiol. *74* : 377 - - 393, 1975.

24162 - **RATHNAM, C.K.M., DAS, V.S.R.** : Inter- and intracellular distribution of carbonic anhydrase, PEP carboxylase and RuDP carboxylase in leaves of *Eleusine coracana*, a C-4 plant. - Z. Pflanzenphysiol. *75* : 360 - 364, 1975.

24163 - **RATHNAM, C.K.M., DAS, V.S.R.** : Biophysical characterization of mesophyll and bundle sheath chloroplasts isolated from the leaves of *Eleusine coracana*, an aspartate-type C-4 plant : Part I - Light absorbing photosynthetic pigment composition. - Indian J. Biochem. Biophys. *12* : 175 - 178, 1975.

24164 - **RATHNAM, C.K.M., EDWARDS, G.E.** : Intracellular localization of certain photosynthetic enzymes in bundle sheath cells of C_4 plants. - Plant Physiol. *56* (Suppl.) : 27, 1975.

24165 - **RATHNAM, C.K.M., EDWARDS, G.E.** : Intracellular localization of certain photosynthetic enzymes in bundle sheath cells of plants possessing the C_4 pathway of photosynthesis. - Arch. Biochem. Biophys. *171* : 214 - 225, 1975.

24166 - **RAU, W.** : Zum Mechanismus der Photoregulation von Morphosen am Beispiel der Carotinoidsynthese. - Ber. deut. bot. Ges. *88* : 45 - 60, 1975.

24167 - **RAUSER, W.E., HORTON, R.F.** : A simple device to measure root growth rates. - J. biol. Educ. *9* : 92 - 95, 1975.

24168 - **RAVEN, C.W., SHROPSHIRE, W.** : Photoregulation of logarithmic fluence-response curves for phytochrome control of chlorophyll formation in *Pisum sativum* L. - Photochem. Photobiol. *21* : 423 - 429, 1975.

24169 - **RAVEN, J.A.** : Algal cells. - In : **BAKER, D.A., HALL, J.L.** (ed.) : Ion Transport in Plant Cells and Tissues. Pp. 125 - 160. North-Holland Publ. Comp., Amsterdam - Oxford, Amer. Elsevier Publ. Comp., New York 1975.

24170 - **RAVEN, J.A., GLIDEWELL, S.M.** : Sources of ATP for active phosphate transport in *Hydrodictyon africanum*. Evidence for pseudocyclic photophosphorylation *in vivo*. - New Phytol. *75* : 197 - 204, 1975.

24171 - **RAVEN, J.A., GLIDEWELL, S.M.** : Effects of CCCP on photosynthesis and on active and passive chloride transport at the plasmalemma of *Hydrodictyon africanum*. - New Phytol. *75* : 205 - 213, 1975.

24172 - **RAVEN, J.A., GLIDEWELL, S.M.** : Photosynthesis, respiration and growth in the shade alga *Hydrodictyon africanum*. - Photosynthetica *9* : 361 - 371, 1975.

24173 - **RAVINSKAYA, A.P.** : Vklyuchenie C^{14} v lishaĭnikovye kisloty pri fotosinteze. [Incorporation of ^{14}C into lichen acids during photosynthesis.] - Fiziol. Rast. *22* : 251 - 255, 1975. [In R, ab : E.]

24174 - **RAVIZZINI, R.A., LESCANO, W.I.M., VALLEJOS, R.H.** : Effect of aurovertin on energy transfer reactions in *Rhodospirillum rubrum* chromatophores. - FEBS Lett. *58* : 285 - 288, 1975.

24175 - **RAWLINS, S.L., RAATS, P.A.C.** : Prospects for high-frequency irrigation. - Science *188* : 604 - 610, 1975. [Ps.]

B24176 - Reaktivnost' Fotosinteticheskogo Apparata. [Reactivity of Photosynthetic Apparatus.] - Izd. kazan. Univ., Kazan' 1975. [In R.]

24177 - **REBEIZ, C.A., MATTHEIS, J.R., SMITH, B.B.** : The net synthesis of protochlorophyll from δ-aminolevulinic acid by isolated plastids. - Plant Physiol. *56* (Suppl.) : 31, 1975.

24178 - **REBEIZ, C.A., MATTHEIS, J.R., SMITH, B.B., REBEIZ, C.C., DAYTON, D.F.** : Chloroplast biogenesis. Biosynthesis and accumulation of Mg-protoporphyrin IX monoester and longer wavelength metalloporphyrins by greening cotyledons. - Arch. Biochem. Biophys. *166* : 446 - 465, 1975.

24179 - **REBEIZ, C.A., MATTHEIS, J.R., SMITH, B.B., REBEIZ, C.C., DAYTON, D.F.** : Chloroplast biogenesis. Biosynthesis and accumulation of protochlorophyll by isolated etioplasts and developing chloroplasts. - Arch. Biochem. Biophys. *171* : 549 - 567, 1975.

24180 - **REBEIZ, C.A., SMITH, B.B., MATTHEIS, J.R., REBEIZ, C.C., DAYTON, D.F.** : Chloroplast biogenesis. Biosynthesis and accumulation of Mg-protoporphyrin IX monoester and other metalloporphyrins by isolated etioplasts and developing chloroplasts. - Arch. Biochem. Biophys. *167* : 351 - 365, 1975.

24181 - **REDMANN, R.E.** : Production ecology of grassland plant communities in western North Dakota. - Ecol. Monogr. *45* : 83 - 106, 1975.[Chl.]

24182 - **REED, D.W., RAVEED, D., REPORTER, M.** : Localization of photosynthetic reaction centers by antibody binding to chromatophore membranes from *Rhodopseudomonas spheroides* strain R26. - Biochim. biophys. Acta *387* : 368 - 378, 1975.

24183 - **REGEHR, D.L., BAZZAZ, F.A., BOGGESS, W.R.** : Photosynthesis, transpiration and leaf conductance of *Populus deltoides* in relation to flooding and drought. - Photosynthetica *9* : 52 - 61, 1975.

24184 - **REGITZ, G., OHAD, I.** : Changes in the protein organization in developing thylakoids of *Chlamydomonas reinhardi* Y-1 as shown by sensitivity to trypsin. - In : **AVRON, M.** (ed.) : Proceedings of the Third International Congress on Photosynthesis. Vol. III. Pp. 1615 - 1625. Elsevier, Amsterdam - Oxford - New York 1975.

24185 - **REHDER, H.** : Phytomasse- und Nährstoffverhältnisse einer alpinen Rasengesellschaft (*Caricetum firmae*). - Verhandl. Ges. Ökol. (Wien) *1975* : 93 - 99, 1975.

24186 - **REICHLE, D.E., O'NEILL, R.V., HARRIS, W.F.**: Principles of energy and material exchange in ecosystems. - In : **VAN DOBBEN, W.H., LOWE-McCONNELL, R.H.** (ed.) : Unifying Concepts in Ecology. Pp. 27 - 43. Dr. W. Junk b.v. Publ., The Hague, Pudoc, Wageningen 1975.

24187 - **REIMER, S., TREBST, A.** : Light-induced conformational changes of the chloroplast thylakoid membrane as indicated by the inactivation of the oxygen evolution system by high internal pH. - Biochem. Physiol. Pflanzen *168* : 225 - - 232, 1975.

24188 - **REǏNGARD, T.A., VOLOVIK, O.I., BERSHTEǏN, B.I., VOLKOVA, N.V., VASILENOK, L.I., DUBROVSKAYA, A.A., ZAITSEVA, N.A., KANIVETS, N.P., MUSHKETIK, L.S., OKANENKO, A.S., OSTROVSKAYA, L.K., POLISHCHUK, A.I., SEMENYUK, I.I., YASNIKOV, A.A.** : O mekhanizme vzaimodeǐstviya piruvatkinazy s fotofosforiliruyushcheǐ sistemoǐ. [Interaction mechanism between pyruvate kinase and phosphorylating system.] - Fiziol. Biokhim. kul't. Rast. *7* : 351 - 355, 444, 1975. [In R, ab : E.]

24189 - **REMENNIKOV, S.M., CHAMOROVSKY, S.K., KONONENKO, A.A., VENEDIKTOV, P.S., RUBIN, A.B.** : High potential oxidation-reduction titration of absorbance changes induced by pulsed laser and continuous light in chromatophores of photosynthesizing bacteria *Rhodospirillum rubrum* and *Ectothiorhodospira shaposhnikovii*. - Studia biophys. *51* : 1 - 13, 1975.

24190 - **REMY, R., BEBEE, G.** : Membrane proteins of higher plant chloroplasts related to photochemical systems and membrane stacking. - In : **AVRON, M.** (ed.) : Proceedings of the Third International Congress on Photosynthesis. Vol. III. Pp. 1675 - 1684. Elsevier, Amsterdam - Oxford - New York 1975.

24191 - **RENGER, G.** : Studies on the reaction mechanism of the oxidizing equivalents produced by system II. - In : **AVRON, M.** (ed.) : Proceedings of the Third International Congress on Photosynthesis. Vol. I. Pp. 127 - 144. Elsevier, Amsterdam - Oxford - New York 1975.

24192 - **RENGER, G.** : The action of 5-chloro-3-tert. butyl-2'-chloro-4'-nitro-salicylanilide and α,α'-bis(hexafluoroacetonyl)aceton on the water-splitting enzyme system Y in spinach chloroplasts. - FEBS Lett. *52* : 30 - 32, 1975.

24193 - **RENGER, G., WOLFF, C.** : Studies on the nature of the primary reactions of photosystem II in photosynthesis. I. The electrochromic 515 nm absorption change as an appropriate indicator for the functional state of the photochemical active centres of system II in DCMU poisoned chloroplasts. - Z. Naturforsch. *30C*: 161 - 171, 1975.

24194 - **RENK, H.** : Hydrology and primary production of phytoplankton in the Baltic. - Pol. Arch. Hydrobiol. *22* : 377 - 397, 1975.

24195 - **RENTHAL, R., LANYI, J.K.** : Light-dependent changes in electrical potential of cell envelope vesicles from *Halobacterium halobium* measured with a cyanine dye. - Biophys. J. *15* (2, Part 2): 68a, 1975.

24196 - **REPKA, J., KOSTREJ, A.** : Dependence of the production process on changes of climatic factors. - Rostlinná Výroba (Praha) *21* : 837 - 844, 1975. [Growth analysis.]

*24197 - **REUCROFT, P.J., SIMPSON, W.H.** : Electrical characteristics of chlorophyll *a* films. - Discuss. Faraday Soc. *51* : 202 - 225, 1971.

24198 - **REUTSKIǏ, V.G., KOZLOVA, Zh.I.** : Primenenie amperometricheskogo metoda dlya opredeleniya skorosti reaktsii Khilla khloroplastov list'ev yachmenya. [Use of the amperometric method for determining the Hill reaction rate of barley leaf chloroplasts.] - Vestsi Akad. Navuk belarus. SSR, Ser. biyal. Navuk *1975* (2): 110 - 113, 142, 1975. [In R.]

24199 - **REUTSKIǏ, V.G., KOZLOVA, Zh.I., SAZONOVA, L.S.** : Fotokhimicheskaya aktivnost' khloroplastov iz list'ev nekotorykh zlakov pri razlichnom rezhime uvlazhneniya pochv. [Photochemical activity of chloroplasts from leaves of some grasses under different soil moisture.] - In : Ėkologo-biologicheskie Issledovaniya Rastitel'nykh Soobshchestv. Pp. 138 - 144, 221. Nauka i Tekhnika, Minsk 1975. [In R.]

24200 - **REYNOLDS, C.S.** : Interrelations of photosynthetic behaviour and buoyancy regulation in a natural population of a blue-green alga. - Freshwater Biol. *5* : 323 - 338, 1975.

24201 - **REYSS, A., PRIOUL, J.L.** : Carbonic anhydrase and carboxylase activities from plants (*Lolium multiflorum*) adapted to different light regimes. - Plant Sci. Lett. *5* : 189 - 195, 1975.

24202 - **REYSSAC, J.** : Évolution quantitative du phytoplancton de la baie du Lévrier de septembre à novembre 1973. - Bull. Mus. nat. Hist. nat. 3e Sér. *328* (Écol. gén. 26) : 69 - 79, 1975. [Chl.]

24203 - **RHODES, B.B., HALL, C.V.** : Effects of CPTA, temperature, and genotype on carotene synthesis in carrot leaves. - HortScience *10* : 22 - 23, 1975.

24204 - **RICARDO, C.P.P.** : Algumas considerações sobre a fisiologia da beterraba-sacarina. [A few considerations on sugar-beet physiology.] - Agron. lusit. *37* : 77 - 93, 1975. [Photosynthates; in Portug., ab : E.]

24205 - **RICHARDS, W.R., WALLACE, R.B., TSAO, M.S., HO, E.** : The nature of a pigment--protein complex excreted from mutants of *Rhodopseudomonas sphaeroides*. - Biochemistry *14* : 5554 - 5561, 1975.

24206 - **RICHTER, E.** : Adenosintriphosphat in Blättern von *Chrysanthemum indicum* L. als Möglichkeit zur Selektion von Pflanzen für die Eignung zur Kultur unter Schwachlichtbedingungen. - Gartenbauwissenschaft *40* : 56 - 59, 1975.

24207 - **RICKMAN, R.W., RAMIG, R.E., ALLMARAS, R.R.** : Modeling dry matter accumulation in dryland winter wheat. - Agron. J. *67* : 283 - 289, 1975.

24208 - **RIDLEY, S.M.** : Interaction of chloroplasts with inhibitors. I. Photodestruction of pigments in mature chloroplasts. - Plant Physiol. *56* (Suppl.) : 11, 1975.

24209 - **RIDLEY, S.M., LAING, J.** : Interaction of chloroplasts with inhibitors. II. Effects of β-carotene inhibition on the conversion of etioplasts to chloroplasts. - Plant Physiol. *56* (Suppl.) : 11, 1975.

24210 - **RIGLER, F.H.** : The concept of energy flow and nutrient flow between trophic levels. - In : **Van DOBBEN, W.H., LOWE-McCONNEL, R.H.** (ed.) : Unifying Concepts in Ecology. Pp. 15 - 26. Dr. W. Junk b.v. Publ., The Hague; Pudoc, Wageningen 1975.

24211 - **RIKHIREVA, G.T., PULATOVA, M.K., CHEKULAEVA, L.N., DRUZHKO, A.B., KOSITSYNA, V.A.** : ĖPR - izuchenie bakteriorodopsinsoderzhashchikh sistem. [ESR study of bacteriorhodopsin-containing systems.] - Biofizika *20* : 938 - 940, 1975. [In R, ab : E.]

24212 - **RIMON, A., OPPENHEIM, A.B.** : Heat induction of the blue-green alga *Plectonema boryanum* lysogenic for the cyanophage SP*lcts* 1. - Virology *64* : 454 - 463, 1975. [Ps.]

24213 - **RIOV, J., BROWN, G.N.** : Extraction, partial purification and properties of ferredoxin-NADP$^+$ reductase from eastern hemlock. - Plant Physiol. *56* (Suppl.): 48, 1975.

24214 - **RIPLEY, E., SAUGIER, B.** : Energy and mass exchange of a native grassland in Saskatchewan. - In : **de VRIES, D.A., AFGAN, N.H.** (ed.) : Heat and Mass Transfer in the Biosphere. Part I. Transfer Processes in Plant Environment. Pp. 311 - 325. Halstead Press, John Wiley, New York 1975.

24215 - **RIPLEY, E.A., REDMANN, R.E.** : Grassland. - In : **MONTEITH, J.L.** (ed.) : Vegetation and the Atmosphere. Vol. 2. Case Studies. Pp. 349 - 398. Academic Press, London - New York - San Francisco 1975. [Ps.]

24216 - **ROBB, J., BUSCH, L., BRISSON, J.D., LU, B.C.** : Ultrastructure of wilt syndrome caused by *Verticillium dahliae*. II. In sunflower leaves. - Can. J. Bot. *53* : 2725 - 2739, 1975. [Chloroplast.]

24217 - **ROBB, J., BUSCH, L., LU, B.C.** : Ultrastructure of wilt syndrome caused by *Verticillium dahliae*. I. In chrysanthemum leaves. - Can. J. Bot. *53* : 901 - 913, 1975. [Chloroplast.]

24218 - **ROBERT, J.-M., PAGÈS, J., PRAT, D.** : Applications de la biométrie cytologique à la définition des stades de développement du *Navicula ostrearia* BORY : incidences de l'évolution pigmentaire sur le verdissement des claires à huîtres. - Physiol. vég. *13* : 225 - 241, 1975.

24219 - **ROBERTS, J., WAREING, P.F.** : An examination of the differences in dry matter production shown by some progenies of *Pinus sylvestris* L. - Ann. Bot. *39* : 311 - 324, 1975. [Growth analysis.]

24220 - **ROBERTS, J.R., RIDDLE, H.C., DAVIES, R.W.** : An improved apparatus for the measurement of oxyger concentration. - Phys. Med. Biol. *20* : 637 - 644, 1975.

24221 - **ROBERTSON, R.N., BOARDMAN, N.K.** : The link between charge separation, proton movement and ATPase reactions. - FEBS Lett. *60* : 1 - 6, 1975.

24222 - **ROBINSON, J.M., GIBBS, M.** : Effects of pH upon the photosynthesis of glycolic acid and the Warburg effect in intact spinach chloroplasts. - Plant Physiol. *56* (Suppl.) : 26, 1975.

24223 - **ROBINSON, J.M., LATZKO, E., GIBBS, M.** : Effect of disalicylidenepropanediamine on the light-dependent reduction of carbon dioxide and glycerate 3-phosphate in intact spinach chloroplasts. - Plant Physiol. *55* : 12 - 14, 1975.

24224 - **ROBINSON, S.J., YOCUM, C.F., IKUMA, H.**: Inhibition of photosynthetic electron transport and phosphorylation by diallate and trifluralin. - Plant Physiol. *56* (Suppl.) : 46, 1975.

24225 - **ROBINSON, S.P., WISKICH, J.T.** : The effects of digitonin on photochemical activities of isolated chloroplasts. - Plant Physiol. *55* : 163 - 167, 1975.

24226 - **ROBINSON, S.P., WISKICH, J.T.** : Effect of digitonin concentration on electron transport, phosphorylation, and proton uptake by subchloroplast particles. - Plant Physiol. *56* : 535 - 539, 1975.

24227 - **ROCKLEY, M.G., WINDSOR, M.W., COGDELL, R.J., PARSON, W.W.** : Picosecond detection of an intermediate in photochemical reaction of bacterial photosynthesis. - Proc. nat. Acad. Sci. USA *72* : 2251 - 2255, 1975.

24228 - **RODIONOV, V.S., KOZUBOV, G.M., TIKHOVA, M.A., SULIMOVA, G.M.** : Izmenenie kontsentratsii glitserolipidov, ul'trastruktury khloroplastov i mitokhondriĭ v list'yakh kartofelya pri zatemnenii. [Effect of darkness on the content of glycerolipids and on the ultrastructure of chloroplasts and mitochondria in the leaves of potato.] - Fiziol. Rast. *22* : 333 - 337, 1975. [In R, ab : E.]

24229 - **RODRIGUEZ, B.P., LAMBETH, V.N.** : Artificial lighting and spacing as photosynthetic and yield factors in winter greenhouse tomato culture. - J. amer. Soc. hort. Sci. *100* : 694 - 697, 1975.

24230 - **RODRIGUEZ, D.B., LEE, T.-C., CHICHESTER, C.O.** : Comparative study of the carotenoid composition of the seeds of ripening *Momordica charantia* and tomatoes. - Plant Physiol. *56* : 626 - 629, 1975.

24231 - **ROGAN, P.G., SMITH, D.L.** : Rates of leaf initiation and leaf growth in *Agropyron repens* (L.) BEAUV. - J. exp. Bot. *26* : 70 - 78, 1975.

24232 - **ROLFE, G.L., BAZZAZ, F.A.** : Effect of lead contamination on transpiration and photosynthesis of loblolly pine and autumn olive. - Forest Sci. *21* : 33 - 35, 1975.

24233 - **ROMANO, J.-C.** : Les adénosines-5'-phosphates (ATP, ADP, AMP) chez des algues phytoplanctoniques. Rôle pour la mesure indirecte de la biomasse. - Compt. rend. Acad. Sci. Paris, Sér. D *281* : 1027 - 1030, 1975.

*24234 - **ROMBACH, J.** : Growth stimulation by cytokinins, thiamine and phytochrome in *Lemna minor* L. in darkness and in light. - Acta bot. neerl. *23* : 348, 1974. [Ps.]

24235 - **RONNENKAMP, R.R., GORZ, H.J., HASKINS, F.A.** : Genetic studies of induced mutants in *Melilotus alba*. IV. Inheritance and complementation of six additional chlorophyll-deficient mutants. - Crop Sci. *15* : 187 - 188, 1975.

24236 - **ROOT, R.A., MILLER, R.J., KOEPPE, D.E.** : Uptake of cadmium - its toxicity, and effect on the iron ratio in hydroponically grown corn. - J. environ. Qual. *4* : 473 - 476, 1975. [Chl.]

24237 - **ROSE, R.J., CRAN, D.G., POSSINGHAM, J.V.** : Changes in DNA synthesis during cell growth and chloroplast replication in greening spinach leaf disks. - J. Cell Sci. *17* : 27 - 41, 1975.

24238 - **ROSEN, D., BARR, R., CRANE, F.L.** : Vanadium compounds and ferrocyanide as ionic redox agents in photosynthesis. - Biochim. biophys. Acta *408* : 35 - 46, 1975.

24239 - **ROSHCHINA, V.D.** : Letuchie komponenty vnutrenneĭ gazovoĭ fazy rasteniĭ. [Volatile components of the internal gas phase in plants.] - In : Fitontsidy. Pp. 78 - 81. Naukova Dumka, Kiev 1975. [In R.]

24240 - **ROSHCHINA, V.D.** : Soderzhanie uglekisloty v gazovoĭ smesi, izvlekaemoĭ iz vegetativnykh tkaneĭ drevesnykh rasteniĭ. [Carbon dioxide content in gas mixture extracted from vegetative tissues of wood plants.] - Nauch. Dokl. vyssh. Shkoly, biol. Nauki *18* (5) : 80 - 85, 1975. [In R.]

24241 - **ROSHCHINA, V.V., AKULOVA, E.A.** : Okislitel'no-vosstanovitel'nye prevrashcheniya tsitokhroma *f* v izolirovannykh khloroplastakh gorokha. [Oxidative-reductive transformations of cytochrome *f* in isolated chloroplasts of pea.] - Fiziol. Rast. *22* : 471 - 478, 1975. [In R, ab : E.]

24242 - **ROSING, J., HARRIS, D.A., SLATER, E.C., KEMP, A., Jr.** : The possible role of tightly bound adenine nucleotides in oxidative and photosynthetic phosphorylation. - J. supramol. Struct. *3* : 284 - 296, 1975.

24243 - **ROSOWSKI, J.R., WILLEY, R.L.** : *Colacium libellae* sp. nov. (*Euglenophyceae*), a photosynthetic inhabitant of the larval damselfly rectum. - J. Phycol. *11* : 310 - 315, 1975.

24244 - **ROSS, J.** : Radiative transfer in plant communities. - In : **MONTEITH, J.L.** (ed.): Vegetation and the Atmosphere. Vol. 1. Principles. Pp. 13 - 55. Academic Press, London - New York - San Francisco 1975.

B24245 - **ROSS, Yu.K.** : Radiatsionnyĭ Rezhim i Arkhitektonika Rastitel'nogo Pokrova. [Radiation Regime and the Architecture of Plant Canopy.] - Gidrometeoizdat, Leningrad 1975. [In R.]

24246 - **ROTA, J.A., MARAVOLO, N.C.** : Transport and mobilization of ^{14}C-sucrose during regeneration in the hepatic, *Marchantia polymorpha*. - Bot. Gaz. *136* : 184 - - 188, 1975.

24247 - **ROTH-BEJERANO, N., LIPS, S.H.** : Glycolate oxidase content of microbodies as affected by nitrate. - Plant Physiol. *55* : 270 - 272, 1975.

24248 - **ROWE, G.T., CLIFFORD, C.H., SMITH, K.L., Jr., HAMILTON, P.L.** : Benthic nutrient regeneration and its coupling to primary productivity in coastal waters. - Nature *255* : 215 - 217, 1975.

24249 - **ROZEMA, J.** : An eco-physiological investigation into the salt tolerance of *Glaux maritima* L. - Acta bot. neerl. *24* : 407 - 416, 1975. [Ps, Chl.]

B24250 - **RUBIN, A.B.** (ed.) : Biofizika Fotosinteza. [Biophysics of Photosynthesis.] - Izd. mosk. Univ., Moskva 1975. [In R.]

24251 - **RUBIN, P.M., McGOWAN, R.E.** : The photo-mediated uptake of glucose-6-phosphate by *Anabaena flos-aquae*. - Plant Physiol. *56* (Suppl.) : 67, 1975.

24252 - **RUCKENBAUER, P.** : Photosynthetic and translocation pattern in contrasting winter wheat varieties. - Ann. appl. Biol. *79* : 351 - 359, 1975.

24253 - **RUDENKO, T.I., KLIMOV, A.A., MAKAROV, A.D.** : Izuchenie konformatsionnykh izmeneniĭ khloroplastov metodom televizionnoĭ lazernoĭ mikroskopii. [Conformations of the chloroplasts studied by television laser microscopy.] - Fiziol. Rast. *22* : 877 - 881, 1975. [In R, ab : E.]

24254 - **RÜDIGER, W.** : Phycobiliproteide. - Ber. deut. bot. Ges. *88* : 125 - 139, 1975.

24255 - **RUFFNER, H.P., KLIEWER, W.M.** : Phosphoenolpyruvate carboxykinase activity in grape berries. - Plant Physiol. *56* : 67 - 71, 1975.

24256 - **RUFNER, R., WITHAM, F.H., COLE, H., Jr.** : Ultrastructure of chloroplasts of *Phaseolus vulgaris* leaves treated with benomyl and ozone. - Phytopathology *65* : 345 - 349, 1975.

*24257 - **RUMBERG, B., SCHRÖDER, H.** : Kinetics and stoichiometry of photophosphorylation. - In : **BRODA, E., LOCKER, A., SPRINGER-LEDERER, H.** (ed.) : Proceedings of the First European Biophysics Congress. Vol. 4. Pp. 57 - 59. Verlag wiener med. Akad., Wien 1971.

24258 - **RURAINSKI, H.J.** : Antagonistic relationships between electron transport and P_{700} in chloroplasts and intact algae. - Z. Naturforsch. *30 C* : 761 - 770, 1975.

24259 - **RUSSELL, W.J., JOHNSON, D.R.** : Translocation patterns in soybeans exposed to $^{14}CO_2$ at four different time periods of the day. - Crop Sci. *15* : 75 - 77, 1975.

24260 - **RUSSELL, W.J., JOHNSON, D.R.** : Carbon-14 assimilate translocation in nodulated and nonnodulated soybeans. - Crop Sci. *15* : 159 - 161, 1975.

24261 - **RUTTER, A.J.** : The hydrological cycle in vegetation. - In : **MONTEITH, J.L.** (ed.) : Vegetation and the Atmosphere. Vol. 1. Principles. Pp. 111 - 154. Academic Press, London - New York - San Francisco 1975. [Stomatal resistance.]

24262 - **RYAN, F.J., BARKER, R., TOLBERT, N.E.** : Inhibition of ribulose diphosphate carboxylase/oxygenase by xylitol 1,5-diphosphate. - Biochem. biophys. Res. Commun. *65* : 39 - 46, 1975.

24263 - **RYAN, F.J., TOLBERT, N.E.** : Ribulose diphosphate carboxylase/oxygenase. 3. Isolation and properties. - J. biol. Chem. *250* : 4229 - 4233, 1975.

24264 - **RYAN, F.J., TOLBERT, N.E.** : Ribulose diphosphate carboxylase/oxygenase. 4. Regulation by phosphate esters. - J. biol. Chem. *250* : 4234 - 4238, 1975.

24265 - **RYAN, F.J., TOLBERT, N.E., BARKER, R.** : Regulation of ribulose diphosphate carboxylase/oxygenase. - Plant Physiol. *56* (Suppl.) : 86, 1975.

24266 - **RYAN, J., STROEHLEIN, J.L., MIYAMOTO, S.** : Sulfuric acid applications to calcareous soils : effects on growth and chlorophyll content of common bermudagrass in the greenhouse. - Agron. J. *67* : 633 - 637, 1975.

24267 - **RYBÁČEK, V.** : The influence of biological age on some growth and reproduction characteristics in three-year old hop plants (*Humulus lupulus* L.). - Rostlinná Výroba (Praha) *21* : 907 - 919, 1975. [Growth analysis.]

24268 - **RYBIN, I.A., KETOVA, L.G.** : Bioėlektricheskaya aktivnost' list'ev kukuruzy i bobov pri raznykh usloviyakh osveshcheniya. [Bioelectric activity of maize and broad bean leaves under different irradiance.] - Fiziol. Biokhim. kul't. Rast. *7* : 599 - 602, 1975. [In R, ab : E.]

24269 - **RYBKINA, G.V., GUSEV, N.A.** : Sravnitel'naya otsenka nekotorykh vozmozhnosteĭ opredeleniya soderzhaniya vody v khloroplastakh. [Comparison of some possibilities of determination of water content in chloroplasts.] - In : Vodoobmen Rasteniĭ pri Neblagopriyatnykh Usloviyakh Sredy. Pp. 223 - 228. Shtiintsa, Kishinev 1975. [In R.]

24270 - **RYĆ, M., LEWAK, S.** : Activity of phosphoenolpyruvate carboxylase in apple seedlings in relation to embryonal dormancy. - Photosynthetica *9* : 299 - 303, 1975.

B24271 - **RYCHNOVSKÁ, M., ŮLEHLOVÁ, B.** : Autökologische Studie der tschechoslowakischen *Stipa*-Arten (Vegetace ČSSR, A8). - Academia, Praha 1975. [Ps.]

24272 - **RYCZKOWSKI, M., SZEWCZYK, E.** : Further research on the photosynthesis in the green developing embryos (dicotyledonous plants). - Z. Pflanzenphysiol. *75* : 175 - 180, 1975.

24273 - **RYDING, S.-O.** : Intercalibration of methods for determining chlorophyll *a* in water. - Vatten *31* : 327 - 332, 1975.

24274 - **RYDING, S.-O.** : Interkalibrering av mätmetoder för bestämning av klorofyll *a*. [Intercalibration of methods for determining chlorophyll *a*.] - Nordforsk Miljövardssekretariatet Publ. *1975* (5) : 1 - 21, 1975. [In Swed.]

24275 - **RYLE, G.J.A., POWELL, C.E.** : Defoliation and regrowth in the graminaceous plant : the role of current assimilate. - Ann. Bot. *39* : 297 - 310, 1975.

24276 - **RYSZKOWSKI, L.** : Energy and matter economy of ecosystems. - In : **Van DOBBEN, W.H., LOWE-McCONNEL, R.H.** (ed.) : Unifying Concepts in Ecology. Pp. 109 - 126. Dr. W. Junk b.v. Publ., The Hague; Pudoc, Wageningen 1975.

24277 - **RYTOVA, N.G.** : Skorost' i prodolzhitel'nost' rosta list'ev u *Phleum pratense* L. (nekotorye aspekty morfogeneza zlakov). [The rate and the duration of leaf growth in *Phleum pratense* L. (some aspects of grass morphogenesis,)] - Bot. Zh. *60* : 1265 - 1277, 1975. [In R, ab : E.]

24278 - **SAEKI, T.** : Distribution of radiant energy and CO_2 in terrestrial communities. - In : **COOPER, J.P.** (ed.) : Photosynthesis and Productivity in Different Environments. Pp. 297 - 322. Cambridge Univ. Press, Cambridge - London - New York - Melbourne 1975.

24279 - **SAGROMSKY, H.** : Chlorophyllbestimmungen mittels eines Aceton-Diäthyläther-Methanol-Petroläther-Gemisches. - Kulturpflanze *23*: 217 - 221, 1975.

24280 - **SAGROMSKY, H.** : Mutants as objects for investigations on the function of chlorophylls. - In : **NASYROV, Yu.S., ŠESTÁK, Z.** (ed.) : Genetic Aspects of Photosynthesis. Pp. 247 - 253. Dr. W. Junk B.V. Publ., The Hague 1975.

24281 - **SAHA, S.** : Photophosphorylation in normal and Tris-washed chloroplasts with 1,4-benzoquinonediimide as the electron acceptor. - Plant Biochem.J. *2* : 71 - - 81, 1975.

24282 - **SAITOH, M., ARAKI, S., SAKURAI, T., OOHUSA, T.** : [Variations in contents of photosynthetic pigments, total nitrogen, total free amino acids and total free sugars in dried lavers obtained at different culture grounds and harvesting times.] - Bull. jap. Soc. Sci. Fish. *41* : 365 - 370, 1975. [In Jap., ab : E.]

24283 - **SAKA, H., MATHUNAKA, S.** : [The growth and photosynthetic activity of artificial chlorophyll mutants in rice plant.] - Proc. Crop Sci. Soc. Jap. *44* : 54 - 60, 1975. [In Jap., ab : E.]

24284 - **SAKAI-IMAMURA, M.** : Fatty acid oxidation and chlorophyll bleaching. - Natural Sci. Rep., Ochanomizu Univ. *26* (2) : 109 - 125, 1975.

24285 - **SAKANISHI, Y., FUKUZUMI, H.** : Effect of temperature on the colouring of leaves in ornamental kale (*Brassica oleracea* L. var. *acephala* DC.). - Bull. Univ. Osaka Pref., Ser. B *27* : 1 - 14, 1975. [Chl.]

24286 - **SAKHAROVA, O.V.** : Skorost' tsiklicheskogo fotofosforilirovaniya izolirovannykh khloroplastov v ontogeneze u razlichnykh vidov i sortov pshenitsy. [Rate of cyclic photophosphorylation of isolated chloroplasts during ontogenesis in various species and cultivars of wheat.] - Byull. vses. nauch.-issled. Inst. Rastenievod. N.I. Vavilova *56* : 73 - 80, 1975. [In R.]

24287 - **SĂLĂGEANU, N.** : Ergebnisse dreijähriger Massenkultur von mikroskopischen Algen auf Laufbändern. - In : **NECHAS, Ĭ.** (ed.) : Izuchenie Intensivnoĭ Kul'tury Vodorosleĭ. Vol. II. Pp. 13 - 24. Třeboň 1975.

24288 - **SALAMON, Z., FRĄCKOWIAK, D.** : Photosensitization in phycoerythrin protein complexes at electrodes. - Photosynthetica *9* : 337 - 339, 1975.

24289 - **SALATENKO, V.N.** : Radiatsionnyĭ rezhim posevov kleshcheviny v usloviyakh orosheniya. [Radiation regime of irrigated castor-bean stand.] - Fiziol. Biokhim. kul't. Rast. *7* : 592 - 598, 1975. [In R, ab : E.]

24290 - **SALCHEVA, G., GRAMATIKOVA, H., FEDINA, I., GUSHTINA, L.** : After-effect of the temperature in the process of hardening winter wheet on the photoassimilation of $^{14}CO_2$ by isolated chloroplasts. - Dokl. bolg. Akad. Nauk *28* : 1101 - 1104, 1975.

24291 - **SALÉ, P.J.M.** : Productivity of vegetable crops in a region of high solar input. IV. Field chamber measurements of French beans (*Phaseolus vulgaris* L.) and cabbages (*Brassica oleracea* L.). - Aust. J. Plant Physiol. *2* : 461 - 470, 1975.

24292 - **SALISBURY, J.L., VASCONCELOS, A.C., FLOYD, G.L.** : Isolation of intact chloroplasts of *Euglena gracilis* by isopycnic sedimentation in gradients of silica. - Plant Physiol. *56* : 399 - 403, 1975.

24293 - **SALLAL, A.-K.J., CODD, G.A.** : The intracellular localization of the glycollate-oxidising enzyme of *Anabaena cylindrica*. - FEBS Lett. *56* : 230 - 234, 1975. [Chl.]

24294 - **SALONEN, K., KOTIMAA, A.-L.** : The determination of dissolved inorganic carbon, a possible source of error in determining the primary production of lake water phytoplankton. - Ann. bot. fenn. *12* : 187 - 189, 1975.

24295 - **SAMARRAI, S.M., MEGLED, F.I.** : The analysis of the yield of Mexican and Arabian varieties of wheat grown in Saudi Arabia. - Proc. Crop Sci. Soc. Jap. *44* : 364 - 369, 1975. [Growth analysis.]

24296 - **SAMISH, Y.B.** : Oxygen build-up in photosynthesizing leaves and canopies is small. - Photosynthetica *9* : 372 - 375, 1975.

24297 - **SAMISH, Y.B., ELLERN, S.J.** : Titratable acids in *Opuntia ficus-indica* L. - J. Range Management *28* : 365 - 369, 1975. [CAM.]

24298 - **SAMORAY, D., HAUSKA, G.** : Differential uncoupling in subchloroplast vesicles of different sizes. - In : **AVRON, M.** (ed.) : Proceedings of the Third International Congress on Photosynthesis. Vol. II. Pp. 1055 - 1066. Elsevier, Amsterdam - Oxford - New York 1975.

24299 - **SANDERS, D.C., PHARR, D.M., KONSLER, T.R.** : Chlorophyll content of a dark green mutant of "Manapal" tomato. - HortScience *10* : 262 - 264, 1975.

*24300 - **SANDERS, J.W., RUIJGROK, T.W., ten BOSCH, J.J.** : Some remarks on the theory of trapping of excitons in the photosynthetic unit. - J. math. Phys. *12* : 534- - 541, 1971.

24301 - **SANDHAUG, A., KJELVIK, S., WIELGOLASKI, F.E.** : A mathematical simulation model for terrestrial tundra ecosystems. - In : **WIELGOLASKI, F.E.** (ed.) : Fennoscandian Tundra Ecosystems. Part 2. Animals and Systems Analysis. Pp. 251 - 266. Springer Verlag, Berlin - Heidelberg - New York 1975. [Simulated Ps.]

24302 - **SANDHU, B.S., HORTON, M.L.** : Diurnal variations in the leaf-water content, leaf-diffusion resistance and net photosynthetic rate of oats, as influenced by water stress. - Indian J. Ecol. *2* : 125 - 131, 1975.

24303 - **SAND-JENSEN, K.** : Biomass, net production and growth dynamics in an eelgrass (*Zostera marina* L.) population in Vellerup Vig, Denmark. - Ophelia *14* : 185 - - 201, 1975.

24304 - **SANDS, R.H., DUNHAM, W.R.** : Spectroscopic studies on two-iron ferredoxins. - Quart. Rev. Biophys. *7* : 443 - 504, 1975.

24305 - **SANDY, J.D., DAVIES, R.C., NEUBERGER, A.** : Control of 5-aminolaevulinate synthetase activity in *Rhodopseudomonas spheroides*. A role for trisulphides. - Biochem. J. *150* : 245 - 257, 1975.

24306 - **SANE, P.V.** : Thermoluminescence studies in photosynthetic organisms. - In : Proceedings of the National Symposium on Thermoluminescence and Its Applications. Pp. 279 - 297. R.R.C., Madras 1975.

24307 - **SANE, P.V., DESAI, T.S., TATAKE, V.G.** : Evidence for a second quencher of chlorophyll fluorescence in spinach leaves. - Indian J. Biochem. Biophys. *12* : 38 - 42, 1975.

24308 - **SAN FELÍU, J.M., MUÑOZ, F.** : Hidrografía y fitoplancton de las costas de Castellón, de septiembre de 1969 a enero de 1971. [Hydrography and phytoplankton of the coast of Castellon, from September 1969 to January 1971.] - Invest. pesq. *39* : 1 - 35, 1975. [Chl; in Span., ab : E.]

24309 - **SANKHLA, N., HUBER, W.** : Regulation of balance between C_3 and C_4 pathway : role of abscisic acid. - Z. Pflanzenphysiol. *74* : 267 - 271, 1975.

24310 - **SANKHLA, N., ZIEGLER, H., VYAS, O.P., STICHLER, W., TRIMBORN, P.** : Eco-physiological studies of Indian arid zone plants. V. A screening of some species for the C_4-pathway of photosynthetic CO_2-fixation. - Oecologia *21* : 123 - 129, 1975.

24311 - **SANTARIUS, K.A.** : Sites of heat sensitivity in chloroplasts and differential inactivation of cyclic and noncyclic photophosphorylation by heating. - J. thermal Biol. *1* : 101 - 107, 1975.

24312 - **SANTARIUS, K.A.** : The effect of heat and solutes on the stability of envelopes and thylakoids of chloroplasts. - Plant Physiol. *56* (Suppl.) : 82, 1975.

24313 - **SANTHANAM, R.** : Aspects of an Indian mangrove forest. - Proc. Int. Symp. Biol. Management Mangroves (Honolulu, Hawaii) *1* : 88 - 95, 1975. [Ps primary production.]

24314 - **SAPEGIN, L.M., DAĬNEKO, M.M.** : Produktyvnist' i struktura fitomasy luchnykh fitotsenoziv. [Productivity and structure of phytomass in meadow biogeocenoses.] - Ukr. bot. Zh. *32* : 291 - 300, 1975. [In Ukr., ab : E.]

24315 - **SAPHON, S., JACKSON, J.B., LERBS, V., WITT, H.T.** : The functional unit of electrical events and phosphorylation in chromatophores from *Rhodopseudomonas sphaeroides*. - Biochim. biophys. Acta *408* : 58 - 66, 1975.

24316 - **SAPHON, S., JACKSON, J.B., WITT, H.T.** : Electrical potential changes, H^+ translocation and phosphorylation induced by short flash excitation in *Rhodopseudomonas sphaeroides* chromatophores. - Biochim. biophys. Acta *408* : 67 - 82, 1975.

24317 - **SAPOZHNIKOV, D.I., POPOVA, I.A., MASLOVA, T.G., KOROLEVA, O.Ya.** : Osobennosti reaktsiĭ violaksantinovogo tsikla, indutsirovannykh blizhnim i dal'nim krasnym svetom. [Reactions of violaxanthin cycle induced by near- and far-red light.] - Fiziol. Rast. *22* : 1108 - 1112, 1975. [In R, ab : E.]

24318 - **SAPOZHNIKOV, D.I., POPOVA, O.F., KAZANSKAYA, Yu.A.** : O reaktsii ėpoksidatsii violaksantinovogo tsikla v zeleneyushchikh prorostkakh kukuruzy. [Epoxidation of violaxanthin cycle in greening maize seedlings.] - Fiziol. Rast. *22* : 718 - - 722, 1975. [In R, ab : E.]

24319 - **SARDA, C., PRIOUL, J.-L., MOYSE, A.** : Structure du limbe et accumulation d'amidon dans les chloroplastes des feuilles d'une Crassulacée, *Bryophyllum daigremontianum* BERGER. - Physiol. vég. *13* : 563 - 577, 1975.

24320 - **SARGENT, D.F., TAYLOR, C.P.S.** : On the respiratory enhancement in *Chlorella pyrenoidosa* by blue light. - Planta *127* : 171 - 175, 1975.

24321 - **SARTI, A.** : Growth and photosynthetic activity of *Lactuca sativa* c.v. *romana* cultivated in three daylight intensities. - In : **CHOUARD, P., de BILDERLING, N.** (ed.) : Phytotronics in Agricultural and Horticultural Research. Phytotronics III. Pp. 142 - 153. Gauthier-Villars, Paris - Bruxelles - Montreal 1975.

*24322 - **SASTRY, Y.S.R., RAO, P.V., LAKSHMINARAYANA, G.** : Bleaching behaviour of chlorophyll substrates under vacuum on exposure to visible light. - Oléagineux *28* : 467 - 470, XXXIX, XLIII, 1973.

24323 - **SATO, T., TSUNO, Y.** : [Studies on CO_2 uptake and CO_2 evolution in each part of crop plants. 1. Photosynthetic rate of corn tassels, sorghum ears and their leaf sheaths.] - Proc. Crop Sci. Soc. Jap. *44* : 281 - 286, 1975. [In Jap., ab : E.]

24324 - **SATO, T., TSUNO, Y.** : [Studies on CO_2 uptake and CO_2 evolution in each part of crop plants. III. Variation of photosynthetic rate in different parts of leaf on rice, corn and sorghum plant.] - Proc. Crop Sci. Soc. Jap. *44* : 389 - - 396, 1975. [In Jap., ab : E.]

24325 - **SATOH, M.** : [The effect of shoot topping on growth and photosynthetic activity of remained leaves in young mulberry plants.] - Sansi-Kenkyu [Acta serol.] *96* : 6 - 13, 1975. [In Jap., ab : E.]

24326 - **SATOH, M., OHYAMA, K.** : Studies on photosynthesis and translocation of photosynthate in mulberry tree. IV. On the translocation of ^{14}C-photosynthetic product in partially defoliated plants. - Proc. Crop Sci. Soc. Jap. *44* : 263 - - 268, 1975.

24327 - **SAUER, K.** : Primary events and the trapping of energy. - In : **GOVINDJEE** (ed.) : Bioenergetics of Photosynthesis. Pp. 115 - 181. Academic Press, New York - San Francisco - London 1975.

24328 - **SAUNDERS, J.A., McCLURE, J.W.** : Phytochrome controlled phenylalanine ammonia lyase in *Hordeum vulgare* plastids. - Phytochemistry *14* : 1285 - 1289, 1975. [Chl.]

24329 - **SAUNDERS, V.A., JONES, O.T.G.** : Detection of two further *b*-type cytochromes in *Rhodopseudomonas spheroides*. - Biochim. biophys. Acta *396* : 220 - 228, 1975.

24330 - **SAVCHENKO, G.E., CHAĬKA, M.T.** : Issledovanie kinetiki temnovogo nakopleniya protokhlorofillida na raznykh stadiyakh razvitiya list'ev yachmenya. [Kinetics of dark protochlorophyllide accumulation in barley leaves at various developmental stages.] - In : **SHLYK, A.A.** (ed.) : Biosintez i Sostoyanie Khlorofillov v Rastenii. Pp. 83 - 103, 245. Nauka i Tekhnika, Minsk 1975. [In R, ab : E.]

24331 - **SAVOVA, N.P., KAREV, K.S.** : Izmenenie na produktivnostta na fotosintezata pri pamuka pod vliyanie na napoyavaneto i toreneto. [Changes in the productivity of photosynthesis in cotton under the influence of irrigation and fertilization.] - Rasteniev. Nauki *12* (2) : 110 - 117, 1975. [In Bulg., ab : E, R.]

24332 - **SAWAMURA, M., NAKASHIMA, M., OSAJIMA, Y.** :[Studies on translocation of ^{14}C--labeled compounds from leaves to fruit in Satsuma Mandarin. (Studies on the quality of citrus fruits. Part III.)] - J. agr. chem. Soc. Jap. *49* : 603 - - 607, 1975. [In Jap., ab : E.]

24333 - **SAXENA, J., SIKKA, H.C., ZWEIG, G.** : Studies on effects of certain quinones. III. Photosynthetic $^{14}CO_2$ incorporation by *Rhodospirillum rubrum*. - Pesticide Biochem. Physiol. *5* : 189 - 195, 1975.

24334 - **SCAWEN, M.D., HEWITT, E.J., JAMES, D.M.** : Preparation, crystallization and properties of *Cucurbita pepo* plastocyanin and ferredoxin. - Phytochemistry *14* : 1225 - 1233, 1975.

24335 - **SCAWEN, M.D., RAMSHAW, J.A.M., BOULTER, D.** : The amino acid sequence of plastocyanin from spinach (*Spinacia oleracea* L.). - Biochem. J. *147* : 343 - 349, 1975.

24336 - **SCHAEDLE, M.** : Tree photosynthesis. - Annu. Rev. Plant Physiol. *26* : 101 - 115, 1975.

24337 - **SCHAEFER, J., STEJSKAL, E.O., BEARD, C.F.** : Carbon-13 nuclear magnetic resonance analysis of metabolism in soybeans labeled by $^{13}CO_2$. - Plant Physiol. *55* : 1048 - 1053, 1975. [Photosynthates.]

24338 - **SCHÄFER, G., TREBST, A., BÜCHEL, K.H.** : 2-anilino-1.3.4-thiadiazole, Hemmstoffe der oxidativen und photosynthetischen Phosphorylierung. - Z. Naturforsch. *30C* : 183 - 189, 1975.

24339 - **SCHANTZ, R., DURANTON, H.** : Les acides aminés de l'Euglène en culture synchrone et pendant le développement du chloroplaste. - In : Les Cycles Cellulaires et leur Blocage chez Plusieurs Protistes. - Coll. int. CNRS. Vol. 240. Pp. 269 - 281. Édit. CNRS, Paris 1975. [Chl.]

24340 - **SCHANTZ, R., SCHANTZ, M.-L., DURANTON, H.** : Changes in amino acid and peptide composition of *Euglena gracilis* cells during chloroplast development. - Plant Sci. Lett. *5* : 313 - 324, 1975.

24341 - **SCHAUB, H., HÖHLER, T., EGLE, K.** : Über den Einfluß des Sauerstoffs auf Stoffproduktion und tagesperiodische Schwankungen der CO_2-Aufnahme. 1. Eine kombinierte Anzuchts- und Meßkammer zur Bestimmung von Photosynthese und Atmung unter definierten Bedingungen. - Photosynthetica *9* : 261 - 267, 1975.

24342 - **SCHEDEL, M., KLEMME, J.-H., SCHLEGEL, H.G.** : Regulation of C_3-enzymes in facultative phototrophic bacteria. The cold-labile pyruvate kinase of *Rhodopseudomonas sphaeroides*. - Arch. Microbiol. *103* : 237 - 245, 1975.

24343 - **SCHEER, H., KATZ, J.J.** : Nuclear magnetic resonance spectroscopy of porphyrins and metalloporphyrins. - In : **SMITH, K.M.** (ed.) : Porphyrins and Metalloporphyrins. Pp. 399 - 524. Elsevier sci. Publ. Comp., Amsterdam - Oxford - New York 1975. [Chl.]

24344 - **SCHEIBE, R., BECK, E.** : Formation of C-4 dicarboxylic acids by intact spinach chloroplasts. - Planta *125* : 63 - 67, 1975.

24345 - **SCHIEWER, U.** : Salztoleranz limnischer Blaualgen. Der Einfluss steigender NaCl--Konzentrationen auf den Stickstoff-, Kohlenhydrat-, Pigmentgehalt und die Kohlenhydratausscheidung. - In : **NEČHAS, Ĭ.** (ed.) : Izuchenie intensivnoĭ Kul'tury Vodorosleĭ. Vol. II. Pp. 85 - 97. Třeboň 1975.

24346 - **SCHIFF, J.A.** : Interactions among cellular compartments in *Euglena* during chloroplast development. - In : **NASYROV, Yu.S., ŠESTÁK, Z.** (ed.) : Genetic Aspects of Photosynthesis. Pp. 63 - 91. Dr. W. Junk B.V. Publ., The Hague 1975.

24347 - **SCHIFF, J.A.** : Photocontrol of chloroplast and cytoplasm during chloroplast devolpment in *Euglena*. - In : Les Cycles Cellulaires et leur Blocage chez Plusieurs Protistes. - Coll. int. CNRS. Vol. 240. Pp. 79 - 93. Édit. CNRS, Paris 1975. [Ps, Chl, Car.]

24348 - **SCHIFF, J.A.** : The control of chloroplast differentiation in *Euglena*. - In : **AVRON, M.** (ed.) : Proceedings of the Third International Congress on Photosynthesis. Vol. III. Pp. 1691 - 1717. Elsevier, Amsterdam - Oxford - New York 1975.

24349 - **SCHINDLER, D.W., FEE, E.J.** : The roles of nutrient cycling and radiant energy in aquatic communities. - In : **COOPER, J.P.** (ed.) : Photosynthesis and Productivity in Different Environments. Pp. 323 - 343. Cambridge Univ. Press, Cambridge - London - New York - Melbourne 1975.

24350 - **SCHLETZ, K.** : Phototaxis bei *Volvox* - Pigmentsysteme der Lichtrichtungsperzeption. - Z. Pflanzenphysiol. *77* : 189 - 211, 1975. [Car.]

24351 - **SCHMID, G.H., RADUNZ, A., MENKE, W.** : The effect of an antiserum to plastocyanin on various chloroplast preparations. - Z. Naturforsch. *30 C* : 201 - 212, 1975.

24352 - **SCHMID, R.** : Deactivation of superoxide dismutase on EDTA-treated chloroplasts. - FEBS Lett. *60* : 98 - 102, 1975.

24353 - **SCHMID, R., CLAUSS, H.** : Multiplication and protein content of chloroplasts of *Acetabularia mediterranea* in blue light after prolonged irradiation with red light. - Protoplasma *85* : 315 - 325, 1975.

24354 - **SCHMID, R., JUNGE, W.** : Current-voltage studies on the thylakoid membrane in the presence of ionophores. - Biochim. biophys. Acta *394* : 76 - 92, 1975.

24355 - **SCHMID, R., JUNGE, W.** : On the influence of extraction and recondensation of CF_1 on the electric properties of the thylakoid membrane. - In : **AVRON, M.** (ed.) : Proceedings of the Third International Congress on Photosynthesis. Vol. II. Pp. 821 - 830. Elsevier, Amsterdam - Oxford - New York 1975.

24356 - **SCHMIDT, B.** : Quenching of chlorophyll-*a* fluorescence by reduced PMS. - FEBS Lett. *58* : 277 - 280, 1975.

24357 - **SCHMIDT, G., LYMAN, H.** : Photocontrol of chloroplast enzyme synthesis in mutant and wild-type *Euglena gracilis*. - In : **AVRON, M.** (ed.) : Proceedings of the Third International Congress on Photosynthesis. Vol. III. Pp. 1755 - 1764. Elsevier, Amsterdam - Oxford - New York 1975.

24358 - **SCHMITT, R.** : New infrared system for measurement of CO_2, CO, H_2O from aircraft. - Atmos. Environ. *9* : 865 - 866, 1975.

24359 - **SCHMITZ, K., SRIVASTAVA, L.M.** : On the fine structure of sieve tubes and the physiology of assimilate transport in *Alaria marginata*. - Can. J. Bot. *53* : 861 - 876, 1975.

24360 - **SCHNARRENBERGER, C.** : Two isoenzymes of glucose-6-phosphate isomerase from spinach leaves. - In : **AVRON, M.** (ed.) : Proceedings of the Third International Congress on Photosynthesis. Vol. II. Pp. 1455 - 1462. Elsevier, Amsterdam - - Oxford - New York 1975.

24361 - **SCHNEEMAN, R., BERG, S.P., KROGMANN, D.W.** : Polylysine and cyanide interactions with electron transport catalysts on the chloroplast membrane. - In : **AVRON, M.** (ed.) : Proceedings of the Third International Congress on Photosynthesis. Vol. I. Pp. 671 - 674. Elsevier, Amsterdam - Oxford - New York 1975.

24362 - **SCHNEEMAN, R., KROGMANN, D.W.** : Polycation interactions with spinach ferredoxin-nicotinamide adenine dinucleotide phosphate reductase. - J. biol. Chem. *250* : 4965 - 4971, 1975.

24363 - **SCHNEIDER, G.W.** : ^{14}C-sucrose translocation in apple. - J. amer. Soc. hort. Sci. *100* : 22 - 24, 1975.

24364 - **SCHNEIDER, H.A.W.** : Chlorophylle : Aspekte der Biosynthese und ihrer Regulation. - Ber. deut. bot. Ges. *88* : 83 - 123, 1975.

24365 - **SCHÖNBOHM, E.** : Der Einfluß von Colchicin sowie von Cytochalasin *B* auf fädige Plasmastrukturen, auf die Verankerung der Chloroplasten sowie auf die orientierte Chloroplastenbewegung. 4. Mitteilung : Zur Mechanik der Chloroplastenbewegung. - Ber. deut. bot. Ges. *88* : 211 - 224, 1975.

24366 - **SCHÖNBOHM, E., HAUPT, W.** : Lichtorientierte Chloroplastenbewegung bei *Mougeotia* spec. - Wiss. Film B *1146* : 1 - 15, 1975.

24367 - **SCHOPF, R., HEISE, K.P., SCHMIDT, B., JACOBI, G.** : Reconstitution of photophosphorylation with thylakoid membranes treated with pyrophosphate. - In : **AVRON, M.** (ed.) : Proceedings of the Third International Congress on Photosynthesis. Vol. II. Pp. 911 - 919. Elsevier, Amsterdam - Oxford - New York 1975.

24368 - **SCHOPFER, P., BAJRACHARYA, D., FALK, H., THIEN, W.** : Phytochrom-gesteuerte Entwicklung von Zellorganellen (Plastiden, Microbodies, Mitochondrien). - Ber. deut. bot. Ges. *88* : 245 - 268, 1975.

24369 - **SCHÖTZ, F.** : Untersuchungen über die Plastidenkonkurrenz bei *Oenothera*. V. Die Stabilität der Konkurrenzfähigkeit bei Verwendung verschiedenartiger mutierter Testplastiden. - Biol. Zentralbl. *94* : 17 - 26, 1975.

24370 - **SCHÖTZ, F., DIERS, L.** : Vergrößerung der Kontaktfläche zwischen Chloroplasten und ihrer cytoplasmatischen Umgebung durch tubuläre Ausstülpungen der Plastidenhülle. - Planta *124* : 277 - 285. 1975.

24371 - **SCHREIBER, U., COLBOW, K., VIDAVER, W.** : Temperature-jump chlorophyll fluorescence induction in plants. - Z. Naturforsch. *30 C* : 689 - 690, 1975.

24372 - **SCHREIBER, U., GROBERMAN, L., VIDAVER, W.** : Portable, solid-state fluorometer for the measurement of chlorophyll fluorescence induction in plants. - Rev. sci. Instrum. *46* : 538 - 542, 1975.

24373 - **SCHREIBER, U., VIDAVER, W.** : Analysis of anaerobic fluorescence decay in *Scenedesmus obliquus*. - Biochim. biophys. Acta *387* : 37 - 51, 1975.

24374 - **SCHRÖDER, H., SIGGEL, U., RUMBERG, B.** : The stoichiometry of non-cyclic photophosphorylation. - In : **AVRON, M.** (ed.) : Proceedings of the Third International Congress on Photosynthesis. Vol. II. Pp. 1031 - 1039. Elsevier, Amsterdam - Oxford - New York 1975.

*24375 - **SCHULZ, H.** : Ein Vergleich an di- und tetraploiden Weidelgrassorten - Ertrag und Carotingehalt in Gefäßversuchen. - Wirtschaftseigene Futter *18* : 257 - 264, 1972.

24376 - **SCHULZE, E.-D., LANGE, O.L., EVENARI, M., KAPPEN, L., BUSCHBOM, U.** : The role of air humidity and temperature in controlling stomatal resistance of *Prunus armeniaca* L. under desert conditions. III. The effect on water use efficiency. - Oecologia *19* : 303 - 314, 1975. [Ps.]

24377 - **SCHULZE, E.-D., LANGE, O.L., KAPPEN, L., EVENARI, M., BUSCHBOM, U.** : The role of air humidity and leaf temperature in controlling stomatal resistance of *Prunus armeniaca* L. under desert conditions. II. The significance of leaf water status and internal carbon dioxide concentration. - Oecologia *18* : 219 - 233, 1975. [Ps.]

24378 - **SCHULZE, E.-D., LANGE, O.L., KAPPEN, L., EVENARI, M., BUSCHBOM, U.** : Physiological basis of primary production of perennial higher plants in the Negev desert. - In : **COOPER, J.P.** (ed.) : Photosynthesis and Productivity in Different Environments. Pp. 107 - 119. Cambridge Univ. Press, Cambridge - London - - New York - Melbourne 1975.

24379 - **SCHUMANN, B., HERRMANN, F., BÖRNER, T., HAGEMANN, R.** : Separation of the Photosystem I chlorophyll-protein complex into several components. - Photosynthetica *9* : 410 - 411, 1975.

24380 - **SCHÜRMANN, P., BUCHANAN, B.B.** : Role of ferredoxin in the activation of sedoheptulose diphosphatase in isolated chloroplasts. - Biochim. biophys. Acta *376* : 189 - 192, 1975.

24381 - **SCHWARTZBACH, S.D., GOLDSTEIN, N.H., SCHIFF, J.A.** : Control of paramylum degradation (PmDeg) in *Euglena gracilis* var. *bacillaris*. - Plant Physiol. *56* (Suppl.) : 33, 1975. [Chloroplast.]

24382 - **SCHWARTZBACH, S.D., SCHIFF, J.A., GOLDSTEIN, N.H.** : Events surrounding the early development of *Euglena* chloroplasts. - Plant Physiol. *56* : 313 - 317, 1975.

24383 - **SCHWARZ, Z.** : Zur Regulation der NADP-abhängigen Glycerinaldehyd-3-phosphat-Dehydrogenase in den Primärblättern von *Phaseolus vulgaris* L. Die Wirkung von Na-Cholat auf die Aktivität des Enzyms *in vitro*. - Biochem. Physiol. Pflanzen *166* : 525 - 536, 1974.

24384 - **SCHWENN, J.D., HENNIES, H.H.** : Enzymes and bound intermediate involved in photosynthetic sulfate reduction of spinach chloroplasts and *Chlorella*. - In : **AVRON, M.** (ed.) : Proceedings of the Third International Congress on Photosynthesis. Vol. I. Pp. 629 - 635. Elsevier, Amsterdam - Oxford - New York 1975.

24385 - **SCOTT, B., GREGORY, R.P.F.** : Properties of protein-chlorophyll complexes from pea (*Pisum sativum* L.) leaves. The organization of chlorophyll. - Biochem. J. *149* : 341 - 347, 1975.

24386 - **SCOTT BURDEN, T., CANVIN, D.T.** : The effect of Mg^{2+} on density of proplastids from the developing castor bean and the use of acetyl-CoA carboxylase activity for their cytochemical identification. - Can. J. Bot. *53* : 1371 - 1376, 1975.

24387 - **SCRUTTON, M.C., FATEBENE, F.** : An assay system for localisation of pyruvate and phosphoenolpyruvate carboxylase activity on polyacrylamide gels and its application to detection of these enzymes in tissue and cell extracts. - Ann. Chem. *69* : 247 - 260, 1975.

24388 - **SEETHARAM, A., NAYAR, K.M.D., HANUMANTHAPPA, H.S.** : Effects of single and combination treatments of *gamma* rays and ethyl methane sulphonate on chlorophyll mutation frequency in rice. - Riso *24* : 179 - 183, 1975.

24389 - **SEGEŤA, V., MILLKOVÁ, J.** : Vliv chladu na kapacitu fotosyntézy a obsah chlorofylu listů okurek. [Chilling effects on the capacity of photosynthesis and on chlorophyll content in cucumber leaves.] - Rostlinná Výroba (Praha) *21* : 409 - - 418, 1975. [In Czech, ab : E, R.]

*24390 - **SEHGAL, P.P., CALHOUN, W.T., HUGHES, C.L., Jr., VICARS, T.M.** : Kinetin and chlorophyll production. - In : Plant Growth Substances 1973. Pp. 1014 - 1021. Hirokawa Publ. Co., Inc., Tokyo 1974.

24391 - **SEIBEL, W., NIERLE, W., EL BAYÂ, A.W.** : Studien über die Wirkung von 2-Chlortrimethylammoniumchlorid (CCC) in der Weizenpflanze. - Fette, Seifen, Anstrichmittel *77* : 20 - 23, 1975. [Ps, Chl, Car.]

24392 - **SEKI, H., YAMAGUCHI, Y., HATAKENAKA, J., ICHIMURA, S.** : Dynamics of organic materials in a natural pool of Shimogamo Hot Spring, Japan. - Arch. Hydrobiol. *75* : 539 - 547, 1975. [Chl.]

24393 - **SEKI, H., YAMAGUCHI, Y., ICHIMURA, S.** : Turnover rate of dissolved organic materials in a coastal region of Japan at summer stagnation period of 1974. - Arch. Hydrobiol. *75* : 297 - 305, 1975. [Chl.]

24394 - **SELGA, M.P., YANEVITS, D.A., RUD', M.S.** : Deĭstvie ul'trafioletovoĭ radiatsii na fotosinteticheskiĭ apparat rasteniĭ i rol' ee v svetokul'ture. [Effect of ultraviolet radiation on the photosynthetic apparatus of plants and its role in light culture.] - In : Biologicheskoe Deĭstvie Ul'trafioletovogo Izlucheniya. Pp. 207 - 208. Nauka, Moskva 1975. [In R.]

24395 - **SELMAN, B.R., JOHNSON, G.L., DILLEY, R.A., VOEGELI, K.K.** : Location of plastocyanin in chloroplast membranes. - In : **AVRON, M.** (ed.) : Proceedings of the Third International Congress on Photosynthesis. Vol. II. Pp. 897 - 909. Elsevier, Amsterdam - Oxford - New York 1975.

24396 - **SELMAN, B.R., JOHNSON, G.L., GIAQUINTA, R.T., DILLEY, R.A.** : Evidence for cyanide and mercury inactivation of endogenous plastocyanin. - J. Bioenerg. *6* : 221 - 231, 1975.

24397 - **SEMENENKO, V.E.** : Metabolitnaya avtoregulyatsiya svetoindutsiruemoĭ beloksinteziruyushcheĭ sistemy i funktsional'noĭ aktivnosti khloroplasta. [Metabolic regulation of light-induced protein synthesizing system and functional activity of chloroplast.] - In : Fotoregulyatsiya Metabolizma i Morfogeneza Rasteniĭ. Pp. 135 - 157, 252, Nauka, Moskva 1975. [In R.]

24398 - **SEMENOVA, G.A., LADYGIN, V.G.** : Ul'trastruktura plastid trekh tipov mutantov *Chlamydomonas reinhardi*, fenotipicheski zheltykh na svetu ili v temnote. [The ultrastructure of plastids of three mutant types of *Chlamydomonas reinhardi* phenotypically yellow in the light or in the darkness.] - Tsitologiya *17* : 1003 - 1008, 1975. [In R, ab : E.]

24399 - **SEMICHAEVSKIĬ, V.D.** : Zakonomernosti obrazovaniya funktsional'no-aktivnykh kompleksov khlorofilla s syvorotochnym al'buminom cheloveka v vodnoĭ srede. [Reguliarities of formation of functionally active complexes of chlorophyll with human serum albumin in aqueous medium.] - Mol. Biol. (Moskva) *9* : 351 - - 360, 1975. [In R, ab : E.]

24400 - **SEMICHAEVSKIĬ, V.D.** : Vliyanie Tritona X-100 na spektral'nye svoĭstva iskusstvennykh khlorofill-belkovykh kompleksov. [Effect of Triton X-100 on spectral properties of artificial chlorophyll-protein complexes.] - Fiziol. Biokhim. kul't. Rast. *7* : 18 - 24, 1975. [In R, ab : E.]

24401 - **SENGER, H.** : Changes in photosynthetic activities in synchronous cultures of *Scenedesmus, Chlorella* and *Chlamydomonas*. - In : Les Cycles Cellulaires et leur Blocage chez Plusieurs Protistes. Coll. int. CNRS. Vol. 240. Pp. 101 - 108. Édit. CNRS, Paris 1975.

24402 - **SENGER, H., BISHOP, N.I., WEHRMEYER, W., KULANDAIVELU, G.** : Development of structure and function of the photosynthetic apparatus during light-dependent greening of a mutant of *Scenedesmus obliquus*. - In : **AVRON, M.** (ed.) : Proceedings of the Third International Congress on Photosynthesis. Vol. III. Pp. 1913- - 1923. Elsevier, Amsterdam - Oxford - New York 1975.

24403 - **SENGER, H., FRICKEL-FAULSTICH, B.** : The regulation of electron flow in synchronized cultures of green algae. - In : **AVRON, M.** (ed.) : Proceedings of the Third International Congress on Photosynthesis. Vol. I. Pp. 715 - 727. Elsevier, Amsterdam - Oxford - New York 1975.

24404 - **SENSER, M., SCHÖTZ, F., BECK, E.** : Seasonal changes in structure and function of spruce chloroplasts. - Planta *126* : 1 - 10, 1975.

24405 - **SENTSOVA, O.Yu., NIKITINA, K.A., GUSEV, M.V.** : Osobennosti kislorodnogo obmena obligatno-fototrofnoĭ sinezelenoĭ vodorosli *Anabaena variabilis* v temnote. [Peculiarities of oxygen exchange in a obligately phototrophic blue-green alga *Anabaena variabilis* in darkness.] - Mikrobiologiya *44* : 283 - 288, 1975. [In R.]

24406 - **SERLIN, R., CHOW, H.-C., STROUSE, C.E.** : The crystal and molecular structure of ethyl chlorophyllide *b* dihydrate at -153°. - J. amer. chem. Soc. *97* : 7237 - - 7242, 1975.

24407 - **SERRUYA, C., BERMAN, T.** : Phosphorus, nitrogen and the growth of algae in Lake Kinneret. - J. Phycol. *11* : 155 - 162, 1975. [Chl.]

24408 - **SERVAITES, J.C., OGREN, W.L.** : Oxygen effects on photosynthesis in soybean mesophyll cells. - Plant Physiol. *56* (Suppl.) : 26, 1975.

24409 - **ŠESTÁK, Z.** : Pigments of plastids and photosynthetic chromatophores. - In : **DEYL, Z., MACEK, K., JANÁK, J.** (ed.) : Liquid Column Chromatography. Pp. 1039- - 1049. Elsevier Publ. Comp., Amsterdam - Oxford - New York 1975.

24410 - **ŠESTÁK, Z., ČATSKÝ, J.** : Bibliography of reviews and methods of photosynthesis - 31, 32, 33, 34. - Photosynthetica *9* : 106 - 119, 223 - 234, 343 - 354, 454 - - 461, 1975.

24411 - **ŠESTÁK, Z., ČATSKÝ, J., SOLÁROVÁ, J., STRNADOVÁ, H., TICHÁ, I.** : Carbon dioxide transfer and photochemical activities as factors of photosynthesis during ontogenesis of primary bean leaves. - In : **NASYROV, Yu.S., ŠESTÁK, Z.** (ed.) : Genetic Aspects of Photosynthesis. Pp. 159 - 166. Dr. W. Junk b.v. Publ., The Hague 1975.

*24412 - **SEVAST'YANOV, V.I., SHUMOV, Yu.S., KOMISSAROV, G.G.** : Fotovol'taicheskiĭ ėffekt v plenkakh pigmentov, kontaktiruyushchikh s ėlektrolitom. III. Raspredelenie potentsiala i vol't-ampernaya kharakteristika dlya besporistykh pigmentirovannykh ėlektrodov v otsutstvie osveshcheniya. [Photovoltaic effect in pigment films which contact with electrolyte. III. Kinetics of potential distributior and voltage characteristics for non-illuminated imporous pigmented electrodes.] - Biofizika *19* : 435 - 439, 1974. [In R, ab : E.]

24413 - **SEYAMA, N., SPLITTSTOESSER, W.E.** : Pigment synthesis in *Cucurbita moschata* cotyledons as influenced by CPTA and several inhibitors. - Plant Cell Physiol. *16* : 13 - 19, 1975.

24414 - **SEYER, P., MARTY, D., LESCURE, A.M., PÉAUD-LENOËL, C.** : Effect of cytokinin on chloroplast cyclic differentiation in cultured tobacco cells. - Cell Differentiation *4* : 187 - 197, 1975. [Chl.]

24415 - **SHADCHINA, T.M.** : Deĭstvie lipoliticheskikh fermentov na aktivnost' fotosistemy I. [Effect of lipolytic enzymes on the activity of photosystem I.] - Fiziol. Biokhim. kul't. Rast. *7* : 142 - 144, 1975. [In R, ab : E.]

24416 - **SHAHAK, Y., CHIPMAN, D.M., SHAVIT, N.** : Differential activity of ADP and ATP analogs as substrates for ATP synthesis and the partial reactions of photophosphorylation. - In : **AVRON, M.** (ed.) : Proceedings of the Third International Congress on Photosynthesis. Vol. II. Pp. 859 - 866. Elsevier, Amsterdam - Oxford - New York 1975.

24417 - **SHAHAK, Y., HARDT, H., AVRON, M.** : Acid-base driven reverse electron flow in isolated chloroplasts. - FEBS Lett. *54* : 151 - 154, 1975.

24418 - **SHAHAK, Y., PICK, U., AVRON, M.** : Energy dependent reverse electron flow in chloroplasts. - In : Proceedings of the Tenth FEBS Meeting. Pp. 305 - 314. Federation of European Biochemical Societies 1975.

*24419 - **SHAĬDUROV, V.S., NARINYAN, S.G.** : Ob ispol'zovanii solnechnoĭ ėnergii vysokogornymi rasteniyami Aragatsa. [Utilization of solar energy by plants of the mountain Aragats.] - Biol. Zh. Arm. *20* : 33 - 39, 1967. [Ps; in R, ab : Arm.]

24420 - **SHALAR', V.M., UNTURA, A.A.** : Nablyudeniya za sutochnoĭ dinamikoĭ pervichnoĭ produktsii fitoplanktona. [Daily dynamics of primary production of phytoplankton.] - Gidrobiol. Zh. *11* (3) : 46 - 48, 1975. [In R.]

24421 - **SHAPIRO, S.L., KOLLMAN, V.H., CAMPILLO, A.J.** : Energy transfer in photosynthesis : pigment concentration effects and fluorescent lifetimes. - FEBS Lett. *54* : 358 - 362, 1975.

24422 - **SHAPOSHNIKOVA, M.G., BINYUKOV, V.I.** : O svyazi mikrovyazkosti mitsell detergenta s fotokhimicheskoĭ aktivnost'yu khlorofilla. [Relation of microviscosity of detergent micelles to photochemical activity of chlorophyll.] - Biofizika *20* : 1129 - 1130, 1975. [In R, ab : E.]

24423 - **SHARMA, H.K., VAIDYANATHAN, C.S.** : 2,3-dihydroxybenzoate 2,3-oxygenase from the chloroplast fraction of *Tecoma stans*. - Phytochemistry *14* : 2135 - 2139, 1975.

24424 - **SHARP, D.D., LIETH, H., WHIGHAM, D.** : Assessment of regional productivity in North Carolina. - In : **LIETH, H., WHITTAKER, R.H.** (ed.) : Primary Productivity of the Biosphere. Pp. 131 - 146. Springer-Verlag, Berlin - Heidelberg - New York 1975.

24425 - **SHARPE, D.M.** : Methods of assessing the primary production of regions. - In : **LIETH, H., WHITTAKER, R.H.** (ed.) : Primary Productivity of the Biosphere. Pp. 147 - 166. Springer-Verlag, Berlin - Heidelberg - New York 1975.

24426 - **SHARPE, P.J.H., DeMICHELE, D.W.** : Computer model of the leaf. - Plant Physiol. *56* (Suppl.) : 86, 1975.

24427 - **SHATILOV, I.S., MAZEIN, V.L.** : Intensivnost' fotosinteza klevera krasnogo pri otritsatel'nykh temperaturakh. [Photosynthetic rate in red clover at temperatures below zero.] - Izv. TSKhA *1975* (4) : 3 - 7, 1975. [In R.]

24428 - **SHATILOV, I.S., POLETAEV, V.V.** : Intensivnost' fotosinteza razlichnykh yarusov list'ev kartofelya, fotosinteticheskiĭ potentsial i urozhaĭ klubneĭ. [Photosynthetic rates of potato leaves of different ages, photosynthetic potential and tuber yield.] - Izv. TSKhA *1975* (2) : 27 - 36, 1975. [In R.]

*24429 - **SHATILOV, I.S., VAULIN, A.V.** : Rol' organov rasteniĭ v formirovanii urozhaya yachmenya na razlichnykh agrofonakh. [Role of plant organs in yield formation in barley under different nutrition.] - Vestn. sel'skokhoz. Nauki *1972* (10) : 19 - 29, 1972. [In R, ab : E, G, R.]

*24430 - **SHATILOV, I.S., VAULIN, A.V.** : Dinamika assimiliruyushcheĭ poverkhnosti i rol' otdel'nykh organov rasteniĭ v formirovanii urozhaya yachmenya. [Dynamics of assimilating foliage area and the role of individual plant organs in yield formation in barley.] - Izv. TSKhA *1972* (1) : 21 - 30, 1972. [Ps; in R, ab : E.]

*24431 - **SHATILOV, I.S., VERBITSKAYA, N.M.** : Fotosinteticheskaya deyatel'nost' mnogoletnikh zlakovykh trav pri senokosnom ispol'zovanii. [Photosynthetic activity of perennial cereal grasses used as aftercrop.] - Izv. TSKhA *1973* (3) : 49 - - 54, 1973. [In R, ab : E.]

24432 - **SHAUKAT, S.S., MOORE, K.G., LOVELL, P.H.** : Some effects of triazines on growth, photosynthesis and translocation of photosynthate in *Pinus* species. - Physiol. Plant. *33* : 295 - 299, 1975.

24433 - **SHEAR, D.B.** : Is the photon in photosynthesis heat or work ? - Biophys. J. *15* (2, Pt. 2) : 115a, 1975.

24434 - **SHEAR, H., WALSBY, A.E.** : An investigation into the possible light-shielding role of gas vacuoles in a planktonic blue-green alga. - Brit. phycol. J. *10* : 241 - 251, 1975. [Ps.]

24435 - **SHEARMAN, R.C., BEARD, J.B.** : Turfgrass wear tolerance mechanisms : I. Wear tolerance of seven turfgrass species and quantitative methods for determining turfgrass wear injury. - Agron. J. *67* : 208 - 211, 1975. [Chl.]

24436 - **SHEEHY, J.E.** : Some optical properties of leaves of eight temperate forage grasses. - Ann. Bot. *39* : 377 - 386, 1975. [Growth analysis.]

24437 - **SHEEHY, J.E., GREEN, R.M., ROBSON, M.J.** : The influence of water stress on the photosynthesis of a simulated sward of perennial ryegrass. - Ann. Bot. *39* : 387 - 401, 1975.

24438 - **SHEEHY, J.E., PEACOCK, J.M.** : Canopy photosynthesis and crop growth rate of eight temperate forage grasses. - J. exp. Bot. *26* : 679 - 691, 1975.

24439 - **SHEIKHOLESLAM, S.N., CURRIER, H.B.** : Turgor pressure gradients and assimilate movement in phloem of *Ecballium elaterium*. - Plant Physiol. *56* (Suppl.) : 18, 1975.

24440 - **SHELDON, R.B., BOYLEN, C.W.** : Factors affecting the contribution by epiphytic algae to the primary productivity of an oligotrophic freshwater lake. - Appl. Microbiol. *30* : 657 - 667, 1975.

24441 - **SHELL, G.S.G., LANG, A.R.G.** : Description of leaf orientation and heliotropic response of sunflower using directional statistics. - Agr. Meteorol. *15* : 33 - - 48, 1975. [Canopy structure.]

24442 - **SHEU-HWA, C.-S., LEWIS, D.H., WALKER, D.A.** : Stimulation of photosynthetic starch formation by sequestration of cytoplasmic orthophosphate. - New Phytol. *74* : 383 - 392, 1975.

24443 - **SHIBATA, H., OCHIAI, H.** : [Studies on δ-amino levulinic acid dehydratase during chloroplast development in radish cotyledons.] - Amino Acid, nucl. Acid [Hakko To Taisha] *32* : 16 - 24, 1975. [In Jap., ab : E.]

24444 - **SHIBLES, R., ANDERSON, I.C., GIBSON, A.H.** : Soybean. - In : **EVANS, L.T.** (ed.): Crop Physiology. Pp. 151 - 189. Cambridge Univ. Press, London - New York 1975. [Ps.]

24445 - **SHIEH, P., TIEN, H.T.** : Can electron conduction occur across bilayer lipid membranes ? - Biophys. J. *15* (2, Pt. 2): 307a, 1975.

24446 - **SHIKHOTOV, V.M., KUCHIN, V.V.** : Izmenenie soderzhaniya askorbinovoĭ kisloty, karotina i khlorofilla v list'yakh vysokogornykh pastbishchnykh rasteniĭ pod vliyaniem gerbitsidov. [Changes in the contents of ascorbic acid, carotene and chlorophyll in leaves of high mountain fodder plants treated with herbicides.] - Agrokhimiya *1975* : 107 - 112, 1975. [In R.]

24447 - **SHILO, M.** : Factors involved in dynamics of algal blooms in nature. - In : **Van DOBBEN, W.H., LOWE-McCONNELL, R.H.** (ed.) : Unifying Concepts in Ecology. Pp. 127 - 132. Dr. W. Junk b.v. Publ., The Hague; Pudoc, Wageningen 1975. [Ps.]

24448 - **SHIMIZU, T., TORIKATA, H., TORII, S.** : [Studies on the effect of crop load on the composition of satsuma mandarin trees. III. Effect of different leaf-to--fruit ratios on the carbohydrate productions of the following year.]- J. jap. Soc. hort. Sci. *43* : 423 - 429, 1975. [In Jap., ab : E.]

24449 - **SHIMSHI, D., EPHRAT, J.** : Stomatal behavior of wheat cultivars in relation to their transpiration, photosynthesis, and yield. - Agron. J. *67* : 326 - 331, 1975.

24450 - **SHIMURA, S., FUJITA, Y.** : Phycoerythrin and photosynthesis of pelagic blue--green alga *Trichodesmium thiebautii* in waters of Kuroshio, Japan. - Mar. Biol. *31* : 121 - 128, 1975.

24451 - **SHIMURA, S., FUJITA, Y.** : Changes in the activity of fucoxanthin-excited photosynthesis in the marine diatom *Phaeodactylum tricornutum* grown under different culture conditions. - Mar. Biol. *33* : 185 - 194, 1975.

24452 - **SHIOMI, N., HORI, S.** : Photosynthetic activity in rice seedlings infected with *Piricularia oryzae* CAVARA. - Annu. Rep. Rad. Center Osaka Prefecture *16* : 65 - 67, 1975.

24453 - **SHIPMAN, L.L., JANSON, T.R., RAY, G.J., KATZ, J.J.** : Donor properties of the three carbonyl groups of chlorophyll *a* : *ab initio* calculations and ^{13}C magnetic resonance studies. - Proc. nat. Acad. Sci. USA *72* : 2873 - 2876, 1975.

24454 - **SHIPMAN, R.H., KAO, I.C., FAN, L.T.** : Single-cell protein production by photosynthetic bacteria cultivation in agricultural by-products. - Biotechnol. Bioeng. *17* : 1561 - 1570, 1975.

*24455 - **SHIRAKAWA, H., WAKAMATSU, K.** : [Light-induced oxidation-reduction reactions of plastocyanin in sonicated chloroplast fragments.] - Sci. Hum. Life *9* (3) : 110 - 115, 1973. [In Jap., ab : E.]

24456 - **SHIROYAMA, T., MILLER, W.E., GREENE, J.C.** : Comparison of the algal growth responses of *Selenastrum capricornutum* PRINTZ and *Anabaena flos-aquae* (LYNGB.) DE BREBISSON in waters collected from Shagawa Lake, Minnesota. - In : **MIDDLEBROOKS, E.J., FALKENBORG, D.H., MALONEY, T.E.** (ed.) : Biostimulation and Nutrient Assessment. Pp. 127 - 148. Utah State Univ., Logan, Utah 1975. [Chl, productivity.]

24457 - **SHIRYAEV, A.I., REǏNGARD, T.A., OSTROVSKAYA, L.K.** : Struktura i funktsii fotosinteticheskikh membran. [Structure and functions of photosynthetic membranes.] - In : Struktura i funktsii Biologicheskikh Membran. Pp. 121 - 139. Nauka, Moskva 1975. [In R.]

24458 - **SHIRYAEV, A.I., REǏNGARD, T.A., POLISHCHUK, A.I., OSTROVSKAYA, L.K.** : Struktura molekulyarnykh agregatov, obladayushchikh vysokoėffektivnym tsiklicheskim fotofosforilirovaniem. [Structure of molecular aggregates highly effective in cyclic photophosphorylation.] - Dokl. Akad. Nauk SSSR *225* : 1453 - 1456, 1975. [In R.]

24459 - **SHLYK, A.A.** : Razvitie issledovaniǐ metabolicheskoǐ geterogennosti fotosinteticheskikh membran. Itogi pervogo ėtapa. [Progress in the studies of metabolic heterogeneity of photosynthetic membranes. Review of the initial steps.] - In: **SHLYK, A.A.** (ed.) : Biosintez i Sostoyanie Khlorofillov v Rastenii. Pp. 104 - - 160, 245 - 246. Nauka i Tekhnika, Minsk 1975. [In R, ab : E.]

24460 - **SHLYK, A.A., KOSTYUK, N.N.** : Vliyanie δ-aminolevulinovoǐ kisloty na nakoplenie protokhlorofillida i obrazovanie khlorofilla v zelenykh list'yakh yachmenya. [Effect of δ-aminolevulinic acid on protochlorophyllide accumulation and chlorophyll formation in green barley leaves.] - Dokl. Akad. Nauk SSSR *225* : 978 - - 980, 1975. [In R.]

24461 - **SHLYK, A.A., PARAMONOVA, T.K.** : Vliyanie ingibirovaniya belkovogo sinteza tsiklogeksimidom na fraktsionirovanie strukturnykh ėlementov khloroplasta. [The effect of protein synthesis inhibition by cycloheximide on the fractionation of chloroplast structural elements.] - Dokl. Akad. Nauk SSSR *225* : 1221 - - 1224, 1975. [In R.]

24462 - **SHLYK, A.A., PARAMONOVA, T.K.** : Vliyanie khloramfenikola na fraktsionnyǐ sostav pigmentno-belkovogo fonda khloroplastov yachmenya. [Effect of chloramphenicol on fractional composition of the pigment-protein pool of barley chloroplasts.] - Fiziol. Rast. *22* : 270 - 276, 1975. [In R, ab : E.]

24463 - **SHLYK, A.A., PRUDNIKOVA, I.V., PARAMONOVA, T.K., SUKHOVER, L.K., MALETS, L.N., GROZOVSKAYA, M.S., MIKHAILOVA, S.A., BALEVA, E.F., NIKOLAEVA, G.N., CHKANIKOVA, R.A., MITSUK, Z.I.** : Vliyanie ingibitorov sul'fgidril'nykh grupp na biosintez protokhlorofillida i khlorofillov *a* i *b*. [Effects of sulfhydryl-group inhibitors on biosynthesis of protochlorophyllide and chlorophylls *a* and *b*.] - In : **SHLYK, A.A.** (ed.) : Biosintez i Sostoyanie Khlorofillov v Rastenii. Pp. 42 - 57, 244. Nauka i Tekhnika, Minsk 1975. [In R, ab : E.]

24464 - **SHLYK, A.A., PRUDNIKOVA, I.V., SAVCHENKO, G.E., AVERINA, N.G., KOSTYUK, N.N., KAMYSHENKO, L.K., VLASENOK, L.I., GAPONENKO, V.I., BALEVA, E.F., PARAMONOVA, T.K., LOSITSKAYA, T.V., VEZITSKIĬ, A.Yu.** : The relationship of chlorophyll biosynthesis to protein and RNA synthesis in greening and green leaves. - In : **NASYROV, Yu.S., ŠESTÁK, Z.** (ed.) : Genetic Aspects of Photosynthesis. Pp. 119- - 132. Dr. W. Junk B.V. Publ., The Hague 1975.

24465 - **SHLYK, A.A., VLASENOK, L.I., AKHRAMOVICH, N.I., VRUBEL', S.V., AKULOVICH, E. M.** : Biosintez khlorofilla *b* v gomogenate zelenykh list'ev v temnote i na svetu. [Biosynthesis of chlorophyll *b* in homogenate of green leaves in the dark and in the light.] - Dokl. Akad. Nauk SSSR *221* : 1234 - 1236, 1975. [In R.]

24466 - **SHNEYOUR, A., AVRON, M.** : Properties of photosynthetic mutants isolated from *Euglena gracilis*. - Plant Physiol. *55* : 137 - 141, 1975.

24467 - **SHNEYOUR, A., AVRON, M.** : A method for producing, selecting, and isolating photosynthetic mutants of *Euglena gracilis*. - Plant Physiol. *55* : 142 - 144, 1975.

24468 - **SHNEYOUR, A., SMILLIE, R.M., RAISON, J.K.** : Biochemical and genetical basis for the temperature sensitivity of photosynthesis and growth in chilling-sensitive plants. - In : **NASYROV, Yu.S., ŠESTÁK, Z.** (ed.) : Genetic Aspects of Photosynthesis. Pp. 349 - 355. Dr. W. Junk B.V. Publ., The Hague 1975.

24469 - **SHOMER-ILAN, A., BEER, S., WAISEL, Y.** : *Suaeda monoica*, a C_4 plant without typical bundle sheaths. - Plant Physiol. *56* : 676 - 679, 1975.

24470 - **SHORTALL, J.G., LIEBHARDT, W.C.** : Yield and growth of corn as affected by poultry manure. - J. Environ. Quality *4* : 186 - 191, 1975. [Growth analysis.]

24471 - **SHOSHAN, V., TEL-OR, E., SHAVIT, N.** : On the role of CF_1 in ATP synthesis and the partial reactions of photophosphorylation. - In : **AVRON, M.** (ed.) : Proceedings of the Third International Congress on Photosynthesis. Vol. II. Pp. 831 - - 838. Elsevier, Amsterdam - Oxford - New York 1975.

*24472 - **SHTINA, E.A., NEKRASOVA, K.A.** : The direct and indirect contribution of soil algae to the primary production of biocenoses. - Ann. Zool. Ecol. anim. *4* (spec. No.) : 37 - 45, 1971.

*24473 - **SHUGART, H.H., GOLDSTEIN, R.A., O'NEILL, R.V., MANKIN, J.B.** : TEEM : a terrestrial ecosystem energy model for forests. - Oecol. Plant. *9* : 231 - 264, 1974. [Ps submodel.]

24474 - **SHUL'GIN, I.A., KLIMOV, S.V., NICHIPOROVICH, A.A.** : Ob adaptivnosti arkhitektoniki rasteniĭ k solnechnoĭ radiatsii. [Adaptation of plant architecture to solar radiation.] - Fiziol. Rast. *22* : 40 - 48, 1975. [In R, ab : E.]

24475 - **SHUMAN, F.R., LORENZEN, C.J.** : Quantitative degradation of chlorophyll by a marine herbivore. - Limnol. Oceanogr. *20* : 580 - 586, 1975.

24476 - **SHUMWAY, L.K., HESS, W.M.** : Ultrastructural changes in white and green tobacco leaf protoplasts during dedifferentiation, and colony formation. - Plant Physiol. *56* (Suppl.) : 38, 1975.

24477 - **SHUTEEV, M.M., POGREBNYAK, M.P., UŠOVICH, A.T., LAVRIK, N.G.** : Khimicheskiĭ sostav vodnykh i pribrezhno-vodnykh rasteniĭ borovykh ozer Altaĭskogo kraya. [Chemical composition of aquatic and coastal-aquatic plants of the borate lakes of the Altai region.] - Sb. nauch. Rabot sib. nauch.-issled. veter. Inst. *24* : 96 - 97, 1975. [Car; in R.]

24478 - **SHUTILOVA, N.I., KLIMOV, V.V., SHUVALOV, V.A., KUTYURIN, V.M.** : Issledovanie fotokhimicheskikh i spektral'nykh svoĭstv subkhloroplastnykh fragmentov FS-2, vysokoochishchennykh ot primesi FS-1. [Photochemical and spectral properties of PS-2 subchloroplast fragments which are highly purified from PS-1 admixtures.] - Biofizika *20* : 844 - 847, 1975. [In R, ab : E.]

24479 - **SHUTILOVA, N.I., KUTYURIN, V.M.** : K voprosu o metode vydeleniya fotokhimicheskski aktivnykh pigment-belkovolipidnykh kompleksov khloroplastov.[Technique of isolation of photochemically active pigment-lipoprotein complexes of chloroplasts.] - Biofizika *20* : 246 - 253, 1975. [In R, ab : E.]

24480 - **SHUVALOV, V.A., KLIMOV, V.V.** : Issledovanie reaktsionnykh tsentrov fotosistemy I khloroplastov metodami dlitel'noĭ lyuminestsentsii, peremennoĭ fluorestsentsii i differentsial'noĭ spektrofotometrii. [Studying of photosystem I reaction centres of chloroplasts by methods of delayed luminescence, variable fluorescence and differential spectrophotometry.] - In : Metody Issledovaniya Fotokhimicheskikh Reaktsiĭ Fotosinteza *in Vitro* i *in Vivo*. Pp. 148 - 159. Pushchino 1975. [In R.]

24481 - **SHUVALOV, V.A., KRASNOVSKIĬ, A.A.** : Izuchenie fotovosstanovleniya kisloroda khloroplastami metodom khemilyuminestsentsii lyuminola i khlorofilla. [Study of oxygen photoreduction in chloroplasts by the method of luminol and chlorophyll chemiluminescence.] - Biokhimiya *40* : 358 - 367, 1975. [In R, ab : E.]

24482 - **SHVALEVA, L.S.** : Soderzhanie khlorofilla i ego belkovo-lipoidnykh kompleksov v list'yakh bobov v svyazi s zasukhoĭ. [Content of chlorophyll and its protein--lipid complexes in broad bean leaves in connection with drought.] - In : Vodoobmen Rasteniĭ pri Neblagopriyatnykh Usloviyakh Sredy. Pp. 92 - 94. Shtiintsa, Kishinev 1975. [In R.]

24483 - **SICHER, R., BISHOP, N.I.** : The effects of high intensity irradiation on an algal mutant deficient in vitamin *E*. - Plant Physiol. *56* (Suppl.) : 10, 1975.

24484 - **SIDDELL, S.G., ELLIS, R.J.** : Protein synthesis in chloroplasts. Characteristics and products of protein synthesis *in vitro* in etioplasts and developing chloroplasts from pea leaves. - Biochem. J. *146* : 675 - 685, 1975.

24485 - **SID'KO, F.Ya., TIKHOMIROV, A.A., ZOLOTUKHIN, I.G., POLONSKIĬ, V.I.** : Rost i razvitie redisa pri osveshchenii dugovoĭ galogenidnoĭ lampoĭ DRIF. [Growth and development of garden radish under illumination with arc halogen lamp DRIF.] - Fiziol. Biokhim. kul't. Rast. *7* : 181 - 184, 1975. [Chl; in R, ab : E.]

24486 - **SIDOROV, A.N.** : Spektral'noe izuchenie kompleksov s perenosom zaryada tetrapirrol'nykh soedineniĭ. [Spectral studies of complexes with the electron transfer in tetrapyrrolic compounds.] - In : Metody Issledovaniya Fotokhimicheskikh Reaktsiĭ Fotosinteza *in Vitro* i *in Vivo*. Pp. 53 - 64. Pushchino 1975. [In R.]

*24487 - **SIEDOW, J.N., CURTIS, V.A., SAN PIETRO, A.** : Studies on photosystem I. I. Relationship of plastocyanin, cytochrome *f* and *P*700. - Arch. Biochem. Biophys. *158* : 889 - 897, 1973.

24488 - **SIEFERMANN, D., YAMAMOTO, H.Y.** : Light-induced de-epoxidation of violaxanthin in lettuce chloroplasts. V. Dehydroascorbate, a link between photosynthetic electron transport and de-epoxidation. - In : **AVRON, M.** (ed.) : Proceedings of the Third International Congress on Photosynthesis. Vol. III. Pp. 1991 - - 1998. Elsevier, Amsterdam - Oxford - New York 1975.

24489 - **SIEFERMANN, D., YAMAMOTO, H.Y.** : NADPH and oxygen-dependent epoxidation of zeaxanthin in isolated chloroplasts. - Biochem. biophys. Res. Commun. *62* : 456 - 461, 1975.

24490 - **SIEFERMANN, D., YAMAMOTO, H.Y.** : Light-induced de-epoxidation of violaxanthin in lettuce chloroplasts. IV. The effects of electron-transport conditions on violaxanthin availability. - Biochim. biophys. Acta *387* : 149 - 158, 1975.

24491 - **SIEFERMANN, D., YAMAMOTO, H.Y.** : Properties of NADPH and oxygen-dependent zeaxanthin epoxidation in isolated chloroplasts. A transmembrane model for the violaxanthin cycle. - Arch. Biochem. Biophys. *171* : 70 - 77, 1975.

24492 - **SIEGEL, M.I., LANE, M.D.** : Ribulose-diphosphate carboxylase from spinach leaves. - In : **COLOWICK, S.P., KAPLAN, N.O.** (ed.) : Methods in Enzymology. Vol. 42. Pp. 472 - 480. Academic Press, New York - San Francisco - London 1975.

24493 - **SIEGELMAN, M.H., RASCHED, I., BÖGER, P.** : Evidence for a 40,000 Dalton species of plastocyanin as the active component. - Biochem. biophys. Res. Commun. *65* : 1456 - 1463, 1975.

24494 - **SIEGENTHALER, P.A., HORAKOVA, J.** : Control of the photosynthetic electron transport by free fatty acids and Mn^2 salts. - In : **AVRON, M.** (ed.) : Proceedings of the Third International Congress on Photosynthesis. Vol. I. Pp. 655 - 664. Elsevier, Amsterdam - Oxford - New York 1975.

24495 - **SIGALAT, C., de KOUCHKOVSKY, Y.** : Fractionnement et caractérisation de l'appareil photosynthétique de l'algue bleue unicellulaire *Anacystis nidulans* I. Obtention de fractions membranaires par "lyse osmotique" et analyse pigmentaire. - Physiol. vég. *13*:243 - 258, 1975.

24496 - **SIGALAT, C., de KOUCHKOVSKY, Y.** : Preparation and properties of photosynthetic fragments of the unicellular blue-green alga *Anacystis nidulans*. - In : **AVRON, M.** (ed.) : Proceedings of the Third International Congress on Photosynthesis. Vol. I. Pp. 621 - 627. Elsevier, Amsterdam - Oxford - New York 1975.

24497 - **SIGGEL, U.** : The control of electron transport by two pH-sensitive sites. - In: **AVRON, M.** (ed.) : Proceedings of the Third International Congress on Photosynthesis. Vol. I. Pp. 645 - 654. Elsevier, Amsterdam - Oxford - New York 1975.

24498 - **SIKORA, Z.** : Management of a collection and studies on the optimal conditions for the cultivation of algae. - Pol. ecol. Stud. *1* : 127 - 136, 1975. [Ps.]

24499 - **SILVERBERG, B.A.** : An ultrastructural and cytochemical characterization of microbodies in the green algae. - Protoplasma *83* : 269 - 295, 1975.

24500 - **SILVERBERG, B.A.** : Some structural aspects of the pyrenoid of the ulotrichalean alga *Stichococcus*. - Trans. amer. microscop. Soc. *94* : 417 - 421, 1975.

24501 - **SILVIUS, J., INGLE, M., BAER, C.H.** : Sulfur dioxide inhibition of photosynthesis in isolated spinach chloroplasts. - Plant Physiol. *56* (Suppl.) : 14, 1975.

24502 - **SILVIUS, J.E., INGLE, M., BAER, C.H.** : Sulfur dioxide inhibition of photosynthesis in isolated spinach chloroplasts. - Plant Physiol. *56* : 434 - 437, 1975.

24503 - **SIMMONS, G.M., Jr.** : Observations on limnetic carbon assimilation rates in Mountain Lake, Virginia : A correction. - Virginia J. Sci. *26* : 135, 1975.

24504 - **ŠIMON, J.** : Production of phytomass in some field crops under irrigation on light soils. - Rostlinná Výroba (Praha) *21* : 845 - 860, 1975.

24505 - **SIMPSON, D.J., BAQAR, M.R., LEE, T.H.** : Unusual ultrastructural features of the chloroplast-chromoplast transformation in *Solanum luteum* fruit. - Aust. J. Plant Physiol. *2* : 235 - 245, 1975.

24506 - **SIMPSON, D.J., BAQAR, M.R., LEE, T.H.** : Ultrastructure and carotenoid composition of chromoplasts of the sepals of *Strelitzia reginae* AITON during floral development. - Ann. Bot. *39* : 175 - 183, 1975.

24507 - **SINCLAIR, T.R., BINGHAM, G.E., LEMON, E.R., ALLEN, L.H., Jr.** : Water use efficiency of field-grown maize during moisture stress. - Plant Physiol. *56* : 245 - 249, 1975. [Ps.]

24508 - **SINCLAIR, T.R., de WIT, C.T.** : Photosynthate and nitrogen requirements for seed production by various crops. - Science *189* : 565 - 567, 1975.

24509 - **SINGH, J.S., LAUENROTH, W.K., STEINHORST, R.K.** : Review and assessment of various techniques for estimating net aerial primary production in grasslands from harvest data. - Bot. Rev. *41* : 181 - 232, 1975.

24510 - **SINGH, N., MISHRA, D.** : The effect of benzimidazole and red light on the senescence of detached leaves of *Oryza sativa* cv. Ratna. - Physiol. Plant. *34* : 67 - 74, 1975. [Chl.]

24511 - **SINGH, R.P., NAIR, K.P.P.** : Defoliation studies in hybrid maize. II. Dry-matter accumulation, LAI, silking and yield components. - J. agr. Sci. *85* : 247 - 254, 1975. [Growth analysis.]

24512 - **SINGH, V.P., BILLORE, S.K.** : Relationship between chlorophyll and energy contents of the *Andropogon* grassland community. - Photosynthetica *9* : 93 - 95, 1975.

24513 - **SIRENKO, L.A.** : Metody kolichestvennogo ucheta rosta vodorosleǐ v kul'ture i vodoeme. [Methods of quantitative study of algae growth in culture and reservoir.] - In : Metody Fiziologo-Biokhimicheskogo Issledovaniya Vodorosleǐ v Gidrobiologicheskoǐ Praktike. Pp. 30 - 50. Naukova Dumka, Kiev 1975. [Chl; in R.]

24514 - **SIRENKO, L.A.** : Diagnostika fiziologicheskogo sostoyaniya kletok vodorosleĭ. [Diagnostics of physiological state of algae cells.] - In : Metody Fiziologo-Biokhimicheskogo Issledovaniya Vodorosleĭ v Gidrobiologicheskoĭ Praktike. Pp. 51 - 74. Naukova Dumka, Kiev 1975. [Oxygen electrode, Winkler method; in R.]

24515 - **SIRENKO, L.A.** : Khromatograficheskoe izuchenie pigmentov poliblefaridovykh vodorosleĭ bez ėkstragirovaniya (po Radchenko i Masyuk [118]). [Chromatographic study of pigments of *Polyblepharidaceae* without extraction (according to Radchenko and Masyuk [118]).] - In : Metody Fiziologo-Biokhimicheskogo Issledovaniya Vodorosleĭ v Gidrobiologicheskoĭ Praktike. Pp. 80 - 81. Naukova Dumka, Kiev 1975. [In R.]

24516 - **SIRENKO, L.A.** : Opredelenie sostoyaniya khlorofilla v kletkakh (po Maslovoĭ [90]). [Determination of chlorophyll state in cells (according to Maslova [90]).] - In : Metody Fiziologo-Biokhimicheskogo Issledovaniya Vodorosleĭ v Gidrobiologicheskoĭ Praktike. Pp. 86 - 87. Naukova Dumka, Kiev 1975. [In R.]

24517 - **SIRENKO, L.A.** : Opredelenie sostoyaniya khlorofilla v kletkakh rasteniĭ (po Sapozhnikovu i Chernomorskomu [130]). [Determination of chlorophyll state in plant cells (according to Sapozhnikov and Chernomorskiĭ [130]).] - In : Metody Fiziologo-Biokhimicheskogo Issledovaniya Vodorosleĭ v Gidrobiologicheskoĭ Praktike. P. 87. Naukova Dumka, Kiev 1975. [In R.]

24518 - **SIRENKO, L.A.** : Vydelenie i ochistka bilikhromoproteidov sinezelenykh vodorosleĭ [140]. [Isolation and purification of bilichrome proteids of blue-green algae[140.]] - In : Metody Fiziologo-Biokhimicheskogo Issledovaniya Vodorosleĭ v Gidrobiologicheskoĭ Praktike. Pp. 88 - 89. Naukova Dumka, Kiev 1975. [In R.]

24519 - **SIRENKO, L.A.** : Vydelenie fikotsianina (po Rotfarb, Godnevu, Gvardiyan [121] . [Phycocyanin isolation (according to Rotfarb, Godnev and Gvardiyan [121]).] - In : Metody Fiziologo-Biokhimicheskogo Issledovaniya Vodorosleĭ v Gidrobiologicheskoĭ Praktike. Pp. 89 - 90. Naukova Dumka, Kiev 1975. [In R.]

24520 - **SIRENKO, L.A.** : Opredelenie kachestvennogo sostava i kolichestva bilikhromoproteinovykh pigmentov (po Okh'lokha, Rafteri [218, 219]). [Determination of qualitative composition and quantity of bilichrome protein pigments (according to O'hEocha and Rafteri [218, 219].] - In : Metody Fiziologo-Biokhimicheskogo Issledovaniya Vodorosleĭ v Gidrobiologicheskoĭ Praktike. Pp. 90 - 91. Naukova Dumka, Kiev 1975. [In R.]

24521 - **SIRENKO, L.A.** : Vydelenie i kolichestvennoe opredelenie karotinoidov i khlorofillov u vodorosleĭ s pomoshch'yu zakreplennykh tonkosloĭnykh khromatogramm (po Khageru i Meĭeru-Bertenratu [197] . [Isolation and quantitative determination of carotenoids and chlorophylls in algae by means of TLC (according to Hager and Meyer-Bertenrath [197]).] - In : Metody Fiziologo-Biokhimicheskogo Issledovaniya Vodorosleĭ v Gidrobiologicheskoĭ Praktike. Pp. 97 - 101. Naukova Dumka, Kiev 1975. [In R.]

24522 - **SIRENKO, L.A., VELICHKO, I.M.** : Ekstraktsiya pigmentov. [Pigment extraction.] - In : Metody Fiziologo-Biokhimicheskogo Issledovaniya Vodorosleĭ v Gidrobiologicheskoĭ Praktike. Pp. 77 - 78. Naukova Dumka, Kiev 1975. [In R.]

24523 - **SIROHI, G.S., GHILDIYAL, M.C.** : Varietal differences in photosynthetic carboxylases & chlorophylls in wheat varieties. - Indian J. exp. Biol. *13* : 42 - - 44, 1975.

24524 - **SIRONVAL, C.** : On plastid states. - In : **AVRON, M.** (ed.) : Proceedings of the Third International Congress on Photosynthesis. Vol. III. Pp. 2153 - 2162. Elsevier, Amsterdam - Oxford - New York 1975.

24525 - **SIVALINGAM, P.M., IKAWA, T., YOKOHAMA, Y., NISIZAWA, K.** : Isolation and examination of biochemical activity of epidermal whole cells from a red alga, *Galaxaura falcata*. - Bull. jap. Soc. sci. Fish. *41* : 183 - 192, 1975.

24526 - **SIVAPALAN, K.** : Photosynthetic assimilation of $^{14}CO_2$ by mature brown stems of the tea plant (*Camelia sinensis* L.). - Ann. Bot. *39* : 137 - 140, 1975.

24527 - **SIVTSEV, M.V.** : Nakoplenie karotinoidov v list'yakh rasteniĭ v svyazi s raznym obespecheniem ikh vodoĭ. [Accumulation of carotenoids in plant leaves in connection with different water supply.] - Nauch. Dokl. vyssh. Shkoly, biol. Nauki *18* (11) : 78 - 83, 1975. [In R.]

24528 - **SIZOV, S.S.** : Izmenenie fiziologo-biokhimicheskikh protsessov v pochkakh muzhskikh i zhenskikh osobeĭ dvudomnykh rasteniĭ v zimniĭ period. [Change of physiological and biochemical processes in buds of male and female speciments in diclinuous plants in winter.] - Nauch. Dokl. vyssh. Shkoly, biol. Nauki *18* (4) : 83 - 88, 1975. [Chl, Car; in R.]

24529 - **SJÖSTRAND, F.S., KRETZER, F.** : A new freeze-drying technique applied to the analysis of the molecular structure of mitochondrial and chloroplast membranes. - J. Ultrastruct. Res. *53* : 1 - 28, 1975.

24530 - **SKEEN, J.N.** : An evaluation of chloroplast pigment characteristics as an indicator of successional status in terrestrial ecosystems. - Amer. Midland Naturalist *94* : 370 - 384, 1975.

*24531 - **SKENE, D.S.** : Chloroplast structure in mature apple leaves grown under different levels of illumination and their response to changed illumination. - Proc. roy. Soc. London B *186* : 75 - 78, 1974.

24532 - **SKLÁDAL, V., KAFKA, K.** : Příspěvek ke studiu fotosyntézy rostlin chmele infračerveným analyzátorem v polních podmínkách. [Studies on photosynthesis in hop plants with infra-red analyzer under field conditions.] - Rostlinná Výroba (Praha) *21* : 373 - 378, 1975. [In Czech, ab : E, R.]

24533 - **SKOKUT, T.A., WU, J.H., DANIEL, R.S.** : Chloroplast ultrastructure in relatior to the preservation of green color in leaves irradiated with high dose of ultraviolet light. - Plant Physiol. *56* (Suppl.) : 64, 1975.

24534 - **SKOROBOGATOVA, R.A., MIRCHINK, T.G.** : Deĭstvie tsitrinina na nekotorye storony obmena veshchestv pshenitsy. [Effect of citrinine on some parts of wheat metabolism.] - Nauch. Dokl. vyssh. Shkoly, biol. Nauki *18* (8) : 79 - 83, 1975. [Chl; in R.]

24535 - **SKRE, O.** : CO_2 exchange in Norwegian tundra plants studied by infrared gas analyzer technique. - In : **WIELGOLASKI, F.E.** (ed.) : Fennoscandian Tundra Ecosystems. Part I. Plants and Microorganisms. Pp. 163 - 183. Springer Verlag, Berlin - Heidelberg - New York 1975.

24536 - **SLAVÍK, B.** : Transpiration rate and intercellular diffusive resistance in the tobacco leaf. - Biol. Plant. *17* : 400 - 404, 1975.

24537 - **SLAVÍK, B.** : Water stress, photosynthesis and the use of photosynthates. - In : **COOPER, J.P.** (ed.) : Photosynthesis and Productivity in Different Environments. Pp. 511 - 536. Cambridge Univ. Press, Cambridge - London - New York - Melbourne 1975.

24538 - **SLAVOV, N.** : Vliyanie na nyakoi agrometeorologichni usloviya v"rkhu natrupvaneto na kaliĭ i tsarevichniya posev. [Influence of some agrometeorological conditions on potassium accumulation in a maize stand.] - Rasteniev. Nauki *12* (4) : 28 - 36, 1975. [Primary production; in Bulg., ab : E, R.]

24539 - **ŚLESAK, E.** : Influence of reducing and oxidating compounds on iron accumulation and chlorophyll content in oat leaves. - Acta Soc. Bot. Pol. *44* : 349 - - 360, 1975.

*24540 - **SLONOV, L.Kh.** : Izmenenie soderzhaniya i sostoyaniya pigmentov v list'yakh raznopolykh osobeĭ yuzhnoĭ konopli v zavisimosti ot usloviĭ pitaniya i vlazhnosti pochvy. [Changes in content and state of pigments in leaves of southern hemp plants of different sex in dependence on nutrition and soil humidity.] - Izv. sev.-kavkaz. nauch. Tsentra vyssh. Shkoly, Ser. estestv. Nauk *1974* (3) : 49 - 52, 1974. [In R.]

*24541 - **SLONOV, L.Kh.** : Pigmenty v list'yakh raznopolykh osobeĭ yuzhnoĭ konopli. [Pigments in leaves of southern hemp plants of different sex.] - In : Voprosy Botaniki (Nal'chik) *1* : 165 - 177, 1974. [In R.]

24542 - **SLUITERS-SCHOLTEN, C.M.T.** : Photosynthesis and the induction of nitrate reductase and nitrite reductase ih bean leaves. - Planta *123* : 175 - 184, 1975.

24543 - **SLUKHAĬ, S.I., SHVEDOVA, O.E.** : Azornyĭ obmen v rasteniyakh kukuruzy pri razlichnoĭ vlagoobespechennosti. [Nitrogen metabolism in maize plants under different water supply.] - Fiziol. Biokhim. kul't. Rast. *7* : 43 - 47, 1975. [Ps; in R, ab : E.]

24544 - **ŠMARDA, J., EBRINGER, L., MACH, J.** : The effect of colicin E2 on the flagellate *Euglena gracilis*. - J. gen. Microbiol. *86* : 363 - 366, 1975. [Chl.]

24545 - **SMILLIE, R., NIELSEN, N.C., HENNINGSEN, K.W., von WETTSTEIN, D.** : Ontogeny and environmental regulation of photochemical activity in chloroplast membranes. - In : **AVRON, M.** (ed.) : Proceedings of the Third International Congress on Photosynthesis. Vol. III. Pp. 1841 - 1860. Elsevier, Amsterdam - Oxford - New York 1975.

24546 - **SMILLIE, R.M., BISHOP, D.G., CONROY, J.** : The implications of genetic, physiological and environmental control of chloroplast ultrastructure to the optimization of photosynthetic activity. - In : **NASYROV, Yu.S., ŠESTÁK, Z.** (ed.) : Genetic Aspects of Photosynthesis. Pp. 305 - 314. Dr. W. Junk B.V. Publ., The Hague 1975.

24547 - **SMIRNOVA, L.F., PALILAVA, A.M., GARDZIEVICH, R.P., SUSHKEVICH, L.U.** : Intėnsiŭnasts' rėaktsii Khila ŭ raslin kukuruzy, yakiya mayots' rozny typ tsytaplazmy. [Rate of Hill reaction in maize plants with different type of cytoplasm.] - Vestsi Akad. Navuk belarus.SSR, Ser. biyal. Navuk *1975* (2) : 48 - - 51, 138, 1975. [In Belorus., ab : R.]

24548 - **SMIRNOVA, L.F., PALILAVA, A.M., SUSHKEVICH, L.U., TSYABUT, L.F.** : Ėlektrontranspartnaya aktyŭnasts' khlaraplastaŭ i pigmentatsyya listsyaŭ gibrydaŭ kukuruzy z roznymi typami tsytaplazmy. [Electron transport activity of chloroplasts and pigmentation of leaves of maize hybrids with different types of cytoplasm.] - Vestsi Akad. Navuk belarus. SSR, Ser. biyal. Navuk *1975* (5) : 41 - 47, 139, 1975. [In Belorus., ab : R.]

*24549 - **SMITH, A.E., WILKINSON, R.E.** : Influence of simazine and atrazine on free fatty acid content in isolated chloroplasts. - Weed Sci. *21* : 57 - 60, 1973. [Ps.]

24550 - **SMITH, A.J.** : Blue-green bacteria : status and photoautotrophic metabolism. - Biochem. Soc. Trans. *3* : 345 - 352, 1975. [Ps.]

24551 - **SMITH, B.B., REBEIZ, C.A.** : Detection of magnesium-protoporphyrin chelatase activity *in vitro*. - Plant Physiol. *56* (Suppl.) : 31, 1975. [Chl.]

24552 - **SMITH, B.N.** : Carbon and hydrogen isotopes of sucrose from various sources. - - Naturwissenschaften *62* : 390, 1975.

24553 - **SMITH, B.N., OLIVER, J., McMILLAN, C.** : Influence of temperature, light intensity and carbon source on $^{13}C/^{12}C$ ratios in plant tissues. - Plant Physiol. *56* (Suppl.) : 25, 1975.

24554 - **SMITH, B.N., ROBBINS, M.J.** : Evolution of C_4 photosynthesis : An assessment based on $^{13}C/^{12}C$ ratios and Kranz anatomy. - In : **AVRON, M.** (ed.) : Proceedings of the Third International Congress on Photosynthesis. Vol. II. Pp. 1579- - 1587. Elsevier, Amsterdam - Oxford - New York 1975.

24555 - **SMITH, B.N., TURNER, B.L.** : Distribution of Kranz syndrome among *Asteraceae*. - Amer. J. Bot. *62* : 541 - 545, 1975.

24556 - **SMITH, C.J.** : Substrate availability in relation to daylength effects on tracheid development in *Picea sitchensis* (BONG) CARR. - Ann. Bot. *39* : 101 - 111, 1975. [Ps.]

24557 - **SMITH, C.J.** : Solvent vapour phototoxicity. - Laboratory Practice *24* (1) : 9 - 11, 1975.

24558 - **SMITH, D., SOBERALSKE, R.M.** : Comparison of the growth responses of spring and summer plants of alfalfa, red clover, and birdsfoot trefoil. - Crop Sci. *15* : 519 - 522, 1975. [Photosynthate transport.]

24559 - **SMITH, D.D., SJOLUND, R.D.** : Photosynthetic activity and membrane polypeptide composition of supergranal chloroplasts from plant tissue cultures containing a viruslike particle. - Plant Physiol. *55* : 520 - 525, 1975.

24560 - **SMITH, K.M.** : General features of the structure and chemistry of porphyrin compounds. - In : **SMITH, K.M.** (ed.) : Porphyrins and Metalloporphyrins. Pp. 3 - 28. Elsevier sci. Publ. Comp., Amsterdam - Oxford - New York 1975. [Chl.]

24561 - **SMITH, K.M.** : Synthesis and preparation of porphyrin compounds. - In :

SMITH, K.M. (ed.) : Porphyrins and Metalloporphyrins. Pp. 29 - 58, Elsevier sci. Publ. Comp., Amsterdam - Oxford - New York 1975. [Chl.]

24562 - **SMITH, P.D.** : The use of *in situ* algal assays to evaluate the effects of sewage effluents on the production of Shagawa Lake phytoplankton. - In : Proceedings : Biostimulation - Nutrient Assessment Workshop. EPA-660/3-75-034. Pp. 143 - 173. National Environm. Res. Center, Corvallis, Oregon 1975. [Chl.]

24563 - **SMITH, R.I.L., STEPHENSON, C.** : Preliminary growth studies on *Festuca contracta* T. KIRK and *Deschampsia antarctica* DESV. on South Georgia. - Brit. Antarct. Surv. Bull. *41-42* : 59 - 75, 1975. [Growth analysis.]

24564 - **SMITH, R.I.L., WALTON, D.W.H.** : A growth analysis technique for assessing habitat severity in tundra regions. - Ann. Bot. *39* : 831 - 843, 1975.

24565 - **SMITH, S.V., KEY, G.S.** : Carbon dioxide and metabolism in marine environments. - Limnol. Oceanogr. *20* : 493 - 495, 1975.

24566 - **SMITH, W.O., Jr.** : The optimal procedures for the measurement of phytoplankton excretion. - Mar. Sci. Commun. *1* : 395 - 405, 1975.

24567 - **SMOLEN, F.** : Illumination in the forest stand at the research site of Báb. - - In : **BISKUPSKÝ, V.** (ed.) : Research Project Báb IBP Progress Report II. Pp. 369 - 380. Veda, Bratislava 1975.

24568 - **SMOLOV, A.P., IGNAT'EV, A.R., POLEVAYA, V.S.** : Stanovlenie funktsiĭ fotosinteticheskogo apparata v kul'ture tkani *in vitro*. [Formation of functions of the photosynthetic apparatus in tissue culture *in vitro*.] - Fiziol. Rast. *22*: 428 - 429, 1975. [In R.]

24569 - **SMOLYAK, L.P., RĖUTSKI, U.R., FILISTOVICH, V.R., KAZLOVA, Zh.I.** : Rolya gruntavykh vod y perarazmerkavanni sonechnaĭ radyyatsyi fitatsenozam. [Effect of underground water on distribution of solar energy by phytocenosis.] - Vestsi Akad. Navuk belarus. SSR, Ser. biyal. Navuk *1975* (1) : 30 - 35, 135, 1975. [In Belorus., ab : R.]

*24570 - **SOEDER, C., STENGEL, E.** : Physico-chemical factors affecting metabolism and growth rate. - In : **STEWART, W.D.P.** (ed.) : Algal Physiology and Biochemistry. Pp. 714 - 740. Blackwell sci. Publ., Oxford - London - Edinburgh - Melbourne 1974. [Ps, Chl.]

24571 - **SOLÓRZANO, L., GRANTHAM, B.** : Surface nutrients, chlorophyll *a* and phaeopigment in some Scottish sea lochs. - J. exp. mar. Biol. Ecol. *20* : 63 - 76, 1975.

24572 - **SOLTERO, R.A., WRIGHT, J.C.** : Primary production studies on a new reservoir; Bighorn Lake-Yellowtail Dam, Montana, U.S.A. - Freshwater Biol. *5* : 407 - 421, 1975.

24573 - **SONG, P.-S., FUGATE, R.D.** : Excited states of biomolecules - III. - Photochem. Photobiol. *22* : 277 - 285, 1975. [Chl.]

24574 - **SONG, S.P., WALTON, P.D.** : Inheritance of leaflet size and specific leaf weight in alfalfa. - Crop Sci. *15* : 649 - 652, 1975.

24575 - **SOROKIN, E.M.** : Kinetic properties of prompt and delayed fluorescence of chlorophyll *a* with λ_{max} = 685 nm when electron transport is blocked on acceptor side of photosystem II. - J. Bioenerg. *7* : 167 - 174, 1975.

24576 - **SOROKIN, E.M.** : Analiz kinetiki perenosa ėlektronov v reaktsionnykh tsentrakh fotosistemy II. [Analysis of the kinetics of electron transfer in photosystem II reaction centres.] - Dokl. Akad. Nauk SSSR *221* : 220 - 222, 1975. [In R.]

24577 - **SOROKIN, E.M.** : Dlitel'nost' fluorestsentsii khlorofilla *a* v kletkakh khlorelly v zavisimosti ot temperatury. [Duration of chlorophyll *a* fluorescence in *Chlorella* cells depending on temperature.] - Biofizika *20* : 941 - 942, 1975. [In R, ab : E.]

*24578 - **SOROKIN, Yu.I., KONOVALOVA, I.W.** : Production and decomposition of organic matter in a bay of the Japan Sea during the winter diatom bloom. - Limnol. Oceanogr. *18* : 962 - 967, 1973. [Ps.]

*24579 - **SOSNOWSKA, J.** : Zbiorowiska Planktonowe Trzech Jezior Mazurskich i Zawartość Chlorofilu w ich Fitoplanktonie. [Plankton Communities of Three Mazurian Lakes and Chlorophyll Content in Their Phytoplankton.] - Monographiae botanicae. Vol. 42. Pp. 1 - 152. Państw. Wydawn. naukowe, Warszawa 1974. [In Pol., ab : E.]

24580 - **SOTO, C., HELLEBUST, J.A., HUTCHINSON, T.C.** : Effect of naphthalene and aqueous crude oil extracts on the green flagellate *Chlamydomonas angulosa*. II. Photosynthesis and the uptake and release of naphthalene. - Can. J. Bot. *53* : 118 - 126, 1975.

24581 - **SOURNIA, A., RICARD, M.** : Phytoplankton and primary productivity in Takapoto Atoll, Tuamotu Islands. - Micronesica *11* : 159 - 166, 1975.

24582 - **SOURNIA, A., RICARD, M.** : Production primaire planctonique dans deux lagons de Polynésie francaise (île de Moorea et atoll de Takapoto). - Compt. rend. Acad. Sci. Paris, Sér. D *280* : 741 - 743, 1975.

24583 - **SPEARING, A.M., KARLANDER, E.P.** : Hill activity in *Chlorella sorokiniana* during cold-induced pigments degradation. - Plant Physiol. *56* (Suppl.) : 11, 1975.

24584 - **SPECHT, J.E., HASKINS, F.A., GORZ, H.J.** : Contents of chlorophylls *a* and *b* in chlorophyll-deficient mutants of sweetclover. - Crop Sci. *15* : 851 - 853, 1975.

24585 - **SPECHT, R.L., BROUWER, Y.M.** : Seasonal shoot growth of *Eucalyptus* spp. in the Brisbane area of Queensland (with notes on shoot growth and litter fall in other areas of Australia). - Aust. J. Bot. *23* : 459 - 474, 1975.

24586 - **SPERLING, J.A.** : Algal ecology of southern Icelandic hot springs in winter. - Ecology *56* : 183 - 190, 1975. [Chl.]

*24587 - **SPERLING, J.A., GRUNEWALD, R.** : Batch culturing of thermophilic benthic algae and phosphorus uptake in a laboratory stream model. - Limnol. Oceanogr. *14* : 944 - 949, 1969.

24588 - **SPIVAK, A.I.** : Fotosintez i vodnyĭ rezhim nekotorykh vidov stepnykh rasteniĭ Yugo-Vostochnogo Zabaĭkal'ya. [Photosynthesis and water relations of some species of steppe plants in South-Eastern Zabaĭkal'e.] - In : Vodnyĭ Obmen v Osnovnykh Tipakh Rastitel'nosti SSSR kak Ėlement Krugovorota Veshchestva i Ėnergii. Pp. 155 - 160. Nauka, Novosibirsk 1975. [In R.]

24589 - **SPODNIEWSKA, I., PIECZYŃSKA, E., KOWALCZEWSKI, A.** : Ecosystem of the Mikołajskie Lake, primary production. - Pol. Arch. Hydrobiol. *22* : 17 - 37, 1975.

24590 - **SPREY, B.** : Membranassoziierte Tubuli während der Chloroplastengenese von *Hordeum vulgare* L. - Protoplasma *84* : 197 - 203, 1975.

24591 - **SPREY, B., LAETSCH, W.M.** : Chloroplast envelopes of *Spinacia oleracea* L. I. Polypeptides of chloroplast envelopes and lamellae. - Z. Pflanzenphysiol. *75*: 38 - 52, 1975.

24592 - **SRIVASTAVA, H.S., JOLLIFFE, P.A., RUNECKLES, V.C.** : Inhibition of gas exchange in bean leaves by NO_2. - Can. J. Bot. *53* : 466 - 474, 1975.

24593 - **SRIVASTAVA, H.S., JOLLIFFE, P.A., RUNECKLES, V.C.** : The effects of environmental conditions on the inhibition of leaf gas exchange by NO_2. - Can. J. Bot. *53* : 475 - 482, 1975.

*24594 - **STADEN, O.L.** : De energiehuishouding van de plantaardige cel. [The energy metabolism of the plant cell.] - Vakbl. Biol. *51* (10) : 244 - 250, 1971. [Ps; in Holl.]

24595 - **STANEV, V.** : Vliyanie mineral'nogo udobreniya i gustoty poseva podsolnechnika na intensivnost' i chistuyu produktivnost' fotosinteza. [Effects of mineral fertilizer and planting density on sunflower photosynthetic rate and net productivity.] - Dokl. SKhA Georgi Dimitrov *8* (3) : 21 - 24, 1975. [In R.]

24596 - **STANHILL, G., MORESHET, S., JURGRAU, M., FUCHS, M.** : The effect of reflecting surfaces on the solar radiation regime and carbon dioxide fixation of a glasshouse rose crop. - J. amer. Soc. hort. Sci. *100* : 112 - 115, 1975. [Ps.]

*24597 - **STANKOVIĆ, Ž.** : Uticaj nedostatka pojedinih jona na dinamiku sadržaja pigmenata hloroplasta pri ozelenjavanju biljaka. [Effect of deficit in some ions on the dynamics of pigment content in chloroplasts during plant greening.] - Zb. Radova prirodno-mat. Fak. Univ. Novom Sadu *4* : 125 - 134, 1974. [In Serbo-croat., ab : E.]

24598 - **STARCK, Z.** : Effect of a limited supply of assimilates on the relationship between their sources and acceptors. - Pol. ecol. Stud. *1* : 81 - 87, 1975.

24599 - **STARODUBTSEV, E.G.** : Faktory, opredelyayushchie sezonnye izmeneniya pervichnoĭ produktsii v severo-zapadnoĭ chasti Tikhogo okeana. [Factors determining the seasonal changes in primary production in the north-west part of Pacific Ocean.] - In :Tr. tikhookeansk. okeanogr. Inst. *9* (Gidrobiologicheskie Issledovaniya v Yaponskom More i Tikhom Okeane) : 15 - 27, 1975. [In R.]

24600 - **STARODUBTSEV, E.G., KAĬGORODOV, N.E.** : Pervichnaya produktsiya i fitoplankton. [Primary production and phytoplankton.] - In : Tr. tikhookeansk. okeanogr. Inst. *9* (Gidrobiologicheskie Issledovaniya v Yaponskom More i Tikhom Okeane): 3 - 8, 1975. [In R.]

24601 - **STARZECKI, W.** : 2.1. Ecophysiological investigation on photosynthetic productivity of leaves of chosen tree species. - Pol. ecol. Stud. *1* : 51 - 63, 1975.

24602 - **STEBBINS, R.L.** : Photographing large trees to show limb structure. - HortScience *10* : 218, 1975.

24603 - **STEINBACK, K.E., GOODENOUGH, U.W.** : Morphological and photosynthetic properties of digitonin-treated chloroplast membranes from the wild-type and *ac-5* strains of *Chlamydomonas reinhardi*. - Plant Physiol. *55* : 864 - 869, 1975.

24604 - **STEINHÜBEL, G.** : Možnosti gravimetrického stanovenia fotosyntézy ihlíc vždyzelených konifer. [Possibilities of gravimetrical determination of photosynthesis in needles of evergreen conifers.] - Ved. Práce výsk. Ústavu les. Hospod., Zvolen *1975* : 227 - 247, 1975. [In Slovak, ab : E, G, R.]

24605 - **STEINHÜBEL, G.** : Non-typical seasonal dynamics of the leaf surface of two plant species in the woodland ecosystem at Báb. - In : **BISKUPSKÝ, V.** (ed.) : Research Project Báb IBP Progress Report II. Pp. 157 - 166. Veda, Bratislava 1975.

24606 - **STEINHÜBEL, G., LEHOTSKÝ, L.** : Rozdielna fotosyntetická aktivita výškových proveniencií smreka, sledovaná gravimetrickou metódou. [Different photosynthetic activity of altitude provenances of spruce observed by the gravimetric method.] - Les. Časopis (Praha) *21* : 3 - 14, 1975. [In Slovak, ab : E, G, R.]

24607 - **STEMLER, A., RADMER, R.** : Source of photosynthetic oxygen in bicarbonate-stimulated Hill reaction. - Science *190* : 457 - 458, 1975.

24608 - **STENGEL, E., SOEDER, C.J.** : Control of photosynthetic production in aquatic ecosystems. - In : **COOPER, J.P.** (ed.) : Photosynthesis and Productivity in Different Environments. Pp. 645 - 660. Cambridge Univ. Press, Cambridge - London - New York - Melbourne 1975.

24609 - **STEPHEN, R., RAI, V., MURTY, K.S.** : Path analysis of grain yield in rice under normal and water-logged conditions. - Riso *24* : 279 - 284, 1975. [Growth analysis.]

24610 - **STERN, W.R.** : Actual and potential photosynthetic production - conclusions. - In : **COOPER, J.P.** (ed.) : Photosynthesis and Productivity in Different Environments. Pp. 661 - 672. Cambridge Univ. Press, Cambridge - London - New York - Melbourne 1975.

24611 - **STEVENS, C.L.R., SCHULTZ, D., Van BAALEN, C., PARKER, P.L.** : Oxygen isotope fractionation during photosynthesis in a blue-green and green alga. - Plant Physiol. *56* : 126 - 129, 1975.

24612 - **STEVENS, S.E., Jr., SIMIC, M.G., RAO, V.S.K.** : X-ray induced inactivation of the capacity for photosynthetic oxygen evolution and nitrate reduction in blue-green algae. - Rad. Res. *63* : 395 - 402, 1975.

24613 - **STEWART, K.D., MATTOX, K.R.** : Comparative cytology, evolution and classification of the green algae with some consideration of the origin of other organisms with chlorophylls *a* and *b*. - Bot. Rev. *41* : 104- 135, 1975.

24614 - **STOCKER, O.** : Prinzipien der Flechtensymbiose. - Flora *164* : 359 - 376, 1975. [Ps.]

24615 - **STOCKING, C.R.** : Iron deficiency and the structure and physiology of maize chloroplasts. - Plant Physiol. *55* : 626 - 631, 1975.

24616 - **STOCKNER, J.G., CLIFF, D.D., MUNRO, K.** : The effects of pulpmill effluent on phytoplankton production in coastal waters of British Columbia. - Fish. mar. Serv. tech. Rep. *578* : 1 - 99, 1975. [Chl.]

24617 - **STOCKNER, J.G., SHORTREED, K.R.S.** : Attached algal growth in Carnation Creek: A coastal rainforest stream on Vancouver Island, British Columbia. - Fish. mar. Serv. tech. Rep. *558* : 1 - 56, 1975. [Chl.]

24618 - **STOCKNER, J.G., SHORTREED, K.R.S.** : Phytoplankton succession and primary production in Babine Lake, British Columbia. - J. Fish. Res. Board Can. *32* : 2413 - 2427, 1975.

*24619 - **STOEV, K.D., DOBREVA, S., SLAVCHEVA, D., GADEVSKA, A.** : Fotosintez list'ev pri issledovanii vinogradnoĭ lozy kak tsel'nogo rasteniya. [Leaf photosynthesis in studying grape-vine as a whole plant.] - In : Fiziologiya Vinogradnoĭ Lozy. (Proc. Symp.). Pp. 407 - 414. Varna 1971. [In R, ab. : F.]

24620 - **STOKES, A.N.** : Proof of a law for calculating absorption of light by cellular suspensions. - Arch. Biochem. Biophys. *167* : 393 - 394, 1975.

24621 - **STOLBOVA, A.V.** : Genetic analysis of light-sensitive mutants of *Chlamydomonas reinhardi*. - In : **NASYROV, Yu.S., ŠESTÁK, Z.** (ed.) : Genetic Aspects of Photosynthesis. Pp. 217 - 223. Dr. W. Junk B.V. Publ., The Hague 1975.

24622 - **STOLOVITSKIĬ, Yu.M., EVSTIGNEEV, V.B.** : Issledovanie pervichnykh fotoprotsessov v molekulakh pigmentov ėlektrometricheskimi metodami. [Studying of primary photoprocesses in pigment molecules by electrochemical methods.] - In : Metody Issledovaniya Fotokhimicheskikh Reaktsiĭ Fotosinteza *in Vitro* i *in Vivo*. Pp. 124 - 137. Pushchino 1975. [In R.]

24623 - **STONE, L.R., KANEMASU, E.T., HORTON, M.L.** : Grain sorghum canopy temperature as influenced by clouds. - Remote Sensing Environ. *4* : 177 - 181, 1975. [Radiation in canopy.]

24624 - **STONER, W.A., MILLER, P.C.** : Water relations of plant species in the wet coastal tundra at Barrow, Alaska. - Arct. alp. Res. *7* : 109 - 124, 1975. [Resistances.]

24625 - **STOY, V.** : 8. Physiology of kernel yield and its application in breeding techniques. Source and sink properties as related to yield in different barley genotypes. - In : Barley Genetics III. Proceedings of the Third International Barley Genetics Symposium. Pp. 641 - 648. Garching 1975. [Ps.]

24626 - **STOY, V.** : Use of tracer techniques to study yield components in seed crops. - In : Tracer Techniques for Plant Breeding. Pp. 43 - 55. Int. at. Energy Agency, Vienna 1975. [Photosynthates.]

24627 - **STOYANOV, Zh.V.** : Dissipatsiya ėnergii i fotosintez pri azotnom i fosfornom pitanii. [Energy dissipation and photosynthesis during nitrogen and phosphorus nutrition.] - Fiziol. Rast. *22* : 625 - 626, 1975. [In R.]

24628 - **STRACK, Z., KARWOWSKA, R., KRASZEWSKA, E.** : The effect of several stress conditions and growth regulators on photosynthesis and translocation of assimilates in the bean plant. - Acta Soc. Bot. Pol. *44* : 567 - 588, 1975.

24629 - **STRAIN, B.R.** : Field measurements of carbon dioxide exchange in some woody perennials. - In : **GATES, D.M., SCHMERL, R.B.** (ed.) : Perspectives of Biophysical Ecology. Pp. 145 - 158. Springer-Verlag, Berlin - Heidelberg - New York 1975.

24630 - **STRASSER, R.J.** : Studies on the oxygen evolving system in flashed leaves. - In : **AVRON, M.** (ed.) : Proceedings of the Third International Congress on Photosynthesis. Vol. I. Pp. 497 - 503. Elsevier, Amsterdam - Oxford - New York 1975.

24631 - **STRASSER, R.J., BUTLER, W.L.** : Absorbance and structure changes during the induction of the oxygen evolution capacity of bean leaves grown in flashes. - - Plant Physiol. *56* (Suppl.) : 9, 1975.

24632 - **STRIBLEY, D.P., READ, D.J.** : Some nutritional aspects of the biology of ericaceous mycorrhizas. - In : **SANDERS, F.E., MOSSE, B., TINKER, P.B.** (ed.) : Endomycorrhizas. Pp. 195 - 207. Academic Press, London - New York - San Francisco 1975. [Photosynthates.]

24633 - **STRNAD, P.** : Phytomass production in cereals cultivated in monoculture. - Rostlinná Výroba (Praha) *21* : 875 - 884, 1975. [Growth analysis.]

24634 - **STROSS, R.G.** : Causes of daily rhythms in photosynthetic rates of phytoplankton. II. Phosphate control of expression in tundra ponds. - Biol. Bull. *149* : 408 - 418, 1975.

24635 - **STUART, A.L., WASSERMAN, A.R.** : Chloroplast cytochrome b_6 : Molecular composition as a lipoprotein. - Biochim. biophys. Acta *376* : 561 - 572, 1975.

24636 - **STUPISHINA, E.A., CHEREZOV, S.N.** : Primenenie IK-spektroskopii dlya izucheniya sostoyaniya vody v rastitel'nykh ob"ektakh. [Use of IR-spectroscopy for studying of water state in plants.] - In : Vodoobmen Rasteniĭ pri Neblagopriyatnykh Usloviyakh Sredy. Pp. 232 - 236. Shtiintsa, Kishinev 1975. [Chloroplast; in R.]

24637 - **STUR, J., MAREK, N., PETKOVA, R.** : Zusammenhänge zwischen lichtinduzierten Änderungen der Sauerstoffkonzentration und des Redoxpotentials in photosynthetisierenden Systemen. - In : **NECHAS, Ĭ.** (ed.) : Izuchenie Intensivnoĭ Kul'tury Vodorosleĭ. Vol. II. Pp. 73 - 82. Třeboň 1975.

24638 - **SUBCHINSKI, V.K., RUUGE, E.K., TIKHONOV, A.N.** : Vzaimodeĭstvie paramagnitnykh zondov s membranami khloroplastov vysshikh rasteniĭ. [Interaction between spin--probes and chloroplast membranes of higher plants.] - Fiziol. Rast. *22* : 882 - 890, 1975. [In R, ab : E.]

24639 - **SUBRAHMANYAM, P., PRABHAKAR, C.S.** : Mobilization and exudation of photo-assimilated ^{14}C by developing pods of groundnut (*Arachis hypogaea* L.). - Plant Soil *43* : 687 - 690, 1975.

24640 - **SUBRAMANIAN, J.** : Electron paramagnetic resonance spectroscopy of porphyrins and metalloporphyrins. - In : **SMITH, K.M.** (ed.) : Porphyrins and Metalloporphyrins. Pp. 555 - 589. Elsevier sci. Publ. Comp., Amsterdam - Oxford - New York 1975. [Chl.]

24641 - **SUCKLING, P.W., DAVIES, J.A., PROCTOR, J.T.A.** : The transmission of global and photosynthetically active radiation within a dwarf apple orchard. - Can. J. Bot. *53* : 1428 - 1441, 1975.

24642 - **SUD'ÏNA, O.G.** : Ėvolyutsiĭnyĭ pidkhid do vyvchennya biokhimichnykh protsesiv na prykladi stanovlennya biosyntezu khlorofilu. [Evolutionary approach to study of biochemical processes as exemplified by the development of chlorophyll biosynthesis.] - Ukr. bot. Zh. *32* : 545 - 562, 672, 1975. [In Ukr., ab : E, R.]

24643 - **SUD'ÏNA, O.G., DOVBYSH, K.P., GOLOD, M.G., FOMISHYNA, R.M.** : Do pytannya pro stan khlorofilazy ta ïogo minlyvist'. [Chlorophyllase state and its variability.] - Ukr. bot. Zh. *32* : 330 - 334, 397, 1975. [In Ukr., ab : E, R.]

24644 - **SUD'ÏNA, O.G., SHNYUKOVA, E.I., KOSTLAN, N.V., MUSHAK, P.O., TUPYK, N.D.** : Osoblyvosti vugletsevogo zhyvlennya *Microcystis aeruginosa* KUETZ. [Aspects of *Microcystis aeruginosa* KUETZ.carbon nutrition.] - Ukr. bot. Zh. *32* : 697 - - 702, 810, 1975. [Ps; in Ukr., ab : E, R.]

24645 - **SUDO, K., ANDO, T.** : [Influence of atmospheric humidity and soil moisture contents on the plant water condition as well as on the growth of sweet pepper and tomato plants.] - Bull. vegetable ornament. Crops Res. Sta., Ser. A *1975* (2) : 49 - 63, 1975. [Growth analysis; in Jap., ab : E.]

24646 - **SUGIMOTO, K.** : [Studies on evapo-transpiration of indica rice plants with reference to crop science I. Relation between variations of transpiration and dry matter production of rice plants.] - Jap. J. trop. Agr. *18* (3) : 131 - - 136, 1975. [Growth analysis; in Jap., ab : E.]

24647 - **SUGIMOTO, K.** : [Studies of evapo-transpiration of indica rice plants with reference to crop science. II. Dry matter production and efficiency of solar energy conversion in the tropical rice plants.] - Jap. J. trop. Agr. *18* (4) : 169 - 174, 1975. [Growth analysis; in Jap., ab : E.]

24648 - **SUHARA, K., TAKEMORI, S., KATAGIRI, M., WADA, K., KOBAYASHI, H., MATSUBARA, H.**: Estimation of labile sulfide in iron-sulfur proteins. - Anal. Biochem. *68* : 632 - 636, 1975. [Ferredoxin.]

24649 - **SULLIVAN, T.P., BRUN, W.A.** : Effect of root genotype on shoot water relations in soybeans. - Crop Sci. *15* : 319 - 322, 1975. [Ps.]

24650 - **SUMIDA, S., YOSHIDA, R., UEDA, M.** : Studies of pesticides effects on *Chlorella* metabolism II. Effect of DCMU on galactolipid metabolism. - Plant Cell Physiol. *16* : 257 - 264, 1975.

24651 - **SUN, A.S., CALVIN, M.** : Stabilization of electron spin resonance probes for photosynthesis studies. - Proc. nat. Acad. Sci. USA *72* : 3107 - 3110, 1975.

24652 - **SUNDBERG, I., NILSHAMMAR-HOLMVALL, M.** : The diurnal variation in phosphate uptake and ATP level in relation to deposition of starch, lipid, and polyphosphate in synchronized cells of *Scenedesmus*. - Z. Pflanzenphysiol. *76* : 270 - 279, 1975.

24653 - **SUNDQVIST, C., KLOCKARE, B.** : Fluorescence properties of protochlorophyllide in flash irradiated dark grown wheat leaves, treated with δ-aminolevulinic acid. - Photosynthetica *9* : 62 - 71, 1975.

24654 - **SUNDQVIST, C., ODENGARD, B., PERSSON, G.** : Light-stimulated accumulation of protochlorophyllide in leaves of different ages treated with δ-aminolevulinic acid.- Plant Sci. Lett. *4* : 89 - 96, 1975.

24655 - **SUSALLA, A.A., MAHLBERG, P.G.** : Plastid organization in phenotypically green leaf tissue of a genetic albino strain of *Nicotiana (Solanaceae)*. - Amer. J. Bot. *62* : 878 - 883, 1975.

24656 - **SUTTON, B.G.** : The path of carbon in CAM plants at night. - Aust. J. Plant Physiol. *2* : 377 - 387, 1975.

24657 - **SUTTON, B.G.** : Glycolysis in CAM plants. - Aust. J. Plant Physiol. *2* : 389 - - 402, 1975.

24658 - **SUTTON, B.G.** : Kinetic properties of phosphorylase and 6-phosphofructokinase of *Kalanchoë daigremontiana* and *Atriplex spongiosa*. - Aust. J. Plant Physiol. *2* : 403 - 411, 1975.

24659 - **SUTTON, B.G.** : Control of glycolysis in succulent plants at night. - In : **MARCELLE, R.** (ed.) : Environmental and Biological Control of Photosynthesis. Pp. 337 - 347. Dr. W. Junk b.v., Publ., The Hague 1975.

24660 - **SUTTON, B.G., TING, I.P.** : Carbohydrate metabolism of cactus. - Plant Physiol. *56* (Suppl.) : 13, 1975.

24661 - **SUZUKI, M.** : Developmental analysis in rice. - In : **MURATA, Y.** (ed.) : JIBP Synthesis. Vol. 11. Pp. 136 - 144. Univ. Tokyo Press, Tokyo 1975.

24662 - **SUZUKI, M., MURATA, Y.** : [Measurement of the efficiency for photosynthetic light energy conversion of rice population under field conditions.] - Proc. Crop Sci. Soc. Jap. *44* : 109 - 113, 1975. [In Jap., ab : E.]

24663 - **SUZUKI, T., UCHIYAMA, M.** : Photoreduction of parathion by spinach chloroplasts. - Chem. pharm. Bull. *23* : 2175 - 2178, 1975.

24664 - **SUZUKI, T., UCHIYAMA, M.** : Photoreduction of parathion by spinach chloroplasts. II. Ferredoxin-independent photoreduction of parathion by heated chloroplasts with an artificial electron donor system. - Chem. pharm. Bull. *23* : 2290 - - 2294, 1975.

24665 - **SUZUKI, T., UCHIYAMA, M.** : Inhibitory effect of parathion on the photosynthetic electron transport system in isolated spinach chloroplasts. - Environm. Contamination Toxicology *14* : 552 - 557, 1975.

24666 - **ŠVIHRA, J., BANAI TÓTH, P.** : Rastovo-produkčný proces ozimnej repky (*Brassica napus* L.) v podmienkach južného Slovenska. [The growth-production process of winter rape (*Brassica napus* L.) under the conditions of Southern Slovakia.] - Rostlinná Výroba (Praha) *21* : 923 - 928, 1975. [Growth analysis; in Slovak, ab : E, R.]

24667 - **SWADER, J.A., CHAN, W.-Y.** : Citric acid enhancement of copper solubility and toxicity in bicarbonate solutions. -Pesticide Biochem. Physiol. *5* : 405 - 411, 1975. [Ps.]

24668 - **SWADER, J.A., KUMAMOTO, J.** : 2-anthraquinonesulfonic acid photosensitized autoxidation of HEPES buffer. - Photochem. Photobiol. *21* : 313 - 315, 1975. [Ps.]

24669 - **SWADER, J.A., STOCKING, C.R., LIN, C.H.** : Light-stimulated absorption of nitrate by *Wolffia arrhiza*. - Physiol. Plant. *34* : 335 - 341, 1975. [Ps.]

24670 - **SWIFT, L.W., SWANK, W.T., MANKIN, J.B., LUXMOORE, R.J., GOLDSTEIN, R.A.** : Simulation of evapotranspiration and drainage from mature and clear-cut deciduous forests and young pine plantation. - Water Resources Res. *11* : 667 - 673, 1975. [Growth analysis.]

24671 - **SYBESMA, C., VINDEVOGEL, G.** : Comparison of spectral shifts induced by ionophores and by strong reductants in chromatophores of photosynthetic bacteria. - In : **AVRON, M.** (ed.) : Proceedings of the Third International Congress on Photosynthesis. Vol. II. Pp. 1171 - 1178. Elsevier, Amsterdam - Oxford - New York 1975.

24672 - **SYDNOR, T.D., FRETZ, T.A., CREAN, D.E., COBBS, M.R.** : Photographic estimation of plant size. - HortScience *10* : 219 - 220, 1975.

24673 - **SYTNYK, K.M., MUSATENKO, L.I., NESTEROVA, A.N., SAVYTS'KA, L.S., BOGDANOVA, T. L.** : Fiziologo-biokhimichni zminy rostuchogo lystka. [Physiological and biochemical changes in the growing leaf.] - Dopovidi Akad. Nauk ukr. RSR, Ser. B - Geol., Geofiz., Khim., Biol. *1975* (11) : 1048 - 1051, 1975. [Ps; in Ukr., ab : E.]

24674 - **SZANIAWSKI, R.K.** : Zarys fizjologii cisa. [Some problems on physiology of yew (*Taxus* sp.).] - In : **BIAŁOBOK, S.** (ed.) : Cis Pospolity - *Taxus baccata* L. Pp. 67 - 77. PWN, Warszawa - Poznań 1975. [Ps; in Pol., ab : E.]

24675 - **SZAREK, S.R., TING, I.P.** : Physiological responses to rainfall in *Opuntia basilaris (Cactaceae)*. - Amer. J. Bot. *62* : 602 - 609, 1975. [Ps.]

24676 - **SZAREK, S.R., TING, I.P.** : Photosynthetic efficiency of CAM plants in relation to C_3 and C_4 plants. - In : **MARCELLE, R.** (ed.) : Environmental and Biological Control of Photosynthesis. Pp. 289 - 297. Dr. W. Junk b.v. Publ., The Hague 1975.

24677 - **SZEICZ, G.** : Instruments and their exposure. - In : **MONTEITH, J.L.** (ed.) : Vegetation and the Atmosphere. Vol. 1. Principles. Pp. 229 - 273. Academic Press, London - New York - San Francisco 1975. [CO_2 content and other bioclimatological parameters.]

*24678 - **SZOKOLAY, A., WILDBRETT, G., BENCZE, K.** : Zur Stabilität von Carotenoidlösungen in Gegenwart insecticider Chlorkohlenwasserstoffe. - Z. Lebensmitt.-Unters. - Forsch. *148* : 22 - 29, 1972.

24679 - **TAGUCHI, S.** : Relationship between photosynthesis and cell size of phytoplankton. - J. Phycol. *11* (Suppl.) : 21, 1975.

24680 - **TAGUCHI, S.** : Phytoplankton and primary production in Futami Bay, Chichi-Jima, Ogasawara Islands. - Bull. Plankton Soc. Jap. *22* : 1 - 10, 1975.

24681 - **TAKABE, T., AKAZAWA, T.** : Further studies on the subunit structure of *Chromatium* ribulose-1,5-bisphosphate carboxylase. - Biochemistry *14* : 46 - 50, 1975.

24682 - **TAKABE, T., AKAZAWA, T.** : The role of sulfhydryl groups in the ribulose-1,5--bisphosphate carboxylase and oxygenase reactions. - Arch. Biochem. Biophys. *169* : 686 - 694, 1975.

24683 - **TAKABE, T., AKAZAWA, T.** : Molecular evolution of ribulose-1,5-bisphosphate carboxylase. - Plant Cell Physiol. *16* : 1049 - 1060, 1975.

24684 - **TAKAHAMA, U., NISHIMURA, M.** : Formation of singlet molecular oxygen in illuminated chloroplasts. Effects of photoinactivation and lipid peroxidation. - Plant Cell Physiol. *16* : 737 - 748, 1975.

24685 - **TAKAHASHI, M., ASADA, K.** : Purification of cytochrome *f* from Japanese-radish leaves. - Plant Cell Physiol. *16* : 191 - 194, 1975.

24686 - **TAKAHASHI, M., IKEDA, T.** : Excretion of ammonia and inorganic phosphorus by *Euphausia pacifica* and *Metridia pacifica* at different concentrations of phytoplankton. - J. Fish. Res. Board Can. *32* : 2189 - 2195, 1975. [Chl.]

24687 - **TAKAHASHI, M., THOMAS, W.H., SEIBERT, D.L.R., BEERS, J., KOELLER, P., PARSONS, T.R.** : The replication of biological events in enclosed water columns. - Arch. Hydrobiol. *76* : 5 - 23, 1975. [Chl.]

24688 - **TAKAKURA, T., GOUDRIAAN, J., LOUWERSE, W.** : A behaviour model to simulate stomatal resistance. - Agr. Meteorol. *15* : 393 - 404, 1975.

24689 - **TAKAMI, S., van BAVEL, C.H.M.** : Numerical experiments on the influence of CO_2 release at ground level on crop assimilation and water use. - Agr. Meteorol. *15* : 193 - 203, 1975.

24690 - **TAKAMIYA, K., NISHIMURA, M.** : Dual roles of ubiquinone as primary and secondary electron acceptors in light-induced electron transfer in chromatophores of *Chromatium* D. - Plant Cell Physiol. *16* : 1061 - 1072, 1975.

24691 - **TAKAMIYA, K.-I., NISHIMURA, M.** : Nature of photochemical reactions in chromatophores of *Chromatium* D. III. Heterogeneity of the photosynthetic units. - Biochim. biophys. Acta *396* : 93 - 103, 1975.

24692 - **TAKEDA, T., YAJIMA, M.** : [An improvement of semiempirical method for estimating the total photosynthesis of the crop population. I. On light-photosynthesis curve of rice leaves.] - Proc. Crop Sci. Soc. Jap. *44* : 343 - 349, 1975. [In Jap., ab : E.]

24693 - **TAKEGAMI, T.** : A study on senescence in tobacco leaf disks I. Inhibition by benzylaminopurine of decrease in protein level. - Plant Cell Physiol. *16* : 407 - 416, 1975.

24694 - **TAKEGAMI, T.** : A study of senescence in tobacco leaf disks II. Chloroplast and cytoplasmic rRNAs. - Plant Cell Physiol. *16* : 417 - 425, 1975.

24695 - **TAKEMOTO, J., BOGORAD, L.** : Subunits of phycoerythrin from *Fremyella diplosiphon* : Chemical and immunochemical characterization. - Biochemistry *14* : 1211 - 1216, 1975.

* 24696 - **TAKEMOTO, J., LASCELLES, J.** : Coupling between bacteriochlorophyll and membrane protein synthesis in *Rhodopseudomonas spheroides*. - Proc. nat. Acad. Sci. USA *70* : 799 - 803, 1973.

24697 - **TALARICO BISIACCHI, L., KOSOVEL, V.** : Ricerche sui pigmenti fotosintetici di *Gracilaria verrucosa* (HUDS.) PAPENFUSS. Dosaggio quantitativo della ficoeritrina. [Photosynthetic pigments from *Gracilaria verrucosa* (HUDS.) PAPENFUSS. Quantitative analysis of phycoerythrin.] - G. bot. ital. *109* : 205 - 219, 1975. [In Ital., ab : E.]

24698 - **TALARICO-BISIACCHI, L., KOSOVEL, V.** : Ricerche ultrastrutturali sulla ficoeritrina di "*Gracilaria verrucosa*" (HUDS.) PAPENFUSS. [Ultrastructural research on phycoerythrin from *Gracilaria verrucosa* (HUDS.) PAPENFUSS.] - Inform. bot. ital. *7* (1) : 21 - 22, 1975. [In Ital., ab : E.]

24699 - **TALLING, J.F.** : Primary production of freshwater microphytes. - In : **COOPER, J.P.** (ed.) : Photosynthesis and Productivity in Different Environments. Pp. 225 - 247. Cambridge Univ. Press, Cambridge - London - New York 1975.

24700 - **TALLING, J.F.** : Primary production of aquatic plants - conclusions. - In : **COOPER, J.P.** (ed.) : Photosynthesis and Productivity in Different Environments. Pp. 281 - 294. Cambridge Univ. Press, Cambridge - London - New York - - Melbourne 1975.

24701 - **TALYSHINSKII, G.M.** : Belkovye fraktsii list'ev iskhodnykh sortov i poluchennykh iz nikh ėksperimental'nykh tri- i tetraploidnykh form shelkovitsy. [Protein fractions of leaves of parental cultivars and resulting experimental tri- and tetraploid forms of mulberry.] - Izv. Akad. Nauk azerb. SSR, Ser. biol. Nauk *1975* (5) : 54 - 60, 1975. [Chloroplasts; in R, ab : Azerb.]

24702 - **TAMBIAN, N.N., MARTIROSYAN, I.A.** : O produktivnosti shtammov khlorelly, vydelennykh iz vodoemov Armyanskoĭ SSR. [Productivity of *Chlorella* strains isolated from water reservoirs of the Armenian S.S.R.] - Sb. nauch. Tr. *6* (Flora, Rastitel'nost' i Rastitel'nye Resursy Armyanskoĭ SSR) : 111 - 114, 1975. [In R, ab : Arm.]

24703 - **TAMBIEV, A.Kh.** : Svechenie vydeleniĭ sine-zelenoĭ vodorosli *Anabaena variabilis* v zavisimosti ot ee fiziologicheskogo sostoyaniya. [Luminescence of exudates of the blue-green alga *Anabaena variabilis* in dependence of its physiological state.] - Nauch. Dokl. vyssh. Shkoly, biol. Nauki *18* (2) : 72 - 74, 1975. [In R.]

24704 - **TAN, G.-Y., DUNN, G.M.** : Stomatal length, frequency, and distribution in *Bromus inermis* LEYSS. - Crop Sci. *15* : 283 - 286, 1975.

24705 - **TANAKA, I.** : The influence of season on the growth of leaf, stem and root. - - In : **MURATA, Y.** (ed.) : JIBP Synthesis. Vol. 11. Pp. 37 - 48. Univ. Tokyo Press, Tokyo 1975.

24706 - **TANAKA, M., HANIU, M., YASUNOBU, K.T., EVANS, M.C.W., RAO, K.K.** : The amino acid sequence of ferredoxin II from *Chlorobium limicola*, a photosynthetic green bacterium. - Biochemistry *14* : 1938 - 1943, 1975.

24707 - **TANAKA, M., HANIU, M., YASUNOBU, K.T., RAO, K.K., HALL, D.O.** : Modification of the automated sequence determination as applied to the sequence determination of the *Spirulina maxima* ferredoxin. - Biochemistry *14* : 5535 - 5540, 1975.

*24708 - **TANAKA, N., NAKANISHI, M., KADOTA, H.** : The excretion of photosynthetic product by natural phytoplankton population in Lake Biwa. - Jap. J. Limnol. *35* (3) : 91 - 98, 1974.

24709 - **TANAKA, Y., KATAYAMA, T.** : Comparative biochemistry of carotenoids in algae - V. Carotenoids in *Rhodomonas baltica* KARSTEN and *Nostoc commune* VANCHER. - Mem. Fac. Fish., Kagoshima Univ. *24* : 127 - 131, 1975.

24710 - **TANG, C.W., ALBRECHT, A.C.** : Chlorophyll-*a* photovoltaic cells. - Nature *254* : 507 - 509, 1975.

24711 - **TANG, C.W., ALBRECHT, A.C.** : Transient photovoltaic effects in metal-chlorophyll-*a*-metal sandwich cells. - J. chem. Phys. *63* : 953 - 961, 1975.

24712 - **TANG, C.W., DOUGLAS, F., ALBRECHT, A.C.** : Pulsed photoconductivity of chlorophyll-*a* films in contact with a nonpolar solution. - J. phys. Chem. *79* : 2723 - 2728, 1975.

24713 - **TANOUE, E., ARUGA, Y.** : Studies on the life cycle and growth of *Platymonas* sp. in culture. - Jap. J. Bot. *20* : 439 - 460, 1975. [Chl.]

24714 - **TĂRÂŢA, G.** : Posibilităţi de notare a intensităţii atacului de plum-pox la prun, prin determinarea conţinutului de clorofilă. [Possibilities of noting the intensity of plum pox attack by determination of chlorophyll content.] - Lucrările ştiinţ. Inst. Cercetări Pomicultură Piteşti *4* : 341 - 347, 1975. [In Roum., ab : E, F, R.]

24715 - **TARCHEVSKIĬ, I.A., CHIKOV, V.I., SULEĬMANOVA, A.Yu., IVANOVA, A.P., ANDRIYANOVA, Yu.E.** : Photosynthesis, assimilation numbers, and photosynthate translocation in wheat of different stalk length. - In : **NASYROV, Yu.S., ŠESTÁK, Z.** (ed.) : Genetic Aspects of Photosynthesis. Pp. 363 - 366. Dr. W. Junk B.V. Publ., The Hague 1975.

24716 - **TAYLOR, O.C., THOMPSON, C.R., TINGEY, D.T., REINERT, R.A.** : Oxides of nitrogen. - In : **MUDD, J.B., KOZLOWSKI, T.T.** (ed.) : Responses of Plants to Air Pollution. Pp. 121 - 139. Academic Press, New York - San Francisco - London 1975. [Ps, Chl.]

24717 - **TAYLOR, R.F., DAVIES, B.H.** : Gas-liquid chromatography of carotenoids and other terpenoids. - J. Chromatogr. *103* : 327 - 340, 1975.

24718 - **TAYLOR, S.E.** : Optimal leaf form. - In : **GATES, D.M., SCHMERL, R.B.** (ed.) : Perspectives of Biophysical Ecology. Pp. 73 - 86. Springer-Verlag, Berlin - - Heidelberg - New York 1975. [Energy balance.]

24719 - **TCHAN, Y.T., ROSEBY, J.E., FUNNELL, G.R.** : A new rapid specific bioassay method for photosynthesis inhibiting herbicides. - Soil Biol. Biochem. *7* : 39 - 44, 1975.

24720 - **TELFER, A., BARBER, J., NICOLSON, J.** : Availability of monovalent and divalent cations within intact chloroplasts for the action of ionophores nigericin and A23187. - Biochim. biophys. Acta *396* : 301 - 309, 1975.

24721 - **TELFER, A., BARBER, J., NICOLSON, J.** : Energy-dependent quenching of chlorophyll *a* fluorescence. Evidence for coupled cyclic electron flow in isolated chloroplasts. - Plant Sci. Lett. *5* : 171 - 176, 1975.

24722 - **TEL-OR, E., AVRON, M.** : Isolation and characterization of a factor which restores the Hill reaction from *Phormidium luridum*. - In : **AVRON, M.** (ed.) : Proceedings of the Third International Congress on Photosynthesis. Vol. I. Pp. 569 - 578. Elsevier, Amsterdam - Oxford - New York 1975.

24723 - **TEL-OR, E., CAMMACK, R., HALL, D.O.** : Immunological comparison of ferredoxins. - FEBS Lett. *53* : 135 - 138, 1975.

24724 - **TEL-OR, E., STEWART, W.D.P.** : Manganese and photosynthetic oxygen evolution by algae. - Nature *258* : 715 - 716, 1975.

24725 - **TENHUNEN, J.D., GATES, D.M.** : Light intensity and leaf temperature as determining factors in diffusion resistance. - In : **GATES, D.M., SCHMERL, R.B.** (ed.) : Perspectives of Biophysical Ecology. Pp. 213- 225. Springer-Verlag, Berlin - Heidelberg - New York 1975.

24726 - **TENNANT, D.** : A test of a modified line intersect method of estimating root length. - J. Ecol. *63* : 995 - 1001, 1975.

24727 - **TERPSTRA, W.** : Investigation of lamellar structure in green leaves and diatoms by means of the study of chlorophyllase activity. - In : **AVRON, M.** (ed.) : Proceedings of the Third International Congress on Photosynthesis. Vol. III. Pp. 2125 - 2130. Elsevier, Amsterdam - Oxford - New York 1975.

24728 - **TERPSTRA, W.** : Chlorophyllase and lamellar structure in *Phaeodactylum tricornutum*. II. Conversion of added chlorophyll into chlorophyllide by small lamellar fragments. - Z. Pflanzenphysiol. *75* : 405 - 414, 1975.

24729 - **TERPSTRA, W., GOEDHEER, J.C.** : Chlorophyllase and lamellar structure in *Phaeodactylum tricornutum*. I. Chlorophyll ⟶ chlorophyllide conversion within the lamellae. - Z. Pflanzenphysiol. *75* : 118 - 130, 1975.

24730 - **TERRY, N., HUSTON, R.P.** : Effects of calcium on the photosynthesis of intact leaves and isolated chloroplasts of sugar beets. - Plant Physiol. *55* : 923 - - 927, 1975.

24731 - **TETLEY, R.M., THIMANN, K.V.** : The metabolism of oat leaves during senescence. IV. The effects of α,α'-dipyridyl and other metal chelators on senescence. - - Plant Physiol. *56* : 140 - 142, 1975.

24732 - **TETT, P., COTTRELL, J.C., TREW, D.O., WOOD, B.J.B.** : Phosphorus quota and the chlorophyll : carbon ratio in marine phytoplankton. - Limnol. Oceanogr. *20* : 587 - 603, 1975.

24733 - **TETT, P., KELLY, M.G., HORNBERGER, G.M.** : A method for the spectrophotometric measurement of chlorophyll *a* and pheophytin *a* in benthic microalgae. - Limnol. Oceanogr. *20* : 887 - 896, 1975.

24734 - **TEW, J., CRESSWELL, C.F., FAIR, P.** : The effect of nitrate and ammonia on the initial carboxylation product and the abundance of microbodies in some C_4 plants and *Hordeum vulgare* L. - In : **AVRON, M.** (ed.) : Proceedings of the Third International Congress on Photosynthesis. Vol. II. Pp. 1249 - 1266. Elsevier, Amsterdam - Oxford - New York 1975. [Assimilation chamber.]

24735 - **THEIDE, B., MENDE, D., WIESSNER, W.** : Structural and photosynthetic differences between autotrophically and photoheterotrophically cultivated *Chlamydobotrys stellata* and their regulation *in vivo*. - In : Les Cycles Cellulaires et leur Blocage chez Plusieurs Protistes. - Coll. int. CNRS 240. Pp. 115 - 121. Edit. CNRS, Paris 1975.

24736 - **THEODÓRSSON, P.** : The study of ^{14}C penetration into filters in primary productivity measurements using double side counting. - Limnol. Oceanogr. *20* : 288 - 291, 1975.

24737 - **THIBAULT, P., MICHEL, A.** : Analysis of photosynthetic induction phenomena in *Zea mays* under anaerobic conditions : ATP synthesis and fluorescence changes. - In : **AVRON, M.** (ed.) : Proceedings of the Third International Congress on Photosynthesis. Vol. II. Pp. 957 - 966. Elsevier, Amsterdam - Oxford - New York 1975.

24738 - **THIEN, W., SCHOPFER, P.** : Control by phytochrome of cytoplasmic and plastid rRNA accumulation in cotyledons of mustard seedlings in the absence of photosynthesis. - Plant Physiol. *56* : 660 - 664, 1975.

24739 - **THOFELT, L.** : Studies on leaf temperature recorded by direct measurement and by thermography. - Acta Univ. upsaliensis *12* : 1 - 143, 1975.

24740 - **THOM, A.S.** : Momentum, mass and heat exchange of plant communities. - In : **MONTEITH, J.L.** (ed.) : Vegetation and the Atmosphere. Vol. 1. Principles. Pp. 57 - 109. Academic Press, London - New York - San Francisco 1975.

*24741 - **THOMANN, R.V., WINFIELD, R.P., DI TORO, D.M.** : Modeling of phytoplankton in Lake Ontario (IFYGL). - In : Proceedings of the 17th Conference on Great Lakes Research. Pp. 135 - 149. Int. Ass. Great Lakes Res. 1974. [Chl.]

24742 - **THOMAS, D.A.** : Stomata. - In : **BAKER, D.A., HALL, J.L.** (ed.) : Ion Transport in Plant Cells and Tissues. Pp. 377 - 412. North-Holland Publ. Comp., Amsterdam - Oxford, Amer. Elsevier Publ. Comp., New York 1975. [Ps.]

*24743 - **THOMAS, E.A.** : Fischsterben in Seeabflüssen durch Hyperphotosynthese. - Verh. int. Verein. theor. angew. Limnol. *18* : 454 - 460, 1972.

24744 - **THOMAS, H.** : The growth responses to weather of simulated vegetative swards of a single genotype of *Lolium perenne*. - J. agr. Sci. *84* : 333 - 343, 1975.

24745 - **THOMAS, H.** : Regulation of alanine aminotransferase in leaves of *Lolium temulentum* during senescence. - Z. Pflanzenphysiol. *74* : 208 - 218, 1975. [Chl.]

24746 - **THOMAS, H., STODDART, J.L.** : Separation of chlorophyll degradation from other senescence processes in leaves of a mutant genotype of meadow fescue (*Festuca pratensis* L.). - Plant Physiol. *56* : 438 - 441, 1975.

24747 - **THOMAS, J.B., van BEEK, J.H.G.M., KLEINEN-HAMMANS, J.W., van ZANTEN, G.A.** : On the effect of pH and ion concentration on chlorophyll spectra *in vivo*. - In : **AVRON, M.** (ed.) : Proceedings of the Third International Congress on Photosynthesis. Vol. III. Pp. 1963 - 1968. Elsevier, Amsterdam - Oxford - New York 1975.

24748 - **THOMAS, J.F., RAPER, C.D., Jr., ANDERSON, C.E., DOWNS, R.J.** : Growth of young tobacco plants as affected by carbon dioxide and nutrient variables. - Agron. J. *67* : 685 - 689, 1975.

24749 - **THOMAS, J.P., O'KELLEY, J.C., HARDMAN, J.K., ALDRIDGE, E.F.** : Flavin as an active component of the photoreversible pigment system of the green alga *Protosiphon botryoides* KLEBS. - Photochem. Photobiol. *22* : 135 - 138, 1975.

24750 - **THOMAS, J.R., KOLODNER, R., TEWARI, K.K.** : Molecular size and the information content of chloroplast DNAs from higher plants. - In : **NASYROV, Yu.S., ŠESTÁK, Z.** (ed.) : Genetic Aspects of Photosynthesis. Pp. 9 - 30. Dr. W. Junk B.V. Publ., The Hague 1975.

24751 - **THOMAS, P., JANAVE, M.T.** : Effects of *gamma* irradiation and storage temperature on carotenoids and ascorbic acid content of mangoes on ripening. - J. Sci. Food Agr. *26* : 1503 - 1512, 1975.

24752 - **THOMAS, S.M., THORNE, G.N.** : Effect of nitrogen fertilizer on photosynthesis and ribulose 1,5-diphosphate carboxylase activity in spring wheat in the field. - J. exp. Bot. *26* : 43 - 51, 1975.

24753 - **THOMPSON, P.J., HABER, A.H., ARGENTIERI, T.M.** : Light regulation of chloroplast development. - Plant Physiol. *56* (Suppl.) : 32, 1975.

24754 - **THOMSON, W.W.** : Effects of air pollutants on plant ultrastructure. - In : **MUDD, J.B., KOZLOWSKI, T.T.** (ed.) : Responses of Plants to Air Pollution. Pp. 179 - 194. Academic Press, New York - San Francisco - London 1975. [Chloroplast.]

24755 - **THOMSON, W.W.** : The structure and function of salt glands. - In : **POLJAKOFF-MAYBER, A., GALE, J.** (ed.) : Plants in Saline Environments. Pp. 118 - 146. Springer-Verlag, Berlin - Heidelberg - New York 1975. [Chloroplast.]

24756 - **THORNBER, J.P.** : Chlorophyll-proteins: Light-harvesting and reaction center components of plants. - Annu. Rev. Plant Physiol. *26* : 127 - 158, 1975.

24757 - **THORNE, S.W., HORVATH, G., KAHN, A., BOARDMAN, N.K.** : Light-dependent absorption and selective scattering changes at 518 nm in chloroplast thylakoid membranes. - Proc. nat. Acad. Sci. USA *72* : 3858 - 3862, 1975.

24758 - **TIESZEN, L.L.** : CO_2 exchange in the Alaskan arctic tundra : seasonal changes in the rate of photosynthesis of four species. - Photosynthetica *9* : 376 - - 390, 1975.

24759 - **TIESZEN, L.L., JOHNSON, D.A.** : Seasonal pattern of photosynthesis in individual grass leaves and other plant parts in arctic Alaska with a portable $^{14}CO_2$ system. - Bot. Gaz. *136* : 99 - 105, 1975.

24760 - **TIKHONOV, A.N., RUUGE, E.K.** : Issledovanie ėlektronnogo transporta v fotosinteticheskikh sistemakh vysshikh rasteniĭ metodom ĖPR. I. Vliyanie predystorii osveshcheniya na kinetiku fotoindutsirovannykh okislitel'no-vosstanovitel'-nykh prevrashcheniĭ *P*700. [ESR study of electron transport in photosynthetic systems of higher plants. I. Effect of preillumination history on the kinetics of photoinduced oxidative-reductive transformations of *P*700.] - Biofizika *20*: 1049 - 1053, 1975. [In R, ab : E.]

24761 - **TIKHONOV, A.N., RUUGE, E.K.** : Issledovanie ėlektronnogo transporta v fotosinteticheskikh sistemakh vysshikh rasteniĭ metodom ĖPR. II. Vliyanie temperatury na kinetiku fotoindutsirovannykh okislitel'no-vosstanovitel'nykh prevrashcheniĭ *P*700. [ESR study of electron transport in photosynthetic systems of higher plants. II. Effect of temperature on the kinetics of photoinduced oxidative-reductive transformations of *P*700.] - Biofizika *20* : 1054 - 1058, 1975. [In R, ab : E.]

24762 - **TIKHONOV, A.N., RUUGE, E.K., SUBCHINSKIĬ, V.K., BLYUMENFEL'D, L.A.** : Issledovanie kinetiki ėlektronnogo transporta i khromaticheskikh perekhodov v izolirovannykh khloroplastakh metodom ĖPR. [Kinetics of electron transport and chromatic transitions in isolated chloroplasts studied by EPR.] - Fiziol. Rast. *22* : 5 - 15, 1975. [In R, ab : E.]

24763 - **TILLEY, L.J., HAUSHILD, W.L.** : Net primary productivity of periphytic algae in the intertidal zone, Duwamish River estuary, Washington. - J. Res. US Geol. Survey *3* : 253 - 259, 1975.

24764 - **TILLEY, L.J., HAUSHILD, W.L.** : Use of productivity of periphyton to estimate water quality. - J. Water Pollut. Cont. Fed. *47* : 2157 - 2171, 1975.

24765 - **TILTON, D.L., BERNARD, J.M.** : Primary productivity and biomass distribution in an alder shrub ecosystem. - Amer. Midland Naturalist *94* : 251 - 256, 1975.

24766 - **TILZER, M.M., GOLDMAN, C.R., de AMEZAGA, E.** : The efficiency of photosynthetic light energy utilization by lake phytoplankton. - Verh. int. Verein. Limnol. *19* : 800 - 807, 1975.

24767 - **TINDALL, D.R., YOPP, J.H., SCHMID, W.E., MILLER, D.M.** : Amino acids of bulk protein, phycocyanin, and soluble fraction of *Aphanothece halophytica*. - Plant Physiol. *56* (Suppl.) : 68, 1975.

24768 - **TISCHNER, R., LORENZEN, H.** : Physiologische Auswirkungen von Hitzeschocks auf synchrone Chlorellen im empfindlichsten Entwicklungsstadium. - Biochem. Physiol. Pflanzen *168* : 233 - 245, 1975.

24769 - **TISHCHENKO, N.N., MAGOMEDOV, I.M.** : Aktivnost' i lokalizatsiya alanin- i aspartataminotransferaz u rasteniĭ s C_4-fotosintezom. [Activity and localization of alanine- and aspartateaminotransferases in plants with C_4-photosynthesis.] - Dokl. Akad. Nauk SSSR *225* : 463 - 465, 1975. [In R.]

24770 - **TITLYANOV, E.A., GLEBOVA, N.T., KOTLYAROVA, L.S.** : Sezonnye izmeneniya v stroenii tallomov *Ulva fenestrata* P. *et* R. [Seasonal changes in the thallus structure of *Ulva fenestrata* P. *et* R.] - Ekologiya *1975* (4) : 36 - 41, 1975. [In R.]

24771 - **TITOVA, E.N., GRIGOR'EV, V.G.** : Regulirovanie raspredeleniya plasticheskikh veshchestv v khmele. [Regulation of the distribution of plastic substances in hop.] - Sel'skokhoz. Biol. *10* : 622 - 623, 1975. [Photosynthates; in R.]

24772 - **TIȚU, H., TĂNASE, V.** : Influența sursei de azot asupra ultrastructurii cloroplastelor din frunzele de porumb (*Zea mays* L.). [Effect of nitrogen source on the ultrastructure of maize leaf chloroplasts.] - Stud. Cercet. Biol., Ser. Biol. veg. *27* : 171 - 175, 1975. [In Roum., ab : E.]

24773 - **TITUS, J., GOLDSTEIN, R.A., ADAMS, M.S., MANKIN, J.B., O'NEIL, R.V., WEILER, P.R., Jr., SHUGART, H.H., BOOTH, R.S.** : A production model for *Myriophyllum spicatum* L. - Ecology *56* : 1129 - 1138, 1975.

24774 - **TJEPKEMA, J.D., YOCUM, C.S.** : Steady-state measurements of respiration rate with an oxygen electrode. - Anal. Biochem. *63* : 341 - 344, 1975.

*24775 - **TKACHUK, K.S., AĖROV, Ǐ.L.** : Vplyv posukhy na optychni vlastyvosti lystkiv ozymoǐ pshenytsi. [Effect of drought on leaf optical properties in winter wheat.]- Dopovidi Akad. Nauk ukr. RSR, Ser. B *1974* : 1122 - 1125, 1153, 1974. [In Ukr., ab : E, R.]

24776 - **TKACHUK, K.S., GULYAEV, B.I., PETRENKO, N.I.** : Vliyanie azotnogo pitaniya na vodoobmen, fotosintez i produktivnost' ozimoǐ pshenitsy. [Influence of nitrogen nutrition on water exchange, photosynthesis and productivity of winter wheat.] - Dokl. Akad. Nauk ukr.SSR, Ser. B *1975* : 650 - 653, 1975. [In R, ab: E.]

24777 - **TODOROVA-TRIFONOVA, A.D.** : Karotinoidni pigmenti pri *Scenedesmus acutus* MEYEN identifitsirani chrez t"nkosloǐna khromatografiya. [Identification of carotenoid pigments in *Scenedesmus acutus* MEYEN by thin-layer chromatography.] - Nauch. Tr. plovdiv. Univ. Paisiǐ Khilendarski, Biol. *13* (5) : 245 - 252, 1975. [In Bulg., ab : E, R.]

24778 - **TODOROVA-TRIFONOVA, A.D., KHAZOVA, Ǐ.V.** : Vliyanie na khloramfenikola i tetratsiklina v"rkhu plastidnite pigmenti na *Scenedesmus acutus* MEYEN kultiviran v ots"stvie na fosfor. [Effect of chloramphenicol and tetracycline on plastid pigments in *Scenedesmus acutus* MEYEN in the absence of phosphorus.] - Nauch. Tr. plovdiv. Univ. Paisiǐ Khilendarski, Biol. *13* (4) : 249 - 260, 1975. [In Bulg., ab : E, R.]

24779 - **TOLBERT, N.E., RYAN, F.J.** : Glycolate biosynthesis by ribulose diphosphate carboxylase/oxygenase. - In : **AVRON, M.** (ed.) : Proceedings of the Third International Congress on Photosynthesis. Vol. II. Pp. 1303 - 1319. Elsevier, Amsterdam - Oxford - New York 1975.

24780 - **TOMBESI, L.** : Rapporti acqua-terreno, evapotraspirazione potenziale, bilanci energetici ed idrologici nell'ambiente climatico di Roma. [Water-soil relationships, potential evapotranspiration, energetic and hydrologic balances in the climatic environment of Rome.] - Ann. Ist. sperim. Nutr. Piante *6* (1): 3 - 75, 1975. [Growth analysis; in Ital., ab : E, F, G.]

24781 - **TOMBESI, L., CALÈ, M.T., FIGLIOLIA, A., IZZA, C., De ROSSI, C.** : Studi sulla nutrizione idrica e minerale del frumento (*Triticum vulgare* cv. Victor). Nota 1. - Effetti esercitati dai principali fattori ambientali su alcuni processi fisiologici. [Studies on water and mineral nutrition of wheat (*Triticum vulgare* cv. Victor). Note 1. - Effects of main environmental factors on some physiological processes.]-Ann. Ist. sperim. Nutr. Piante *6* (5) : 3 - 27, 1975. [Ps; in Ital., ab : E, F, G, Span.]

24782 - **TOMBESI, L., GIANCARLINI, G., FAVOLA, G., MORETTI, R., BARONI, R.** : Energetic and hydrologic balances and automation of an irrigation plant. - Ann. Ist. sperim. Nutr. Piante *6* (2) : 3 - 31, 1975. [Radiation in canopy.]

24783 - **TOMKIEWICZ, M., CORKER, G.** : Chlorophyll cation radical in phospholipid vesicles. - In : **AVRON, M.** (ed.) : Proceedings of the Third International Congress on Photosynthesis. Vol. I. Pp. 265 - 272. Elsevier, Amsterdam - Oxford - New York 1975.

24784 - **TOMKIEWICZ, M., CORKER, G.A.** : Chlorophyll sensitized charge separation in phospholipid vesicles. - Photochem. Photobiol. *22* : 249 - 256, 1975.

24785 - **TONECKI, J.** : Changes of respiration intensity and chlorophyll content in needles of Norway spruce (*Picea abies* L. KARST) seedlings treated with 2,4,5-T and dalapon. - Acta agrobot. *28* : 177 - 195, 1975.

24786 - **TORBICKI, H., RENK, H.** : Utilization of Gibbs triangle in the study of the primary production dependence on phytoplankton biomass and insolation. - Pol. Arch. Hydrobiol. *22* : 399 - 412, 1975.

24787 - **TRACHTENBERG, C.H., McCLOUD, D.E.** : Net photosynthesis of peanut leaves at varying light intensities and leaf ages. - Soil Crop Sci. Soc. Florida Proc. *35*: 54 - 55, 1975.

24788 - **TRAVIS, D.M., STEWART, K.D., WILSON, K.G.** : Nuclear and cytoplasmic chloroplast mutants induced by chemical mutagens in *Mimulus cardinalis* : Genetics and ultrastructure. - Theor. appl. Genet. *46* (2) : 67 - 77, 1975.

24789 - **TREBST, A.** : Electron flow across the thylakoid membrane. Native and artificial energy conserving sites in photosynthetic electron flow. - In : **AVRON, M.** (ed.) : Proceedings of the Third International Congress on Photosynthesis. Vol. I. Pp. 439 - 448. Elsevier, Amsterdam - Oxford - New York 1975.

24790 - **TREBST, A., REIMER, S., HAUSKA, G.** : Electron and proton shuttles across the photosynthetic membrane. - In : **QUAGLIARIELLO, E., PAPA, S., PALMIERI, F., SLATER, E.C., SILIPRANDI, N.** (ed.) : Electron Transfer Chains and Oxidative Phosphorylation. Pp. 343 - 350. North-Holland Publ. Comp., Amsterdam 1975.

24791 - **TREBST, A., WIETOSKA, W.** : Wirkungsmechanismus und Struktur-Aktivitätsbeziehungen des Aminotriazinon-Herbizids Metribuzin. Hemmung des photosynthetischen Elektronentransports von Chloroplasten durch Metribuzin. - Z. Naturforsch. *30 C*: 499 - 504, 1975.

24792 - **TREFFRY, T.** : Developmental changes in the surface properties of chloroplast membranes. - Planta *126* : 11 - 17, 1975.

24793 - **TREFFRY, T., ALBERTSSON, P.-Å.** : Studies of etioplast and developing chloroplast membranes using polymer two-phase systems. - In : **AVRON, M.** (ed.) : Proceedings of the Third International Congress on Photosynthesis. Vol. III. Pp. 1661 - 1666. Elsevier, Amsterdam - Oxford - New York 1975.

24794 - **TREGUBENKO, M.Ya., FILIPPOV, G.L., VISHNEVSKIĬ, N.V.** : Fotosinteticheskiĭ potentsial kukuruzy pri uluchshenii usloviĭ vyrashchivaniya. [Photosynthetic potential of maize under improved growth conditions.] - Byull. vses. nauch.-issled. Inst. Kukuruzy *1972* (2 [25]) : 13 - 16, 1972. [Growth analysis; in R.]

24795 - **TREHARNE, K.J., NELSON, C.J.** : Effect of growth temperature on photosynthetic and photo-respiratory activity in tall fescue. - In : **MARCELLE, R.** (ed.) : Environmental and Biological Control of Photosynthesis. Pp. 61 - 69. Dr. W. Junk b.v. Publ., The Hague 1975.

*24796 - **TREIBS, A.** : On the chromophores of porphyrin systems. - Ann. New York Acad. Sci. *206* : 97 - 115, 1973. [Chl.]

24797 - **TREICHEL, S.** : Crassulaceensäurestoffwechsel bei einem salztoleranten Vertreter der *Aizoaceae : Aptenia cordifolia*. - Plant Sci. Lett. *4* : 141 - 144, 1975.

24798 - **TRENBATH, B.R., ANGUS, J.F.** : Leaf inclination and crop production. - Field Crop Abstr. *28* : 231 - 244, 1975.

*24799 - **TRENCH, R.K.** : Nutritional potentials in *Zoanthus sociathus (Coelenterata, Anthozoa)*. - Helgoländer wiss. Meeresuntersuch. *26* : 174 - 216, 1974. [Ps.]

24800 - **TRENCH, R.K.** : Of "leaves that crawl": functional chloroplasts in animal cells. - Symp. Soc. exp. Biol. *29* (Symbiosis) : 229 - 265, 1975.

24801 - **TRETYAK, T.V., OKANENKO, A.S.** : Vliyanie zasukhi na vodnyĭ rezhim u razlichnykh po ėkologicheskomu proiskhozhdeniyu sortov sakharnoĭ svekly. [Effect of drought on water regime in varieties of sugar beet differing by ecological origin.] - Fiziol. Biokhim. kul't. Rast. *7* : 35 - 42, 1975. [Chloroplast; in R, ab : E.]

24802 - **TRIBE, M.A., ERAUT, M.R., SNOOK, R.K.** : Photosynthesis. - Basic Biology Course. Vol. 6. Cambridge Univ. Press. Cambridge - London - New York - Melbourne 1975.

24803 - **TRIFONOVA, I.S.** : Pokazateli produktivnosti fitoplanktona v ozerakh raznogo tipa. [Indices of phytoplankton productivity in lakes of various type.] - In: Osnovy Bioproduktivnosti Vnutrennikh Vodoemov Pribaltiki. Pp. 184 - 185. Vil'nyus 1975. [In R.]

24804 - **TROUGHTON, J.H.** : Photosynthetic mechanisms in higher plants. - In : **COOPER, J.P.** (ed.) : Photosynthesis and Productivity in Different Environments. Pp. 357 - 391. Cambridge Univ. Press, Cambridge - London - New York - Melbourne 1975.

24805 - **TROUGHTON, J.H.** : Light level and the mean speed of translocation in *Zea mays* leaves. - In : **MARCELLE, R.** (ed.) : Environmental and Biological Control of Photosynthesis. Pp. 373 - 385. Dr. W. Junk b.v. Publ., The Hague 1975.

24806 - **TROUGHTON, J.H., CARD, K.A.** : Temperature effects on the carbon-isotope ratio of C_3, C_4 and Crassulacean-acid-metabolism (CAM) plants. - Planta *123* : 185 - - 190, 1975.

24807 - **TROUGHTON, J.H., CARD, K.A.** : Application of carbon isotope fractionation by plants to ecosystem investigations. - What's new Plant Physiol. *7* (9) : 1 - 5, 1975.

24808 - **TROUGHTON, J.H., FORK, D.C., CHANG, F.H.** : Environmental effects on the membrane associated electron transport reactions of photosynthesis. - In : **MARCELLE, R.** (ed.) : Environmental and Biological Control of Photosynthesis. Pp. 387 - 403. Dr. W. Junk b.v. Publ., The Hague 1975.

24809 - **TROXLER, R.F., BROWN, A.S.** : Metabolism of δ-aminolevulinic acid in red and blue-green algae. - Plant Physiol. *55* : 463 - 467, 1975.

24810 - **TROXLER, R.F., FOSTER, J.A., BROWN, A.S., FRANZBLAU, C.** : The α and β subunits of *Cyanidium caldarium* phycocyanin : properties and amino acid sequences at the amino terminus. - Biochemistry *14* : 268 - 274, 1975.

24811 - **TRUKHAN, Ė.M.** : Kvantovoėlektronnyĭ podkhod k izucheniyu pervichnoĭ stadii fotosinteza. [Quantum electron approach to studying primary stages of photosynthesis.] - Probl. kosm. Biol. *31* (Upravlenie Fiziologicheskimi Protsessami i Ikh Modelirovanie) : 235 - 256, 1975. [In R.]

24812 - **TRUKHAN, Ė.M., DERYABKIN, V.N., KIREEV, V.B.** : Ėksperimental'noe issledovanie fizicheskikh kharakteristik pervichnoĭ stadii fotosinteza. [Experimental studies of physical characteristics of primary stage of photosynthesis.] - Probl. kosm. Biol. *31* (Upravlenie Fiziologicheskimi Protsessami i Ikh Modelirovanie): 256 - 272, 1975. [In R.]

24813 - **TRUKHAN, Ė.M., KIREEV, V.B.** : Ob odnoĭ kooperativnoĭ modeli pervichnoĭ stadii fotosinteza. [Co-operative model of primary stage of photosynthesis.] - Probl. kosm. Biol. *31* (Upravlenie Fiziologicheskimi Protsessami i Ikh Modelirovanie): 273 - 293, 1975. [In R.]

24814 - **TRÜPER, H.G.** : The enzymology of sulfur metabolism in phototrophic bacteria - a review. - Plant Soil *43* : 29 - 39, 1975.

24815 - **TSANG, M.L.-S., SCHIFF, J.** : Studies of sulfate utilization by algae 14. Distribution of adenosine-3'-phosphate-5'-phosphosulfate (PAPS) and adenosine--5'-phosphosulfate (APS) sulfotransferases in assimilatory sulfate reducers. - Plant Sci. Lett. *4* : 301 - 307, 1975. [Ps.]

24816 - **TSANG, M.L.-S., SCHIFF, J.A.** : Two patterns of assimilatory sulfate reduction in photosynthetic and non-photosynthetic organisms. - Plant Physiol. *56* (Suppl.) : 36, 1975.

24817 - **TSEL'NIKER, Yu.L.** : Vliyanie intensivnosti sveta na chislo i razmery khloroplastov u drevesnykh porod. [Effect of irradiance on the number and dimensions of chloroplasts in trees.] - Fiziol. Rast. *22* : 262 - 269, 1975. [In R, ab : E.]

24818 - **TSEL'NIKER, Yu.L.** : Vliyanie intensivnosti sveta na opticheskie svoĭstva khloroplastov i tkaneĭ list'ev drevesnykh porod. [Effect of illuminance on optical properties of chloroplasts and leaf tissues of trees.] - Fiziol. Rast. *22* : 695 - 701, 1975. [In R, ab : E.]

24819 - **TSUBO, Y.** : Formation of colorless cells in algae. - In : **TOKIDA, J., HIROSE, H.** (ed.) : Advance of Phycology in Japan. Pp. 180 - 193. Dr. W. Junk b.v. Publ., The Hague 1975. [Chloroplast.]

24820 - **TSUNO, Y.** : [The influence of transpiration upon the photosynthesis in several crop plants.] - Proc. Crop Sci. Soc. Jap. *44* : 44 - 53, 1975. [In Jap., ab : E.]

24821 - **TSUNO, Y., SATO, T., MIYAMOTO, H., HARADA, N.** : [Studies on CO_2 uptake and CO_2 evolution in each part of crop plants. II. Photosynthetic activity in the leaf sheath and ear of rice plant.] - Proc. Crop Sci. Soc. Jap. *44* : 287 - - 292, 1975. [In Jap., ab : E.]

24822 - **TSUNODA, S., YAMAGUCHI, K.** : Regression of grain yield on leaf area index of rice. - In : **MURATA, Y.** (ed.) : JIBP Synthesis. Vol. 11. Pp. 145 - 149. Univ. Tokyo Press, Tokyo 1975.

24823 - **TSUSHIMOTO, G., KIKUCHI, T., ISHIDA, M.R., MATSUBARA, T.** : A simple and rapid method of cell disruption for chloroplast isolation from *Euglena* cells. - Annu. Rep. Res. Reactor Inst. Kyoto Univ. *8* : 121 - 123, 1975.

24824 - **TU, J.C.** : Localization of infectious soybean mosaic virus in mottled soybean seeds. - Microbios *14* : 151 - 156, 1975. [Chl.]

24825 - **TUMERMAN, L.** : Comparative phenomena in photosynthesis. - In : **AVRON, M.** (ed.) : Proceedings of the Third International Congress on Photosynthesis. Vol. I. Pp. 195 - 198. Elsevier, Amsterdam - Oxford - New York 1975.

24826 - **TUNSTALL, B.R., CONNOR, D.J.** : Internal water balance of brigalow (*Acacia harpophylla* F. MUELL.) under natural conditions. - Aust. J. Plant Physiol. *2* : 489 - 499, 1975. [Ps.]

24827 - **TUQUET, C., GUILLOT-SALOMON, T.** : Modifications dela composition en lipides polaires des plastes isolés de feuilles d'Orge étiolées soumises à l'action d'éclair répétés. - Compt. rend. Acad. Sci. Paris, Sér. D *280* : 2853 - 2856, 1975.

24828 - **TUQUET, C., LUBAC, M. de** : Action de la phospholipase *A* sur l'enveloppe des chloroplastes d'Epinard. - Compt. rend. Acad. Sci. Paris, Sér. D *281* : 1313- - 1316, 1975.

24829 - **TURČIĆ, M., CANIĆ, V.** : Razdvajanje karotenoida iz lishća spanaća khromatografijom na tankom sloju skroba. [Separation of spinach leaf carotenoids by thin-layer chromatography on starch.] - Zbor. prir. Nauke, Matitsa srpska *49* : 152 - 158, 1975. [In Serbian, ab : E.]

24830 - **TURGEON, R., WEBB, J.A.** : Leaf development and phloem transport in *Cucurbita pepo* : carbon economy. - Planta *123* : 53 - 62, 1975. [Ps.]

24831 - **TURGEON, R., WEBB, J.A., EVERT, R.F.** : Ultrastructure of minor veins in *Cucurbita pepo* leaves. - Protoplasma *83* : 217 - 232, 1975. [Phloem transport.]

24832 - **TURISHCHEVA, M.S.** : Chloroplast DNA-membrane complex in *Pisum sativum*. - In : **NASYROV, Yu.S., ŠESTÁK, Z.** (ed.) : Genetic Aspects of Photosynthesis. Pp. 1 - - 8. Dr. W. Junk B.V. Publ., The Hague 1975.

24833 - **TÜRK, R., WIRTH, V.** : Über die SO_2-Empfindlichkeit einiger Moose. - Bryologist *78* : 187 - 193, 1975. [Ps.]

24834 - **TÜRK, R., WIRTH, V.** : The pH dependence of SO_2 damage to lichens. - Oecologia *19* : 285 - 291, 1975. [Ps.]

24835 - **TURNER, B.D.** : Energy flow in arboreal epiphytic communities. An empirical model of net primary productivity in the alga *Pleurococcus* on larch trees. - Oecologia *20* : 179 - 188, 1975.

24836 - **TURNER, J.F., TURNER, D.H.** : The regulation of carbohydrate metabolism. - Annu. Rev. Plant Physiol. *26* : 159 - 186, 1975.

24837 - **TURNER, N.C., JARVIS, P.G.** : Photosynthesis in Sitka spruce (*Picea sitchensis* (BONG.) CARR. IV. Response to soil temperature. - J. appl. Ecol. *12* : 561 - - 576, 1975.

24838 - **TYLER, J.E.** : The *in situ* quantum efficiency of natural phytoplankton populations. - Limnol. Oceanogr. *20* : 976 - 980, 1975.

24839 - **TYREE, M.T.** : Some inconsistencies that arise when the Canny-Phillips model of phloem translocation is applied to known ^{14}C-assimilate profiles in plant stems and petioles. - Can. J. Bot. *53* : 1128 - 1131, 1975.

24840 - **TYSZKIEWICZ, E., ROUX, E.** : Effects of low temperatures on formation and conservation of high energy state (X_e) appearing in spinach chloroplasts during the light step of the two-stage phosphorylation : I. - Formation of X_e below 0 °C. II. - Conservation of X_e between -30 °C and -196 °C. - Biochem. biophys. Res. Commun. *65* : 1400 - 1408, 1975.

24841 - **UCHIJIMA, Z.** : Dry matter production of crops in relation to climatic conditions. - In : **MURATA, Y.** (ed.) : JIBP Synthesis. Vol. 11. Pp. 86 - 104. Univ. Tokyo Press, Tokyo 1975.

24842 - **UEHARA, Y., OGAWA, T., SHIBATA, K.** : Effects of abscisic acid and its derivatives on stomatal closing. - Plant Cell Physiol. *16* : 543 - 546, 1975. [Stomatal resistance.]

24843 - **ULIK, T.M.** : Adaptation of *Cyanidium caldarium* to variation in light intensity and CO_2 stress. - J. Phycol. *11* (Suppl.) : 15, 1975.

24844 - **UMRIKHINA, A.V., BUBLICHENKO, N.V., BEGICHEV, V.N., KRASNOVSKIĬ, A.A.** : Fotoobrazovanie svobodnykh radikalov v sisteme bakteriokhlorofill-*p*-benzokhinon. [Photoformation of free radicals in the system bacteriochlorophyll-*p*-benzoquinone.] - Dokl. Akad. Nauk SSSR *221* : 974 - 977, 1975. [In R.]

*24845 - **UPITIS, V.V., NOLLENDORF, A.F., PAKALNE, D.S.** : Maloizuchennye mikroėlementy v kul'ture khlorelly. Vanadiĭ. [Little known microelements in *Chlorella* culture. Vanadium.] - Latv. PSR Zinatnu Akad. Vestis [Izv. Akad. Nauk latv. SSR] *1974* (11) : 15 - 22, 1974. [Chl; in R, ab : E.]

24846 - **URIBE, E.G.** : Protection of chloroplasts from energy-dependent inactivation of photophosphorylation by EDTA. - In : **AVRON, M.** (ed.) : Proceedings of the Third International Congress on Photosynthesis. Vol. II. Pp. 1073 - 1079. Elsevier, Amsterdam - Oxford - New York 1975.

24847 - **USMANOV, P.D., ABDULLAEV, H.A., USMANOVA, O.V., SOKHIBNAZAROV, Sh.** : Mutation variability of chloroplasts in *Arabidopsis thaliana* (L.) HEYNH. - In : **NASYROV, Yu.S., ŠESTÁK, Z.** (ed.) : Genetic Aspects of Photosynthesis. Pp. 189 - 201. Dr. W. Junk B.V. Publ., The Hague 1975.

24848 - **USMANOV, P.D., ABDULLAEV, Kh.A., PINKHASOV, Yu.I., BIKASIYAN, G.R.** : Genetika, struktura i funktsiya plastid pestrolistnykh rasteniĭ arabidopsisa i khlopchatnika. [Genetics, structure and function of plastids in *Arabidopsis* and cotton plants with variegated leaves.] - Genetika *11* (4) : 22 - 29, 1975. [In R, ab : E.]

24849 - **USTIMENKO, G.V., POPOV, V.P., VASIL'EVA, V.N.** : Fotosinteticheskaya deyatel'nost' rasteniĭ vigny v usloviyakh razlichnogo zagushcheniya. [Photosynthetic activity of cowpea plants under varying density.] - Sel'skokhoz. Biol. *10* : 936 - 938, 1975. [Growth analysis; in R.]

24850 - **USUDA, H., KANAI, R., MIYACHI, S.** : Carbon dioxide assimilation and photosystem II deficiency in bundle sheath strands isolated from C_4 plants. - Plant Cell Physiol. *16* : 485 - 494, 1975.

24851 - **UTSUNOMIYA, H., YAMAGATA, M., DOI, Y.** : [Scanning electron microscopy of the endosperm of cereal crops. 3. Starch cell layer of white-core rice.] - Bull. Fac. Agr. Yamaguti Univ. *26* : 1 - 18, 1975. [Photosynthates; in Jap., ab : E.]

24852 - **UTSUNOMIYA, H., YAMAGATA, M., DOI, Y.** : [Scanning electron microscopy of the endosperm of cereal crops. 4. Starch cell layer of imperfect grain of rice (non-glutinous) and glutinous rice.] - Bull. Fac. Agr. Yamaguti Univ. *26* : 19 - 44, 1975. [Photosynthates; in Jap., ab : E.]

24853 - **VÁCLAVÍK, J.** : Comparison of the changes in net photosynthetic CO_2 uptake and water vapour efflux during leaf ontogenesis with the differences between the leaves according to their descending insertion level. - Biol. Plant. *17* : 411 - 415, 1975.

24854 - **VAKLINOVA, S., GOUSHINA, L.** : Effect of nitrate and ammonia nitrogen on carboanhydrase activity in plants with C-3 and C-4 type fixation of CO_2. - Dokl. bolg. Akad. Nauk *28* : 399 - 402, 1975.

24855 - **VALANNE, N.** : Peripheral structures of plastids and ultrastructural localization of acid phosphatase and succinic dehydrogenase in a variegated *Betula pubescens* mutant. - Can. J. Bot. *53* : 1072 - 1077, 1975.

24856 - **VALLEJOS, R.H., ANDREO, C.S., RAVIZZINI, R.A.** : Divalent-cation ionophores and Ca^{2+} transport in spinach chloroplasts. - FEBS Lett. *50* : 245 - 249, 1975.

24857 - **VALLESPINÓS, F., ESTRADA, M.** : Nitrógeno particulado en la región del NW. de África. Distribución y relación con otros parámetros. [Particulate nitrogen in the upwelling zone of NW Africa. Distribution and relationships with other parameters.] - Result. Exped. cient. Buque oceanogr. "Cornide de Saavedra" *4* : 131 - 143, 1975. [Chl.] ; in Span.]

24858 - **VALLE-TASCÓN, S. DEL, GIMÉNEZ-GALLEGO, G., RAMÍREZ, J.M.** : Light-dependent ATP formation in a non-phototrophic mutant of *Rhodospirillum rubrum* deficient in oxygen photoreduction. - Biochem. biophys. Res. Commun. *66* : 514 - 519, 1975.

24859 - **VALLE-TASCÓN, S. DEL, RAMÍREZ, J.M.** : Origin of the ATP formed during the light-dependent oxygen uptake catalyzed by *Rhodospirillum rubrum* chromatophores. - Z. Naturforsch. *30C* : 46 - 52, 1975.

24860 - **VAMBUTAS, V., BERTSCH, W.** : Delayed light studies on photosynthetic energy conversion. VIII. Evidence from millisecond emission of chloroplasts for two adenylate binding sites on membrane-bound coupling factor, CF_1. - Biochim. biophys. Acta *376* : 169 - 179, 1975.

24861 - **VAN, T.K., GARRARD, L.A.** : Effect of UV-B radiation on net photosynthesis of some C_3 and C_4 crop plants. - Soil Crop Sci. Soc. Florida Proc. *35* : 1 - 3, 1975.

*24862 - **VAN BAALEN, C.** : Growth, photosynthetic, and respiratory rates of the microalgae. - In : **LASKIN, A.I., LECHEVALIER, H.A.** (ed.) : Handbook of Microbiology. Vol. 4. Pp. 21 - 28. CRC Press, Cleveland, Ohio 1974.

24863 - **VAN BAVEL, C.H.M.** : A behavioral equation for leaf carbon dioxide assimilation and a test of its validity. - Photosynthetica *9* : 165 - 176, 1975.

24864 - **VAN BEST, J.A., DUYSENS, L.N.M.** : Reactions between primary and secondary acceptors of Photosystem II in *Chlorella pyrenoidosa* under anaerobic conditions, as studied by chlorophyll *a* fluorescence. - Biochim. biophys. Acta *408* : 154 - 163, 1975.

24865 - **VANDEN DRIESSCHE, T.** : Chloroplast functions are influenced by morphactins. - Biochem. Physiol. Pflanzen *168* : 543 - 551, 1975.

24866 - **VANDEN DRIESSCHE, T.** : Circadian rhythm in the Hill reaction of *Acetabularia*. - In : **AVRON, M.** (ed.) : Proceedings of the Third International Congress on Photosynthesis. Vol. I. Pp. 745 - 751. Elsevier, Amsterdam - Oxford - New York 1975.

24867 - **VANDEN DRIESSCHE, T.** : Circadian modulation of structure and function in *Acetabularia*. - In : Les Cycles Cellulaires et leur Blocage chez Plusieurs Protistes. - Coll. int. CNRS 240. Pp. 33 - 40. Edit. CNRS, Paris 1975.

24868 - **VANDEN DRIESSCHE, T., HAYET, M.** : Circadian rhythm of photosynthesis as influenced by anti-auxin. - Protoplasma *83* : 181 - 182, 1975.

24869 - **VANDERMEULEN, D.L., GOVINDJEE** : Anthroyl stearate as a fluorescent probe of a chloroplast membrane energized state related to photophosphorylation. - Biophys. J. *15* (2, Pt. 2) : 277a, 1975.

24870 - **VANDERMEULEN, D.L., GOVINDJEE** : Anthroyl stearate : A fluorescent probe for chloroplasts. - In : **AVRON, M.** (ed.) : Proceedings of the Third International

Congress on Photosynthesis. Vol. II. Pp. 1095 - 1105. Elsevier, Amsterdam - Oxford - New York 1975.

24871 - **VANDERMEULEN, D.L., GOVINDJEE** : Interactions of fluorescent analogs of adenine nucleotides with coupling factor protein isolated from spinach chloroplasts. - FEBS Lett. *57* : 272 - 275, 1975.

24872 - **VAN DIE, J., TAMMES, P.M.L.** : Phloem exudation from monocotyledonous axes. - - In : **ZIMMERMANN, M.H., MILBURN, J.A.** (ed.) : Transport in Plants I. Phloem Transport. (Encyclopedia of Plant Physiology N.S., Vol. 1.). Pp. 196 - 222. Springer-Verlag, Berlin - Heidelberg - New York 1975. [Photosynthates.]

*24873 - **VAN DIE, J., VONK, C.R., TAMMES, P.M.L.** : Studies on phloem exudation from *Yucca flaccida* HAW. XII. Rate of flow of ^{14}C-sucrose from a leaf to the wounded inflorescence top. Evidence for a primary origin of the major part of the exudate sucrose. - Acta bot. neerl. *22* : 446 - 451, 1973.

24874 - **VAN DUIJVENDIJK-MATTEOLI, M.A., DESMET, G.M.** : On the inhibitory action of cadmium on the donor side of Photosystem II in isolated chloroplasts. - Biochim. biophys. Acta *408* : 164 - 169, 1975.

24875 - **VAN GINKEL, G.** : Detailed action spectra of photophosphorylation. - In : **AVRON, M.** (ed.) : Proceedings of the Third International Congress on Photosynthesis. Vol. II. Pp. 1121 - 1130. Elsevier, Amsterdam - Oxford - New York 1975.

24876 - **VAN GORKOM, H.J.** : Identification of the primary reactants in photosystem II. - In : **AVRON, M.** (ed.) : Proceedings of the Third International Congress on Photosynthesis. Vol. I. Pp. 159 - 162. Elsevier, Amsterdam - Oxford - New York 1975.

24877 - **VAN GORKOM, H.J., PULLES, M.P.J., WESSELS, J.S.C.** : Light-induced changes of absorbance and electron spin resonance in small photosystem II particles. - Biochim. biophys. Acta *408* : 331 - 339, 1975.

24878 - **VAN HUMMEL, H.C., HULSEBOS, T.J.M., WINTERMANS, J.F.G.M.** : Biosynthesis of galactosyl diglycerides by non-green fractions from chloroplasts. - Biochim. biophys. Acta *380* : 219 - 226, 1975.

24879 - **VAN HUMMEL, H.C., WINTERMANS, J.F.G.M.** : Observations on the biosynthesis of galactolipids in spinach leaves. - In : **AVRON, M.** (ed.) : Proceedings of the Third International Congress on Photosynthesis. Vol. III. Pp. 2029 - 2037. Elsevier, Amsterdam - Oxford - New York 1975.

24880 - **VAN KEULEN, H., LOUWERSE, W., SIBMA, L., ALBERDA, T.** : Crop simulation and experimental evaluation - a case study. - In : **COOPER, J.P.** (ed.) : Photosynthesis and Productivity in Different Environments. Pp. 623 - 643. Cambridge Univ. Press, Cambridge - London - New York - Melbourne 1975.

24881 - **VAN RENSEN, J.J.S.** : Effects of N-(phosphomethyl)glycine on photosynthetic reactions in *Scenedesmus* and in isolated spinach chloroplasts. - In : **AVRON, M.** (ed.) : Proceedings of the Third International Congress on Photosynthesis. Vol. I. Pp. 683 - 687. Elsevier, Amsterdam - Oxford - New York 1975.

24882 - **VAN RENSEN, J.J.S.** : Lipid peroxidation and chlorophyll destruction caused by diquat during photosynthesis in *Scenedesmus*. - Physiol. Plant. *33* : 42 - 46, 1975.

24883 - **VAN SAMBEEK, J.W., PICKARD, B.G.** : Mediation of rapid electrical, catabolic, photosynthetic, & transpirational changes by Ricca's factor : Studies on intact plants. - Plant Physiol. *56* (Suppl.) : 81, 1975.

24884 - **VAN SAMBEEK, J.W., ULBRIGHT, C.R., PICKARD, B.G.** : Mediation of rapid electrical, catabolic, photosynthetic & transpirational changes by Ricca's factor : Studies on excised leaves. - Plant Physiol. *56* (Suppl.) : 81, 1975.

24885 - **VAN STEVENINCK, M.E., VAN STEVENINCK, R.F.M.** : Evidence for structural units in chloroplast thylakoids. - Protoplasma *86* : 381 - 389, 1975.

24886 - **VARFOLOMEEV, S.D., ZAĬTSEV, S.V., IL'INA, M.D., BEREZIN, I.V.** : Kinetika i mekhanizm inaktivatsii izolirovannykh khloroplastov. [Kinetics and mechanism of inactivation of isolated chloroplasts.] - Mol. Biol. (Moskva) *9* : 893 - - 902, 1975. [In R, ab : E.]

24887 - **VARLET GRANCHER, C.** : Variation et estimation de l'énergie d'origine solaire reçue sur des plans d'inclinaison et d'azimut variables. - Ann. agron. *26* : 245 - 264, 1975.

24888 - **VASIL, I.K., GILES, K.L.** : Induced transfer of higher plant chloroplasts into fungal protoplasts. - Science *190* : 680, 1975.

24889 - **VAULIN, A.V.** : Ob ispol'zovanii assimilyatsionnykh kamer pri opredelenii fotosinteza v polevykh usloviyakh. [Use of assimilation chambers for determination of photosynthesis in field conditions.] - Sel'skokhoz. Biol. *10* : 434 - - 437, 1975. [In R, ab : E.]

24890 - **VECHAR, A.S., BYKAVA, L.M.** : Rolya flavinavykh zluchėnnyaŭ u svetaadchuval'nykh rėaktsyyakh raslin. [Role of flavin compounds in light sensitive reactions in plants.] - Vestsi Akad. Navuk belarus.SSR, Ser. biyal. Navuk *1975* (2) : 59 - 64, 139, 1975. [Chloroplast movements; in Beloruss., ab : R.]

24891 - **VECHER, A.S., KLINGER, Yu.E., RESHETNIKOV, V.N.** : Adenozintrifosfataznaya i pirofosfataznaya aktivnost' izolirovannykh yader i khloroplastov prorostkov rzhi s razlichnoĭ ploidnost'yu. [ATPase and pyrophosphatase activities in isolated nuclei and chloroplasts of rye seedlings of different ploidity.] - Dokl. Akad. Nauk belorus. SSR *19* : 648 - 651, 1975. [In R.]

24892 - **VECHER, A.S., MAS'KO, A.A., KOVAL'CHUK, R.A.** : Osnovnye komponenty khimicheskogo sostava membran khloroplastov rzhi. [Basic components of chemical composition of rye chloroplast membranes.] - Dokl. Akad. Nauk belorus. SSR *19* : 1132 - 1134, 1975. [In R.]

24893 - **VECHER, A.S., MAS'KO, A.A., RESHETNIKOV, V.N.** : Raspredelenie nukleinovykh kislot v strukturnykh ėlementakh khloroplastov. [Distribution of nucleic acids in structural elements of chloroplasts.] - In : Pitanie i Obmen Veshchestv u Rasteniĭ. Pp. 3 - 6. Nauka i Tekhnika, Minsk 1975. [In R.]

24894 - **VECHER, A.S., MAS'KO, A.A., RESHETNIKOV, V.N.** : Issledovanie sostava i svoĭstv belkov khloroplastov di- i tetraploidnykh rasteniĭ. [Composition and properties of proteins of chloroplasts of di- and tetraploid plants.] - In : Fiziologo--Biokhimicheskie Aspekty Rosta i Razvitiya Rasteniĭ. Pp. 3 - 6, 167. Nauka i Tekhnika, Minsk 1975. [In R.]

24895 - **VEDERNIKOV, V.I.** : Zavisimost' assimilyatsionnogo chisla i kontsentratsii khlorofilla "*a*" ot produktivnosti vod v razlichnykh temperaturnykh oblastyakh mirovogo okeana. [Dependence of assimilation number and chlorophyll *a* concentration on water productivity in different temperature areas of the world ocean.] - Okeanologiya *15* : 703 - 707, 1975. [In R, ab : E.]

24896 - **VEDERNIKOV, V.I., KOBLENTS-MISHKE, O.I., SUKHANOVA, I.N., KARABASHEV, G.S., FISHER, Ya.** : Sravnenie vertikal'nogo izmeneniya kolichestva vzvesi, khlorofilla, fitoplanktona i intensivnosti lyuminestsentsii pigmentov v ekvatorial'nom i peruanskom raĭonakh vostochnoĭ Patsifiki. [Comparison of vertical changes in quantities of particulate matter, chlorophyll, phytoplankton, and the intensity of pigment luminescence in the equatorial and peruvian regions of eastern Pacific.] - Tr. Inst. Okenologii Akad. Nauk SSSR *102* (Ėkosistemy Pelagiali Tikhogo Okeana) : 165 - 174, 1975. [In R, ab : E.]

24897 - **VELICHKO, I.M.** : Izvlechenie pigmentov iz vodorosleĭ s plazmaticheskoĭ obolochkoĭ (po Masyuk i Radchenko [91, 92]). [Pigment extraction from algae with plasmatic membrane (according to Masyuk and Radchenko [91, 92]).] - In : Metody Fiziologo-Biokhimicheskogo Issledovaniya Vodorosleĭ v Gidrobiologicheskoĭ Praktike. Pp. 79 - 80. Naukova Dumka, Kiev 1975. [In R.]

24898 - **VELICHKO, I.M.** : Opredelenie kolorimetricheskim metodom summarnogo soderzhaniya khlorofillov v ėkstraktakh iz vodorosleĭ. [Colorimetric determination of total chlorophyll content in algae extracts.] - In : Metody Fiziologo-Biokhimicheskogo Issledovaniya Vodorosleĭ v Gidrobiologicheskoĭ Praktike. Pp. 81 - 83. Naukova Dumka, Kiev 1975. [In R.]

24899 - **VELICHKO, I.M.** : Razdelenie pigmentov s pomoshch'yu bumazhnoĭ khromatografii. [Pigment separation by means of paper chromatography.] - In : Metody Fiziologo-Biokhimicheskogo Issledovaniya Vodorosleĭ v Gidrobiologicheskoĭ Praktike. Pp. 83 - 85. Naukova Dumka, Kiev 1975. [In R.]

24900 - **VELICHKO, I.M.** : Razdel'noe opredelenie pigmentov s pomoshch'yu reaktsii Krausa (polevoĭ metod [23]). [Pigment separation among immiscible solvents by means of Kraus reaction (field method [23]).] - In : Metody Fiziologo-Biokhimicheskogo Issledovaniya Vodorosleĭ v Gidrobiologicheskoĭ Praktike. Pp. 85 - - 86. Naukova Dumka, Kiev 1975. [Chl, Car; in R.]

24901 - **VELTHUYS, B.R.** : Binding of the inhibitor NH_3 to the oxygen-evolving apparatus of spinach chloroplasts. - Biochim. biophys. Acta *396* : 392 - 401, 1975.

24902 - **VELTHUYS, B.R.** : Flash number dependent luminescence of isolated spinach chloroplasts at different pH's, low temperature and in the presence of NH_4Cl. - In : **AVRON, M.** (ed.) : Proceedings of the Third International Congress on Photosynthesis. Vol. I. Pp. 93 - 100. Elsevier, Amsterdam - Oxford - New York 1975.

24903 - **VELTHUYS, B.R., AMESZ, J.** : Temperature and preillumination dependence of delayed fluorescence of spinach chloroplasts. - Biochim. biophys. Acta *376* : 162 - 168, 1975.

24904 - **VELTHUYS, B.R., VISSER, J.W.M.** : The reactivation of EPR signal II in chloroplasts treated with reduced dichlorophenol-indophenol : evidence against a dark equilibrium between two oxidation states of the oxygen evolving system. - FEBS Lett. *55* : 109 - 112, 1975.

24905 - **VERDUIN, J.** : Rate of carbon dioxide transport across air-water boundaries in lakes. - Limnol. Oceanogr. *20* : 1052 - 1053, 1975.

24906 - **VERDUIN, J.** : Photosynthetic rates in Lake Superior. - Verh. int. Verein. Limnol. *19* : 689 - 693, 1975.

24907 - **VERMA, S.B., ROSENBERG, N.J.** : Accuracy of lysimetric, energy balance, and stability-corrected aerodynamic methods of estimating above-canopy flux of CO_2. - Agron. J. *67* : 699 - 704, 1975.

24908 - **VERMEGLIO, A., MATHIS, P.** : Light-induced absorption changes in spinach chloroplasts/ A comparative study at -50° and -170 °C. - In : **AVRON, M.** (ed.) : Proceedings of the Third International Congress on Photosynthesis. Vol. I. Pp. 323 - 334. Elsevier, Amsterdam - Oxford - New York 1975.

24909 - **VERNON, L.P., KLEIN, S.M.** : Nature of plant chlorophylls *in vivo* and their associated proteins. - Ann. New York Acad. Sci. *244* : 281 - 296, 1975.

24910 - **VERNOTTE, C., BRIANTAIS, J.-M., ARMOND, P., ARNTZEN, C.J.** : Preillumination effects on chloroplast structure and photochemical activity. - Plant Sci. Lett. *4* : 115 - 123, 1975.

24911 - **VERNOTTE, C., BRIANTAIS, J.M., ARNTZEN, C.J.** : Comparison of excitons transfers changes induced by cations and by adaptation in states I and II. - In : **AVRON, M.** (ed.) : Proceedings of the Third International Congress on Photosynthesis. Vol. I. Pp. 183 - 193. Elsevier, Amsterdam - Oxford - New York 1975.

24912 - **VEZITSKIĬ, A.Yu., RUDOĬ, A.B.** : Prevrashcheniya khlorofillovykh pigmentov v postėtiolirovannykh list'yakh raznykh vidov rasteniĭ na ranneĭ stadii zeleneniya. [Conversions of chlorophyll pigments in postetiolated leaves of plants of different species at early greening stage.] - In : **SHLYK, A.A.** (ed.) : Biosintez i Sostoyanie Khlorofillov v Rasteniі. Pp. 58 - 82, 244 - 245. Nauka i Tekhnika, Minsk 1975. [In R, ab : E.]

24913 - **VIDOVIČ, J.** : A device for the measurements of radiation in stands with automatically moving sensors. - Biol. Plant. *17* : 75 - 78, 1975.

24914 - **VIERKE, G., MÜLLER, M.** : On the kinetics and mechanism of spontaneous intramolecular reduction of the central metal ion in K_2[Mn(IV)-2-α-hydroxyethyl isochlorin e_4] acetate in aqueous alkaline solutions and its relation to the binding sites of manganese in photosynthesis. - Z.Naturforsch. *30C* : 327 - 332, 1975.

24915 - **VIGNES, D., CALMÉS, J.** : Quelques modifications physico-chimiques et physiologiques liées à la sénescence des feuilles. - Physiol. Plant. *33* : 188 - 193, 1975.

24916 - **VIRGIN, H.I.** : *In vivo* absorption spectra of protochlorophyll$_{650}$ and protochlorophyll$_{636}$ within the region 530-700 nm. - Photosynthetica *9* : 84 - 92, 1975.

24917 - **VISSER, J.W.M., RIJGERSBERG, C.P.** : Reversibility of light-induced reactions of photosystem I and II at temperatures below 200 °K. - In : **AVRON, M.** (ed.): Proceedings of the Third International Congress on Photosynthesis. Vol. I. Pp. 399 - 408. Elsevier, Amsterdam - Oxford - New York 1975.

24918 - **VITOLA, Ā., GUBARE, G., KRISTKALNE, S., KREICBERGS, O., SELGA, M.** : Augu spējas pielāgoties pazeminātai apgaismojuma intensitātei atkarībā no minerālās barošanās apstākliem. [Plant adaptability to low illuminance depending on mineral nutrition.]-In : Tautsaimniecībā Derīgo Augu Agrotehnika. Pp. 123 - - 132. Zinātne, Riga 1975. [Ps; in Latv., ab : R.]

24919 - **VIVEKANANDAN, M., GNANAM, A.** : Physiological studies on isolated mesophyll cells of *Canna edulis* KER. - Indian J. exp. Biol. *13* : 30 - 33, 1975.

24920 - **VIVEKANANDAN, M., GNANAM, A.** : Studies on the mode of action of aminotriazole in the induction of chlorosis. - Plant Physiol. *55* : 526 - 531, 1975.

24921 - **VIVEKANANDAN, M., GNANAM, A.** : Studies on the mechanism of action of amitrole on chloroplast development. - Indian J. Biochem. Biophys. *12* : 374 - 378, 1975.

24922 - **VIVEKANANDAN, M., GNANAM, A.** : Studies on the effect of aminotriazole on chloroplast development in *Phaseolus radiatus* L. - Curr. Sci. *44* : 842 - 845, 1975.

24923 - **VLASENOK, L.I., AKHRAMOVICH, N.I., VRUBEL', S.V., AKULOVICH, E.M.** : Raspredelenie khlorofillov *a* i *b* pri mnogokratnom fragmentirovanii granal'nykh lamell khloroplastov. [Distribution of chlorophylls *a* and *b* upon multiple fragmentation of chloroplast grana lamellae.] - In : **SHLYK, A.A.** (ed.) : Biosintez i Sostoyanie Khlorofillov v Rastenii. Pp. 183 - 196, 246 - 247. Nauka i Tekhnika, Minsk 1975. [In R, ab : E.]

24924 - **VOGELMANN, H., GHAHREMANI, B., WAGNER, F.** : Preparation of porphobilinogen and uroporphyrin III from δ-aminolaevulinic acid by pretreated cells of *Chromatium vinosum*. - Europe. J. appl. Microbiol. *2* : 19 - 28, 1975.

24925 - **VOLGER, H.G., HEBER, U.** : Cryoprotective leaf proteins. - Biochim. biophys. Acta *412* : 335 - 349, 1975. [Chloroplast.]

24926 - **VOLKOV, A.G., LOZHKIN, B.T., BOGUSLAVSKIĬ, L.I.** : Perenos protonov i ėlektronov cherez bisloĭnye membrany i granitsu razdela faz dekan/voda v prisutstvii khlorofilla. [Proton and electron transfer through bilayer membranes and decane/water interface in the presence of chlorophyll.] - Dokl. Akad. Nauk SSSR *220* : 1207 - 1210, 1975. [In R.]

24927 - **VOLLENWEIDER, R.A.** : Input-output models. With special reference to the phosphorus loading concept in limnology. - Schweiz. Z. Hydrol. *37* : 53 - 84, 1975. [Chl, primary production.]

24928 - **VOLODARSKIĬ, N.I., BYSTRYKH, E.E.** : Funktsional'naya aktivnost' fotosinteticheskogo apparata rasteniĭ podsolnechnika pod vliyaniem zasukhi. [Functional activity of photosynthetic apparatus of sunflower plants during drought.] - Sel'skokhoz. Biol. *10* : 716 - 721, 1975. [In R, ab : E.]

24929 - **VOOKOVÁ, B.** : Analysis of chlorophylls in some soil algae at Báb. - In : **BISKUPSKÝ, V.** (ed.) : Research Project Báb IBP Progress Report II. Pp. 233 - 237. Veda, Bratislava 1975.

24930 - **VOROB'EVA, L.M., KRASNOVSKIĬ, A.A., KAYUPOVA, G.A.** : Svetovye prevrashcheniya pigmentnogo kompleksa *Anacystis nidulans*. [Light-induced changes of pigment complex of *Anacystis nidulans*.] - Fiziol. Rast. *22* : 16 - 26, 1975. [In R, ab : E.]

24931 - **VOSKRESENSKAYA, N.P.** : Printsipy fotoregulirovaniya metabolizma rasteniĭ i regulyatornoe deĭstvie krasnogo i sinego sveta na fotosintez. [Principles of photoregulation of plant metabolism and regulatory action of red and blue light on photosynthesis.] - In : Fotoregulyatsiya Metabolizma i Morfogeneza Rasteniĭ. Pp. 16 - 36, 249. Nauka, Moskva 1975. [In R.]

24932 - **VOSKRESENSKAYA, N.P., POLYAKOV, M.A.** : Enhancement of CO_2-gas exchange in leaves of *Convallaria majalis* L., its action spectrum and light saturation. Plant Sci. Lett. *5* : 333 - 338, 1975.

24933 - **VOSKRESENSKAYA, N.P., POYARKOVA, N.M., KHODZHIEV, A., DROZDOVA, I.S.** : Regulatory action of blue light on the activity of carboxylating enzymes and enzymes of the glycollate pathway in broad bean and maize plants. - In : **NASYROV, Yu. S., ŠESTÁK, Z.** (ed.) : Genetic Aspects of Photosynthesis. Pp. 167 - 175. Dr. W. Junk B.V. Publ., The Hague 1975.

24934 - **VOWINCKEL, T., OECHEL, W.C., BOLL, W.G.** : The effect of climate on the photosynthesis of *Picea mariana* at the subarctic tree line. 1. Field measurements. - Can. J. Bot. *53* : 604 - 620, 1975.

24935 - **VOZNESENSKIĬ, V.L.** : Maksimal'nye intensivnosti fotosinteza rasteniĭ yugo-vostochnykh Karakumov. [Maximal values of photosynthetic rate of south-east Karakum desert plants.] - Bot. Zh. *60* : 992 - 1000, 1975. [In R.]

24936 - **VOZNYAK, V.M., KIM, V.A., EVSTIGNEEV, V.B.** : Issledovanie metodom ĖPR mekhanizma fotookisleniya khlorofilla *a* nitrosoedineniyami. [ESR studies of the mechanism of photooxidation of chlorophyll *a* by nitro-compounds.] - Biofizika *20* : 406 - 410, 1975. [In R, ab : E.]

24937 - **VOZNYAK, V.M., KIM, V.A., EVSTIGNEEV, V.B.** : Fotokhimicheskaya generatsiya anion-radikalov bakteriokhlorofilla i bakteriofeofitina. [Photochemical generation of anion-radicals of bacteriochlorophyll and bacteriopheophytin.] - Zh. prikl. Spektroskop. *23* : 54 - 59, 1975. [In R.]

24938 - **VREDENBERG, W.J.** : The kinetic of light-induced changes in the electrical potential measured across the thylakoid membranes of intact chloroplasts. - In: **AVRON, M.** (ed.) : Proceedings of the Third International Congress on Photosynthesis. Vol. II. Pp. 929 - 939. Elsevier, Amsterdam - Oxford - New York 1975.

24939 - **VREDENBERG, W.J., TONK, W.J.M.** : On the steady-state electrical potential difference across the thylakoid membranes of chloroplasts in illuminated plant cells. - Biochim. biophys. Acta *387* : 580 - 587, 1975.

24940 - **VRKOČ, F.** : The effect of cultural practices and conditions of locality on growth-dynamics and productivity of main field crops. - Rostlinná Výroba (Praha) *21* : 817 - 823, 1975. [Growth analysis.]

*24941 - **VRUBLEVSKAYA, K.G., ZAĬTSEVA, T.A., PLOTNIKOVA, A.N., MANDEL', T.E.** : Vliyanie kachestva sveta na protsessy zapasaniya ėnergeticheskikh veshchestv v rasteniyakh. [Effect of light quality on processes of energetic substances formation in plants.] - In : Tretiĭ Vsesoyuznyĭ Biokhimicheskiĭ S"ezd. Referaty Nauchnykh Soobshcheniĭ. Vol. 1. P. 147. Riga 1974. [Ps; in R.]

24942 - **VYAS, N.L., VYAS, L.N.** : Distribution of chlorophyll and carotenoids in different canopy strata of a dry deciduous tropical forest at Udaipur, India. - Photosynthetica *9* : 241 - 245, 1975.

24943 - **WABER, J., SAKAI, W.S.** : Further studies of the ultrastructure of D_2O grown winter rye. - Protoplasma *84* : 273 - 281, 1975.

24944 - **WADA, K., HASE, T., MATSUBARA, H.** : Evolutionary information involved in primary structures of chloroplast-type ferredoxins. - J. Biochem. *78* : 637 - 639, 1975.

24945 - **WADA, K., HASE, T., TOKUNAGA, H., MATSUBARA, H.** : Amino acid sequence of *Spirulina platensis* ferredoxin : a far divergency of blue-green algal ferredoxins. - FEBS Lett. *55* : 102 - 104, 1975.

24946 - **WAGENER, K.** : Kinetic isotope effects of oxygen in photosynthesis and respiration. - In : **GOLDBERG, E.D.** (ed.) : The Nature of Seawater. Pp. 433 - 451. Dahlem Konferenzen, Berlin 1975.

24947 - **WAGGONER, P.E.** : Micrometeorological models. - In : **MONTEITH, J.L.** (ed.) : Vegetation and the Atmosphere. Vol. 1. Principles. Pp. 205 - 228. Academic Press, London - New York - San Francisco 1975. [Ps.]

24948 - **WAGNER, G.J., SIEGELMAN, H.W.** : Large-scale isolation of intact vacuoles and isolation of chloroplasts from protoplasts of mature plant tissues. - Science *190* : 1298 - 1299, 1975.

24949 - **WAIDYANATHA, U.P. de S., KEYS, A.J., WHITTINGHAM, C.P.** : Effects of carbon dioxide on metabolism by the glycollate pathway in leaves. - J. exp. Bot. *26*: 15 - 26, 1975.

24950 - **WAIDYANATHA, U.P. de S., KEYS, A.J., WHITTINGHAM, C.P.** : Effects of oxygen on metabolism by the glycollate pathway in leaves. - J. exp. Bot. *26* : 27 - 32, 1975.

*24951 - **WAKAMATSU, K.** : [Isolation of the new fluorescent substances from spinach leaves.] - Sci. hum. Life *9* : 1 - 7, 1971. [Ps; in Jap., ab : E.]

24952 - **WALK, R.-A., METZNER, H.** : Reinigung und Charakterisierung von Chloroplasten -Carbonat- Dehydratase (Isoenzym I) aus Blättern von *Lactuca sativa*. - Hoppe-Seyler's Z. physiol. Chem. *356* : 1733 - 1741, 1975.

24953 - **WALKER, K.F.** : The seasonal phytoplankton cycles of two saline lakes in central Washington. - Limnol. Ocenogr. *20* : 40 - 53, 1975.

24954 - **WALKER, N.A.** : Uncoupling in particles and intact chloroplasts by amines and nigericin - a discussion of the role of swelling. - FEBS Lett. *50* : 98 - 101, 1975.

24955 - **WALL, J.D., GEST, H.** : Prospects for the molecular biology of photosynthetic bacteria. - In : **AVRON, M.** (ed.) : Proceedings of the Third International Congress on Photosynthesis. Vol. II. Pp. 1179 - 1188. Elsevier, Amsterdam - Oxford - New York 1975.

24956 - **WALLEN, D.G., CARTIER, L.D.** : Molybdenum dependence, nitrate uptake and photosynthesis of freshwater plankton algae. - J. Phycol. *11* : 345 - 349, 1975.

24957 - **WALLENTINUS, I.** : Primary production of macroalgae measured by the ^{14}C method. - Merentutkimuslait. Julk./Havsforskningsinst. Skr. *239* : 72 - 77, 1975.

24958 - **WALLES, B., HUDÁK, J.** : A comparative study of chloroplast morphogenesis in seedlings of some conifers (*Larix decidua, Pinus sylvestris* and *Picea abies*). - Stud. forest. suec. *127* : 1 - 22, 1975.

24959 - **WALLES, B., HUDÁK, J.** : Etioplast and chromoplast development in the lycopenic mutant of maize. - J. submicroscop. Cytol. *7* : 325 - 334, 1975.

24960 - **WALSH, G.E.** : Utilization of energy by primary producers in four ponds in northwestern Florida. - In : Proceedings : Biostimulation - Nutrient Assessment Workshop. EPA-660/3-75-034. Pp. 249 - 274. National Environm. Res. Center, Corvallis, Oregon 1975.

24961 - **WALTON, D.W.H., GREENE, D.M., CALLAGHAN, T.V.** : An assessment of primary production in a sub-antarctic grassland on South Georgia. - Brit. Antarct. Surv. Bull. *41/42* : 151 - 160, 1975.

24962 - **WANG, W., BOYNTON, J.E., GILLHAM, N.W., GOUGH, S.** : Genetic control of chlorophyll biosynthesis in *Chlamydomonas* : analysis of a mutant affecting synthesis of δ-aminolevulinic acid. - Cell *6* : 75 - 84, 1975.

24963 - **WANKA, F.** : Possible role of the pyrenoid in the reproductional phase of the cell cycle of *Chlorella*. - In : Les Cycles Cellulaires et leur Blocage chez Plusieurs Protistes. - Coll. int. CNRS 240. Pp. 131 - 136. Édit. CNRS, Paris 1975.

24964 - **WARD, A.K., WETZEL, R.G.** : Some effects of sodium on blue-green algal growth. - J. Phycol. *11* (Suppl.) : 14, 1975. [Ps, Chl.]

24965 - **WARD, A.K., WETZEL, R.G.** : Sodium : some effects on bluegreen algal growth. - J. Phycol. *11* : 357 - 363, 1975. [Ps, Chl.]

24966 - **WARD, B.** : The non-light-dependent reduction of 2,6-dichlorophenolindophenol by cells of the blue-green alga *Anacystis nidulans*. - Can. J. Microbiol. *21* : 419 - 422, 1975.

24967 - **WAREING, P.F., PATRICK, J.** : Source-sink relations and the partition of assimilates in the plant. - In : **COOPER, J.P.** (ed.) : Photosynthesis and Productivity in Different Environments. Pp. 481 - 499. Cambridge Univ. Press, Cambridge - London - New York - Melbourne 1975.

24968 - **WARRINGTON, I.J., MITCHELL, K.J.** : The suitability of three high intensity lamp sources for plant growth and development. - J. agr. Eng. Res. *20* : 295 - 302, 1975. [Growth analysis.]

24969 - **WASSINK, E C.** : Photosynthesis and productivity in different environments - conclusions. - In : **COOPER, J.P.** (ed.) : Photosynthesis and Productivity in Different Environments. Pp. 675 - 687. Cambridge Univ. Press, Cambridge - London - New York - Melbourne 1975.

24970 - **WATANABE, I.** : Transformation factor from carbon dioxide net assimilation to dry weight in crops. I. Soybean. - Proc. Crop Sci. Soc. Jap. *44* : 68 - 72, 1975.

24971 - **WATANABE, I.** : Transformation factor from carbon dioxide net assimilation to dry weight in crops. II. Peanut. - Proc. Crop Sci. Soc. Jap. *44* : 403 - 408, 1975.

24972 - **WATANABE, I.** : Transformation factor from carbon dioxide net assimilation to dry weight in crops. III. Rice. - Proc. Crop Sci. Soc. Jap. *44* : 409 - 413, 1975.

24973 - **WATLING-PAYNE, A.S., SELWYN, M.J.** : Decrease of proton permeability of CF_1--deficient chloroplast particles by triphenyltin. - FEBS Lett. *58* : 57 - 61, 1975.

24974 - **WATSON, B.T.** : The influence of low temperature on the rate of translocation in the phloem of *Salix viminalis*. - Ann. Bot. *39* : 889 - 900, 1975. [Photosynthates.]

24975 - **WATT, K.E.F.** : Critique and comparison of biome ecosystem modeling. - In : **PATTEN, B.C.** (ed.) : Systems Analysis and Simulation Ecology. Vol. III. Pp. 139 - 152. Academic Press, New York - San Francisco - London 1975.

24976 - **WEBB, W.L.** : Dynamics of photoassimilated carbon in Douglas fir seedlings. - Plant Physiol. *56* : 455 - 459, 1975.

24977 - **WEBB, W.L.** : The distribution of photoassimilated carbon and the growth of Douglas-fir seedlings. - Can. J. Forest Res. *5* : 68 - 72, 1975.

24978 - **WEBER, A.** : Chlorophylle und Carotinoide der *Chaetophorineae (Chlorophyceae; Ulotrichales)* 2. Der Einfluss unterschiedlicher Stickstoffkonzentrationen auf die Pigmentgarnitur und die Morphogenese der Grünalge *Fritschiella tuberosa* IYENGAR. - Arch. Microbiol. *102* : 45 - 52, 1975.

24979 - **WEBER, J.A., GATES, D.M.** : The effects of light intensity and pH on photosynthesis in *Elodea densa*. - Plant Physiol. *56* (Suppl.) : 12, 1975.

24980 - **WEBSTER, G.L., BROWN, W.V., SMITH, B.N.** : Systematics of photosynthetic carbon fixation pathways in *Euphorbia*. - Taxon *24* : 27 - 33, 1975.

*24981 - **WEEDON, B.C.L.** : Some recent studies on carotenoids and related compounds. - Pure appl. Chem. *35* : 113 - 130, 1973.

24982 - **WEGMANN, K., PRISTAVU, N.** : The role of the nitrogen source in the carbon metabolism of *Dunaliella*. - In : **AVRON, M.** (ed.) : Proceedings of the Third International Congress on Photosynthesis. Vol. II. Pp. 1525 - 1531. Elsevier, Amsterdam - Oxford - New York 1975.

24983 - **WEHRMEYER, W., SCHNEIDER, H.** : Elektronenmikroskopische Untersuchungen zur reversiblen Veränderung der Chloroplastenfeinstruktur von *Rhodella violacea* bei Stickstoffmangel. - Biochem. Physiol. Pflanzen *168* : 519 - 532, 1975.

24984 - **WEIDNER, M., NORDHORN, G., KREMER, B.P., KÜPPERS, U.** : Untersuchungen zur Photosynthese bei *Giffordia mitchellae*. - Z. Pflanzenphysiol. *76* : 423 - 443, 1975.

24985 - **WEILAND, R.T., NOBLE, R.D., CRANG, R.E.** : Photosynthetic and chloroplast ultrastructural consequences of manganese deficiency in soybean. - Amer. J. Bot. *62* : 501 - 508, 1975.

24986 - **WEIMBERG, R.** : Effect of growth in highly salinized media on the enzymes of the photosynthetic apparatus in pea seedlings. - Plant Physiol. *56* : 8 - 12, 1975.

24987 - **WEINBERG, M.B., CASTELFRANCO, P.A.** : Effect of EPTC on plastid membrane constituents in germinating cucumber cotyledons. - Weed Sci. *23* : 185 - 187, 1975. [Chl.]

24988 - **WEINMANN, R., KREEB, K.** : CO_2-Gaswechsel von Sklerophyllen im nördlichen Gardasee-Gebiet. - Ber. deut. bot. Ges. *88* : 205 - 210, 1975.

24989 - **WEISE, G., GNAUCK, A.** : Toxische Schwermetalleffekte bei submersen Wasserpflanzen. - Wiss. Z. tech. Univ. Dresden *24* : 1459 - 1460, 1975. [Ps.]

24990 - **WEISE, G., GNAUCK, A.H.** : Toxikologische Untersuchungen an submersen Makrophyten in Fliessgewässern auf der Grundlage Infrarotgasanalytischer Messungen (IRGA-Methodik). - Environ. Protect. Eng. *1* : 137 - 149, 1975.

24991 - **WEISS, C.** : The molecular orbital theory of chlorophyll. - Ann. New York Acad. Sci. *244* : 204 - 213, 1975.

24992 - **WEISS, C.M.** : Field evaluation of the algal assay procedure on surface waters of North Carolina. - In : **MIDDLEBROOKS, E.J., FALKENBORG, D.H., MALONEY, T.E.** (ed.) : Biostimulation and Nutrient Assessment. Pp. 29 - 76. Utah State Univ., Logan, Utah 1975. [Chl, productivity.]

24993 - **WEISSENBÖCK, G.** : Aktivitätsverlauf der Phenylalanin-, Tyrosin-Ammonium-Lyase (PAL,TAL) und Chalkon-Flavanon-Isomerase im Vergleich zur C-Glycosylflavon--Akkumulation im wachsenden Hafersproß (*Avena sativa* L.) bei Belichtung und Dunkelheit. - Z. Pflanzenphysiol. *74* : 226 - 254, 1975. [Chl.]

24994 - **WEISSENBÖCK, G., EFFERTZ, B.** : Entwicklungs- und lichtabhängige Akkumulation von C-Glycosylflavonen im Haferkeimling (*Avena sativa* L.). - Z. Pflanzenphysiol. *74* : 298 - 326, 1975. [Chl.]

24995 - **WELLBURN, A.R.** : δ-aminolaevulinic acid formation in greening *Avena* laminae. - Phytochemistry *14* : 699 - 701, 1975.

24996 - **WELLBURN, A.R.** : Succinyl coenzyme *A* and porphobilinogen formation in isolated etio-chloroplasts and greening laminae. - Phytochemistry *14* : 1171 - 1173, 1975.

24997 - **WELLBURN, A.R., COBB, A.H.** : Developmental changes in the levels of SDS-extractable polypeptides in developing plastid membranes. - In : **AVRON, M.** (ed.): Proceedings of the Third International Congress on Photosynthesis. Vol. III. Pp. 1647 - 1659. Elsevier, Amsterdam - Oxford - New York 1975.

24998 - **WELLER, D., DOEMEL, W., BROCK, T.D.** : Requirement of low oxidation-reduction potential for photosynthesis in a blue-green alga (*Phormidium* sp.). - Arch. Microbiol. *104* : 7 - 13, 1975.

24999 - **WERDAN, K., HELDT, H.W., MILOVANCEV, M.** : The role of pH in the regulation of carbon fixation in the chloroplast stroma. Studies on CO_2 fixation in the light and dark. - Biochim. biophys. Acta *396* : 276 - 292, 1975.

25000 - **WERGIN, W.P., POTTER, J.R.** : The effects of fluometuron on the ultrastructural development, chlorophyll accumulation and photosynthetic competence in developing velvetleaf seedlings. - Pestic. Biochem. Physiol. *5* : 265 - 279, 1975.

25001 - **WERTHMÜLLER, K., SENGER, H.** : Changes in the photosynthetic apparatus of autotrophic, mixotrophic and photo-heterotrophic synchronized cultures of *Scenedesmus*.- In : **AVRON, M.** (ed.) : Proceedings of the Third International Congress on Photosynthesis. Vol. III. Pp. 1969 - 1976. Elsevier, Amsterdam - Oxford - New York 1975.

25002 - **WESSELS, J.S.C., BORCHERT, M.T.** : Studies on subchloroplast particles. Similarity of grana and stroma photosystem I and the protein composition of photosystem I and photosystem II particles. - In : **AVRON, M.** (ed.) : Proceedings of the Third International Congress on Photosynthesis. Vol. I. Pp. 473 - 484. Elsevier, Amsterdam - Oxford - New York 1975.

25003 - **WESTLAKE, D.F.** : Primary production of freshwater macrophytes. - In : **COOPER, J.P.** (ed.) : Photosynthesis and Productivity in Different Environments. Pp. 189 - 206. Cambridge Univ. Press, Cambridge - London - New York - Melbourne 1975.

25004 - **WETTERMARK, G., STYMNE, H., BROLIN, S.E., PETERSSON, B.** : Substrate analyses in single cells. I. Determination of ATP. - Anal. Biochem. *63* : 293 - 307, 1975.

25005 - **WETZEL, R.G.** : Primary production. - In : **WHITTON, B.A.** (ed.) : River Ecology.

Pp. 230 - 247. Blackwell sci. Publ., Oxford 1975.

25006 - **WHATLEY, F.R.** : Chloroplasts. - In : **WOLSTENHOLME, G.E.W., FITZSIMONS, D.W.** (ed.) : Energy Transformation in Biological Systems. CIBA Foundation Symp. No. 31. Pp. 41 - 61. Elsevier, Amsterdam 1975.

25007 - **WHATLEY, J.M.** : The occurrence of a peripheral reticulum in plastids of the gymnosperm, *Welwitschia mirabilis*. - New Phytol. *74* : 215 - 220, 1975.

25008 - **WHATLEY, J.M.** : Chloroplast structure in coiled and uncoiling croziers of *Pilularia globulifera*. - New Phytol. *74* : 413 - 420, 1975.

25009 - **WHEELER, C.T., LAWRIE, A.C.** : 35. Nitrogen fixation in root nodules of alder and pea in relation to the supply of photosynthetic assimilates. - In : **NUTMAN, P.S.** (ed.) : Symbiotic Nitrogen Fixation in Plants. International Biological Programme, Vol. 7. Pp. 497 - 509. Cambridge Univ. Press, Cambridge - London - New York - Melbourne 1975.

25010 - **WHITE, W.S., WETZEL, R.G.** : Nitrogen, phosphorus, particulate and colloidal carbon content of sedimenting seston of a hard-water lake. - Verh. int. Verein. Limnol. *19* : 330 - 339, 1975. [Chl.]

25011 - **WHITESIDE, W.F., VANDEMARK, J.S., SPLITTSTOESSER, W.E.** : Changes in various constituents of onion as influenced by two light regimes. - HortScience *10* : 18 - 20, 1975. [Chl.]

25012 - **WHITMARSH, J., LEVINE, R.P.** : Fluorescence polarization measurements excited by horizontally polarized light. - In : **AVRON, M.** (ed.) : Proceedings of the Third International Congress on Photosynthesis. Vol. I. Pp. 223 - 228. Elsevier, Amsterdam - Oxford - New York 1975.

25013 - **WHITTAKER, R.H., LIKENS, G.E.** : The biosphere and man. - In : **LIETH, H., WHITTAKER, R.H.** (ed.) : Primary Productivity of the Biosphere. Pp. 305 - 328. Springer-Verlag, Berlin - Heidelberg - New York 1975. [Chl, primary production.]

25014 - **WHITTAKER, R.H., LIKENS, G.E., LIETH, H.** : Scope and purpose of this volume. - In : **LIETH, H., WHITTAKER, R.H.** (ed.) : Primary Productivity of the Biosphere. Pp. 3 - 5. Springer-Verlag, Berlin - Heidelberg - New York 1975. [Primary productivity.]

25015 - **WHITTAKER, R.H., MARKS, P.L.** : Methods of assessing terrestrial productivty. - In : **LIETH, H., WHITTAKER, R.H.** (ed.) : Primary Productivity of the Biosphere. Pp. 55 - 118. Springer-Verlag, Berlin - Heidelberg - New York 1975.

25016 - **WHITTLE, S.J., CASSELTON, P.J.** : The chloroplast pigments of the algal classes *Eustigmatophyceae* and *Xanthophyceae*. I. *Eustigmatophyceae*. - Brit. phycol. J. *10* : 179 - 191, 1975.

25017 - **WHITTLE, S.J., CASSELTON, P.J.** : The chloroplast pigments of the algal classes *Eustigmatophyceae* and *Xanthophyceae*. II. *Xanthophyceae*. - Brit. phycol. J. *10* : 192 - 204, 1975.

25018 - **WIEBE, H.H.** : Photosynthesis in wood. - Physiol. Plant. *33* : 245 - 246, 1975.

25019 - **WIĘCKOWSKI, S.** : Metabolism of chloroplast pigments and photosynthetic activity at different phases of leaf ontogenesis. - Pol. ecol. Stud. *1* : 33 - 40, 1975.

25020 - **WIEGERT, R.G.** : Simulation models of ecosystems. - Annu. Rev. Ecol. Syst. *6* : 311 - 338, 1975.

25021 - **WIELAND, N.K., BAZZAZ, F.A.** : Physiological ecology of three codominant successional annuals. - Ecology *56* : 681 - 688, 1975. [Ps.]

25022 - **WIELGOLASKI, F.E.** : Comparative productivity of ecosystems : an introduction. - In : **Van DOBBEN, W.H., LOWE-McCONNELL, R.H.** (ed.) : Unifying Concepts in Ecology. Pp. 65 - 66. Dr. W. Junk b.v. Publ., The Hague; Pudoc, Wageningen 1975.

25023 - **WIELGOLASKI, F.E.** : Primary production of tundra. - In : **COOPER, J.P.** (ed.) : Photosynthesis and Productivity in Different Environments. Pp. 75 - 106. Cambridge Univ. Press, Cambridge - London - New York - Melbourne 1975.

25024 - **WIELGOLASKI, F.E.** : Principles in the use of wide-scale models of tundra data

- In : **WIELGOLASKI, F.E.** (ed.) : Fennoscandian Tundra Ecosystems. Part 2. Animals and Systems Analysis. Pp. 245 - 250. Springer Verlag, Berlin - Heidelberg - New York 1975. [Photosynthates.]

25025 - **WIELGOLASKI, F.E., KJELVIK, S.** : Energy content and use of solar radiation of Fennoscandian tundra plants. - In : **WIELGOLASKI, F.E.** (ed.) : Fennoscandian Tundra Ecosystems. Part I. Plants and Microorganisms. Pp. 201 - 207. Springer Verlag, Berlin - Heidelberg - New York 1975.

25026 - **WIERSMA, J.V., BAILEY, T.B.** : Estimation of leaflet, trifoliolate, and total leaf areas of soybean. - Agron. J. *67* : 26 - 30, 1975.

B25027 - **WIEßNER, W.** : Bioenergetik bei Pflanzen. - VEB Gustav Fischer Verlag, Jena 1975.

25028 - **WIGNARAJAH, K., JENNINGS, D.H., HANDLEY, J.F.** : The effect of salinity on growth of *Phaseolus vulgaris* L. I. Anatomical changes in the first trifoliate leaf. - Ann. Bot. *39* : 1029 - 1038, 1975. [Palisade structure.]

25029 - **WILD, A., RÜHLE, W., GRAHL, H.** : The effect of light intensity during growth of *Sinapis alba* on the electron-transport and the noncyclic photophosphorylation. - In : **MARCELLE, R.** (ed.) : Environmental and Biological Control of Photosynthesis. Pp. 115 - 121. Dr. W. Junk b.v. Publ., The Hague 1975.

25030 - **WILDNER, G.F., HAUSKA, G.A.** : Localization and function of cytochrome 552 in *Euglena gracilis*. - In : **AVRON, M.** (ed.) : Proceedings of the Third International Congress on Photosynthesis. Vol. I. Pp. 525 - 534. Elsevier, Amsterdam - Oxford - New York 1975.

25031 - **WILKINSON, J.F., BEARD, J.B.** : Anatomical responses of "Merion" Kentucky bluegrass and "Pennlawn" red fescue at reduced light intensities. - Crop Sci. *15* : 189 - 194, 1975. [Chloroplast.]

25032 - **WILKINSON, J.F., BEARD, J.B., KRANS, J.V.** : Photosynthetic-respiratory responses of "Merion" Kentucky bluegrass and "Pennlawn" red fescue at reduced light intensities. - Crop Sci. *15* : 165 - 168, 1975.

25033 - **WILLARD, J.M., GIBBS, M.** : Fructose-diphosphate aldolase from blue-green algae. - In : **COLOWICK, S.P., KAPLAN, N.O.** (ed.) : Methods in Enzymology. Vol. 42. Pp. 228 - 234. Academic Press, New York - San Francisco - London 1975.

25034 - **WILLENBRINK, J., RANGONI-KÜBBELER, M., TERSKY, B.** : Frond development and CO_2--fixation in *Laminaria hyperborea*. - Planta *125* : 161 - 170, 1975.

25035 - **WILLERT, D.J. von** : Stomatal control, osmotic potential and the role of inorganic phosphate in the regulation of the Crassulacean acid metabolism in *Mesembryanthemum crystallinum*. - Plant Sci. Lett. *4* : 225 - 229, 1975.

25036 - **WILLERT, D.J. von** : Die Bedeutung des anorganischen Phosphats für die Regulation der Phosphoenolpyruvat Carboxylase von *Mesembryanthemum crystallinum* L. - Planta *122* : 273 - 280, 1975.

25037 - **WILLIAMS, D.W., CHANCELLOR, W.J.** : Irrigated agricultural production response to constraints in energy-related inputs. - Trans. ASAE *18* : 459 - 466, 1975. [Modelling production.]

25038 - **WILLIAMS, G.J., III, LAZOR, R., YOURGRAU, P.** : Temperature adaptations in the Hill reaction of altitudinally and latitudinally diverse populations of *Verbascum thapsus* L. - Photosynthetica *9* : 35 - 39, 1975.

25039 - **WILLIAMS, J.H., WILSON, J.H.H., BATE, G.C.** : The growth of groundnuts (*Arachis hypogaea* L. cv. Makulu Red) at three altitudes in Rhodesia. - Rhodesia J. agr. Res. *13* : 33 - 43, 1975. [Growth analysis.]

25040 - **WILLIAMS, M.W., ARAKAWA, E.T., BIRKHOFF, R.D., HAMM, R.N., SCHWEINLER, H.C., MacRAE, R.A.** : Optical properties of chloroplasts and red blood cells in the vacuum uv. - Rad. Res. *61* : 185 - 190, 1975.

25041 - **WILLIAMS, P.F.** : Growth of broad beans infected by *Botrytis fabae*. - J. hort. Sci. *50* : 415 - 424, 1975. [Growth analysis.]

25042 - **WILLIAMS, R.F.** : The shoot apex, leaf growth and crop production. - J. aust. Inst. agr. Sci. *41* : 18 - 26, 1975. [Growth analysis.]

25043 - **WILLIAMS, R.J.P.** : Proton-driven phosphorylation reactions in mitochondrial and chloroplast membranes. - FEBS Lett. *53* : 123 - 125, 1975.

25044 - **WILLISON, J.H.M.** : The relatiosnhip of plastic deformation in freeze-etching to the orientation of a protein particle. - J. Microscopy *105* : 81 - 85, 1975.

25045 - **WILSON, D.** : Leaf growth, stomatal diffusion resistances and photosynthesis during droughting of *Lolium perenne* populations selected for contrasting stomatal length and frequency. - Ann. appl. Biol. *79* : 67 - 82, 1975.

25046 - **WILSON, D.** : Stomatal diffusion resistances and leaf growth during droughting of *Lolium perenne* plants selected for contrasting epidermal ridging. - Ann. appl. Biol. *79* : 83 - 94, 1975.

25047 - **WILSON, D.** : Variation in leaf respiration in relation to growth and photosynthesis of *Lolium*. - Ann. appl. Biol. *80* : 323 - 338, 1975.

25048 - **WILSON, J.R.** : Influence of temperature and nitrogen on growth, photosynthesis and accumulation of non-structural carbohydrate in a tropical grass, *Panicum maximum* var. *trichoglume*. - Neth. J. agr. Sci. *23* : 48 - 61, 1975.

25049 - **WILSON, J.R.** : Comparative response to nitrogen deficiency of a tropical and temperate grass in the interrelation between photosynthesis, growth, and the accumulation of non-structural carbohydrate. - Neth. J. agr. Sci. *23* : 104 - - 112, 1975.

25050 - **WILTENS, J., SCHREIBER, U., VIDAVER, W.** : Reversible effects of desiccation on the photosynthetic apparatus of an intertidal red algae. - Plant Physiol. *56* (Suppl.) : 20, 1975.

25051 - **WINDSOR, M.W., ROCKLEY, M.G., COGDELL, R.J., PARSON, W.W.** : Picosecond flash photolysis and spectroscopy and kinetics of intermediates in bacterial photosynthesis. - In : **JOUSSOT-DUBIEN, J.** (ed.) : Lasers in Physical Chemistry and Biophysics. Pp. 369 - 378. Elsevier sci. Publ. Comp., Amsterdam 1975.

25052 - **WINKLER, E., GAMPER, L., SCHWIENBACHER-MASCOTTI, M.** : Die Stoffproduktion verschiedener Maissorten im Zentralalpenraum (Brixen - Innsbruck - Rinn) in Abhängigkeit vom Temperatur- und Niederschlagsfaktor. - Veröff. Museum Ferdinand. *55* : 253 - 292, 1975.

25053 - **WINTER, D.F., BANSE, K., ANDERSON, G.C.** : The dynamics of phytoplankton blooms in Puget Sound, a fjord in the northwestern United States. - Mar. Biol. *29* : 139 - 176, 1975.

25054 - **WIRTH, V., TÜRK, R.** : Zur SO_2-Resistenz von Flechten verschiedener Wuchsform. - Flora *164* : 133 - 143, 1975.

25055 - **WITHERS, N., HAXO, F T.** : Chlorophyll c_1 and c_2 and extraplastidic carotenoids in the dinoflagellate, *Peridinium foliaceum* STEIN. - Plant Sci. Lett. *5* : 7 - 15, 1975.

25056 - **WITHERS, N.W., HAXO, F.T.** : Pigments of the binucleate marine dinoflagellate *Peridinium foliaceum* STEIN. - J. Phycol. *11* (Suppl.) : 15, 1975.

25057 - **WITT, H.T.** : Primary acts of energy conservation in the functional membrane of photosynthesis. - In : **GOVINDJEE** (ed.) : Bioenergetics of Photosynthesis. Pp. 493 - 554. Academic Press, New York - San Francisco - London 1975.

25058 - **WIUM-ANDERSEN, S.** : The influence of the zooplankton anaesthetising substance Physostigmine salicylicum on photosynthesis. - Arch. Hydrobiol. *76* : 379 - - 383, 1975.

25059 - **WOJCIESKA, U., ŚLUSARCZYK, M.** : Dystrybucja produktów fotosyntezy w źdźbłach długo- i krótkosłomych pszenic ozimych. [Distribution of photosynthetic products in ahort and long culms of winter wheats.] - Acta agrobot. *28* : 263 - 273, 1975. [In Pol., ab : E.]

25060 - **WOJCIESKA, U., SZCZYPA, E.** : Transport asymilatów wytworzonych w poszczególnych organach roślin zbożowych w różnych fazach wzrostu i rozwoju. [Translocation of photosynthates produced in different organs of cereal plants at various stages of growth and development.] - Pamiet. puławski *62* : 51 - 67, 1975. [In Pol., ab : E, R.]

25061 - **WOJCIESKA, U., WOLSKA, E.** : Wpływ chlorku chlorocholiny na aktywność fotosyntetyczną i transport asymilatów do korzeni żyta. [The effect of CCC on photosynthetic activity and photosynthate transport to rye roots.] - Pamiet. puławski *62* : 23 - 37, 1975. [In Pol., ab : E, R.]

25062 - **WOLFF, C., GLÄSER, M., WITT, H.T.** : Studies on the photochemical active chlorophyll-a_{II} in system II of photosynthesis. - In : **AVRON, M.** (ed.) : Proceedings of the Third International Congress on Photosynthesis. Vol.I. Pp. 295 - - 305. Elsevier, Amsterdam - Oxford - New York 1975.

25063 - **WOLIŃSKA, D.** : Starzenie się chloroplastów roślin wyższych. [Ageing of chloroplasts of higher plants.] - Wiadom. bot. *19* : 165 - 180, 1975. [In Pol.]

B25064 - **WOLKEN, J.J.** : Photoprocesses, Photoreceptors, and Evolution. - Academic Press, New York - San Francisco - London 1975. [Ps.]

25065 - **WOLPERT, J.S., ERNST-FONBERG, M.L.** : A multienzyme complex for CO_2 fixation. - Biochemistry *14* : 1095 - 1102, 1975. [Purification.]

25066 - **WOLPERT, J.S., ERNST-FONBERG, M.L.** : Dissociation and characterization of enzymes from a multienzyme complex involved in CO_2 fixation. - Biochemistry *14*: 1103 - 1107, 1975.

25067 - **WONG, W., SACKETT, W.M., BENEDICT, C.R.** : Isotope fractionation in photosynthetic bacteria during carbon dioxide assimilation. - Plant Physiol. *55* : 475- - 479, 1975.

25068 - **WOOD, F.E., CUSANOVICH, M.A.** : The reaction of *Rhodospirillum rubrum* cytochrome c_2 with iron hexacyanides. - Bioinorg. Chem. *4* : 337 - 352, 1975.

25069 - **WOOD, K.G.** : Photosynthesis of *Cladophora* in relation to light and CO_2 limitation; $CaCO_3$ precipitation. - Ecology *56* : 479 - 484, 1975.

25070 - **WOOD, P.M., BENDALL, D.S.** : The kinetics and specificity of electron transfer from cytochromes and copper proteins to *P*700. - Biochim. biophys. Acta *387* : 115 - 128, 1975.

25071 - **WOODWARD, J., MERRETT, M.J.** : Induction potential for glyoxylate cycle enzymes during the cell cycle of *Euglena gracilis*. - Europe. J. Biochem. *55* : 555 - 559, 1975.

25072 - **WOOLHOUSE, H.W.** : Membrane structure and transport problems considered in relation to phosphorus and carbohydrate movements and regulation of endotrophic mycorrhizal associations. - In : **SANDERS, F.E., MOSSE, B., TINKER, P.B.** (ed.): Endomycorrhizas. Pp. 209 - 239. Academic Press, London 1975.

25073 - **WOOLHOUSE, H.W., BATT, T.** : The nature and regulation of senescence in plastids. - In : **SUNDERLAND, N.** (ed.) : Perspectives in Experimental Biology. Vol. 2. Botany. Pp. 163 - 175. Pergamon Press, Oxford - New York - Toronto - Sydney - Paris - Braunschweig 1975.

25074 - **WOOLLEY, P.** : Models for metal ion function in carbonic anhydrase. - Nature *258* : 677 - 682, 1975.

25075 - **WOŹNY, A., SZWEYKOWSKA, A.** : Effect of cytokinins and antibiotics on chloroplast development in cotyledons of *Cucumis sativus*. - Biochem. Physiol. Pflanzen *168* : 195 - 209, 1975.

25076 - **WRAIGHT, C.A., COGDELL, R.J., CLAYTON, R.K.** : Some experiments on the primary electron acceptor in reaction centres from *Rhodopseudomonas sphaeroides*. - Biochim. biophys. Acta *396* : 242 - 249, 1975.

25077 - **WRISCHER, M., LJUBESIC, N., DEVIDÉ, Z.** : Ultrastructural studies of plastids in leaves of *Fraxinus excelsior* L. var. *aurea* (WILLD). - J. Microscop. Biol. cellulaire *23* : 105 - 112, 1975.

25078 - **WRISCHER, M., LJUBEŠIĆ, N., DEVIDÉ, Z.** : Transformation of plastids in the leaves of *Acer negundo* L. var. *odessanum* (H.ROTHE). - J. Cell Sci. *18* : 509 - - 518, 1975.

25079 - **WU, J.H., SKOKUT, T.A.** : Retardation of ultraviolet light accelerated senescence by visible light or by benzyladenine in *Nicotiana glutinosa* leaves. - Plant Physiol. *56* (Suppl.) : 75, 1975. [Chl.]

25080 - **WUJEK, D.E., CAMBURN, K.E., ANDREWS, H.T.** : An ultrastructural study of pyrenoids in *Leptosiropsis torulosa*. - Protoplasma *86* : 263 - 268, 1975.

25081 - **WYDRZYNSKI, T., GOVINDJEE** : A new site bicarbonate effect in photosystem II of photosynthesis : Evidence from chlorophyll fluorescence transients in spinach chloroplasts. - Biochim. biophys. Acta *387* : 403 - 408, 1975.

25082 - **WYDRZYNSKI, T., GOVINDJEE** : Bicarbonate effect on fluorescence transients in Tris-washed and heat treated chloroplasts with various electron donors. - Biophys. J. *15* (2, Pt. 2) : 222a, 1975.

25083 - **WYDRZYNSKI, T., GROSS, E.L., GOVINDJEE** : Effects of sodium and magnesium cations on the "dark"- and light-induced chlorophyll *a* fluorescence yields in sucrose-washed spinach chloroplasts. - Biochim. biophys. Acta *376* : 151 - 161, 1975.

25084 - **WYDRZYNSKI, T., ZUMBULYADIS, N., SCHMIDT, P.G., GOVINDJEE** : Water proton relaxation as a monitor of membrane-bound manganese in spinach chloroplasts. - Biochim. biophys. Acta *408* : 349 - 354, 1975.

25085 - **WYDRZYNSKI, T., ZUMBULYADIS, N., SCHMIDT, P.G., GUTOWSKY, H.S., GOVINDJEE** : Proton relaxation and charge accumulation during oxygen evolution in photosynthesis. - Proc. nat. Acad. Sci. USA *73* : 1196 - 1198, 1975.

25086 - **YAJIMA, M., TAKEDA, T.** : [An improvement of semiempirical method for estimating the total photosynthesis of the crop population. II. On light attenuation and vertical distribution of P_{max} in rice plant population.] - Proc. Crop Sci. Soc. Jap. *44* : 350 - 356, 1975. [In Jap., ab : E.]

25087 - **YAKOVLEV, A.P., OVCHINNIKOVA, M.F.** : Ul'trastruktura i fotokhimicheskaya aktivnost' khloroplastov gibridnoĭ i inbrednoĭ kukuruzy. [Ultrastructure and photochemical activity of chloroplasts of hybrid and inbred maize.] - Nauch. Dokl. vyssh. Shkoly, biol. Nauki *18* (4) : 112 - 116, 1975. [In R.]

25088 - **YAKUBOVA, M.M., RUBIN, A.B., KHRAMOVA, G.A., MATORIN, D.N.** : Hill reaction and delayed fluorescence in mutants of *Gossypium hirsutum*. - In : **NASYROV, Yu.S., ŠESTÁK, Z.** (ed.) : Genetic Aspects of Photosynthesis. Pp. 263 - 269. Dr. W. Junk B.V. Publ., The Hague 1975.

25089 - **YAKUSHEV, B.I., MOVCHUN, A.V.** : Vliyanie travyanistoĭ rastitel'nosti na rost i dinamiku uglevodov v organakh sosny obyknovennoĭ. [Effect of herbaceous population on the growth and carbohydrate dynamics in roots of *Pinus sylvestris*.] - In : Ėkologo-biologicheskie Issledovaniya Rastitel'nykh Soobshchestv. Pp. 102 - 106. Nauka i Tekhnika, Minsk 1975. [In R.]

25090 - **YAKUSHKINA, N.I., DULIN, A.F.** : Osobennosti vliyaniya fitogormonov na protsess fotofosforilirovaniya v khloroplastakh yachmenya. [The features of the phytohormones effect on photophosphorylation in barley chloroplasts.] - Nauch. Dokl. vyssh. Shkoly, biol. Nauki *18* (4) : 78 - 82, 1975. [In R.]

25091 - **YAKUSHKINA, N.I., PUSHKINA, G.P.** : Izmenenie intensivnosti fotofosforilirovaniya v prorostkakh kukuruzy pod vliyaniem gibberellina i kinetina. [Effect of gibberellin and kinetin on the rate of photophosphorylation in maize seedlings.] - Fiziol. Rast. *22* : 1132 - 1137, 1975. [In R, ab : E.]

25092 - **YAMADA, M., NAKAMURA, Y.** : Fatty acid synthesis by spinach chloroplasts II. The path from PGA to fatty acids. - Plant Cell Physiol. *16* : 151 - 162, 1975.

25093 - **YAMAMOTO, H.Y., CHENCHIN, E.E., YAMADA, D.K.** : Effect of chloroplast lipids on violaxanthin de-epoxidase activity. - In : **AVRON, M.** (ed.) : Proceedings of the Third International Congress on Photosynthesis. Vol. III. Pp. 1999 - - 2006. Elsevier, Amsterdam - Oxford - New York 1975.

25094 - **YAMAMOTO, T., TONOMURA, Y.** : pH jump-induced phosphorylation of adenosine diphosphate in thylakoidal membranes. Dependence of the rate on pH and concentrations of substrates. - J. Biochem. (Tokyo) *77* : 137 - 146, 1975.

25095 - **YAMASHITA, T.** : [Metabolism of glycine through the glycolate pathway and carbon dioxide production in mulberry leaves.] - J. sericult. Sci. Jap. *44* : 294 - 300, 1975. [In Jap., ab : E.]

25096 - **YAMASHITA, T., TOMITA, G.** : Comparative study of the reactivation of oxygen evolution in chloroplasts inhibited by various treatments. - Plant Cell Physiol. *16* : 283 - 296, 1975.

25097 - **YAMASHITA, T., TSUTSUMI, M., YOSHINARI, S.** : [Changes in the activities of enzyme related to carbon dioxide fixation and glycolate pathway with the age of mulberry leaves.] - J. sericult. Sci. Jap. *44* : 1 - 6, 1975. [In Jap., ab: E.]

25098 - **YAMAZAKI, S., TAKISAWA, H., TAMAURA, Y., HIROSE, S., INADA, Y.** : Inhibition of adenosine triphosphatase activity of chloroplast coupling factor (CF_1) by troponin component, TN-I. - FEBS Lett. *56* : 248 - 251, 1975.

25099 - **YASNIKOV, A.A., BERSHTEĬN, B.I., VOLKOVA, N.V., VASILENOK, L.I., VOLOVIK, O. I., DUBROVSKAYA, A.A., ZAĬTSEVA, N.A., KANIVETS, N.P., MUSHKETIK, L.S., OKANENKO, A.S., OSTROVSKAYA, L.K., POLISHCHUK, A.I., REĬNGARD, T.A.** : Dva mekhanizma svetozavisimogo transporta protona v khloroplastakh rasteniĭ. [Two mechanisms of the light-dependent transport of proton in chloroplasts of plants.] - Dokl. Akad. Nauk SSSR *224* : 1449 - 1452, 1975. [In R.]

25100 - **YASNIKOV, A.A., BERSHTEĬN, B.I., VOLKOVA, N.V., VASILENOK, L.I., VOLOVIK, O. I., ZAĬTSEVA, N.A., KANIVETS, N.P., MUSHKETIK, L.S., OKANENKO, A.S., OSTROVSKAYA, L.K., PETRENKO, S.S., POLISHCHUK, A.I., REĬNGARD, T.A., SEMENYUK, I.I.**: Regulation by pyruvate kinase and phosphatase of inorganic phosphate incorporation during photophosphorylation. - In : **NASYROV, Yu.S., ŠESTÁK, Z.** (ed.) : Genetic Aspects of Photosynthesis. Pp. 287 - 293. Dr. W. Junk B.V. Publ., The Hague 1975.

25101 - **YASNIKOV, A.A., VOLKOVA, N.V., ZAĬTSEVA, N.A., KANIVETS, N.P., VASILENOK, L. I., MUSHKETIK, N.S., BERSHTEĬN, B.I., OKANENKO, A.S., OSTROVSKAYA, L.K., REĬNGARD, T.A.** : Light-dependent transport of protons in maize chloroplasts. - Photosynthetica *9* : 311 - 317, 1975.

25102 - **YASNIKOVA, E.A.** : Pigmenty pochek zimuyushchikh rasteniĭ i ikh uchastie v fotokhimicheskikh reaktsiyakh. [Pigments of hibernating plant buds and their participation in photochemical reactions.] - Fiziol. Biokhim. kul't. Rast. *7* : 603 - 606, 1975. [In R, ab : E.]

25103 - **YIM, Y.-J.** : Dry matter production and leaf area index of herb community in central Korea. - Korean J. Bot. *18* (3) : 87 - 91, 1975.

25104 - **YOCH, D.C., ARNON, D.I., SWEENEY, W.V.** : Characterization of two soluble ferredoxins as distinct from bound iron-sulfur proteins in the photosynthetic bacterium *Rhodospirillum rubrum*. - J. biol. Chem. *250* : 8330 - 8336, 1975.

25105 - **YOCUM, C.F., NELSON, N., RACKER, E.** : A combined procedure for preparation of plastocyanin, ferredoxin and CF_1. - Prep. Biochem. *5* : 305 - 317, 1975.

25106 - **YOCUM, C.S., LOMMEN, P.W.** : Mesophyll resistances. - In : **GATES, D.M., SCHMERL, R.B.** (ed.) : Perspectives of Biophysical Ecology. Pp. 45 - 54. Springer Verlag, Berlin - Heidelberg - New York 1975.

25107 - **YORDANOV, I., ZEINALOV, Y., STAMENOVA, M.** : Influence of post-action of high temperatures on photosynthetic activity, composition of lamellar proteins and spectral characteristics of pigment systems I and II. - Biochem. Physiol. Pflanzen *168* : 567 - 573, 1975.

25108 - **YOSHIDA, F., KOHNO, H., KAIDO, A.** : The mineral nutrition of cultured chlorophyllous cells of tobacco. (III) Kinetic studies of H_2PO_4 and SO_4 uptakes by the cells. - Bull. Fac. Agr., Tamagawa Univ. *15* : 10 - 21, 1975.

25109 - **YOSHIDA, K., GOTOH, K.** : [Translocation and distribution of ^{14}C-assimilates related to stem termination habits in soybeans.] - Proc. Crop Sci. Soc. Jap. *44* : 185 - 193, 1975. [In Jap., ab : E.]

25110 - **YOSHIDA, M., SONE, N., HIRATA, H., KAGAWA, Y., TAKEUCHI, Y., OHNO, K.** : ATP synthesis catalyzed by purified DCCD-sensitive ATPase incorporated into reconstituted purple membrane vesicles. - Biochem. biophys. Res. Commun. *67* : 1295 - 1300, 1975.

25111 - **YOSHIDA, T., MOSS, D.N., RASMUSSON, D.C.** : Effect of stomatal frequency in barley on photosynthesis and transpiration. - Bull. Kyushu agr. exp. Sta. *18*: 71 - 80, 1975.

25112 - **YOUSEF, Y.A., PADDEN, T.J., GLOYNA, E.F.** : Diurnal changes in radionuclides uptake by phytoplankton in small scale ecosystems. - Water Res. *9* : 181 - 187. 1975. [Ps.]

25113 - **YU, W., ALFANO, R.R., SEIBERT, M.** : Fluorescent kinetics of chlorophyll in Photosystem I and Photosystem II enriched fractions of spinach. - Biophys. J. *15* (2, Pt. 2): 222a, 1975.

25114 - **YU, W., HO, P.P., ALFANO, R.R., SEIBERT, M.** : Fluorescent kinetics of chlorophyll in Photosystem I and II enriched fractions of spinach. - Biochim. biophys. Acta *387* : 159 - 164, 1975.

25115 - **YURKEVICH, I.D., GOLOD, D.S., PARFENOV, V.I.** : Voprosy biogeotsenologii i biologicheskaya produktivnost' elovykh lesov Belorussii. [Problems of biogeocoenology and biological productivity of spruce forests in Belorussia.] - - In : Ėkologo-biologicheskie Issledovaniya Rastitel'nykh Soobshchestv. Pp. 14 - 27. Nauka i Tekhnika, Minsk 1975. [In R.]

*25116 - **YUSHKOV, P.I., KULIKOV, N.V.** : Deĭstvie khronicheskogo *gamma*-oblucheniya na molodye seyantsy sosny obyknovennoĭ. [Effect of chronical *gamma* irradiation on young seedlings of pine.] - Tr. Inst. Ėkol. Rast. Zhivot. *67* : 291 - 293, 1970. [Photosynthates; in R.]

25117 - **YUSUFOV, A.G., ASHUROVA, O.B.** : Starenie ukorenennykh list'ev. [Aging of rooted leaves.] - Fiziol. Rast. *22* : 741 - 746, 1975. [Chl; in R, ab : E.]

25118 - **ZABKA, G.G., CHATURVEDI, S.N.** : Water conservation in *Kalanchoe blossfeldiana* in relation to carbon dioxide dark fixation. - Plant Physiol. *55* : 532 - 535, 1975.

25119 - **ZABLEN, L., WOESE, C.R.** : Procaryote phylogeny IV : Concerning the phylogenetic status of a photosynthetic bacterium. - J. mol. Evol. *5* : 25 - 34, 1975.

25120 - **ZABLEN, L.B., KISSIL, M.S., WOESE, C.R., BUETOW, D.E.** : Phylogenetic origin of chloroplast and prokaryotic nature of its ribosomal RNA. - Proc. nat. Acad. Sci. USA *72* : 2418 - 2422, 1975.

25121 - **ZACHLEDER, V., DOUCHA, J., BERKOVÁ, E., ŠETLÍK, I.** : The effect of synchronizing dark period on populations of *Scenedesmus quadricauda*. - Biol. Plant. *17*: 416 - 433, 1975. [Ps, Chl.]

25122 - **ZAKRZHEVSKIĬ, D.A., KALASHNIKOV, Yu.E., SINYAKOVA, R.S., KUTYURIN, V.M.** : K voprosu o roli ėkzogennogo vodoroda pri fotoreduktsii vodorosleĭ. [Role of exogenous hydrogen in photoreduction of algae.] - Dokl. Akad. Nauk SSSR *222* : 493 - 496, 1975. [Ps; in R.]

25123 - **ZALENSKY, O.V.** : Potential photosynthesis of Central Asian desert plants. - In: **COOPER, J.P.** (ed.) : Photosynthesis and Productivity in Different Environments. Pp. 129 - 132. Cambridge Univ. Press, Cambridge - London - New York - Melbourne 1975.

25124 - **ZAMAZOVA, L.M., KRENDELEVA, T.E.** : Ob al'ternativnykh putyakh transporta ėlektronov, indutsiruemogo fotosistemoĭ I. [Alternative ways of electron transport induced by photosystem I.] - Nauch. Dokl. vyssh. Shkoly, biol. Nauki *18* (4) : 49 - 55, 1975. [In R.]

25125 - **ZASLONKIN, V.P.** : Ottok plasticheskikh veshchestv iz list'ev gorokha i vliyanie na ėtot protsess molibdena. [Efflux of photosynthates from pea leaves as affected by molybdenum.] - Nauch. Tr. orlov. (shatilovskoĭ) obl. sel'.-khoz. opyt. Sta. Lisitsyna *7* : 297 - 306, 1975. [In R.]

25126 - **ZAVITKOVSKI, J., SALMONSON, B.J.** : Effects of *gamma* radiation on biomass production of ground vegetation under broadleaved forests of northern Wisconsin. - Rad. Bot. *15* : 337 - 348, 1975.

25127 - **ZAVODNIK, N.** : Effects of temperature and salinity variations on photosynthesis of some littoral seaweeds of North Adriatic sea. - Bot. mar. *18* : 245 - - 250, 1975.

25128 - **ZAZZERINI, A.** : Fotosintezi e respirazione in due cultivars di peperone infettate von *Verticillium dahliae* KLEB. [Photosynthesis and respiration in two cultivars of *Capsicum* infected with *Verticillium dahliae* KLEB.] - Ann. Fac. Agr. Univ. Studi Perugia *30* : 157 - 165, 1975. [In Ital., ab : E.]

25129 - **ZBYTNIEWSKI, Z., DREWA, G., PAUTSCH, F.** : Effect of detergents and of phospho-gypsum on the oxygen and chlorophyll *a* levels and on the dry weight of the residue of brackish water under laboratory conditions. - Merentutkimuslaitoksen Julk / Havsforskningsinst. Skr. *239* : 100 - 104, 1975.

25130 - **ZDANOWSKI, B., BNIŃSKA, M., KORYCKA, A., SOSNOWSKA, J.** : The effect of mineral fertilization on ecosystem structure and functioning in lakes of different trophic type. Part I. The effect of lake fertilization on changes in chemical composition of water and macrophytes, chlorophyll content and primary production of pelagic zone. - Pol. Arch. Hydrobiol. *22* : 217 - 232, 1975.

25131 - **ZEÏNALOV, Yu., PETKOVA, R.** : Effect of temperature rise on the oxygen flash yields and on the oxygen burst under continuous irradiance in isolated spinach chloroplasts. - Photosynthetica *9* : 288 - 292, 1975.

25132 - **ZELDIN, M.H., COHEN, C.E., BEN-SHAUL, Y., SCHIFF, J.A.** : Measurement *in vivo* of light-induced spectroscopic changes of protochlorophyll(ide) and chlorophyll(ide) in *Euglena*. - Plant Physiol. *56* (Suppl.) : 33, 1975.

25133 - **ZELDIN, M.H., SCHIFF, J.A.** : Absorption changes in extracts from *Euglena gracilis* var. *bacillaris* mutant W_3BUL on blue illumination. - Plant. Physiol. *56* (Suppl.) : 33, 1975.

25134 - **ZELENSKIĬ, M.I., MOGILEVA, G.A.** : Ob otsenke sostoyaniya fotosinteticheskogo apparata rasteniĭ po fotokhimicheskoĭ aktivnosti khloroplastov. [Evaluation of state of the photosynthetic apparatus of plants using the photochemical activity of chloroplasts.] - Byull. vses. nauch.-issled. Inst. Rastenievod. N.I. Vavilova *56* : 31 - 36, 1975. [In R.]

25135 - **ZELENSKIĬ, M.I., MOGILEVA, G.A.** : Potentsiometricheskoe issledovanie kinetiki reaktsii Khilla : zavisimost' ot kontsentratsii ferritsianida. [Effect of ferricyanide concentration on kinetics of Hill reaction studied by potentiometry.] - Fiziol. Rast. *22* : 636 - 646, 1975. [In R, ab : E.]

25136 - **ZELITCH, I.** : Improving the efficiency of photosynthesis. - Science *188* : 626-- 633, 1975.

25137 - **ZELITCH, I.** : Environmental and biological control of photosynthesis : general assessment. - In : **MARCELLE, R.** (ed.) : Environmental and Biological Control of Photosynthesis. Pp. 251 - 262. Dr. W. Junk b.v. Publ., The Hague 1975.

25138 - **ZELITCH, I.** : Pathways of carbon fixation in green plants. - Annu. Rev. Biochem. *44* : 123 - 145, 1975.

25139 - **ZELITCH, I.** : Getting more CO_2 into food. - Nature *256* : 90 - 91, 1975. [Photorespiration.]

25140 - **ZEMÁNEK, M.** : Vliv retardantu CCC a antitranspirantu TAG na vodní provoz a tvorbu výnosu jarní pšenice. [The effect of retardant CCC and antitranspirant TAG on water balance and yield formation in spring wheat.] - Rostlinná Výroba (Praha) *21* : 379 - 392, 1975. [In Czech, ab : E, R.]

25141 - **ZEN'KEVICH, E.I., LOSEV, A.P., GURINOVICH, G.P.** : Migratsiya ènergii èlektronnogo vozbuzhdeniya v smeshannykh assotsiatakh khlorofilla i ego proizvodnykh. [Migration of electron excitation energy in mixed associates of chlorophyll and its derivatives.] - Mol. Biol. (Moskva) *9* : 516 - 523, 1975. [In R, ab : E.]

25142 - **ZHEREBCHUK, L.K., OLEVINSKAYA, Z.M.** : Vliyanie gibberellina na soderzhanie khlorofilla i karotinoidov v list'yakh zdorovogo i porazhennogo X-virusom kartofelya. [Effect of gibberellin on content of chlorophyll and carotenoids in potato leaves of healthy and X-virus affected plants.] - Fiziol. Biokhim. kul't. Rast. *7* : 86 - 91, 1975. [In R, ab : E.]

25143 - **ZHIGALOVA, T.V., GAVRILENKO, V.F.** : Fiziologicheskaya aktivnost' khloroplastov sortov pshenitsy razlichnoĭ produktivnosti pod vliyaniem nekotorykh faktorov. [Physiological activity of chloroplasts of wheat cultivars with different productivity under the effect of some factors.] - Sel'skokhoz. Biol. *10* : 832 - 840, 1975. [In R, ab : E.]

*25144 - **ZHIZNEVSKAYA, G.Ya.** : Uchastie metallov - mikroèlementov medi, zheleza i molibdena - v formirovanii i funktsionirovanii fotosinteticheskogo apparata zelenogo lista. [Participation of metals - trace elements copper, iron and molybdenum - in the formation and functioning of the photosynthetic apparatus of a green leaf.] - In : Mikroèlementy - Regulyatory Zhiznedeyatel'nosti i Produktivnosti Rasteniĭ. Pp. 7 - 36. Zinatne, Riga 1971. [In R.]

*25145 - **ZHIZNEVSKAYA, G.Ya., MUTUSKIN, A.A.** : Med' i zhelezo list'ev bobovykh v fazu obrazovaniya generativnykh organov. [Copper and iron in leaves of *Fabaceae* in the phase of formation of generative organs.] - Agrokhimiya *1973* (10) : 97 - - 102, 1973. [Ferredoxin, plastocyanin; in R.]

25146 - **ZHMURKO, L.G., BOBYR, A.D.** : Deyaki fiziologo-biokhimichni osoblyvosti vybirnoï fitotoksychnosti tryazyniv na foni virusnoï infektsiï. [Certain physiological and biochemical peculiarities in selective phytotoxicity of triazines against background of viral infection.] - Mikrobiol. Zh. *37* (2) : 197 - 201, 1975. [Chl; in Ukr., ab : E, R.]

25147 - **ZHUKOVA, G.Ya.** : Problema proiskhozhdeniya i èvolyutsii plastid v svete dannykh èmbriologii rasteniĭ. [Problem of origin and evolution of plastids as elucidated by the data of plant embryology.] - Bot. Zh. *60* : 713 - 738, 1975. [In R.]

25148 - **ZIEGLER, I., LIBERA, W.** : The enhancement of CO_2 fixation in isolated chloroplasts by low sulfite concentrations and by ascorbate. - Z. Naturforsch. *30C*: 634 - 637, 1975.

25149 - **ZIEMAN, J.C.** : Quantitative and dynamic aspects of the ecology of turtle grass, *Thalassia testudinum*. - In : Estuarine Research. Vol. I. Chemistry, Biology and the Estuarine System. Pp.541 - 562. Academic Press, New York - San Francisco - London 1975. [Production.]

25150 - **ZIEMAN, J.C.** : Seasonal variation of turtle grass, *Thalassia testudinum* KÖNIG, with reference to temperature and salinity effects. - Aquatic Bot. *1* : 107 - - 123, 1975. [Production.]

25151 - **ZILINSKAS, B., GOVINDJEE** : Silicomolybdate and silicotungstate mediated DCMU--insensitive Photosystem II reaction : electron flow, chlorophyll *a* fluorescence, and delayed light emission changes. - Biophys. J. *15* (2, Pt. 2) : 224a, 1975.

25152 - **ZILINSKAS, B.A., GOVINDJEE** : Silicomolybdate and silicotungstate mediated dichlorophenyldimethylurea-insensitive photosystem II reaction. Electron flow, chlorophyll *a* fluorescence and delayed light emission changes. - Biochim. biophys. Acta *387* : 306 - 319, 1975.

B25153 - **ZIMMERMANN, M.H., MILBURN, J.A.** (ed.) : Transport in Plants I. Phloem Transport. - Springer-Verlag, Berlin - Heidelberg - New York 1975. [Photosynthates.]

25154 - **ZINGMARK, R.G., MILLER, T.G.** : The effects of mercury on the photosynthesis and growth of estuarine and oceanic phytoplankton. - In : **VERNBERG, F.J.** (ed.): Physiological Ecology of Estuarine Organisms. Vol. 3. Pp. 45 - 57. Belle W. Baruch Library mar. Sci., Columbia, S.C. 1975.

25155 - **ZINKIEWICZ, E., SKIBA, T., BURSZEWSKA-SAMIEC, H.** : Wstępne badania nad możliwością zastosowania szybkiej metody izotopowej do porównania intensywności fotosyntezy liści lucerny. [Preliminary investigations on the possibility of applying a rapid isotopic method for comparing photosynthetic rate in alfalfa leaves.] - Pamięt. puławski *62* : 7 - 22, 1975. [In Pol., ab : E, R.]

*25156 - **ZOLOTOV, V.I., FEVRALEV, V.S.** : Produktivnost' fotosinteza gibridov kukuruzy v svyazi s udobreniem i gustotoĭ rasteniĭ. [Photosynthetic efficiency of maize hybrids in relation to fertilizer application and plant density.] - Byul. vses. nauch.-issled. Inst. Kukuruzy (Dnepropetrovsk) *25* : 29 - 32, 1972. [In R.]

25157 - **ZSOLNAY, A.** : Total labile carbon in the euphotic zone of the Baltic Sea as measured by BOD. - Mar. Biol. *29* : 125 - 128, 1975. [Chl.]

*25158 - **ŻURBICKI, Z.** : Atmospheric electricity and plant nutrition. - Acta horticult. *29* : 413 - 427, 1973. [Ps.]

25159 - **ZÜRRER, H., BACHOFEN, R.** : Austauschreaktionen zwischen ATP and Pyrophosphat in isolierten Chloroplasten. - Bull. schweiz. bot. Ges. *85* : 85 - 95, 1975.

25160 - **ZURZYCKI, J.** : Adjustment processes of the photosynthetic apparatus to light conditions, their mechanism and biological significance. - Pol. ecol. Stud. *1* : 41 - 49, 1975.

25161 - **ZURZYCKI, J.** : Improvements of the microrespirometric method for the measurements of photosynthesis. - Pol. ecol. Stud. *1* : 93 - 94, 1975.

E R R A T A

Ref. no.	For	Read
Volume 1, Part 2		
5674	4674	5674
Volume 5, Part 1		
19346	Foto intetIcheskogo Foto inteza	FotosintetIcheskogo Fotosinteza
19367	FotointetIcheskaya	FotosintetIcheskaya
Volume 5, Part 2		
p.181 last line	--	BAKRI, M.D.L. 15490

Authors' names are presented in the form in which they appear in the respective publication. The names from papers published in Cyrillic characters are transcribed as shown on p. III of this volume. Alternative spellings and forms of the name of the same author are usually cross-indexed. The numbers in *italics* refer to publications in which the respective author acts as an editor.

A

AARONSON, S. 21505
ABDULLAEV, H.A. 24847
see ABDULLAEV, Kh.A.
ABDULLAEV, Kh.A. 24848
see ABDULLAEV, H.A.
ABDULLAEVA, S.K. 22534
ABILOV, Z.K. 21506, 22514
ABRAMYAN, L.Kh. 21700
ABUTALYBOV, M.G. 21506
ACKER, S. 21507, 21922, 24041
ACKERSON, L.C. 22408
ACKERSON, R.C. 21508
ADABRA-MICHANOL, Y. 21509
ADAMCZAK, B. 23646
ADAMS, G.E. *23292*
ADAMS, M. 22321
ADAMS, M.S. 21510, 23596, 23745, 23992, 24773
ADDY, N.D. 23582
ADEDIPE, N.O. 21511-3
ADOLFSEN, R. 21514-5
AÊROV, Ÿ.L. 24775
AFGAN, N.H. *23368*, *24004*, *24214*
AGEE, B.A. 24090
AGHION, J. 22235
AIKAZYAN, V.Ts. 21516
AIKIN, W.J. 21517
AIMI, R. 21518
AIRAPETYAN, R.B. 21642
AITKEN, A. 21519
AKAZAWA, T. 21618-9, 23840-1, 23927, 24681-3
AKHMEDOV, Yu.K. 21520
AKHRAMOVICH, N.I. 24465, 24923
AKINOLA, J.O. 21521-2
AKIYAMA, T. 21523-6, 23897
AKOBUNDU, I.O. 21527
AKOYUNOGLOU, G. 21528, 21604, 23531
AKSENOVICH, A.V. 21529
AKULOVA, A.E. 22894
AKULOVA, E.A. 21530-1, 23709, 23742-3, 23750, 24241
AKULOVICH, E.M. 24465, 24923
AKULOVICH, N.K. 21532
ALAM, M.I. 21533
ALBERDA, T. 24880
ALBERTE, R.S. 21534-7, 21925
ALBERTSSON, P.-Å. 23310, 24793
ALBITSKAYA, O.N. 22054
ALBRECHT, A.C. 24710-2
ALBUZIO, A. 23974
ALDERFER, R.G. 21538
ALDRIDGE, E.F. 24749
ALEǏNIKOV, I.M. 23340
ALEKSEEV, V.A. 21539-41
ALESHIN, A.D. 21542
ALFANO, R.R. 25113-4
AL-HASAN, R.H. 21543
ALI, A. 22662
ALIEV, D.A. 21544
ALIEV, K.A. 21545
ALIEV, Z.Sh. 22514
ALINA, B.A. 22422
ALLAKHVERDOV, B.L. 21546
ALLEN, J.F. 21547-8
ALLEN, L.H.Jr. 21549-50, 24507
ALLEN, M.J. 21551
ALLESSIO, M.L. 21552-3
ALLEWELDT, G. 21554-5
ALLMARAS, R.R. 21556, 24207
ALOFE, C.O. 21557
ALSCHER, R. 21558
ALSOP, W.R. 21559
ALTMAN, A. 21560
AL'ZHANOVA, R.M. 21561
AMANOV, M.A. 23561
AMBARD-BRETTEVILLE, F. 22264
AMBLER, R.P. 21562, 23635
AMBROSAŬ, A.L. 21563
AMBROSOV, A.L. 21564
AMEMIYA, A. 23352
AMERKHANOVA, M.B. 21565
AMESZ, J. 21566-8, 24903
AMEZAGA, A.de 24766
AMILENI, A.R. 22083
AMIRDZHANOV, A.G. 21569
ANANYAN, A.A. 21570-1
ANDERSEN, J.M. 21825
ANDERSEN, W. 21572
ANDERSEN, W.R. 21573
ANDERSON, C.E. 24748
ANDERSON, G.C. 25053
ANDERSON, I.C. 24444
ANDERSON, J.M. 21574-5
ANDERSON, L.E. 21576-80, 22274, 22977, 23949
ANDERSON, L.L. 22069-70
ANDERSON, M.C. 21581
ANDERSON, O.R. 21582, 23527-8
ANDERSON, R. 21510, 23992
ANDERSON, R.E. 21583-4
ANDERSON, R.J. 21585
ANDO, T. 24645
ANDRE, C. 21586
ANDRÉ, M. 23376
ANDREEVA, T.F. 21587

B

BAALEN, C. van see VAN BAALEN, C.
BABA, S. 21641
BABA, Y. 21641
BABAYAN, R.S. 21642
BABCOCK, G.T. 21643-6, 21817-9, 22582
BABUSHKIN, L.N. 21647-8
BACCARINI-MELANDRI, A. 21649-50, 23615, 24095
BACH, B. 23754
BACHOFEN, R. 21651, 21780, 23484, 24034-5, 25159
BADGER, M.R. 21592
BAER, C.H. 24501-2
BAGIYAN, L.G. 23964
BAHL, J. 21653-4
BAHR, J.T. 22940
BAIER, D. 21655
BAILEY, T.B. 23985, 25026
BAIN, J.M. 22033
BAIS, R. 21656
BAJRACHARYA, D. 24368
BAKER, C.H. 21657-8, 22162
BAKER, D.A. *23479*, *23847*, *24169*, *24742*
BAKER, D.N. 21535, 22749, 23587, 23603-4
BAKER, E.F.I. 21659
BAKER, J.B. 21660
BAKER, N.R. 21661-4
BAKER, T.S. 21665
BAKHRAMDZHANOVA, N. 21757
BALANDREAU, J.P. 21666
BALASHOV, S.P. 23432
BALDRY, C.W. 22119-20
BALDY, C. 21667
BALEVA, E.F. 24463-4
BALL, E. 23480-1, 24055
BALTSCHEFFSKY, H. 21668
BALTSCHEFFSKY, M. 21669, 21777, 22947
BAMBERG, S. 22379
BAMBERG, S.A. 21670
BAMBERGER, E.S. 21671-3
BANAI TÓTH, P. 24666
BANASIK, J. 21674
BANERJI, D. 21675
BANNER, R.E. 24044
BANSE, K. 25053
BAQAR, M.R. 24505-6
BARABAL'CHUK, K.A. 21676
BARANINA, I.I. 21677
BARANOV, A.A. 21678
BARANOWSKA, H. 22516
BARANSKIǏ, P.I. 23339
BARBER, J. 21679, 21720, 22766, 23656, 24720-1
BARCKHAUS, R.H. 21680
BARDEN, J.A. 21640
BARICA, J. 21681-2
BARKER, R. 24262, 24265
BARMORE, C.R. 21683
BARNABAS, A.D. 21684
BARNES, D.K. 21960
BARNES, R.L. 23605
BAR-NUN, S. 21685
BARONI, R. 24782
BAROOVA, S.R. 21686
BARR, R. 21687-9, 24238
BARROW, J.R. 22043
BARSKIǏ, E.L. 21690
see BARSKY, E.L.
BARSKY, E.L. 21691
see BARSKIǏ, E.L.
BARTA, A.L. 21692
BARTOŠ, J. 21693
BARTSCH, R.G. 21562, 23563, 23635
BASANTANI, H.T. 22844
BASIOUNY, F.M. 21694
BASNIZKY, J. 21695
BASSHAM, J.A. 21558, 22077, 23370, 24059
BASSI, P.K. 21696
BASZYŃSKI, T. 21697-8, 23227
BATE, G.C. 22128-9, 25039
BATES, J.W. 21699
BATIČ, F. 23556
BATIKYAN, G.G. 21700
BATLLE, A.M.Del.C. 21701
BATT, T. 21702, 25073
BATTERSBY, A.R. 21703
BATZLI, G.O. 21704
BAUER, H. 21705
BAUER, P.-J. 22753
BAUER, R. 21706, 22794
BAULD, J. 21707
BAUMGARTNER, A. 23578
BAVEL, C.H.M.van see VAN BAVEL, C.H.M.
BAXTER, J. 22150
BAZIER, R. 22135
BAZZAZ, F.A. 21708-9, 21995, 24183, 24232, 25021
BEADLE, C.L. 23295
BEALE, S.I. 21710
BEARD, C.F. 24337
BEARD, J.B. 24435, 25031-2
BEARDALL, J. 21711, 22552
BEARDEN, A.J. 21584, 21712, 23518
BEARDSELL, M.F. 21713
BEBEE, G. 24190
BECHER, B.M. 21714-6
BECK, E. 21717, 22320, 24344, 24404
BECKER, J.F. 21718
BECKER, K. 21719, 23400-1
BEDDARD, G.S. 21720-1
BEDNÁŘOVÁ, E. 22829
BEEK, J.H.G.M.van see VAN BEEK, J.H.G.M.
BEER, S. 21722, 24469
BEERS, J. 24687
BEEVERS, H. 22982, 23927
BEGICHEV, V.N. 24844
BEHNKE, H.-D. 21723
BEǏDEMAN, I.N. *23679*
BEISENHERZ, W.W. 21724
BEKASOVA, O.D. 21725-6, 23016, 23996
BEKINA, R.M. 21727-8, 23491
BELETSKAYA, D.K. 22606
BELETSKIǏ, Yu.D. 21729

BOGGESS, W.R. 24183
BOGOMOLNIĬ, R.A. 21834, 23458
BOGORAD, L. 21600-1, 21835-6, 22562, 24695
BOGUSLAVSKIĬ, L.I. 21837, 24926
BOGUSPAEV, K.K. 21565
BÖHM, H. 22757-8
BÖHME, H. 21838-9, 22809
BOHR, R. 23646
BOĬCHENKO, E.A. 21840, 22305-6
BOKANI, A. 23754
BOKÁNY, A. see BOKANI, A.
BOKOVAYA, M.M. 21841
BOLHÀR-NORDENKAMPF, H.R. 21842
BOLL, W.G. 24934
BOLTON, J.R. 21843, 22376, 22682, 23600-1
BOMMEGOWDA, A. 23220
BOMSEL, J.L. 21844
BONCH-OSMOLOVSKAYA, E.A. 21691
BONDARENKO, V.I. 21845
BONEN, L. 21846
BONNEMAIN, J.L. 21847
BONNER, H.S. 22289
BONNETT, H.T. 21848
BONOTTO, S. 22277
BONTE, J. 21849
BONZI, L.M. 21850
BONZON, M. 21851-2
BOOTE, K.J. 21550
BOOTS, R.S. 24773
BOOYSEN, P. deV. 21853
BORCHERT, M.T. 25002
BORCHERT, R. 21854
BORG, D.C. 22242
BORISEVICH, G.P. 21855
BORISOV, A.Yu. 21690, 21856
BORISOVA, I.G. 21857
BORNEFELD, T. 21858-9
BÖRNER, T. 21860, 22643, 24379
BORNKAMM, R. 21861-2
BORODINA, S.M. 24104
BORODKIN, S.O. 22976
BORSHCHEVSKAYA, T.N. 21678
BORZENKOVA, R.A. 21863
BOSCH, J.J.ten see TEN BOSCH, J.J.
BOSCHETTI, A. 21931, 22317
BOTEY SERRA, J. 21864
BOTHE, H. 21865
BOUCHER, F. 21866
BOUCHON, J. 21633
BOUGES-BOCQUET, B. 21750, 21867-8
BOULTER, D. 22529, 24335
BOUMA, D. 21869
BOUNIAS, M. 21870
BOURDU, R. 21871-2, 22624
BOUREAU, M. 23376
BOURQUE, D.P. 21873
BOUTHELIER, V. 21971
BOVEY, F. 23874-5
BOWES, B.G. 22114
BOWES, G. 21874-5
BOX, E. 21876
BOYD, C.E. 21877
BOYER, J.S. 21878, 22415, 22830-1, 22974, 23673
BOYER, P.D. 21879
BOYER, Y. 21880-2
BOYLE, J.E. 21883
BOYLEN, C.W. 24440
BOYNTON, J.E. 22117, 24962
BRACKENHOFER, H. 23422
BRADBEER, J.W. 21616, 21884-5, 22594, 23689
BRADBURY, I.K. 22584
BRADY, C.J. 21886
BRADY, R.A. 21887
BRAND, J.J. 21888-9, 22164
BRÄNDLE, E.P.O. 21890
BRANDT, A.B. 21891-4
BRANTON, D. 21895
BRAUM, J.G. 21896
BRAVDO, B. 21625, 21897
BREBNER, J. 23538
BRETON, J. 21718, 21898-900, 22394
BREUZÉ, G. 21901, 22748
BREWSTER, J.L. 23864
BREZEANU, A. 21902
BŘEZINA, V. 22362
BRIANTAIS, J.-M. 21612-3, 24910-1
BRIN, G.P. 23201-2
BRINKHUIS, B.H. 21903
BRISSON, J.D. 24216
BRITTON, G. 21904, 22616
BRITZ, S.J. 21905-7
BROCK, T.D. 21707, 21908-11, 23501, 24998
BROCKMANN, H.Jr. 22549
BRODA, E. 21912-3, B21914, *23478*, *24257*
BRODY, S.S. 22063
BROGÅRDH, T. 21916
BROLIN, S.E. 25004
BRONISZ, D. 23646
BRONNER, F. *22233*
BROOKSBANK, P. 22577
BROUERS, M. 21918
BROUWER, R. 21919
BROUWER, Y.M. 24585
BROWN, A.P. 22405
BROWN, A.S. 21920, 24809-10
BROWN, D.H. 21699
BROWN, G.N. 22224, 23616-7, 24213
BROWN, H.E. 21921
BROWN, J. 21922
BROWN, J.A. 21923
BROWN, J.S. 21924-5, 22450-1
BROWN, K.W. 22964
BROWN, R.H. 21926, 22110, 22489, 23072, 23452
BROWN, T.J. 23963
BROWN, W.V. 21926-7, 24980
BROWNING, G. 21928-9
BRUCE, D. 21930
BRUCKERHOFF, D.N. 22762
BRÜGGER, M. 21931
BRUICE, T.C. 23563
BRULFERT, J. 21932
BRUN, W.A. 22084, 24649

C

D

E

ETIENNE, A.L. 22359-60, 23365
ETTL, H. 22361-2
EUGSTER, C.H. 21940
EVANS, E.H. 22363-4
EVANS, L.T. *21951*, *22282*, 22365-9, *22367-9*, *22419*, *23587*, *23695*, *23736*, *23975*, *24444*
EVANS, M.C.W. 21983, 22370-6, 22658, 23799, 24706
EVANS, N. 22377
EVANS, T.A. 22378
EVENARI, M. 21695, 22379-80, 23015, 23297-300, 24376-8
EVERS, A.K. 22381
EVERT, R.F. 24831
EVSTIGNEEV, V.B. 21725-6, 22382-6, 22567-8, 22621-2, 23089, 23100, 23720, 24622, 24936-7
EYLES, J.C. 22387

F

FABBRI, E. 21649-50
FABBRI, F. 21850, 22388
FABRIS, G.L. 22389
FAIR, P. 24734
FAJER, J. 22242, 22390
FAJSZI, Cs. 22161
FALK, H. 22391, 22888, 24368
FALKENBORG, D.H. *22592*, *24456*, *24992*
FALKOWSKI, P.G. 22392
FALUDI-DÁNIEL, Á. 22393-5, 22815-6, 23746
FAN, L.T. 24454
FANICA-GAIGNIER, M. 22093-4
FARAH, S.M. 22396
FARINEAU, J. 22397-400
FARINEAU, N. 22401
FARMER, R.E.Jr. 22402
FARQUHAR, G.D. 22403
FARRAR, J.F. 22404
FARRINGTON, P. 22593
FATEBENE, F. 24387
FAUSET, C.R. 22405
FAVOLA, G. 24782
FEDINA, I. 24290
FEDINA, I.S. 22406
FEDOROV, V.D. 22407
FEE, E.J. 24349
FEHER, G. 22198, 22408, 23886
FEHÉR, M. 22409
FEHÉR, M.T. 22221
FEICK, R. 23826
FEIERABEND, J. 22410
FEIGE, G.B. 22411-2
FEKETE, G., 22413-4
FELDER, E.M. *23292*
FELEKI, Z. 21604
FELLOWS, R.J. 22415
FELTON, R.H. 22242
FENCHEL, T. 21815
FENNA, R.E. 22416
FERGUSON, J.F. 22511
FERRARI, I. 22417
FERREE, D.C. 22418, 22661
FETISOVA, Z.G. 21690
FEVRALEV, V.S. 25156
FIALA, M. 23809
FICK, G.W. 22419
FIEDLER, U. 22420
FIELD, G.F. 23070
FIGLIOLIA, A. 21974, 24781
FIKSIŃSKI, K. 22460
FILIPPOV, G.L. 22421, 24794
FILIPPOVICH, I.I. 22422-4
FILISTOVICH, V.R. 24569
FINAKOV, G.Z. 21892
FINENKO, Z.Z. 21775
FIOLET, J.W.T. 22425-7
FIRSTATER, E. 21649
FISCHER, K. 23633
FISCHER, K.S. 22428-30
FISCHER, R.A. 22431
FISCHEROVÁ, H. 22432
FISCUS, E.L. 21534
FISHER, D.B. 23937-8
FISHER, J.A. 22761
FISHER, N.M. 21929
FISHER, N.S. 22433
FISHER, R.W. 22434
FISHER, Ya. 24896
FITE, D.G. 21864
FITZGERALD, M.P. 23415
FITZSIMONS, D.W. *25006*
FLEMING, A.A. 22435
FLIEGE, R. 22732
FLINT, R.W. 22436
FLOROVA, N.B. 23830-1
FLOWER-ELLIS, J.G.K. 22437
FLOYD, G.L. 22535, 24292
FLUHR, R. 22438-40
FOCK, H. 23329
FOCK, H.A. 23509
FOCKE, R. 22441
FOGG, G.E. 21543, 22442
FOMISHYNA, R.M. 24643
FONG, F. 22906
FONG, F.K. 22443-5, 23144
FONTVIEILLE, D. 22446
FOOKES, C.J.R. 22095
FORD, E.D. 22447
FORD, M.A. 22448, 22960
FORDE, B.J. 22449
FORESTER, J. 22790
FORK, D.C. 21774, 22450-1, 23728-30, 24808
FORMAN, A. 22390
FORNASIERO-BARONI, R. 23607
FORSTER, G.R. 24051
FORTI, G. 22452, 22933-6
FOSTER, J.A. 21920, 24810
FOSTER, J.M. 22453, 22858
FOTT, J. 22454
FOUASSIN, A. 22278
FOUSOVÁ, S. 21637, 22455
FOWLER, C.F. 22456, 23148

GRAY, E.A. 22591
GRAY, J.C. 22049-50, 23254, 23256
GREEF, J.A. de see DE GREEF, J.A.
GREEN, R.M. 24437
GREENE, D.M. 24961
GREENE, J.C. 22592, 24456
GREENWOOD, E.A.N. 22593
GREGORY, P. 22594
GREGORY, R.P.F. 22595-6, 24154, 24385
GREGSON, K. 21801, 22597
GREPPIN, H. 21851-2
GRESHAM, C.A. 22598
GRIDNĖVA, N.V. 23180
GRIFFITH, O.H. 22174
GRIFFITHS, M. 22599
GRIFFITHS, W.T. 22600-2, 23704
GRIGORA, M.Yu. 23935
GRIGOR'EV, V.G. 24771
GRIME, J.P. 22603
GRIMME, L.H. 22604-5, 23248-9, 24117
GRINENKO, V.V. 22606
GRINTAL', A.R. 22607
GRISHINA, G.S. 22608
GRITS, M.G. 22988
GROBERMAN, L. 24372
GRODZINSKI, B. 22609
GRODZINSKIĬ, A.M. 22610-1
GROGAN, C.O. *22984*
GROLLMAN, A.P. 24010
GROMET-ELHANAN, Z. 21797, 22612-3, 23362-3
GRONEBAUM-TURCK, K. 22614
GROOTH, B.G. de see de GROOTH, B.G.
GROSS, E.L. 21608, 22193-5, 22615, 24102-3, 25083
GROSS, J.A. 22616
GROUZIS, J.-P. 22286
GROVES, R.H. 22617
GROZOVSKAYA, M.S. 23960, 24463
GRUMBACH, K.H. 22618-9, 23402, 23404
GRUNERT, S. 22466
GRUNEWALD, R. 24587
GRZESIAK, S. 23547
GUARDINO, V. 22620
GUBARE, G. 23222, 24918
GUDAUSKAS, R.T. 23424
GUDKOV, N.D. 22621-2
GUÉRIN-DUMARTRAIT, E. 22623, 23345
GUERN, M. 22624
GUERRIER, D. 21932
GUIKEMA, J.A. 22625
GUILLAUME, E. 21881-2
GUILLOTIN, J. 22626
GUILLOT-SALOMON, T. 22627, 24827
GUKASYAN, I.A. 23032
GUKASYAN, L.A. 21700
GULAYA, N.K. 22628
GULYAEV, B.I. 23887, 24776
GUN-AAZHAV, T. 22629
GUPTA, P.K. 22630
GUPTA, S.K. 23952, 24040
GUREVICH, G.M. 24158
GURICHEVA, N.P. 22631
GURINOVICH, G.P. 22632-3, 25141
GURU, B.C. 22184
GUSEV, M.V. 24405
GUSEV, N.A. 24269
GUSHCHINA, L.M. 22055
GUSHTINA, L. 24290
GUTELMACHER, B.L. 22634
GUTIERREZ, M. 22635, 22840
GUTOWSKY, H.S. 25085
GUTSER, R. 22636

H

HABER, A.H. 24753
HÄDER, D.-P. 22637
HADLEY, G. 24114
HADLEY, M. 21954
HADLEY, N.F. *22316*, *23298*, *23692*
HAEDER, H.E. 22638-9
HAEHNEL, W. 22640
HAFF, L. 22641
HAGAN, R.M. 22187
HÄGELE, W. 22642
HAGEMAN, R.H. 21875, 23282-3, 23504
HAGEMANN, R. 21860, 22643, 24379
HAGER, A. 22644
HAGER, H. 22645
HAGIN, R.D. 22279
HAINES, E.B. 22646
HAISMAN, D.R. 22647
HALÁSZ, N. 22815-6
HALES, B.J. 22648, 23437
HALEVY, A.H. 24063
HALL, A.E. 22102, 22649-53
HALL, C.A.S. 22654
HALL, C.V. 24203
HALL, D.O. 21583, 21591, *21622*, *21982*, *22370-2*, 22655-60, *22655-6*, *22710*, 22720, 22979, 23473-4, 24707, 24723
HALL, F.R. 22418, 22661
HALL, J.L. *23479*, *23847*, *24169*, *24742*
HALL, R. 22662
HALL, S.M. 22663
HALLDALL, P. 22664, 23834
HÄLLGREN, J.E. 22665
HALLIWELL, B. 22666
HALPERN, S. 21872
HALSEY, Y.D. 22667
HALVA, E. 22668
HAMAKOGA, M. 23579
HAMILTON, P.L. 24248
HAMLIN, L. 22669
HAMM, R.N. 25040
HAMMER, U.T. 22389
HAMMES, G.G. 21987-9
HAMMOND, J.H. 22581
HAMMOND, L.C. 21550
HAMPP, R. 22670-2
HANAN, J.J. 21517
HANCOCK, J.G. 23713
HANDLEY, J.F. 25028
HANEY, A.W. 21709

HUNT, J.F. 22279
HUNT, I.V. 22849
HUNT, L.A. 23493
HUNT, R. 22603, 22850
HUNTER, R.B. 22263
HURDUC, N. 23739-40
HURME, H. 23027
HUSS, K. 22665
HUSTON, R.P. 24730
HUSZÁR, J. 22851
HUTCHINSON, T.C. 24580
HUTH, W. 22852
HUTSON, K.G. 22853
HUTTON, M.J. 23522
HUZULÁK, J. 22854
HWANG, K.E. 23802
HYER, R.C. 21986

I

IANCU, M. 22855
IBRAHIM, I.K.A. 22966
ICHIKAWA, T. 22856-7, 22883
ICHIMURA, S. 24392-3
IDSO, S.B. 22453, 22858
IGARASHI, Y. 23142-3
IGNATENKO, I.V. 22053
IGNAT'EV, A.R. 24068, 24568
IHLENFELDT, M.J.A. 22859
IIJIMA, T. 22860
IIZUKA, S. 22861
IKAWA, T. 23843, 24525
IKEDA, T. 24686
IKEGAMI, I. 22862
IKENAGA, T. 22863
IKUMA, H. 24224
IKUSHIMA, T. 22864
ILANI, A. 23533
ILANI, S. 22356
IL'INA, M.D. 21690, 24886
ILLIK, M. 23754
ILLYES, Gh. 22865
ILMAVIRTA, V. 22866-7
IL'YUSHCHENKO, T.E. 21947
IMAI, A. 23051
IMAI, H. 22868-9
IMAMALIEV, A.I. 22870
IMURA, T. 22871
INABA, T. 22872
INADA, K. 22873
INADA, Y. 22874, 25098
INAYAMA, M. 22897
INCOLL, L.D. 23295, 23447
INCROPERA, F.P. 22875
INGLE, M. 24501-2
INGLE, R.K. 22876
INNIS, G.S. 22877
INOSAKA, M. 22878-9
INOUE, K. 22880
INOUE, Y. 22856-7, 22881-4, 23132, 23874
INOUYE, M. 23885
IOFFE, N.T. 23510
IRIYAMA, K. 21597
IRVINE, J.E. 22885
ISAACSON, R.A. 22408, 23886
ISAAKIDOU, J. 22886, 23957
ISANGALIN, F.Sh. 22887
ISENRING, H.-P. 22391, 22888
ISHIDA, M.R. 24823
ISO, N. 23182
ISRAELSTAM, G.F. 21533
ITIER, B. 24005
ITO, K. 22878-9
ITOH, S. 22889
IVANCHENKO, V.M. B22890, 22891
IVANISHCHEV, V.N. 23625
IVANKINA, N.G. 22892, 23859
IVANOV, A.F. 22893
IVANOV, B.N. 21531, 22894
IVANOV, N.P. 24076
IVANOVA, A.P. 24715
IVANOVA, S.B. 24050
IVANOVICH, V.A. 24078
IVNITSKAYA, I.N. 23012
IWAI, S. 22869
IWAKI, H. 22895-6
IWAKIRI, S. 22880, 22897
IWAMURA, T. 22898
IWANIJ, V. 22899
IWATA, T. 23669
IZAWA, S. 22900
IZVOSHCHIKOV, V.P. 22901
IZZA, C. 21974, 24781

J

JABBEN, M. 22902
JACKSON, A.H. 22377
JACKSON, J.B. 22903-4, 24315-6
JACOBI, G. 22905, 23195-8, 24367
JACOBSON, B.S. 22906
JACQUARD, P. 22907
JACQUES, G. 23809
JACQUES, G.L. 22908-9
JACQUINOT, L. 22910
JADHAV, S.J. 22911
JAGANNATH, M.K. 23220
JAGENDORF, A.T. 22912-3, 23906
JÄGER, E. 21765
JAGER, J.M. de see DE JAGER, J.M.
JAHN, O.L. 22914
JAHNKE, L.S. 22915
JAHREN, A. 23105
JAKRLOVÁ, J. 22916
JAKUCS, P. 22917
JAMES, D.M. 24334
JAMES, G.B. 23295
JANA, M.K. 21603
JANA, P.K. 22524
JANÁK, J. 24409
JANAVE, M.T. 24751
JANÍČEK, J. 22829
JANK, H.-W. 21602
JANKOVIĆ, M.M. 22918

K

L

LI, B.D. 23393
LI, E.H. 23394
LI, Y.-S. 23395-6
LIAAEN-JENSEN, S. 21940, 23855
LIBERA, W. 23397, 25148
LICHTENTHALER, H.K. 21719, 21959, 22618-9, 22673, 23115-6, 23398-404, 24032
LIDDLE, M.J. 23405
LIDSTER, P.D. 23406
LIEBHARDT, W.C. 24470
LIEN, S. 22842, B23407
LIEN, T. 22465
LIER, J.B. 23277
LIETH, H. *21876*, *21955*, *22654*, 23408-12, *23408-12*, *B23413*, *23414*, *23741*, 24424, *24424-5*, *25013-5*, 25014
LIFSHITZ, Y. 21996-7
LIKENS, G.E. 23414, 25013-4
LILLEY, R.McC. 23415-6
LIMAR', R.S. 23172, 23417
LIMPÁROVÁ, M. 23418
LIN, C.H. 23419, 24669
LIN, L. 23420-1
LINDEN, J.C. 23422
LINDER, S. 22948
LINDQUIST, P. 23527-8
LINDSAY, D.C. 23423
LINDSAY, J.G. 24098
LINDSEY, D.W. 23424
LINSCOTT, D.L. 22279
LINSER, H. 22968, 23425
LINT, P.J.A.L. de see DE LINT, P.J.A.L.
LINVILL, D.E. 23426
LIPKIND, B.I. 22534, 23199
LIPS, S.H. 22980-1, 23014, 23427, 24247
LIPSCHULTZ, C.A. 22497
LIPSKAYA, A.A. 24050
LIPSKAYA, G.A. 23428-9
LISTER, G.R. 22086-7
LITTAN, A. 24064
LITTLE, C.H.A. 23430-1
LITTLETON, E.J. 21802
LITVIN, F.F. 21733, 22305-6, 23205-6, 23432-4
LITVINENKO, L.G. 23339, 23435
LIVAK, K.L. 24125
LJUBEŠIĆ, N. 25077-8
LLAMBIAS, E.B.C. 21701
LLOYD, E.J. 23797
LOACH, K. 23431
LOACH, P.A. 23436-7
LOBODA, N.I. 23438, 23792
LOCKER, A. *23478*, *24257*
LOEBLICH, A.R.III. 23439
LOEBLICH, L.A. 23440
LOFTUS, M.E. 23441, 23590
LOHONYAI, N. 23442
LÖHR, E. 23443
LOKUTSIEVSKAYA, L.K. 21968
LOMMEN, P.V. 23444
LOMMEN, P.W. 25106
LONDON, R.E. 23445
LONERGAN, T.A. 23446
LONG, S.P. 23447
LOOMIS, R.S. 22419, 23448
LOOS, E. 23449
LOPATA, W.-D. 23450
LÓPEZ GORGÉ, J. 22081, 23336-7
LORD, J.M. 23451-4, 23597
LORENC, M. 23778
LORENZEN, C.J. 24475
LORENZEN, H. 23794, 24768
LORIMER, G.H. 21592, 22522, 23455
LOSEV, A.P. 22633, 25141
LOSITSKAYA, T.V. 24464
LOSSOW, K. 23456
LOUASON, G. 22131
LOUGUET, P. 21849
LOUWERSE, W. 23457, 24688, 24880
LOVELL, P.H. 24432
LOVENBERG, W. *23708*
LOVETT, J.V. 23338
LOVEYS, B.R. 21799, 23218-9, 24086
LOWE-McCONNEL, R.H. *23871*, *24186*, *24210*, *24276*, *24447*, *25022*
LOZHKIN, B.T. 21837, 24926
LOZIER, R.H. 23458
LOZOVA, G.I. 23459-61
LU, B.C. 24216-7
LUBAC, M. de 24828
LUBAŃSKA, G. 23862-3
LUCAS, W.J. 23462-3
LUDFORD, P. 22157
LUDLOW, C.J. 23464
LUDLOW, M.M. 23465, 23813
LUDWIG, J.A. 23466
LUDWIG, L.J. 22038-9, 23467
LUEBS, R.E. 23468
LUGANSKAYA, A.N. 23469
LUISIER, J.-L. 22888
LUKASHEV, E.P. 23139, 23470
LUKPANOV, Zh.L. 21732
LUK'YANOVA, L.M. 23471
LUMPKIN, O. 23472
LUMSDEN, J. 23473-4
LÜNING, K. 22261, 23475
LUNNEY, C.A. 23476
LUPP, G. 23633
LURIE, S. 22107, 23477
LUŚCIŃSKA, M. 23646
LÜTTGE, U. 22946, 23478-81, 24055
LÜTZ, C. 23482-3
LUTZ, H.U. 21651, 21780, 23484, 24034
LUTZ, M. 23485, 24065
LUUKKANEN, O. 22686
LUXMOORE, R.J. 24670
LUZHNOVA, M.I. 23486
LUZZANA, M.R. 23487
LYAKHNOVICH, Ya.P. 23488-9
LYMAN, H. 23490, 24357
LYSENKO, G.G. 21728, 23491
LYTTLETON, J.W. 23492
LYUBENKO, M.O. 23683

M

N

O

Q

R

ROGERS, L.J. 21589, 22202, 22853, 23944
ROGOZIŃSKA, E. 22516
ROLFE, G.L. 21708, 21995, 24232
ROLLIN, H. 23628
ROMANO, J.-C. 24233
ROMANOVA, E.N. 22053
ROMBACH, J. 24234
ROMODAN, V.N. 23684
RONNENKAMP, R.R. 24235
ROOT, R.A. 24236
ROSA, L. 22452
ROSARIO, E.J. del 24002
ROSE, R.J. 24086, 24237
ROSEBY, J.E. 24719
ROSEN, D. 24238
ROSEN, J. 24047
ROSENBERG, N.J. 23367-8, 24907
ROSHCHINA, V.D. 24239-40
ROSHCHINA, V.V. 24241
ROSING, J. 24242
ROSOWSKI, J.R. 24243
ROSS, J. 24244
see ROSS, Yu.K.
ROSS, R.T. 21585
ROSS, Yu.K. B24245
see ROSS, J.
ROSSI, C. de 24781
ROSSWALL, T. *22053, 23304*
ROTA, J.A. 24246
ROTEM, J. 22269
ROTH, D. 21765, 24033
ROTH, H.D. 23290
ROTH-BEJERANO, N. 23014, 24247
ROTTENBERG, H. 24042
ROTTENBURG, T. 23005
ROUSSAUX, J. 22401
ROUSSEAU, K. 22242
ROUX, E. 21899-900, 22624, 23361, 24840
ROWE, A. 23597
ROWE, G.T. 24248
ROWE, P.R. 23071
ROWLAND, A.O. 22727
ROWLEY, J.A. 22449
ROY, G. 23538
ROZEMA, J. 24249
RUBIN, A.B. 21855, 23215, 23470, 23525, 23970-1, 24189, B24250, 25088
RUBIN, B.A. 21727, 22513, 23491, 23831
RUBIN, L.B. 23970-1
RUBIN, P.M. 24251
RUCKENBAUER, P. 24252
RUD', M.S. 24394
RUDENKO, T.I. 24253
RÜDIGER, W. 23178-9, 24254
RUDOI, A.B. 24912
RUFFNER, H.P. 24255
RUFNER, R. 24256
RÜHLE, W. 25029
RUIJGROK, T.W. 24300
RUMBERG, B. 24257, 24374
RUNECKLES, V.C. 22181, 24592-3
RUNNING, S.W. 22255
RURAINSKI, H.J. 24258
RUSSEL, W.J. 24259-60
RUTTEN, P. 23628
RUTTER, A.J. 24261
RUUGE, É.K. 23241, 24638, 24760-2
RUYTER, A. de see DE RUYTER, A.
RŮŽIČKA, J. 22420
RUZIEVA, R.Kh. 21530-1, 22894
RYAN, F.J. 24262-5, 24779
RYAN, J. 24266
RYBÁČEK, V. 24267
RYBIN, I.A. 24268
RYBKINA, G.V. 23028, 24269
RYĆ, M. 24270
RYCHNOVSKÁ, M. B24271
RYCZKOWSKI, M. 24272
RYDING, S.-O. 24273-4
RYLE, G.J.A. 24275
RYSZKOWSKI, L. 24276
RYTOVA, N.G. 24277

S

SAAKOV, V.S. 21678
SAAKYAN, M.A. 21642
SABAD, Ya. 23364
see SZABAD, J.
SABATER, B. 23555
SABHARWAL, P.S. 23048
SACKETT, W.M. 25067
SADDLER, H.D.W. 23589
SADO, T. 23758
SAEKI, T. 24278
SAFORD, W.E. 21954
SAGER, R. 23885
SAGHER, D. 24130
SAGROMSKY, H. 24279-80
SAHA, S. 24281
SAHU, G. 24142
SAI, P.K. 22462
SAIJO, Y. 23659
SAITO, T. 23182
SAITO, Y. 21523, 22769
SAITOH, M. 24282
SAKA, H. 24283
SAKAI, W.S. 24943
SAKAI-IMAMURA, M. 24284
SAKAMOTO, E. 22884
SAKANISHI, Y. 24285
SAKANO, K. 23256
SAKHAROVA, O.V. 24286
SAKURAI, T. 24282
SĂLĂGEANU, N. 24287
SALAI, L. 23364
see SZALAY, L.
SALAMON, Z. 24288
SALATENKO, V.N. 24289
SALCHEVA, G. 24290
SALDANA, G. 21921
SALE, P.J.M. 24291
SALINGER, S. 21862
SALISBURY, J.L. 24292

T

U

V

X

Y

Z

This index contains a selection of primary items chosen according to their importance in photosynthesis research and to their relevance and occurrence. The word "Photosynthesis" is not regarded as a main theme, but partial processes, photosynthetic parameters and the factors affecting photosynthesis are listed. The processes and other characteristics are summarized into several main themes when presented in combination with individual factors, *e.g.* carbon fixation pathways, electron transport chain, chlorophyll, carotenoids, gas exchange, photorespiration, algae productivity, ecosystem productivity and canopy functioning (including photosynthate translocation and distribution), resistances to CO_2 and water vapour transfer,*etc.*

Several items from branches related to photosynthesis research were also chosen for convenience, *e.g.* dealing with respiration, plant growth and development, water relations, anatomy, *etc.* These items contain only references to papers within the scope of this bibliography.

A

Algae chlorophylls see Chlorophylls *c*, *d*

Algae, CO_2 and O_2 exchange see Gas exchange in algae

Algae, depth distribution in reservoirs
21682, 21726, 21742, 21805, B21829, 21936, 21972, 22155, 22158, 22256-7, 22268, 22389, 22417, 22446, 22453, 22488, 22493-4, 22517, 22531, 22578, 22628, 22646, 22742, 22745-6, 22866, 22921, 22943-4, 23016, 23065, 23087, 23098, 23133, 23269, 23314, 23377, 23387, 23441, 23456, 23593, 23606, 23824, 23931-2, 24051, 24058, 24061, 24140, 24144, 24194, 24200, 24248, 24349, 24440, 24572, 24578-9, 24616, 24618, 24680, 24708, 24743, 24763, 24766, 24857, 24896, 24906, 24953, 25053, 25130, 25157

Algae in cosmonautics B22052

Algae life cycles see Ontogeny of algae

Algae mass cultures productivity (*cf.* also Algae and photosynthetic bacteria, cultivation) 21735, 21789, 23229, 24119, 24287, 24454, 24702

Algae photosynthesis and production
21707, 21742, 21772, 21805, B21829, 21952-3, 22000, 22132, 22155, 22211, 22248, 22256-8, 22268, 22287, 22292-3, 22494, 22538, 22628, 22898, 22943-4, 23074, 23188-9, 23225, 23278, 23301, 23377, 23686, 23824, 24118, 24144, 24308, 24349, 24420, 24440, 24562, 24586, 24644, 24687, 24708, 24786, 24838, 24906, 24957, 24960

Algae, primary productivity in reservoirs (*cf.* also Chlorophyll as measure of production of algae and water reservoirs)
21707, 21731, 21955, 22169, 22604, 22634, 23164, 23268, 23414, 23534, 23569, 23644, 23685, 23765, 24062, 24447, 24503, 24608, 24617, 24699-700, 24736, 24969, 25003, 25005, 25121

Algae, primary productivity, methods (*cf.* also O_2 determination (other than O_2 electrode); O_2 electrode)
21742, 21952, 21955, 21958, 22143, 22177, 22577, 22592, 22654, 22757-8, 22921, 22944, 23414, 24087, 24294, 24456, 24514, 24732, 24786, 24957, 25005

Algae synchronous cultures see Algae and photosynthetic bacteria, cultivation; Ontogeny of algae ...

Altitude see Pressure, altitude ...

Amino acids see Proteins, amino acids, nucleic acids ...

δ-Aminolaevulinic acid see Chlorophyll biosynthesis ...

Amphistomatous leaf, gas exchange in (*cf.* also Leaf epidermis, stomata) 22960

Anaerobic atmosphere see N_2, anaerobic atmosphere ...

Antibiotics and carbon fixation pathways
21616, 21702, 21724, 22018, 22117, 22192, 23451, 23480, 23715, 24016

Antibiotics and carotenoids 21642, 24778

Antibiotics and chlorophyll
21505, 21642, 21679, 21724, 21741, 22107, 22225, 22317, 22338, 22469, 22889, 23025, 23095, 23108-9, 23224, 23272, 23396, 23528, 23851, 23882, 23969, 24016, 24076, 24397, 24443, 24462, 24464, 24534, 24544, 24745, 24778, 25019, 25075

Antibiotics and chloroplast (chromatophore)
21545, 21863, 21931, 22083, 22117, 22469, 22729, 23224, 23637, 24076, 24348, 24354, 24462

Antibiotics and electron transport chain
21611, 21669, 21859, 21938, 21997, 22018, 22154, 22179, 22499, 22501, 22536, 22612-3, 22773, 22809, 22894, 22903-4, 22923, 22970, 23124, 23127, 23199, 23362, 23627, 23704, 23718, 23721, 23851, 24193, 24241, 24315-6, 24354, 24757, 24869-70, 25090

Antibiotics and gas exchange 22018, 22834, 22946, 23641, 23718

B

C

C_4 pathway of carbon fixation
21575, 21580, 21652, 21711, 21739, 21773, 21785, 21810, 21814, B21833, 21926, 21941, 21967, 22071, 22075, 22119-20, 22150, 22156, 22203, 22249, 22328-9, 22368, 22398, 22556, 22580, 22635, 22697-8, 22700, 22833-5, 22837-40, 22868-9, 22958, 22965, 22983, 23001, 23032, 23170, 23231, 23323, 23343, 23374, 23397, 23714, B23767, 23786, 23877, 23902, 23926, 23929, 23973, 23986, 24008, 24074, 24160-1, 24164-5, B24250, 24270, 24310, 24344, 24469, 24552, 24554-5, 24675-6, 24734, 24742, 24769, 24804, 24806-8, 24980, 25113, 25138

C_3, C_4, CAM pathways, comparison see Carbon metabolic types ...

Calibration of infra-red analyser see Infra-red gas analyser ...

Caloric values see Calorimetry ...

Calorimetry 22173, 22258, 22297, 22570, 22765, 22778, 22918-9, 23137, 23408, 23630, 24153, 25025

Calorimetry, methods 23408, B24024

Calvin-Benson cycle see C_3 pathway of carbon fixation

CAM 21580, 21652, 21884-5, 21932, 22098-9, 22152, 22203, 22239, 22318, 22951, 22958, 22962-3, 23011, 23013, 23121, 23300, 23374, 23379, 23480-1, 23539, 23610, 23714, 23764, 23786-7, 23926, 23928-30, 24126, 24297, 24319, 24554, 24656-9, 24675-6, 24742, 24797, 24804, 24806, 24808, 24980, 25035-6, 25118

Canopy, CO_2 profiles
21549, 21581, 21633, 21801, 21803, 22593, 22597, 22645, 22797, 22984, 23578, 23700, 23814, 24004, 24015, 24215, 24278, 24296, 24689, 24947

Canopy density, thickness
21521, 21524, 21542, 21826, 21950, 22007, 22184, 22396, 22428-30, 22447, 22844, 22910, 22985, 23067, 23221, 23233, 23313, 23426, 23738, 23896, 23945, 24080, 24115, 24229, 24291, 24441, 24511, 24595, 24849, 24887, 25156

Canopy, horizontal structure
21633, 21761, 22854, 23039, 23412, 23856, 24093, 24215, 24244, 24474

Canopy, leaf age 21928, 22065, 22097, 22606, 24278, 25086

Canopy, leaf angles
21523, 21633, 21761, 22885, 22910, 23059, 23117, 23346, 23814, 24215, 24244, 24441, 24474, 24758, 24798, 24880, 24887

Canopy microclimate and macroclimate
22271, 22680, 22880, 22984, 23304, 23426, 23814, 24004, 24740, 24782, 24947

Canopy photosynthesis and PhAR profile (*cf.* also Canopy, radiation profile)
21538, 22302, 22775, 23007, 23856, 23910, 24289, 24474, 24662, 24758, 24880, 25086

Canopy photosynthesis, direct measurement
21538, 21802, 21841, 22040, B22052, 22123, 22896, 23007, 23443, 23894-5, 24438, 24880, 25015

Canopy photosynthesis, energy balance
21802-3, 22144, 22680, 22765, 22941, 23592, 24004-5, 24057, 24110, 24474, 24512, 24689, 24740, 24780, 24782, 24907

Canopy photosynthesis, mass and momentum balance
21802, 22144, 22765, B23407, 23592, 23814, 24004, 24689, 24740, 24907

Canopy photosynthesis, model see Model ...

Canopy, radiation distribution; reflection, transmission, absorption, albedo, *etc.* (*cf.* also Canopy, radiation profile)
21523, 21539, 21541, 21667, 21822, 22123, 22554, 22271, 22302, 22593, 22606, 22728, 22895, 22985, 23003-4, 23039, 23070, 23117, 23367-8, 23426, 23692, 23783, 23856, 24005, 24131, 24215, 24244, 24276, 24278, 24289, 24474, 24569, 24623, 24641, 24798, 24887, 24913

Carotenoids in model systems 21614, 21855, 23340, 23510, 23533

Carotenoids in photosynthesis mechanism
21807, 21892, 21954, 21986, 22020, 22101, 22134, 22154, 22289, 22322, 22450, 22580-1, 22619, 22644, 22711, 22774, 22904, 23169, 23292-3, 23421, 23946, 24133, 24136, 24189, 24209, 24316, 24318, 24451, 24488-91, 25057, B25064, 25093

Carotenoids in physiology of photosynthesis 23752, 24019, 24068, 24315, 24348, 24451

Carotenoids in seeds and fruits 21571, 22323, 22914, 23277, 23324-5, 23575, 23878, 23890, 24069, 24071, 24092, 24129, 24230, 24751

Carotenoids luminescence *in vitro* 23340

Chamber, assimilation see Assimilation chamber

Chemiosmotic hypothesis, proton transport in chloroplast
21531, 21607-8, 21631-2, 21687, 21745, 21752, 21766, 21843-4, 21879, 21890, 21912, 21997, 22017, 22103, 22146, 22179, 22215, 22233, 22324, 22425-6, 22456, 22491, 22528, 22536, 22583, 22586, 22612-3, 22625, 22657, 22704, 22706, 22721-2, 22773-4, 22894, 22900, 22904, 22912-3, 22946, 22969-72, 22996, 23183-4, 23192, 23208, 23362, 23434, 23449, 23591, 23615, 23647, 23656, 23662, 23726, 23842, 23847, 23891, 23921, B24024, 24028, 24042, 24081, 24099, 24169, B24176, 24221, 24311, 24316, 24355, 24367, 24374, 24405, 24416-7, 24497, 24525, 24545, 24671, 24790, 24856, 24870, 24938, 24973, 25006, 25030, 25043, 25057, 25099, 25101, 25143, 25152

Chlorobium chlorophyll see Chlorophylls, *Chlorobium*

Chlorophyll absorption spectra *in vitro*
21726, 21775, 21796, 21891, 21918, 22072, 22161, 22207, 22235, 22242, 22278, 22378, 22445, 22458, 22534, 22633, 22642, 22959, 23044, 23089, 23097, 23340, 23342, 23344-5, 23348-9, 23353, 23361, 23388-9, 23433, 23450, 23472, 23667, 23812, 23857, 23936, 23966, 24065, 24180, B24250, 24273, 24327, 24357, 24400, 24486, 24733, 24749, 24784, 24800, 24937, 24991, 25016-7, B25064, 25141

Chlorophyll absorption spectra *in vivo*
21507, 21532, 21606, 21678, 21685, 21691, 21714, 21733-4, 21743, 21834, 21889, 21898, 21906-7, 21918, 21970, 22020, 22058, 22079, 22087, 22134, 22138, 22182, 22222, 22247-8, 22275, 22326, 22333-4, 22393, 22395, 22450-1, 22458, 22461, 22483, 22509, 22558, 22580, 22595, 22600-1, 22605, 22615, 22623, 22642, 22664, 22780, 22799, 22814, 22862, 22902, 22973, 23000, 23022, 23029, 23043, 23097-8, 23110, 23132, 23139, 23205, 23208, 23278, 23344-5, 23384, 23402, 23488, 23515, 23527, 23565, 23667, 23820, 23823, 23826-7, 23839, 23874, 23876, 23879-80, 23884, 23892, 23916, B23933, 23965-7, 24041, 24158, 24190, 24227, 24283, 24327, 24345, 24385, 24398, 24434, 24466, 24495-6, 24631, 24727-9, 24747, 24756-7, 24796, 24818, 24908-9, 24916, 24930, 25006, 25040, 25107, 25132-3

Chlorophyll and its products determination, column chromatography
21622, 22166, 22340, 22377, 22477, 22553, 24409, 24800

Chlorophyll and its products determination, electrophoresis and other methods
21605, 21775, 22471, 24756

Chlorophyll and its products determination, *in vivo*
21758, 22213, 22605, 22748, 22799, 22862, 22988, 23057, 23111, 23344, 23421, 23538, 24205, 24479, 24516-7, 25132

Chlorophyll and its products determination, paper chromatography, thin-layer chromatography 21622, 22278, 22323, 22549, 22928, 23054, 23383-4, 23400, 23450, 23781, 24521, 24978, 25016-7, 25055

Chlorophyll and its products determination, spectral methods
21758, 22021, 22199, 22298, 22329, 22340, 22549, 22871, 22929, 23016, 23441, 23565, 23812, 23971, 23996, 24179, 24273-4, 24279, 24372, 24513, 24733, 24898-9

Chlorophyll and production of higher plants
21699, 21760, 21862, 22012, 22435, 22775, 23091, B23407, 23410, 23570, 23831, 24040, 24512, 24530, 24942, 25013, 25015, 25123

Chlorophyll as measure of production of algae and water reservoirs
21543, 21681-2, 21726, 21742, 21776, 21805, 21953, B22052, 22121, 22287, 22353, 22357, 22392, 22417, 22453-4, 22464, 22488, 22493, 22517, 22531, 22538, 22551, 22578, 22592, 22646, 22742, 22756, 22764, 22807, 22858, 22898, 22943, 22976, 23016-7, 23087, 23096, 23098, 23107, 23133, 23188-9, 23269-70, 23275, 23278, 23284-5, 23301, 23314-6, B23407, 23441, 23542, 23569, 23590, 23593, 23595, 23663, 23684, 23702, 23759, 23809, 23824, 23844, 23947, 23963, 24051, 24058, 24060-2, 24090, 24118, 24140, 24194, 24200, 24202, 24308, 24349, 24392-3, 24456, 24475, 24562, 24570-2, 24579, 24581-2, 24586, 24616-8, 24680, 24686-7, 24699, 24708, 24732, 24741, 24763-4, 24786, 24803, 24857, 24895-6, 24927, 24953, 24957, 24964, 24992, 25003, 25010, 25053, 25129-30, 25157

Chlorophyll biosynthesis and precursors
21506, 21528, 21532, 21534, 21558, 21575, 21642, 21664, 21678, 21701, 21703, 21710, 21741, 21831, 21872, 21918, 21922, 22001, 22011, 22067, 22094, 22181, 22265, 22290, 22307, 22317, 22322, 22335, 22337, 22341-2, 22391, 22438-40, 22467, 22469, 22472-3, 22506, 22600-2, 22619, 22670, 22691, 22735, 22789, 22799-800, 22813-6, 22827, 22888, 22902, 22911, 22990, 23035, 23095, 23101, 23108-9, 23277, 23291, 23312, 23398, 23400, 23436, 23451, 23527-8, 23531, 23564, 23581, 23614, 23619, 23639, 23642, 23674, 23697, 23704, B23767, 23768, 23781, 23812, 23823, 23829, 23831, 23874-5, 23879-83, 23913, 23915, 23924, 23951, 23960, 23968, 24019, 24109, 24117, 24158, 24168, 24177-80, 24209, 24305, 24330, 24340, 24348, 24364, 24381, 24398, 24400, 24402, 24413, 24459-60, 24463-5, 24561, 24590, 24597, 24642, 24653-4, 24696, 24727, 24756, 24792, 24800, 24912, 24916, 24924, 24962, 24986-7, 24991, 24995-6, 25000, B25064, 25121, 25132-3, 25144

Chlorophyll chemical structure
21615, 22072, 22134, 22242, 22340, 22378, 22408, 22416, 22444-5, 22642, 22887-8, 22927, 22949, 23301, 23583, 23857, 24217, 24327, 24343, 24406, 24453, 24560, 24640, 24991

Chlorophyll complexes *in vitro*
21614, 21917-8, 21969, 22072, 22161, 22235, 22483, 22534, 23144, 23358, 23360-1, 23364, 23388-9, 23584, 23769, 24385, 24399-400, 24422, 25141

Chlorophyll complexes *in vivo*
21507, 21528, 21532, 21534, 21536-7, 21558, 21574-5, 21599, 21601, 21733, 21745, 21768, 21796, 21860, 21900, 21918, 21922-5, 21963, 22058, 22079, 22113-4, 22182, 22194-5, 22213, 22221-2, 22230, 22243, 22265, 22275-6, 22289, 22322, 22326, 22338, 22393-5, 22408, 22416, 22443, 22462, 22514, 22580, 22596, 22600-1, 22615, 22623, 22642-3, 22773, 22780, 22814, 22988, 23000, 23052, 23082, 23095, 23103, 23110-1, 23163, 23205, 23344, 23347, 23428-9, 23434, 23460-1, 23488, 23498, 23562, 23580, 23753, B23767, 23768, 23823, 23825-6, 23830, 23857, 23860, 23874, 23879, 23892, 23909, 23916, B23933, 23954, 23998, 24041, 24077, 24154, 24158, 24190, 24205, 24208, 24221, B24250, 24293, 24306-7, 24327, 24348, 24364, 24374, 24379, 24459, 24462, 24479, 24497, 24653, 24729, 24747, 24756, 24808, 24876, 24909, 24916, 24962, 25057, B25064, 25107

Chlorophyll degradation
21796, 21857, 21886, 21889, 21896, 21961, 21969, 22068, 22102, 22235, 22300, 22312, 22323, 22342, 22377, 22390, 22408, 22468, 22499, 22534, 22581, 22633, 22647, 22692, 22864, 22905, 22914, 22959, 23016, 23019, 23029, 23044, 23050, 23066, 23089, 23189, 23191, 23249, 23301, 23324-5, 23331, 23342, 23429, 23450, 23470, 23535-6, 23599, 23630, 23702, 23809, 23878, 23924, 23993, 24037, 24058, 24065, 24089, 24129, 24208, 24227, 24284, 24400, 24475, 24562, 24571, 24583, 24643, 24680, 24693, 24714, 24729, 24731, 24746, 24796, 24819, 24876, 24882, 24920, 24937, 24995, 25000, 25077-8

Chlorophyll delayed light emission, luminescence *in vitro*
21636, 23204-6, 23340, 23364, 23381, 23434, 23715, 25141

Chlorophyll delayed light emission, luminescence *in vivo*
21679, 21970, 21993, 22106-7, 22125, 22291, 22359-60, 22482, 22559, 22585, 22683, 22708, 22766, 22795, 22856-7, 22882-3, 22889, 22956-7, 23066, 23139, 23205-6, 23215, 23326-8, 23434, 23477, 23485, 23538, 23576, 23622, 23726, 23851, 23964, B24250, 24306, 24327, 24480-1, 24575, 24703, 24791, 24812, 24860, 24901-3, 25088, 25151-2

Chlorophyll energetic states *in vitro* (*cf*. also Chlorophyll in model systems)
21721, 22046, 22174, 22384-6, 22534, 22642, 22871, 22887, 23130-1, 23359, 23364, 23378, 23433, 23533, 24197, 24327, 24573, 24622, 24783, 25141

Chlorophyll energetic states *in vivo*
21607, 21816, 21856, 22020, 22022, 22057, 22088, 22125, 22134, 22182, 22193, 22275-6, 22288, 22299, 22443, 22458, 22558, 22580-1, 22632, 22642, 23046, 23130, 23206, 23434, 23437, 23520, 23631, 23874, 23954, 23966, 24158, B24250, 24306, 24327, 24480, 24496, 24811, 24813

Chlorophyll energetics model see Model ...

Chlorophyll, enzymes of synthesis and degradation (other than chlorophyllase)
21701, 21703, 21710, 22093-4, 22275-6, 22307, 22341, 22439-40, 22670, 22672, 23035, 23095, 23312, 24364, 24443

Chlorophyll fluorescence *in vitro*
21597, 21636, 21721, 21917-8, 22160, 22235, 22278, 22480, 22483, 22534, 22664, 23012, 23130, 23153, 23185, 23205, 23327, 23348, 23359, 23433, 23831, 23970, B24250, 24327, 24399-400, 24421

Chlorophyll fluorescence *in vivo*
21506, 21567, 21597, 21600, 21607-8, 21664, 21679, 21690, 21706, 21718, 21720, 21733, 21744, 21749, 21767, 21774, 21790, 21810, 21825, 21856, 21898-9, 21901, 21918, 21924, 21962-5, 21993, 22020, 22044, 22062, 22078-9, 22088, 22090-1, 22182, 22214, 22264-5, 22275-7, 22291, 22298-300, 22325, 22328-9, 22332, 22359, 22405, 22426, 22458, 22462-3, 22483, 22514, 22527, 22558, 22580-1, 22605, 22615, 22629, 22632, 22642, 22683, 22721, 22748, 22766, 22794, 22815-6, 22857, 22933-4, 22936, 22955-7, 22973, 23017, 23021, 23023-4, 23031, 23042-3, 23102-3, 23118, 23120, 23132, 23140-1, 23153, 23184-5, 23187, 23192, 23205, 23208-9, 23240-1, 23323, 23327, 23394-6, 23402, 23434, 23441, 23446, 23586, 23616-8, 23622, 23639, 23647, 23656, 23726, 23728-30, B23767, 23768, 23794, 23830, 23843, 23916, B23933, 23934, 23954-7, 23970-1, 24041, 24103, 24133, 24158, 24179, B24250, 24307, 24327, 24330, 24356, 24371-3, 24401-3, 24417, 24421, 24451, 24478, 24480, 24496, 24575, 24577, 24631, 24653, 24721, 24728, 24735, 24737, 24756, 24792, 24825, 24864, 24870, 24874, 24910-1, 24917, 24930, 25002, 25012, 25050, 25076, 25081-3, 25107, 25113-4, 25151-2

Chlorophyll forms see Chlorophyll complexes

Chlorophyll in flowers 24506

Chlorophyll in model systems (*cf*. also Chlorophyll energetic states *in vitro*)
21614, 21721, 21837, 21969, 21991, 22046, 22072, 22161, 22174, 22207, 22262, 22386, 22458, 22480, 22534, 22568, 22621-2, 22751, 22798, 22871, 22915, 23100, 23130, 23144, 23202-3, 23290, 23340, 23348-9, 23358, 23360, 23381, 23438, 23469, 23533, 23566, 23584, 23630, 23792, 23830, 23857, B24250, 24399, 24412, 24422, 24710-2, 24783-4, 24844, 24926, 24936, 25141

Chlorophyll in photosynthesis mechanism
21643, 21646, 21662, 21664, 21690, 21807, 21816, 21843, 21892, 21899, B21914, 21954, 21963, 21991, 22020, 22046, 22106, 22125, 22134, 22193, 22198, 22221, 22243, 22288-9, 22299, 22390, 22443, 22450, 22458, 22558, 22568, 22580, 22583, 22596, 22640, 22748, 22933, 22970, 22972, 23000, 23046, 23101-2, 23118, 23141, 23200, 23215, 23292, 23327, 23347, 23402, B23407, 23434, 23437, 23558, 23566, 23574, 23584, 23720, 23761, 23825-6, 23857, 23876, 23916, 23934, 23955, 23957, 23966, 24133, 24154, 24191, 24227, B24250, 24280, 24306-7, 24327, 24374, 24548, 24575, 24622, 24811-3, 24876, 24914, 24931, 25051, 25057, B25064

Chlorophyll in physiology of photosynthesis
21662, 21697, 21721, 21776, 21793, 21842, B21914, 22010, 22085-6, 22089, 22184, 22192, 22413, 22453, 22623, 22815, 22848, 22851, 23242, 23258, 23278, 23470, 23739, 23754, 23780, 23845, 23867, 23895-7, 24019, 24148, 14197, 24280, 24321, 24323, 24348, 24378, 24401, 24715, 24804, 24821, 24882

Chlorophyll in seeds and fruits
22323, 22468, 22911, 22914, 23277, 23319, 23324-5, 23395, 23406, 23829-30, 23878, 23890, 24071, 24129, 24299, 24824

Chlorophyll luminescence see Chlorophyll delayed light emission ...

Chlorophyll, methods see Chlorophyll and its products determination

Chlorophyll number see Chlorophyll in physiology of photosynthesis

Chlorophyll unit see Photosynthetic (chlorophyll) unit

Chlorophyllase 21683, 23599, 24361, 24643, 24727-9

Chlorophyllase and other enzymes of chlorophyll synthesis and degradation, methods 23599

Chlorophylls *a*, *b* and their ratio
21509, 21520, 21534, 21536-7, 21564, 21575, 21595, 21606, 21608, 21613, 21626, 21629, 21642, 21662, 21678, 21697, 21726, 21734, 21743, 21747, 21760, 21767, 21772, 21773, 21825, 21842, 21872, 21880, 21894, 21923, 21925, 21948-9, 21974, 22056, 22067, 22079, 22086, 22197, 22213, 22248, 22252, 22265, 22290, 22323, 22326, 22338, 22350, 22394, 22397, 22413, 22435, 22466-7, 22476, 22516, 22518, 22527, 22534, 22555-6, 22565, B22572, 22573, 22618-20, 22624, 22673, 22694, 22793, 22800, 22805, 22815-6, 22818, 22841, 22851, 22862, 22926, 22928, 22959, 22967, 22973, 22980-1, 22988, 23057, 23080, 23083, 23086, 23092, 23149, 23193, 23200, 23226, 23248, 23259, B23262, 23263, B23264, 23269, 23272, 23288, 23350, 23356, 23383-4, 23404, B23407, 23417, 23428-9, 23443, 23460, 23463-4, 23488-9, 23515, 23526, 23528, 23536, 23550, 23560, 23562, 23567, 23570, 23583, 23586, 23590, 23638, 23642, 23687, 23718, 23751-3, 23753, B23767, 23771, 23775, 23779, 23794, 23804, 23810-1, 23820, 23822, 23835, 23876, 23883, 23890, 23916, 23960, 23964, 23973, 24008, 24019, 24040-1, 24068, 24075-8, 24084, 24086, 24111, 24122, 24127, 24148, 24163, 24190, 24226, 24228, 24280, 24284, 24299, 24308, 24328, 24364, 34411, 24414, 24421, 24457-9, 24462-4, 24479, 24482, 24505, 24523-4, 24530, 24540-1, 24545, 24548, 24560, 24584, 24597, 24603, 24615, 24654, 24713, 24715, 24775, 24785, 24818, 24824, 24879, 24892, 24912, 24915, 24923, 24929, 24942, 24984-5, 25002, 25016-7, 25019, B25064, 25102, 25117, 25152

Chlorophylls *c*, *d*
21595, 21726, 21775, 22095, 22538, 22767, 22865, 22926-8, 22930, 23016, 23269, 23393, 23590, 24233, 24308, 24451, 24882, 25055-6

Chlorophylls, *Chlorobium* 24796

Chloroplast see also Thylakoid; Stroma of chloroplast; Pyrenoid; Ribosome of chloroplast; Phycobilisome

Chloroplast and cell counting methods 24253, 24513

Chloroplast and chromatophore biopotentials
21518, 22047, 22253, 22332, 22587, 22769, 22819, 22892, 22971, 23230, 23317, 23859, 23972, 24026, 24197, 24355, 24445, 24812, 24938-9

Chloroplast and chromatophore chemical composition (*cf.* also Lipids...; Proteins...)
21545, 21551, 21574-5, 21653, 21685, 21728, 21747, 21768, 21792, 21817, 21825, 21835, 21931, 21996, 22002-3, 22032, 22056, 22060, 22076, 22135, 22145, 22149, 22195, 22215, 22237, 22280, 22285, 22322, 22422, 22469, 22619-20, 22641, 22688, 22694, 22712, 22729, 22737-41, 22779, 22781, 22783, 22810, 22820-1, 22967, 23009, 23228, 23281, 23373, 23459, 23461, 23473, 23482, 23490, 23543, 23621, 23630, 23660, 23681, 23690, 23706, 23726, 23730, B23767, 23788-9, 23821-2, 23836, 23852, 23870, 23880, 23935, 23969, 24035, 24049-50, 24086, 24095, 24103, 24112, 24162, 24178, 24180, 24182, 24184, 24190, 24228, 24253, 24319, 24340, 24348, 24353, 24357, 24368, 24379, 24381, 24397, 24423, 24468, 24549, 24559, 24591, 24694, 24701, 24735, 24738, 24750, 24793, 24827, 24832, 24855, 24865, 24878-9, 24890, 24892-4, 24920, 24997, 25029, B25064, 25084, 25096, 25147

Chloroplast and chromatophore dimensions
21663, 21851-2, 21863, 21871-2, 22117, 22148, 22290, 22467, 22496, 22565, B22572, 22573, 22624, 22870, 22924-5, 23034, 23092, 23350, 23392, 23549, B23767, B23933, 24075-6, 24088, 24196, 24218, 24253, 24315, 24340, 24347, 24817, 24847, 25008, B25064

Chloroplast and chromatophore distribution in cell see Chloroplast and chromatophore number and distribution

Chloroplast and chromatophore fragments
21572, 21578, 21599-600, 21606, 21613, 21632, 21685, 21690, 21743, 21745, 21825, 21888, 21922, 21983, 21996, 22079, 22164, 22243-4, 22289, 22326, 22345, 22349-50, 22374-5, 22406, 22462, 22483, 22501, 22514, 22539, 22596, 22605, 22626, 22667, 22705, 22708, 22748, 22773-4, 22780-1, 22856, 22906, 22913, 22922-3, 22980-1, 23056-7, 23093, 23110, 23120, 23122, 23127, 23139--41, 23169, 23187, 23242, 23310, 23345, 23373, 23421, 23473, 23515-6, 23586, 23642, 23655, 23724, 23730, 23742, 23852, 23860, 23876, 23916, B23933, 23934-5, 24008, 24099, 24116, 24165, 24190, 24226, 24298, 24385, 24395, 24415, 24423, 24443, 24455, 24457-9, 24462, 24478-80, 24487, 24493, 24495-6, 24591, 24603, 24728, 24876-7, 24879, 24909, 24923, 24954, 25002, 25006, 25019, 25030, 25057, 25083, 25099, 25113-4

Chloroplast and chromatophore number and distribution
21663, 21747, 21863, 21871-2, 21974, 22082, 22313, B22572, 22573, 22624, 22978, 22988, 23034, 23071, 23092, 23350, 23689, B23767, B23933, 24075-6, 24086, 24100, 24196, 24237, 24353, 24467, 24499, 24817-9, 24847, 24985, 25031

Chloroplast and chromatophore replication, ontogeny
21528, 21653, 21747, 21850, 21871-2, 22011, 22068, 22083, 22147, 22180, 22193, 22237, 22277, 22307, 22312-3, 22323, 22338, 22361-2, 22401, 22423, 22469, 22492, 22602, 22692, 22800, 22816, 22945, 23094-5, 23109, 23129, 23279, 23319, 23353, 23369, 23482-3, 23527, 23642, 23674, 23676-7, 23689, 23766, B23767, 23835, 23882, 23903, 23919, 23969, 23987, 24086, 24157, 24184, 24209, 24237, 24340, 24346-8, 24382-3, 24506, 24545-6, 24590, 24735, 24792-3, 24867, 24958-9, 24996-7, 25000, 25008, 25075, 25147

Chloroplast and chromatophore volume changes
22017, 22232, 22526, 22536, 22557, 22560, 22820-1, 22891, 23726, 23847, 24078, 24136, 24187, 24867, 25160

Chloroplast, isolated, carbon fixation in see Carbon fixation in isolated chloroplasts ...

Chloroplast, isolated, gas exchange by
21717, 21818, 22140, 22452, 22536, 22666, 22721, 22833-4, 22980-1, 23183, 23195, 23492, 23715, 23841, 24501, 24631

Chloroplast isolation
21565, 21716, 22111, 22304, 22370, 22463, 22590, 22626, 22836, 22891, 22996, 23357, 23415, 23705, B23933, 24149, 24292, 24823, 24948

Chloroplast, localization of electron transport chain in thylakoid see Electron transport chain localization in thylakoid

Chloroplast movements
21676, 21905-7, 22702-3, 24166, 24365-6, 24770, 24890, 25160

Chloroplast outer membrane
21546, 21612-3, 21651, 21718, 21755, 21844, 21871, 22033, 22114, 22223, 22281, 22338, 22536, 22694, 22732, 22891, 22967, 23209, 23353, 23507, 23755, 23924, 24049, 24370, 24828, 24997, 25137

Chloroplast proteins (and other photosynthetic proteins), methods
21565, 21600, 21685, 21860, 22780, 23281, 23321, 23482, 23498, 23821, 24379

Chloroplast ultrastructure (*cf.* also Chloroplast outer membrane; Stroma..; Thylakoid..)
21546, 21582, 21595, 21601, 21606-8, 21610, 21651, 21662, 21664, 21700, 21756, 21767, 21778, 21791, 21825, 21848, 21850-2, 21871, 21873, 21899, 21978, 21984, 22009, 22033, 22117, 22147, 22175-6, 22193, 22200, 22223-4, 22233, 22240-1, 22277, 22312-3, 22322, 22328, 22336, 22355, 22393-5, 22415, 22422-4, 22449, 22467, 22473, 22492, 22538, 22642, 22647, 22698, 22740, 22743, 22783, 22793, 22800, 22815-7, 22890, 22926, 22945, 22975, 23031, 23093, 23108-9, 23129, 23280, 23310, 23319, 23322-3, 23325, 23350, 23357, 23366, 23369, 23392, 23402, 23450, 23528, 23530, 23575, 23631, 23642, 23647, 23670, 23674, 23680-1, 23689, 23705, 23755, B23767, 23769, 23773, 23820, 23822, 23849, 23883, 23903-5, 23919, 23924, 23969, 23987, 24030, 24067,

Competition in ecosystem 21667, 21845, 21862, B22358, 22447, 22907, 23338, 24614

Conductance for transfer of gases see Resistance ...

"Contribution" of individual organs see Biomass distribution and redistribution; Photosynthate translocation ...

Correlations within plant 21786, 22015

Cosmic radiation see Ionizing radiation ...

Coupling factor 1 see ATPase ...

Cover, vegetative see Canopy ...; Ecosystem ...

Crassulacean Acid Metabolism see CAM

Cultivar differences, carbon fixation pathways 22050, 22975, 24523

Cultivar differences, carotenoids
21571, 22178, 22696, 22860, 23149, 23193, 23526, 23810-1, 24375, 24892

Cultivar differences, chlorophyll
21513, 21949, 22012, 22085, 22178, 22300, 22696, 22851, 22911, 23149, 23193, 23288, 23526, 23570, 23810-1, 24019, 24108, 24285, 24523, 24714-5, 24892

Cultivar differences, chloroplast (chromatophore) 24892, 25087

Cultivar differences, ecosystem productivity and canopy functioning
21513, 21570, 21949-50, 22212, 22260, 22725, 22755, 22879, 23117, 23544, 23734-5, 23763, 23777, 23779, 23795, 23862, 24080, 24142, 24252, 24363, 24375, 24609, 24646-7, 24798, 25059, 25125

Cultivar differences, electron transport chain
22298, 22409, 23672, 23944, 24286, 24821, 25087, 25134

Cultivar differences, gas exchange
21570, 21677, 21967, 22042, 22085, 22212, 22298, 22678, 22725-6, 22796, 22851, 22952, 22966, 23172, 23763, 23779, 23795-6, 23922, 24142, 24252, 24523, 24715, 24787

Cultivar differences, photorespiration 23796

Cultivar differences, resistances to CO_2 and water vapour transfer
21529, 24449, 24649

Cultivar differences, respiration 22952, 23796, 25137

Cultivation of algae and photosynthetic bacteria see Algae and photosynthetic bacteria, cultivation; Algae mass cultures productivity

Cuticular CO_2 and O_2 exchange 23873

Cuticular resistance see Resistance, cuticular

Cytochromes
21562, 21567-8, 21611, 21613, 21638, 21669, 21689, 21773, 21807, 21810, 21825, 21839, 21923-4, 21938, 21965, 21970, 21993, 21996, 22021, 22063, 22141, 22145-6, 22269, 22322, 22325, 22478, 22499, 22501, 22525, 22580-1, 22588, 22605, 22640, 22648, 22667, 22683, 22698, 22759, 22774, 22794, 22809-12, 22935, 22972, 22983, 23057, 23123-4, 23126, 23148, 23173, 23200, 23253, 23261, 23276, 23292, 23335, 23421, 23515, 23519, 23548, 23574, 23580, 23611, 23647, 23704, 23724, B23767, 23843, 23847, 24027, 24095-7, 24163, 24189, 24221, 24241, B24250, 24281, 24327, 24329, 24340, 24403, 24418, 24466, 24487, 24631, 24635, 24690-1, 24723, 24808, 24879, 24908-9, 25006, 25030, B25064, 25068, 25070, 25076

Cytochromes, methods 22021, 23253, 23276, 24635, 24685

D

Dark CO_2 fixation
21783, 22537, 22556, 23013, 23099, 23121, 23210, 23300, 23333, 23379, 23610, 23786, 23926, 24083, 24999, 25018

Data recording and processing 21801, 22597, 23834

Decapitation see Defoliation ...

Defoliation, decapitation, ear and root removal, effect on carotenoids 22581

Defoliation, decapitation, ear and root removal, effect on chlorophyll
22006, 23529, 23756, 24127

Defoliation, decapitation, ear and root removal, effect on ecosystem productivity and canopy functioning
21522, 21634, 21786, 22096, 22352, 22663, 23499, 23774, 23776, 23895, 24127, 24275, 24326, 24511, 24598, 24744, 24820, 24970

Defoliation, decapitation, ear and root removal, effect on gas exchange
21538, 21634, 21786, 22006, 22210, 22952, 23218-9, 23430, 23756, 23962, 24275, 24598, 24820

Defoliation, decapitation, ear and root removal, effect on resistances to CO_2 and water vapour transfer 23218, 23430

Defoliation, decapitation, ear and root removal, effect on respiration
22952, 23962, 24275

Defoliation, methods 24275

Deuterium oxide, tritium oxide 21641, 24943

Development, leaf, plant see Leaf (and plant) development and ageing

Dew see Precipitation, dew ...

Dew measurement see Aerodynamic methods ...

Dichroisms determination (methods and results)
21715, 21898, 21900, 21917, 22113-4, 22394, 22593, 22596, 22979, 23055, 23707, 23909, 23998, 24154, 24327, 24385, 24784, 24810

Differentiation of tissues see also Leaf (and plant) development and ageing; Ontogeny.

Differentiation of tissues and carotenoids 21872

Differentiation of tissues and chlorophyll 21872, 21925, 23048

Differentiation of tissues and chloroplast (chromatophore) 21872, 23490

Differentiation of tissues and ecosystem productivity and canopy functioning
21882, 22015, 23468, 24267

Differentiation of tissues and electron transport chain 21925

Differentiation of tissues and gas exchange 21593, 21696

Diffusive resistance see Resistance ...

Diurnal changes (biological clock) in algae productivity
21972, 21990, 22292, 22493, 22578, 22654, 22858, 22866, 22943, 23759, 24144, 24420, 24634, 25112

Diurnal changes (biological clock) in carbon fixation pathways and parameters
22038, 22303, 24126, 24656

Diurnal changes (biological clock) in carotenoids 22303, 23471

Diurnal changes (biological clock) in chlorophyll 22303, 23441, 24194, 24285

Diurnal changes (biological clock) in chloroplast (chromatophore) 21905, 22744, 24867

Diurnal changes (biological clock) in ecosystem productivity and canopy functioning
21549, B22052, 22108, 22116, 22144, 22999, 23592, 23802, 23910, 23945, 23983, 24143, 24214, 24244, 24259, 24438, 24441, 24567, 24641, 24656, 24689, 24887, 24907

E

Ecotypes, geographical types, and resistances to CO_2 and water vapour transfer 22284

Ecotypes, geographical types, and respiration 22950

Ecotypes, geographical types, carotenoids in 23752-3, 24530

Ecotypes, geographical types, chlorophylls in 21709, 22992, 23752-3, 24530

Efficiency, photosynthetic see Irradiance and gas exchange, analysis of light curves

Electron transport chain activity
21646, 21697, 21744, 21825, 21915, 22106, 22142, 22151, 22164, 22215, 22243, 22427, 22514, 22526, 22567, 22576, 22605, 22683, 22706, 22720, 22800, 22821, 22882, 22973, 23043, 23148, 23395, 23794, 23851, 23920, 24225, 24496, 24664, 24684, 24762, 24791, 24909, 25002, 25102

Electron transport chain activity, methods 21744, 23876

Electron transport chain components see Cytochromes; Ferredoxin...; Ferredoxin-NADP reductase; NADP...; O_2 evolution...; Photosystems; Plastocyanin; Quinones...

Electron transport chain components and photorespiration 23981-2

Electron transport chain localization in thylakoid
21535, 21575, 21646, 21778, B21979, 22017, 22100, 22346, 22354, 22528, 22580, 22644, 22657, 22704-6, 22969-71, 23093, 23631, B23767, 24095, 24097, 24136, 24327, 24395, 24471, 24491, 24789-90, 25006, 25030

Electron transport chain model see Model ...

Electron transport chain, serological analysis
21613, 21743, 21777-8, 21792, 22217, 22574, 23253, 23577, 23624, 23798, 24095, 24097, 24130, 24135, 24182, 24351, 24471, 24860, 25030

Emerson effect, Blinks effect 21638, 22580, 23724, 23989, 24724, 24932

Energy balance, leaf B22358, 22875

Energy utilization, plant and ecosystem
21629, 21832, 21973, B21979, 22036, 22123, 22185, 22297, 22302, 22543, 22750, 22755, 22765, 22829, 23003, 23106, 23145, 23147, 23423, 23814, 23862, 23925, 24229, 24419, 24438, 24647, 24780, 24782, 24863, 24969, 25023, 25025

Enzymes and carbon fixation pathways 23504

Enzymes and chlorophyll 21943, 23140-1, 23951, 24305

Enzymes and chloroplast (chromatophore)
21551, 21857, 22739-41, 23742-3, 24328, 24360, 24423, 24828, 24855

Enzymes and electron transport chain
21551, 23014, 23139, 23141, B23407, 23934, 24184, 24415, 24874

Enzymes and gas exchange 23504

Enzymes and photorespiration 22347, 22942, 23231, 23981-2

Enzymes and respiration 21814, 23231, 24657, 24659, 24795

Enzymes of carbon fixation pathways other than RuBPC, PEPC and malic enzyme
21576, 21578-80, 21655, 21672-3, 21702, 21724, 21814, 21840, 21933, 21937, 21942, 22018, 22054, 22081, 22150, 22157, 22239, 22250, 22274, 22303, 22635, 22671, 22695, 22697, 22699, 22713, 22785-6, 22802, 22838, 22958, 22977, 22980-1, 23030-1, 23033, 23060, 23062, 23072, 23336-7, 23370, 23397, 23419, 23586, 23674, B23767, 23788, 23797, 23868, 23949, 23974, 24002, 24008, 24106, 24112, 24161, 24164-5, 24223, 24255, 24342, 24346, 24352, 24357, 24360, 24380, 24383, 24386, 24397, 24658, 24682, 24734, 24769, 24804, 24931, 24933, 24984, 25033, 25065-6, 25073, 25097, 25101, 25138

Enzymes of carbon fixation pathways other than RuBPC, PEPC, malic enzyme, malate dehydrogenase, methods 21579, 21937, 22081, 22699, 22713, 22802, 23868, 24002, 25033, 25065-6

Enzymes of carotenoids synthesis and degradation see Carotenoids, enzymes ...

Enzymes of chlorophyll synthesis and degradation see Chlorophyll, enzymes ...

Enzymes of electron transport chain, methods 21617

Enzymes of photorespiration see Photorespiration enzymes

Epidermis see Leaf epidermis ...

EPR, NMR (methods and results)
21583, 21617, 21643-6, 21816-9, 21970, 21981, 21983, 22060, 22088, 22090, 22126-7, 22146, 22373, 22375-6, 22390, 22459, 22561, 22642, 22682, 22708, 22887, 23055, 23057, 23076, 23084, 23089, 23139, 23192, 23215, 23253, 23437, 23445, 23472, 23546, 23583, 23600-1, 23655, 23729-30, 23799, 23966, 24211, 24304, 24327, 24337, 24344, 24453, 24638, 24640, 24651, 24760-2, 24783-4, 24844, 24877, 24904, 24917, 24936-7, 25019

Ethylene see Gases, organic ...

Evolution see Phylogeny

Excitation resistance see Resistance, carboxylation and excitation

Exhaust gases see Pollution of air ...

Exposure chamber see Assimilation chamber

Extension growth, leaf dimensions
21533, 21853, 21878, 21882, 21928-9, 22007, 22012, 22015, 22097, 22113, 22228, 22335, 22401, 22485, 22516, 22663, 22701, 22727, 22844, 22859, 22885, 22999, 23223, 23423, 23431, 23465-6, 23493, 23500, 23521, 23587, 23604, 23645, 23650, 23669, 23695, 23779, 23813, 23950, 23961, 23975, 23984, 24000, 24076, 24085, 24123, 24231, B24271, 24444, 24628, 24673, 24718, 24744, 24770, 25042, 25045-6, 25059

Extraction of pigments see Pigments ...

Exudation of photosynthates see Photosynthate translocation ...

F

Fatty acids see Lipids, fatty acids ...

Ferredoxin, ferredoxin-NADP reductase, methods
21589, 22372, 23055, 23577, 24213, 24334, 24648, 24707, 25105

Ferredoxin, flavoproteins, rubredoxin
21530, 21547, 21548, 21583-4, 21589, 21591, 21611, 21620, 21638, 21668, 21737-8, 21743, 21865, 21935, 21937, B21979, 21980-1, 21983, 22051, 22126, 22346-7, 22373-4, 22588, 22658-60, 22669, 22693, 22736, 22773, 22853, 22979, 23057, 23124, 23150, 23272, 23517, 23525, 23563, 23577, 23600-1, 23708, 23719-20, B23767, 23798-9, 23847, 24041, 24074, 24223, 24304, 24327, 24362, 24380, 24418, 24615, 24648, 24651, 24663-4, 24706, 24720, 24723, 24784, 24875, 24917, 24944-5, 25006, B25064, 25091, 25104-5, 25124, 25145

Ferredoxin-NADP reductase, pteridines
21638, 21743, 22027, 23184, 23261, 23272, B23767, 23798, 24213

Flashes of light see Irradiation, flash

Flavoproteins see Ferredoxin ...

Flooding and carotenoids 23682

Flooding and chlorophyll 23682

Flooding and ecosystem productivity and canopy functioning 24609

Flooding and electron transport chain B24024

Flooding and gas exchange 22830, 24183

Fluorescence, methods 22021, 22664, 23146, 23153, 23538, 23616, 25114

Fluorine see Pollution of air ...

Foliage see Canopy ...

Fraction I protein see Ribulose 1,5-bisphosphate carboxylase

Frost (hardiness) see Temperature, low ...

Fungus diseases see Phytopathological effects ...

Fusicoccin see Growth regulators ...

G

Gas exchange in algae
21711, 21738, 21770, 21772, 21808, 21903, 21966, 21976, B22052, 22092, 22104, 22109, 22137, 22211, 22265, 22287, 22303-4, 22363-4, 22387, 22407, 22433, 22488, 22490, 22494, 22522, 22623, 22673, 22717, 22764, 22790-1, 22828, 22842, 22846-8, 22861, 22944, 22961, 23018, 23043, 23099, 23133, 23165, 23225, 23439, 23446, 23475, 23537, 23552-3, 23571-2, 23606, 23659, 23685, 23718, 23729, 23884, 24013, 24021, 24029, 24043, 24053, 24072, 24134, 24172, 24346-8, 24401-2, 24405, 24434, 24483, 24525, 24565, 24570, 24608, 24612, 24637, 24667, 24679, 24708, 24724, 24735, 24800, 24862, 24865, 24867, 24956, 24964-5, 24979, 24984, 24998, 25001, 25127

Gas exchange in isolated chloroplasts see Chloroplast , isolated, gas exchange by

Gas exchange, model see Model ...

Gas exchange of other organs than leaf 24759

Gases, organic, and carotenoids 22323

Gases, organic, and chlorophyll 21683, 22323

Gases, organic, and ecosystem productivity and canopy functioning 24470

Gases, organic, and electron transport chain 22346

Gases, organic, and gas exchange 22434, 22830-1, 23151

Gasometric methods, generally 22295, 22830, 22901, 23155, 23306, 23559, 24863, 25015

Gasometric system, conditioning of air 23443, 23651, 23727, 23834, 23979, 24151

Gasometric system, open 22152-3, 22189, 23306, 23443, 23457, 23559, 23727, 23988, 24592

Gasometric system, semiclosed and closed
22005, 22252, 22308, 22344, 22819, 23049, 23306, 23559, 23652, 23834, 24291, 25032

Genetics and ecosystem productivity and canopy functioning
22516, 23220, 23740, 23795, 23814, 23897, 24219, 24260, 24375, 24574, 24625

Genetics of carbon fixation pathways 22049-50, 23256, 23754, 23769, 23797, 23814

Genetics of carotenoids 23193, 23675, 23739, 24203, 24375, 24540, 24548

Genetics of chlorophyll
21960, 22435, 22624, 22735, 22911, 23193, 23331, 23350, 23594, 23629, 23739, 23754, 23768, 23897, 24022, 24540-1, 24548, 24788

Genetics of chloroplast (chromatophore)
21575, 21871, 22554, 22624, 22643, 23071, 23279, 23350, 23490, 23766, B23767, 23905, 24346, 24370, 24397, 24655, 24788, 24819, 25087, 25147

Genetics of electron transport chain
22870, 23672, 23814, 24468, 24547-8, 24891, 25087

Genetics of gas exchange
21637, 21967, 22455, 22487, 22624, 22870, 22985, 23352, B23540, 23629, 23769, 23795-6, 23814, 24614, 25047, 25111, 25156

Genetics of photorespiration 23629, 25137

Genetics of respiration 22624, 25047

Glycollate metabolism see Photorespiration ...

Glyoxysome see Peroxisome ...

Granum see Thylakoid ...

Gravimetric determination of photosynthesis see Dry-matter production ...

Gross photosynthetic rate
21526, 21659, 21782, 22038, 22252, 22494, 22563, 22678, 22896-7, 23013, 23091, 23145, 23278, 23298, B23407, 23464, 23553, 23603, 23632, 23894, 23994, 24029, 24082, 24110, 24122, 24193, 24323-4, 24438, 24692, 24880, 24915

Growth analysis, methods
21795, 21802, 21882, 22404, 22681, 22716, 23155, 23234, 23784, 24143, 24425, 24509, 24564, 24602, 24672, 24726, 24971-2

Growth analysis, net assimilation rate, leaf area ratio, relative growth rate
21523-5, 21621, 21803, 21853, 21919, 21951, 22074, 22076, 22112, 22123, 22260, 22273, 22314, 22369, 22380, 22396, 22402, 22414, 22419, 22430-1, 22449, 22479, 22516, B22572, 22593, 22603, 22676, 22787, 22806, 22829, 22855, 22885, 22916, 22974, 22984, 22998, 23002, 23037-8, 23063-4, 23106, 23156-7, 23181, 23220, 23223, 23234-5, 23251, 23305, 23443, 23448, 23499, 23502, 23557, 23578, 23582, 23650, 23661, 23695, 23712, 23732, 23734-6, 23763, 23776-7, 23813-4, 23864, 23893, 23895-7, 23925, 23975, 24014, 24020, 24115, 24143, 24186, 24219, 24267, 24275, 24291, 24295, 24303, 24321, 24432, 24438, 24444, 24474, 24504, 24509, 24537, 24563, 24585, 24605, 24610, 24646-7, 24661, 24666, 24705, 24748, 24759, 24794-5, 24798, 24822, 24830, 24841, 24849, 25039, 25041-2, 25047, 25049, 25052, 25103

Growth analysis, specific leaf area, leaf area index, leaf area duration
21521, 21523-5, 21556, 21621, 21658, 21667, 21686, 21761, 21793, 21802, 21809, 21853-4, 21919, 21929, 21950-1, 22030, 22074, 22076, 22097, 22108, 22118, 22185, 22246, 22260, 22273, 22282, 22331, 22343, 22352, 22367-9, 22396, 22402, 22414, 22419, 22430, 22448, 22516, 22524, 22589, 22593, 22676, 22725, 22734, 22775-6, 22806, 22829, 22854, 22885, 22897, 22910, 22960, 22984, 22998, 23002, 23007, 23038-9, 23059, 23091, 23145, 23156-7, 23168, 23220-1, 23234-5, 23251, 23304, 23367, 23410, 23426, 23443, 23468, 23493, 23499, 23545, 23557, 23578, 23582, 23587, 23645, 23653, 23669, 23695, 23732, 23736-7, 23784, 23805, 23814, 23819, 23856, 23893, 23895-7, 23923, 23925, 23945, 24003, 24019-20, 24080, 24093, B24132, 24186, 24214-5, 24219, 24249, 24267, B24271, 24291, 24295, 24321, 24324, 24430, 24436-8, 24470, 24474, 24509, 24511, 24563, 24569, 24574, 24605, 24609, 24625, 24633, 24646-7, 24661, 24668, 24670, 24689, 24705, 24752, 24758, 24773, 24780-1, 24794-5, 24798, 24822, 24841, 24880, 24940, 25015, 25039, 25041, 25049, 25052, 25086, 25103

Growth regulators and carbon fixation pathways
21933, 22786, 23219, 23974, 24016, 25063

Growth regulators and carotenoids
21509, 21520, 21959, 22401, 23080, 23425, 23526, 23675, 23807, 23974, 24391, 25142

Growth regulators and chlorophyll
21509, 21520, 21536, 21701, 21781, 21948, 21959, 21961, 22067-8, 22290, 22401, 22468, 22472, 22571, 22735, 22785, 23019, 23025, 23080, 23356, 23425, 23526, 23535-6, 23638, 23674, 23883, 23974, 24037, 24084, 24127, 24168, 24390-1, 24414, 24464, 24510, 24745-6, 24753, 25075, 25079, 25142

Growth regulators and chloroplast (chromatophore)
21536, 22401, 22703, 22785, 24368, 24414, 24694, 24865, 24931, 25063, 25075

I

Infra-red radiation, effect on photosynthetic parameters see Irradiance, spectral composition ...; Temperature, high ...

Inhibitors of electron transport chain (*cf.* also Pesticides ...; Antibiotics ...) 21528, 21533, 21559, 21567, 21576, 21627, 21638, 21643-5, 21649-50, 21671, 21679, 21687, 21689, 21691, 21706, 21728, 21744, 21746, 21763-4, 21766, 21774, 21800, 21804, 21806, 21817, 21819, 21825, 21839, 21859, 21883, 21907, 21970, 21997, 22013, 22079, 22105, 22106-7, 22141-2, 22145, 22154, 22179, 22192, 22202, 22214, 22217, 22240, 22243-4, 22309-10, 22325, 22330, 22349-50, 22359-60, 22397-8, 22400, 22405, 22426-7, 22434, 22473, 22478, 22490-1, 22501, 22525-8, 22536, 22540-1, 22548, 22559, 22561, 22574-6, 22582-3, 22612, 22637, 22669, 22683, 22689, 22691, 22704, 22748, 22773, 22794, 22801, 22811--2, 22819-20, 22824, 22834-5, 22837, 22872, 22900, 22906, 22934, 22946, 22955, 22957, 22961, 22995, 23005, 23008, 23032, 23043, 23066, 23076-7, 23114, 23118-9, 23122, 23132, 23150, 23184-5, 23187, 23199, 23208, 23215, 23219, 23253, 23272-3, 23309, 23317, 23322, 23326-7, 23343, 23365, 23396, 23404, 23446, 23451, 23477, 23479, 23484, 23486, 23490-1, 23495, 23506-7, 23518-9, 23538, 23541, 23548, 23551, 23573, 23591, 23599, 23615, 23618, 23622, 23647, 23656, 23660, 23715, 23718, 23722, 23724, 23728, 23730, 23762, 23808, 23837, 23840, 23847, 23851, 23883, 23884, 23888-9, 23898, B23933, 23944, 23954-5, 23957, 23960, 23970, 23999, 24011, 24013, B24024, 24032, 24035, 24042, 24136, 24160, 24169, 24171, B24176, 24184, 24187, 24191, 24193, 24208, 24212, 24223-5, 24238, 24241, 24251, 24258, 24293, 24307, 24315, 24318, 24320, 24338, 24344, 24346, 24351, 24373, 24396, 24403, 24413, 24417-8, 24463, 24467, 24481, 24490, 24496-7, 24525, 24542, 24545, 24559, 24575, 24650, 24665, 24669, 24719, 24721, 24735, 24737, 24790-1, 24846, 24850, 24859, 24869-70, 24874-5, 24876-7, 24881, 24901-4, 24910-1, 24938, 24949, 24998, 25000, 25012, 25059, 25061-2, 25071, 25081, 25088, 25091, 25096, 25124, 25151-2

Insertion see Ontogeny ...

Intercellular spaces, CO_2 concentration inside 21757, 21767, 21773, 22041, 22652, 23928, 24239-40

Intermediates of carbon fixation pathways, methods 22710, 25067

Intracellular resistance see Resistance, intracellular (mesophyll)

Ionizing radiation (gamma, X, cosmic, *etc.*)and carbon fixation pathways 25116

Ionizing radiation (gamma, X, cosmic, *etc.*) and carotenoids 24751

Ionizing radiation (gamma, X, cosmic, *etc.*) and chlorophyll 21603, 21831, 22911, 24141, 24388

Ionizing radiation (gamma, X, cosmic, *etc.*) and ecosystem productivity and canopy functioning 23763, 25126

Ionizing radiation (gamma, X, cosmic, *etc.*) and gas exchange 23850, 24612

Ionizing radiation (gamma, X, cosmic, *etc.*) and respiration 23850, 24001

Irradiance, compensation see Compensation irradiance

Irradiance (PhAR) and algae productivity 21735, 22211, 22268, 22436, 22464, 22493, 22531, 22858, 22867, 22943, 23096--7, 23099, 23133, 23387, 23554, 23664, 23824, 24349, 24440, 24616-7, 24680, 24766, 24786, 24953, 24960, 25053, 25121

Irradiance (PhAR) and biliproteins 23083

Irradiance (PhAR) and carbon fixation pathways 21576, 21616, 21711, 21724, 21810, 21975, 22157, 22274, 22398, 22507, 22530, 22552, 22868, 22905, 22980-1, 23013, 23060, 23195-7, 23286, 23419, 23451, 23154, 23668, 23772, 23938, 24016, 24160, 24201, 24265, 24348, 24380, 24397, 24553, 24708, 24843

Irradiance (PhAR) and carotenoids 21509, 22401, 22450, 22507, 22518, 22619, 22644, 22841, 23116, 23393, 23471, 23804, 24129, 24146, 24228, 24317-8, 24451, 24489, 24621, 25077-8

K

L

Leaf optical properties (*cf.* also Carotenoids absorption spectra *in vivo*;Chlorophyll absorption spectra *in vivo*)
21629, 21695, 22247, 22296, 22302, 22333-5, 22508, 23368, 23523, 23692, 23992, 24436, 24775, 24818

Leaf resistance see Resistance for water vapour ...; Resistance, stomatal ...

Leaf, sun- and shade leaf see Leaf anatomy

Leaf temparature (methods and results)
21569, 21695, 21773, 22505, 22507, 22652, 22875, 23297, 23368, 23447, 23791, 23992, 24045, 24627, 24718, 24739, 24889, 24934

Leaf temperature measurement see Leaf temperature

Leaf thickness
21633, 21801, 22039, 22082, 22084, 22157, 22509, 22516, 22787, 22885, 22937, 23521, 23787, 23848, 23896, 24076, 24299, 24411, 24601

Light see Irradiance ...; Canopy, radiation ...

Lighting system see Irradiation, illumination ...

Linear dichroism see Dichroisms ...

Lipids, fatty acids, and carbon fixation pathways 22411, 23036

Lipids, fatty acids, and chlorophyll 24284

Lipids, fatty acids, and chloroplast
21575, 21653-4, 22002-3, 22620, 22673, 22694, 22712, 22779, 22820, 23116, 23228, 23323, 23402, 23459, 23482, 23730, B23933, 24468, 24827, 24878-9, B25064, 25092-3

Lipids, fatty acids, and electron transport chain 22711, 23227, 23935, 24494

M

"Maintenance" respiration see Respiration, "growth" and "maintenance"

Malate dehydrogenase, methods see Malic enzyme, malate dehydrogenase, methods

Malic enzyme
21580, 21814, 22054, 22157, 22239, 22381, 22397, 22399, 22635, 22697-8, 22983, 22997, 23025, 23030-1, 23060, 23231, 23289, 23586, 23797, 23877, 23938, 24008, 24122, 24126, 24164-5, 24311, 24615, 24734, 24795, 24804, 25065-6

Malic enzyme, malate dehydrogenase, methods 22381, 25065-6

Mass culture of algae see Algae mass cultures ...

Mehler reaction see Photosystem 1 activity ...

Membrane transport of CO_2 21916, 23848, 24426

Mesophyll resistance see Resistance, intracellular (mesophyll)

Microbody see Peroxisome

Microelements see Mineral elements (other than N, P, K) ...

Mineral elements (N, P, K) and algae productivity
21735, 22132, 22417, 22551, 22976, 23301, 23606, 23932, 24248, 24349, 24687, 24764, 25130

Mineral elements (N, P, K) and biliproteins 22496

Mineral elements (N, P, K) and carbon fixation pathways
21672, 21785, 22150, 22552, 22685, 22837, 23333, 23386, 23480, 23660, 24064, 24442, 24734, 24752, 24854, 24982, 25009, 25035-6

Mineral elements (N, P, K) and carotenoids
21697, 22518, 22565, 23375, 23550, 23853, 24145-6, 24540, 24597, 24778, 24978

Mutants, chloroplast (chromatophore) in
21613, 21825, 21860, 22078, 22117, 22393, 22395, 22643, 22737, 22800, 23031, 23129, 23265-6, 23279-80, 23512, 23636, 23746-7, 23820-2, 23849, 23885, 23919, 24067, 24357, 24369, 24381-2, 24398, 24402, 24603, 24696, 24788, 24847, 24855, 24959

Mutants, electron transport in
21575, 21807, 21888, 21935, 22078-9, 22117, 22145, 22325, 22498, 22501, 22842, 22957, 23404, B23407, 23548, 23822, 24041, 24190, 24280, 24283, 24329, 24351, 24466-7, 24545, 24603, 24858, 24955, 25088

Mutants, gas exchange in
21677, 21771, 22265, 22393, 22842, 23278, 23718, 23845, 23849, 24280, 24283, 24466, 24483, 24848

Mutants, photosynthetic, isolation and selection 23513

N

N_2, anaerobic atmosphere, and gas exchange 21572, 23764

NAD see NADP, NAD

NADP, NAD 21530, 21548, 21611, 21638, 21673, 21737, 21743, 21837, 22051, 22134, 22164, 22347, 22601, 22659, 22706, 22786, 22821, 23078, 23146, 23170, 23201, 23231, 23261, 23351, 23370, 23591, 23724, 23726, 23798, 23934, 24172, 24187, 24242, 24258, 24384, 24415, 24468, 24489, 24491, 24651, 24665, 24790, 24804, 24875, 25006, 25030, 25124

NADP, NAD , methods 22921

Net assimilation rate see Growth analysis, net assimilation rate ...

Net photosynthetic rate see Gas exchange ...

Nitrogen see N_2 ..., Mineral elements (N, P, K)

NMR see EPR, NMR ...

Nuclear magnetic resonance see EPR, NMR ...

Nucleic acids see Proteins, amino acids, nucleic acids ...

O

O_2 and biliproteins 24405

O_2 and carbon fixation pathways
21592, 21618, 21652, 21883, 22397, 22530, 23452, 23841, 23877, 23926-7, 24222, 24408

O_2 and chlorophyll 22068, 22093, 22230, 22259, 23613, 23825, 23993, 24405, 24930

O_2 and ecosystem productivity and canopy functioning 22479, 24123-5

O_2 and electron transport chain
21611, 22104, 22236, 22345, 22721, 22934-5, 23261, 23632, 24099, 24329, 24859

O_2 and gas exchange
21711, 21717, 21800, 21812, B21914, 21926, 22031, 22066, 22070-1, 22086, 22131, 22152, 22251-2, 22305, 22521, 22442, 22544, 22608, 22649, 22796, 22833, 22935, 23218, 23329, 23444, 23493, B23540, 23764, 23796, 23814, 23841, 23926, 23928, 23994, 24082-3, 24296, 24405, 24447, 24593, 24611-2, 24804, 24946, 24998

Ontogeny of leaf, insertion level, and chlorophyll (continued)
23835, 23896, 23896, 23959, 23964, 24076, 24236, 24284-5, 24321, 24330, 24404, 24411, 24485, 24523, 24541, 24584, 24643, 24654, 24693, 24746, 24792, 24915-6, 24920, 24994, 25019, 25061, 25063, 25117

Ontogeny of leaf, insertion level, and chloroplast
21661, 21663, 22313, 22627, 23265-6, 23392, B23767, 24076, 24269, 24533, 25019, 25063, 25073, 25075

Ontogeny of leaf, insertion level, and ecosystem productivity and canopy functioning
21511-2, 21540, 21552, 21684, 21786, 21802, 21842, 21847, 22004, 22547, 22678, 22725, 22776, 22897, 23628, 23912, 23990, 24326, 24598, 24830

Ontogeny of leaf, insertion level, and electron transport chain
21544, 21561, 21970, 23428, B24176, 24336, 24411, 24915, 24928

Ontogeny of leaf, insertion level, and gas exchange
21517, 21538, 21590, 21629, 21633, 21661-2, 21786, 21802, 21842, 21869, 21970, 22004, 22014, 22025, 22031, 22226-8, 22678, 22725, 22776, 22786-7, 22796, 22836, 22840, 22897, 22960, 22962, 23045, 23049, 23134, 23257, 23431, 23465, 23493, 23539, B23540, 23643, 23694, 23706, 23756-7, 23779, 23783-4, 23835, 23896, 23922, 24138, 24183, 24323, 24336, 24411, 24428, 24523, 24532, 24592, 24596, 24601, 24692, 24759, 24787, 24821, 24830, 24853, 24880, 24915, 24919, 24931, 24985, 25018, 25034, 25061, 25073, 25097, 25155

Ontogeny of leaf, insertion level, and photorespiration
22226, 22228, 22542, 22796, 23171, 24915

Ontogeny of leaf, insertion level, and resistances to CO_2 and water vapour transfer
21538, 22226, 22787, 22962, 22964, 23134, 23431, 23757, 24915

Ontogeny of leaf, insertion level, and respiration
21590, 21842, 22226-8, 22786, 22796, 23091, 23257, 24821, 24830, 24919, 25047, 25073

Ontogeny of plant and carbon fixation pathways
21639, 22530, 23927, 23973, 24336, 24675, 24752, 24986

Ontogeny of plant and carotenoids 21571, 21642, 21870, 21921, 23086, 23810-1

Ontogeny of plant and chlorophyll
21642, 21870, 21948, 22247, 22295, 22512, 23086, 23258, 23287, 23810-1, 23913, 23973, 24084, 24752

Ontogeny of plant and chloroplast 21873

Ontogeny of plant and ecosystem productivity and canopy functioning
21557, 21667, 22034-5, 22428-9, 22431, 22547, 22593, 22771, 22776, 22823, 22844, 22960, 22732, 23734-5, 23740, 23862, 23912, 24084, 24115, 24125, 24291, 24626

Ontogeny of plant and electron transport chain 21544, 23086, 24286, 24336

Ontogeny of plant and gas exchange
21542, 21878, 21967, 22014, 22075, 22284, 22776, 22836, 22960, 23171-2, 23297, 23330, 23376, 23431, 23443, 23493, 23784, 23845, 23907, 24084, 24272, 24280, 24759, 25111, 25155

Ontogeny of plant and photorespiration 23973

Ontogeny of plant and resistances to CO_2 and water vapour transfer 23696, 23757

Ontogeny of plant and respiration 22563, 23443, 23587, 23973, 24084

Optical properties, leaf see Leaf optical properties

Oscillations, short-term fluctuations, steady and non-steady state, in gas exchange
22582, 22935, 22999, 23609, 23649, 23840, 24449

Oscillations, short-term fluctuations, steady and non-steady state, in resistances to CO_2 and water vapour transfer 22403, 23609

Osmotically active substances and algae productivity 23099

Osmotically active substances and carbon fixation pathways
22412, 22609, 23480-1, 25035

Osmotically active substances and chlorophyll 23523, 24713, 24727-8

Osmotically active substances and chloroplast (chromatophore)
21755, 22062, 22536, 23995, 24036, 24311

Osmotically active substances and ecosystem productivity and canopy functioning
21909, 22085, 23952

Osmotically active substances and electron transport chain
21755, 23415, 23842, B24176, 24720

Osmotically active substances and gas exchange
21755, 21908-10, 22321, 22938, 23099, 23523

Osmotically active substances and photorespiration 22609

Osmotically active substances and respiration 22484

Osmotically active substances, use for water stress induction 21908-10

Oxygen see O_2

Ozone see Pollution of air ...

P

P680 21643, 21645-6, 21662, 22243-4, 22580, 23518, 23954, 23957, 24041, 24133, 24307, 24804, 24917, 25057, 25062, 25151

P700, *P750*, *P890*, *etc.*
21528, 21534, 21536, 21567-8, 21638, 21644, 21662, 21687-9, 21743, 21773, 21810, 21816, 21825, 21843, 21866, 21922-3, 21925, 21959, 22079, 22090, 22134, 22141, 22219, 22288-9, 22354, 22374, 22390, 22393, 22498, 22501, 22561, 22580, 22588, 22794, 22816, 22862, 23020, 23022, 23041, 23043, 23046-7, 23056-7, 23076, 23084, 23102, 23111, 23120, 23140, 23184, 23187, 23239, 23241, 23347, 23420, 23470, 23515-6, 23518, 23552, 23573, 23586, 23600-1, 23638, 23724, 23729-30, B23767, 23798, 23825, 23847, 23876, 23916, 23954, 23966, 24096, 24163, 24189, 24208, 24238, 24241, 24258, 24327, 24478-80, 24487, 24496, 24631, 24671, 24691, 24756, 24760-2, 24789, 24804, 24909, 25002, 25050, 25057, 25062, 25070

Paramagnetic oxygen analyser see O_2 determination

Paramagnetic resonance see EPR, NMR

PEP carboxylase (PEPC) see Phosphoenolpyruvate carboxylase

Peroxisome, glyoxysome, microbody
21535, 21957, 22009, 22071, 22111, 22410, 22906, 22980-2, 23011, 23195, 23310, 23427, 23714, 24036, 24247, 24319, 24426, 24499, 24734, 25000, 25007

Pesticides, herbicides and algae productivity 22550

Pesticides, herbicides and carbon fixation pathways 22183, 22192, 22398

Pesticides, herbicides and carotenoids
22240, 22636, 23248-9, 23964, 24446, 24678, 24921-2

Pesticides, herbicides and chlorophyll
21679, 21732, 21781, 21842, 22266, 22339, 22555, 22670, 22804-5, 22911, 23165, 23226, 23248-9, 23576, 23955, 23957, 23964, 23967, 24048, 24432, 24446, 24549, 24785, 24882, 24902, 24920-2, 24987, 25000, 25146

Pesticides, herbicides and chloroplast (chromatophore)
21527, 22240. 22339, 22550, 23867, 24920-2, 25000

Pesticides, herbicides and ecosystem productivity and canopy functioning
21779, 21781, 22185, 22331, 22356, 23236, 23907, 24432

Pesticides, herbicides and electron transport chain
21527, 21559, 22202, 23165, 23242, 23486, 23657, 23837, 23941-2, 23944, 24032, 24193, 24208, 24241, 24490, 24663, 24665, 24791, 24881, 24921

Pesticides, herbicides and gas exchange
21640, 21779, 22240, 22349, 22356, 22397, 22418, 22433, 22535, 22550, 22790, 22872, 22946, 23165, 23194, 23242, 23249, 23311, 23548, 23625, 23657, 23867, 23907, B24176, 24432, 24719, 24881-2, 24919-21, 25000, 25122

Pesticides, herbicides and photorespiration 23150, 23867, 24949-50

Pesticides, herbicides and respiration 23165, 23311, 23867, 24048, 24785

Petiole see Stem, petiole

pH, effect on algae productivity 21789

pH, effect on biliproteins 21725, 23532

pH, effect on carbon fixation pathways
21592, 21652, 21655, 21711, 21875, 22523, 22732, 22835, 22837, 22940, 23195-6, 23198, 23336-7, 23927, 24222, 24255, 24262, 24352, 24360, 24658, 24843, 24984, 24999, 25036

pH, effect on carotenoids 22644, 24489

pH, effect on chlorophyll
22068, 22269, 22426, 22647, 22808, 22843, 23618, 23656, 23892, 24497, 24747, 24757, 24860, 24902, 24911

pH, effect on chloroplast (chromatophore) 21631, 23995, 24355, 25094

pH, effect on electron transport chain
21588, 21631, 21691, 21745-6, 21888, 21890, 21997, 22063, 22164, 22202, 22233, 22325, 22349-50, 22456, 22525, 22536, 22567, 22575-6, 22586, 22594, 22708, 22774, 22801, 22811, 22824, 22900, 22913, 22972, 22996, 23123, 23183-4, 23192, 23362-3, 23449, 23555, 23591, 23615, 23721, 23798, 23851, 24032, 24042, 24081, 24098, 24136, B24176, 24187, 24213, 24257, 24281, 24316, 24367, 14374, 24417-8, 24497, 24637-8, 24757, 24790, 24869-70, 24886, 24911, 24938, 25006, 25094, 25110, 25143

pH, effect on gas exchange
22054, 22452, 22511, 22536, 22578, 22665, 22732, 22828, 22836, 22848, 22935, 23058, 23462-3, 23596, 23840, 23843, 24834, 24919, 24979, 25069

PhAR see Irradiance ...; Canopy, radiation ...

Phosphoenolpyruvate carboxylase
21722, 21736, 21739, 21782-3, 21814, 21874, 21932, 21941, 22054, 22077, 22119-20, 22150, 22203, 22239, 22251-2, 22381, 22397, 22406, 22556, 22609, 22635, 22697-8, 22727, 22835, 22837-8, 22905, 22951, 22958, 22997, 23030-1, 23060, 23072, 23195, 23231, 23274, 23289, 23343, 23374, 23382, 23397, 23481, 23610, 23632, 23754, B23767, 23772, 23797, 23814, 23877, 23926, 23973, 24008, 24074, 24100, 24122, 24126, 24161-2, 24164-5, 24201, 24255, 24270, 24309, 24344, 24469, 24523, 24525, 24615, 24657, 24797, 24804, 24933, 24984, 24986, 25035-6, 25065-6, 25138

Phosphoenolpyruvate carboxylase, methods 22381, 23343, 24309, 24387, 25065-6

Phosphorus see Mineral elements (N, P, K)

Photoperiod and carbon fixation pathways 21932, 22366, 22725, 23539, 24126

Photoperiod and carotenoids 23417

Photoperiod and chlorophyll 21734, 23035, 23417, 23528, 23564, 23861

Photoperiod and chloroplast (chromatophore) 21851-2, 22978

Photoperiod and ecosystem productivity and canopy functioning
21882, 22725, 23712, 23815, 24625

Photoperiod and electron transport chain 23005

Photoperiod and gas exchange
21696, 22366, 22726, 23371, 23712, 23764, 24556, 24625, 25118

Photophosphorylation, cyclic
21528, 21530, 21559, 21563, 21588, 21611, 21613, 21688, 21698, 21744, 21888, 21913, B21914, 22017, 22051, 22134, 22233, 22392, 22527, 22540-1, 22566, 22576, 22669, 22706-7, 22711, 22812, 22834, 22842, 22894, 22922-3, 23057,

Photophosphorylation, cyclic (continued)
23209, 23242, 23273, 23351, 23415, 23624, 23647, 23660, 23668, 23837, 23866, 23888, 23920, 23940, 23944, B24024, 24032, 24048, 24099, 24136, 24159, B24176, 24226, 24286, 24311, 24355, 24416, 24458, 24502, 24669, 24721, 24730, 24789, 24791, 24856, 24869, 24875, 25006, 25030, 25143

Photophosphorylation in photosynthetic bacteria see Photosynthetic bacteria, photophosphorylation

Photophosphorylation mechanism see Chemiosmotic hypothesis

Photophosphorylation, methods 21858, 21982, 23183, 23186, 24149, 24351, 24869

Photophosphorylation, model see Model ...

Photophosphorylation, non-cyclic
21514-5, 21528, 21530, 21548, 21559, 21561, 21563, 21588, 21607-8, 21611, 21613, 21687-8, 21727, 21744-5, 21753, 21766, 21859, 21888, 21912-3, B21914, 21996, 22017, 22025, 22038, 22134, 22216-7, 22286, 22426-7, 22498, 22513, 22540-1, 22566, 22568, 22574-5, 22580, 22586, 22625, 22657, 22659, 22669, 22689-90, 22711, 22720, 22722, 22724, 22766, 22801, 22834, 22870, 22891-2, 22894, 22900, 22906, 22913, 22946, 23005, 23008, 23066, 23079, 23118, 23163, 23173, 23201, 23209, 23215, 23242, 23273, 23351, 23434, 23486, 23506-7, 23541, 23555, 23624, 23647, 23660, 23718, 23723, 23762, 23847, 23888-9, 23898, 23920, B23933, 23944, 23997, 24011, 24024, 24032, 24035, 24091, 24095, 24104, 24107, 24136, 24159-60, B24176, 24188, 24224, 24242, B24250, 24257, 24281, 24298, 24311-2, 24338, 24351,24367, 24374, 24411, 24416, 24457, 24466, 24471, 24502, 24568, 24721-2, 24730, 24737, 24789, 24840, 24846, 24856, 24859-60, 24870, 24875, 24941, 24951, 24954, 24999, 25029-30, B25064, 25090-1, 25094, 25100, 25102, 25124

Photophosphorylation, pseudocyclic
22548, 22707, 22833, 23273, 23847, 23888, 23944, 24170, 24351

Photoreduction see H_2 evolution ...

Photorespiration enzymes
21711, 21957, 22071, 22410, 22542, 22940, 22980-2, 23427, 24122, 24247, 24933

Photorespiration, metabolic cycles
21543, 21587, 21618-9, 21711, 21717, 21728, 21785, 21800, 22070, 22110-1, 22192, 22310, 22320, 22347, 22442, 22542, 22548, 22609, 22649, 22942, 23072--3, 23198, 23309, 23329, 23455, 23467, 23509, 23511, 23714, 23877, 23973, 23981-2, 24064, 24113, 24222, 24293, 24408, 24779, 24804, 24949-50, 24969, 25071, 25091, 25097, 25136, 25138-9, 25160

Photorespiration rate
21535, 21659, 21897, 21926, 21967, 21994, 22005, 22038, 22066, 22152, 22228, 22521, 22649, 23073, 23171, 23329, 23493, 23634, 23796, 23994, 24072, 24122, 24264, 24336, 25137, 25160

Photosynthate translocation and distribution
21511-2, 21535, 21552-4, 21557, 21560, 21593, 21630, 21634, 21684, 21692, 21740, 21762, 21786, 21847, 21853, 21878, 21883, 21951, 21966, 21999, 22004, 22037-8, 22064, 22096, 22123, 22137, 22196, 22212, 22229, 22351, 22365, 22367-9, 22428-9, 22470, 22489, 22547, 22638, 22663, 22678, 22687, 22714, 22723, 22732, 22754, 22770-2, 22777, 22826, 22845, 22872, 22878-9, 22910, 22932, 22965, 22974, 22985, 23032, 23045, 23073, 23085, 23175, 23212, 23220, 23236-7, 23243, 23251, 23332, 23372, 23385, 23431, 23587, 23603, 23649, 23694, 23696, 23713, 23756, 23773-4, 23803, 23817, 23902, 23908, 23912, 23937, 23958, 23975, 23990, 24000-1, 24023, 24088, 24105, 24114, 24120, 24124, 24142, 24150, 24204, 24246, 24252, 24259-60, 24275, 24301, 24326, 24332, 24337, 24359, 24363, 24432, 24439, 24444, 24448, 24508, 24526, 24556, 24558, 24598, 24626, 24628, 24632, 24639, 24656, 24708, 24715, 24765, 24771, 24799, 24805, 24830, 24839, 24851-2, 24872-3, 24965, 24967, 24969, 24974, 24976-7, 25059-61, 25072, 25089, 25109, 25125, B25153

Proteins, amino acids, nucleic acids,and electron transport chain 22243, 23495

Proteins, amino acids, nucleic acids,and gas exchange
21699, 22228, 23766, 24138, 25121

Protochlorophyll(ide) see Chlorophyll biosynthesis

Proton transport in chloroplast see Chemiosmotic hypothesis

Pteridines see Ferredoxin-NADP reductase

Pyrenoid 22743, 23366, 24243, 24500, 25080

Q

Quantum yield and requirement
21585, 21638, 21662, 21690, 21727, 21750, 21834, 21891, 21893, B21914, 21993, 22134, 22174, 22236, 22310, 22580, 22722, 22957, 23103, 23147-8, 23209, 23365, B23407, 23673, 23965, 24133, 24315, 24401-2, 24433, 24576, 24804, 24811, 24838, 24904, 25057

Quantum yield and requirement, methods see Quantum yield ...

Quinones in photosynthesis
21568, 21631, 21638, 21646, 21697, 21721, 21752, 21804, 21810, 21867, 21954, 21959, 22100, 22141, 22325, 22354, 22525-8, 22580, 22588, 22618-9, 22626, 22637, 22667, 22673, 22706, 22759, 22794, 22812, 22934, 22969, 22972, 23076, 23116, 23129, 23184-5, 23200, 23215, 23272, 23290, 23292, 23398, 23400-4, 23552, 23566, 23574, 23655, 23687, 23730, 23886, 23921, B23933, 24028, 24032, 24133, 24163, 24281, 24333, 24346, 24403, 24490, 24497, 24583, 24691, 24784, 24789, 24864, 24876-7, 24921, 25006, 25029, 25057, 25076, 25099

R

Radiation in canopy see Canopy, radiation ...

Radiation, light see Irradiance ...

Rain, precipitation measurement see Aerodynamic methods ...

Reaction centres see *P*680; *P*700 ...

Recycling of CO_2 inside cell and leaf 21897, 22066, 22397, 23171, 24264, 25136

Relative growth rate see Growth analysis, net assimilation rate ...

Relative water content see Water saturation deficit

Resistance, carboxylation and excitation 22749, 23609, 23994, 24100, 24201, 24411, 24537, 24915

Resistance, cuticular 23634, 23703, 24624

Resistance, intracellular (mesophyll)
22039, 22041, 22075, 22157, 22209, 22284, 22314, 22321, 22484, 22649, 22652, 22686, 22787, 22875, 22937, 22938, 22950, 22960, B23006, 23430-1, 23609, 23653, 23696, 23773, 23780, 23791, 23887, 23994, 24100, 24201, B24271, 24411, 24537, 24592-3, 24675-6, 24730, 24776, 24837, 25106, 25111

Resistance, leaf boundary layer
21538, 21748, 22075, 22209, 22284, 22316, 22584, 22589, 22649, 22652, 22749, 22938, 23296, 23493, 23846, 24100, 24536, 24593, 24676, 24863, 24915

Resistance, stomatal (and intercellular) (*cf.* also Resistances for water vapour ...)
21538, 21549-50, 21713, 21748, 21787-8, 21810, 21813, 21887, 21968, 22039,

Ribulose 1,5-bisphosphate oxygenase see Ribulose 1,5-bisphosphate carboxylase ...; Photorespiration enzymes

Root removal see Defoliation ...

Root, underground part, and carbon fixation pathways 23333

Root, underground part, and chlorophyll 23771

Root, underground part, and ecosystem productivity and canopy functioning
21692, 22348, 22547, 22714, 23168, 23271, 23332, 24185, 24303

Root, underground part, and gas exchange
21609, 21666, 21692, 21786, 22938, 23786, 24837

Rooted leaves, gas exchange in 22782

RuBP carboxylase, RuBPC see Ribulose 1,5-bisphosphate carboxylase

Rubredoxin see Ferredoxin ...

S

Saccharides and algae productivity 25121

Saccharides and carbon fixation pathways
21639, 21661, 21739, 21784, 21798, 21942, 22080, 22250, 22411, 22449, B22572, 22608, 22697, 22828, 22868-9, 22975, 23073, 23210-1, 23214, 23379, 23422, 23467, 24000, 24059, 24161, 24262, 24319, 24382, 24404, 24442, 24552, 24652, 24659, 24836

Saccharides and carotenoids 22673

Saccharides and chlorophyll
22337, 22673, 22735, 22789, 23951, 23959, 24397, 24985

Saccharides and chloroplast (chromatophore)
22176, 22673, 22783, 23681, 23924, 24381, 24476, 24865, 24985

Saccharides and ecosystem productivity and canopy functioning 21853, 22229, 23984

Saccharides and electron transport chain 22673, 23726, 23917, 23989, 24568

Saccharides and gas exchange 22673, 23649, 24397, 24568

Saccharides and photorespiration 21619

Salinity of soil and carbon fixation pathways
21736, 22484, 23274, 23786, B24070, 24986, 25035

Salinity of soil and chlorophyll 22482, 23523, 24249, 24986

Salinity of soil and chloroplast (chromatophore) 24755

Salinity of soil and ecosystem productivity and canopy functioning
21508, 22485, 22937, 23113, B24070, 24204, 24249, 24470, 24628, 24986

Salinity of soil and electron transport chain 23274, B24070

Salinity of soil and gas exchange
21508, 21699, 22344, 22484-5, 22791, 22965, 23112-3, 23522-3, B24070

Salinity of soil and resistances to CO_2 and water vapour transfer 22484, B24070

Salinity of soil and respiration 22484, 24001

Samples for pigment determination see Pigments ...

Seasonal changes in algae productivity
21681, 21742, B21829, 21953, 21990, 22121, 22132, 22143, 22155, 22177, 22220, 22287, 22294, 22353, 22389, 22417, 22436, 22453-4, 22488, 22493, 22500, 22517, 22531, 22538, 22551, 22578, 22592, 22628, 22646, 22745-6, 22756, 22807, 22858, 22867, 22898, 22976, 22986, 22991, 23017, 23040, 23088, 23188-90, 23267, 23275, 23301, 23315-6, 23334, 23377, 23456, 23505, 23542, 23554, 23595, 23646,

T

Transpiration and photosynthesis
21670, 21713, 21765, 21773, 21995, 22042, 22075, 22112, 22171, 22210, 22284, 22314, 22343-4, B22358, 22507, 22651, 22686, 22831, 22951, 23112-3, 23444, 23628, 23666, 23707, 23775, 23784, 23900, B24132, 24175, 24376, 24588, 24675-6, 24689, 24718, 24776, 24820

Tritium oxide see Deuterium oxide, tritium oxide ...

U

Ubiquinones see Quinones ...

Ultraviolet radiation see Irradiance, spectral composition ...

Uncouplers of electron transport chain (*cf.* also Antibiotics and electron transport chain)
21531, 21559, 21588, 21632, 21745, 21766, 21797, 21859, 21959, 21997, 22103, 22106, 22136, 22154, 22392, 22425, 22452, 22491, 22548, 22625, 22689, 22705-7, 22766, 22801, 22903, 22913, 22933, 22936, 23066, 23184, 23192, 23480, 23484, 23541, 23617, 23641, 23722, B23933, 23989, 24011, B24024, 24034-5, B24176, 24187, 24192, 24224-5, 24298, 24367, 24418, 24471, 24497, 24525, 24720, 24790, 24860, 24869, 24911, 24954, 24999, 25929, 25096

V

Virus diseases see Phytopathological effects ...

Vitamin K_3 see Quinones ...

Volume changes in leaf and other organs
21787, 22187, 22351, 22855, 22920, 22999, 24269, 24801

W

Warburg effect see O_2 and gas exchange

Water, heavy see Deuterium oxide, tritium oxide ...

Water saturation deficit
21550, 21556, 21713, 21765, 21823, 21861, 22074, 22097, 22118, 22283-4, 22314-6, 22327, 22348, 22380, 22419, B22572, 22584, 22676, 22752, 22920, 22937, 23090, 23263, B23264, 23295-6, 23299, B23302, 23307-8, 23330, 23468, 23587, 23608-9, 23665, 23696, 23813, 23887, 24063, B24271, 24302, 24377, 24389, 24444, 24537, 24624, 24645, 24775, 24801, 25045, 25140

Water splitting mechanism see O_2 evolution mechanism and kinetics

Wind (air-flow rate) and ecosystem productivity and canopy functioning
21633, 21928, 23304, 23346, 24744

Wind (air-flow rate) and gas exchange 22487, 23444, 23727

Wind (air-flow rate) and resistances to CO_2 and water vapour transfer
22584, 22762, 23846

Wind measurement see Aerodynamic methods ...

X

X-rays see Ionizing radiation ...

Xanthophylls
21509, 21564, 21642, 21663, 21678, 21719, 21743, 21747, 21891, 21940, 22086, 22323, 22465, 22580, 22618-9, 22694, 23115, 23169, 23200, 23259, 23399, 23471, 23568, 23781, 24092, 24148, 24163, 24317-9, 24413, 24479, 24489, 24541, 24777, 24892, 24909

Xanthophylls of algae
21595, 21678, 22166, 22241, 22450-1, 22465, 22538, 22599, 22616, 22767, 22865, 22926, 22930, 23213, 23393, 23489, 23953, 24233, 24350, 24451, 24709, 24829, 24978, 25016-7, 25055-6

Xanthophylls of photosynthetic bacteria
21904, 22101, 22167, 22626, 22679, 22954, 23421, 24056, 24092, 24550

Xerophytes see Drought ...; Temperature, high ...

Y

Yield formation see Biomass distribution ...; Photosynthate translocation ...

Z

Zeaxanthin see Carotenoids ...

This index contains a selection of plant genera and types interesting as experimental material for physiological, ecological and agricultural studies. Latin scientific names of plant genera and English names of plant groups and types are the main items which present the reference numbers.

A

Abies 21629, 21633, 21705, 21902, 22009, 22086, 22778, 23044, 23082, 23091, 23430-1, 23500, 23818, 23830, 24057, 24336, 24716

Acer 21705, 21808-9, 22139, 22165, 22171, 22190-1, 22647, 22734, 22854, 22893, 22973, 23422, 23530, 23551, 23605, 23679, 24336, 24528, 24601, 24716, 24817-8, 25019, 25078

Acetabularia 21598-601, 21785, 21890, 22089, 22280, 22417, 23018, 23230, 23626-7, 23697, B23767, 23969, 24353, 24865-8

Aesculus 24817-8

Agave 22316, 22508, 24872

Alder see *Alnus*

Alfalfa see *Medicago*

Algae (*cf.* also *Acetabularia*, A. blue-green, A. brown, A. green, A. red, *Anabaena*, *Anacystis*, *Ankistrodesmus*, *Chlamydomonas*, *Chlorella*, *Chrysophyta*, Diatoms, *Dinoflagellates*, *Dunaliella*, *Euglena*, *Nostoc*, *Porphyridium*, *Scenedesmus*, *Ulva*) 21546, 21575, 21595, 21681, 21707, 21726, 21772, 21776, 21805, 21829, 21883, 21910, 21925, 21952, 22048, 22121, 22238, 22257, 22268-9, 22292-4, 22357, 22387, 22389, 22450, 22454, 22458, 22493-4, 22538, 22545, 22592, 22599, 22628, 22634, 22644, 22654, 22743, 22745-6, 22756, 22763, 22791, 22807, 22846-7, 22867, 22898, 22921, 22929, 22943, 22963, 22976, 23016-7, 23088, 23091, 23096, 23099, 23133, 23188-9, 23210, 23241, 23267, 23269-70, 23275, 23301, 23314-6, 23334, 23370, 23393, B23407, 23456, 23462-3, 23505, 23542, 23590, 23595, 23606, 23622, 23644, 23647, 23659, 23667, 23686, 23707, 23726, 23744, 23824, 23844, 23931-2, 23953, 24051, 24059-60, 24062, 24090, 24202, 24233, 24274, 24276, 24282, 24287, 24306, 24308, 24407, 24409, 24503, 24513-7, 24521, 24554, 24562, 24566, 24571-2, 24578-9, 24582, 24589, 24599-600, 24634, 24679-80, 24699, 24708-9, 24732-3, 24741, 24763, 24766, 24786, 24803, 24816, 24835, 24838, 24862, 24897-8, 24906, 24953, 24969, 24992, 25005, 25016-7, 25053, 25058, 25129, 25138, 25154

Algae, blue-green (*cf.* also *Anabaena*, *Anacystis*, *Nostoc*) 21547, 21562, 21572, 21575, 21617, 21620, 21628, 21682, 21774, 21805, 21830, 21836, 21846, 21908, 21911-2, 21920, 21925, 21981, 22054, 22104-5, 22121, 22155, 22225, 22294, 22349-50, 22363, 22442, 22453-4, 22461, 22464, 22490, 22497, 22515, 22517, 22521, 22538, 22546, 22592, 22637, 22642, 22644, 22658-60, 22664-5, 22693, 22745-6, 22763, 22767, 22803, 22853, 22867, 22876, 22961, 23010, 23083, 23097-8, 23107, 23110-1, 23128, 23153, 23165, 23178, 23268, 23275, 23322, 23383-4, B23407, 23434, 23453, 23456, 23474, 23505, 23546, 23606, 23728-30, 23750, 23788, 23824, 23843, 23868, 23870, 23914, 23931-2, 23954, 23969, 24061, 24072, 24118-9, 24134, 24154, 24169, 24200, 24212, B24250, 24254, 24258, 24273-4, 24308, 24345, 24421, 24434, 24450, 24472, 24498, 24518-20, 24550, 24572, 24579, 24586-7, 24589, 24611-2, 24618, 24644, 24648, 24695, 24707, 24722-4, 24756, 24767, 24809-10, 24816, 24843, 24909, 24929, 24944-5, 24960, 24964-5, 24982, 24998, 25005, 25033, 25112

Algae, brown 21562, 21903, 21945, 22166, 22256, 22258, 22261, 22451, 22462, 22538, 22607, 22644, 22660, 22687, 22767, 22803, 22927, 23040, 23074, 23211, 23213, 23335, B23407, 23475, 23534, 23606, 23728, 23827, 23941, 24073, 24154, 24169, 24409, 24728-9, 24816, 24957, 24984, 25034, 25127

Algae, green (*cf.* also *Acetabularia, Ankistrodesmus, Chlamydomonas, Chlorella, Dunaliella, Scenedesmus, Ulva*)
21584, 21610, 21620, 21737, 21755-6, 21767, 21769, 21805-6, 21871, 21896, 21905-7, 21925, 21945, 21978, 22121, 22256, 22277, 22294, 22305, 22337, 22361, 22364, 22442, 22450-1, 22453-4, 22462, 22515, 22538, 22548, 22578, 22591-2, 22642, 22644, 22659-60, 22702-3, 22717, 22745-6, 22757-8, 22764, 22767, 22790, 22794, 22803, 22828, 22846, 22866, 22928, 22965, 22991, 23040, 23042, 23097-8, 23107, 23165, 23212, 23229, 23366, 23456, 23474, 23485, 23534, 23537, 23554, 23567, 23596, 23606, 23685, 23728, 23824, 23843, 23847, 23868, 23876, 23931, 23987, 24029, 24053, 24072, 24169-72, B24250, 24273, 24304, 24350, 24365-6, 24409, 24420, 24456, 24472, 24498, 24500, 24572, 24579, 24589, 24608, 24611, 24613, 24617-8, 24683, 24713, 24723, 24735, 24743, 24749, 24770, 24808, 24909, 24929, 24944, 24957, 24978, 25003, 25005-6, 25069, 25080, 25112, 25127

Algae, red. (*cf.* also *Porphyridium*)
21562, 21575, 21737, 21846, 21871, 21925, 21945, 22091-2, 22137, 22411, 22495-7, 22515, 22538, 22546, 22642, 22644, 22659-60, 22767, 22790, 22861, 23040, 23152, 23177-8, 23182, 23214, 23344-5, B23407, 23445, 23474, 23571-2, 23632, 23843, 23868, 23954, 24169, 24254, 24409, 24525, 24545, 24697-8, 24809, 24816, 24957, 24983, 25006, 25050, 25070, 25127

Allium 22484, 22508, 22719, 22747, 24527, 24716, 25011

Alnus 21705, 22086, 22333, 22685, 24336, 24765, 25009

Alpine plants 23167, 23303-4, 23545, 25015

Amaranthus 22033, 22183, 22249, 22328, 22489, 22620, 22698, 22768, 22863, 22958, 23032-3, 23694, 23714, B23767, 23977, 24546, 24716

Anabaena 21519, 21589, 22305-6, 22478, 22561, 22665, 23043, 23058, 23110-1, 23724, 23884, 24013, 24251, 24293, 24405, 24456, 24703

Anacystis 21859, 21865, 21889, 22274, 22407, 22558, 22609, 22859, 23187, 23390, 23730, 23969, 24011, 24073, 24447, 24495-6, 24930, 24966

Ananas 22098, 22152, 23786

Ankistrodesmus 21865, 22324, 22550, B23407, 24273, 24579

Antirrhinum 22148-9, 23129, 23624, 24135-6, 24351

Apple see *Malus*

Apricot see *Armeniaca*

Aquatic macrophytes (*cf.* also *Elodea, Phragmites, Typha*)
21847, 21973, 22000, 22297, 22305, 22731, 22625, 23188-9, 23476, 23596, 23632, 23646, 23683, 23814, 23819, 23859, B24024, 24099, 24303, 24440, 24477, 24700, 24773, 24969, 24979, 25003, 25005, 25103, 25130, 25149-50

Arabidopsis 21870, 22204-5, 22432, B23767, 23845, 24847-8

Arachis 21550, 22352, 22685, 23039, 24424, 24639, 24787, 24820, 24861, 24971, 25039

Arbor vitae see *Thuja*

Armeniaca 22747, 23297

Artichoke see *Cynara*

Ash see *Fraxinus*

Asparagus 22252, 22747

Aspen see *Populus*

Atriplex 21652, 21748, 21773-4, 21810-4, 21825, 22033, 22156, 22249, 22318, 22328, 22484, 22544, 22698, 22983, 23033, 23113, 23692, 23714, B23767, 24657-8, 24804, 24806, 25018

Avena 21916, 21948, 22028, 22068, 22107, 22130, 22156, 22348, 22566, 22611, 22615, 22672, 22700, 22719, 23063, 23157, 23355, 23395-6, 23482-3, 23486, 23524, 23586, 23588, 23716-7, 23788, 23859, 24001, 24199, 24284, 24302, 24424, 24508, 24539, 24564, 24716, 24725, 24731, 24750, 24792-3, 24861, 24870, 24907, 24940, 24993-7, 25136

Avocado see *Persea*

B

Bacteria, photosynthetic (*cf.* also *Chlorobium*, *Chromatium*, *Halobacterium*, *Rhodopseudomonas*, *Rhodospirillum*)
21690, 21703, 21815, 21829, 21843, 21912, 21938, 22020-1, 22055, 22090-1, 22154, 22167, 22188, 22267-8, 22390, 22458, 22499, 22515, 22580, 22644, 22658-60, 22679, 22695, 22767, 22903, 22913, 22931, 23020, 23022, 23120, 23124, 23142-3, 23161, 23203, 23312, 23347, 23391, B23407, 23434, 23437, 23474, 23484, 23501, 23525, 23615, 23667, 23870, 23914, 23965-6, 24031, 24098, 24189, 24227, 24306, 24327, 24367, 24409, 24454, 24459, 24671, 24682-3, 24723, 24784, 24814, 24955, 25051, 25057, 25067, 25119

Bamboo see *Bambussa*

Bambussa B24024

Banana see *Musa*

Barley see *Hordeum*

Bean see *Phaseolus*

Beech see *Fagus*

Bermuda grass see *Cynodon*

Beta 21563, 21676, 21765, 21779, 21782, 21841, 21847, 22007, 22028, 22066, 22108, 22123, 22171, 22245, 22318, 22345, 22347, 22419, 22462, 22487, 22565, 22647, 22651, 22666, 22711, 22719, 22721, 22775-6, 22806, 22829, 23002-3, 23170, 23221, 23252, 23259, 23330, 23448, 23461, 23549, 23588, 23625, 23648-9, 23714, 23726, 23737, B23767, 23773, 23814, 23887, 23925, 23990, 24033, 24196, 24204, 24442, 24468, 24504, 24537, 24552, 24557, 24643, 24683, 24716, 24730, 24801, 24812, 24839, 24894, 24840, 25137-8, 25158

Betula 21705, 21760-1, 22686, 22734, 22893, 22948, 23257, 23500, 23521, 23666, 23679, 24057, 24240, 24601, 24739, 24817-8, 24855, 25018, 25025

Birch see *Betula*

Blackberry see *Rubus*

Blueberry see *Vaccinium*

Bluegrass see *Poa*

Brassica 21841, 21921, 22123, 22647, 22660, 22741, 22747. 22797, 23101, 23223, 23356-7, 23457, 23528, 23577, 23580, 23691, 23798, 23876, 23920, 23952, 23972, 24092, 24285, 24291, 24468, 24564, 24666, 24716, 24842, 24925, 25158

Colocynthis 23913

Coniferous plants (*cf.* also *Abies, Cedrus, Larix, Picea, Pinus, Pseudotsuga, Taxus, Thuja, Tsuga*)
21705, 22073, 22086, 22224, 22255, 23082, 23091, 23135, 23257, 23318, 23666, 23943, 24001, 24372, 24604, 25007

Corn see *Zea*

Cornelian cherry see *Cornus*

Cornus 21809, 22028, 22190, 22719, 23605

Cotton see *Gossypium*

Cottonwood see *Populus*

Cowberry see *Vaccinium*

Cowpea see *Vigna*

Crabgrass see *Digitaria*

Crataegus 21809

Cucumber see *Cucumis*

Cucumis 21703, 21732, 21957, 22051, 22082, 22313, 22401, 22797, 22872, 22897, 22978, 22999, 23222, 23442, 23527, 23581, 23690-1, 23716-7, 23853, 23972, 24130, 24177-80, 24389, 24468, 24557, 24918, 24987, 25075, 25158

Cucurbita 21673, 21791, 21847, 22110, 22734, 23025, 23528, 23575, 23750, 24334, 24413, 24830-1

Currant see *Ribes*

Cyanobacteria see Algae, blue-green

Cynara 21695

Cynodon 21508, 22953, 23245, 24266

Cyperus 22660, 23033, 23657

D

Dactylis 21621, 22030, 22919, 23061, 23375, 24436, 24438

Dallis grass see *Paspalum*

Daucus 21841, 21948, 22636, 22713, 22719, 22747, 23026, 23425, 23448, 23632, 23675, 23807, 23951, 24203, 24716, 25136

Deciduous trees and shrubs (*cf.* also *Acer, Aesculus, Alnus, Amygdalus, Armeniaca, Betula, Carpinus, Carya, Castanea, Cerasus, Citrus, Cornus, Crataegus, Eucalyptus, Fagus, Fraxinus, Hevea, Hibiscus, Juglans, Malus, Mangifera, Morus, Olea, Persea, Persica, Pirus, Populus, Prunus, Quercus, Ribes, Robinia, Rubus, Salix, Sambucus, Sorbus, Syringa, Tamarix, Tilia, Ulmus, Vitis*)
22016, 22028, 22086, 22492, 22734, 23499, 23599, 23605, 23814, 24093, 24147, 24153, 24239, 24336, 24423, 24670, 24808, 24826, 25126

Desert plants and ecosystems 21670, 22379, 22544, 22951, 23297-300, 23466, 23658, 23692, 23752-3, 24378, 24629, 24935

Dewberry see *Rubus*

Diatoms 21543, 21582, 21682, 21711, 21775, 22109, 22158, 22287, 22357, 22362, 22387, 22433, 22442, 22451, 22453-4, 22488, 22515, 22531, 22538, 22550, 22552, 22559, 22578, 22580, 22592, 22644, 22660, 22745-6, 22763, 22790, 22807, 22848, 22865--7, 22927, 22991, 23107, 23128, 23268, 23275, 23377, 23441, 23456, 23505, 23542, 23569, 23606, 23632, 23808-9, 23824, 23843, 23931-2, 23940, 23981-2, 24111, 24118, 24218, 24233, 24274, 24308, 24381-2, 24440, 24451, 24472, 24475, 24572, 24579, 24581, 24589, 24616-8, 24727, 24732, 24816, 24929, 24956, 25005, 25112

Digitaria 21874, 21999, 22502, 22698, 22833-4, 22837-8

Dinoflagellatae, Dinophyceae 21966, 22241, 22744, 22926, 22930, 23087, 23439, 23441, 23611, 24816, 25004, 25055-6

Dogwood see *Cornus*

Douglas fir see *Pseudotsuga*

Dunaliella 21736, 21909, 23440, 23941, 24030, 24982

E

Egg plant see *Solanum*

Elaeis 22123, 23448

Elder see *Sambucus*

Elm see *Ulmus*

Elodea 22721, 22892, 23188, 24546, B25064

Endive see *Cichorium*

Equisetum 22543, 22658, 22660, 22979, 23870

Ericaceae see Heath plants and communities

Eucalyptus 23634, 24336, 24585

Euglena 21562, 21575, 21703, 21741, 21804-5, 21888, 21925, 21935, 21976, 21992, 22107, 22110-1, 22121, 22159, 22192, 22214, 22264, 22303-5, 22307, 22312, 22442, 22453, 22462, 22469, 22521, 22535, 22596, 22616, 22644, 22660, 22711, 22729, 22745-6, 22767, 22827, 22864, 22933, 23095, 23224, 23354, 23384, B23407, 23446, 23451, 23453-4, 23456, 23490, 23569, 23577, 23597, 23606, 23623, 23638--9, 23728, B23767, 23824, 23843, 23851, 23868, 23915, 23954, 23968-9, 24027, 24067, 24097, 24128, 24130, 24154, 24243, 24273, 24292, 24304, 24308, 24339--40, 24346-8, 24357, 24397, 24409, 24459, 24466-7, 24544, 24579, 24608, 24613, 24723, 24819, 24823, 25030, B25064, 25065-6, 25071, 25112, 25120, 25132-3

Euphorbia 22249, 22328, 23986, 24980

Evergreen plants see Sempervirent plants

F

Fagopyrum 22028

Fagus 21633, 21705, 21902, 22065, 22208, 22254, 22619, 23091, 23138, 23818, 24601

Ferns 22016, 22147, 22434, 22475, 22644, 22965, 23460, 23464, 23661, 24013, 24371, 25008, 25025

Fescue see *Festuca*

Festuca 21853, 21926, 22668, 22918, 23061, 23136, 23375, 23795-7, 24435-6, 24438, 24563, 24746, 24795, 24961, 25031-2

Ficus 22619, 23846, 24557, 25117

Fig see *Ficus*

Fir see *Abies*

Flax see *Linum*

Forage crops (*cf.* also *Brassica*, Grasses, Leguminous plants, *Lupinus*, *Medicago*, *Trifolium*, *Vicia*, *Vigna*, *etc.*)
21667, 22123, 22279, B23767, 24446, 24584

Forest (including undergrowth) plants and ecosystems (*cf.* also Coniferous plants, Deciduous trees and shrubs, Ferns, *Fragaria*, Grasses, Heath plants and communities, Lichens, Liverworts, Medicinal plants, Mosses, *Sphagnum*, *Vaccinium*, *etc.*)
21760, 21825, 22271-3, 22302, 22327, 22447, 22505, 22573, 22645, 22747, 22832, 22841, 23135, 23138, 23232, 23234, 23313, B23407, 23412, 23534, 23578, 23582, 23605, 23741, 23911, 23943, 24150, 24278, 24313, 24473, 24530, 24567, 24835, 24942

Fountain-grass see *Pennisetum*

Foxtail millet see *Setaria*

Fragaria 22719, 22783, 23371, 23716

Fraxinus 22189-91, 22492, 22647, 22893, 23243, 23521, 23551, 23829, 24240, 24336, 24817-8, 25077

Fruit plants and trees (*cf.* also *Ananas*, *Armeniaca*, *Cerasus*, *Citrullus*, *Citrus*, *Cocos*, *Cucumis*, *Cucurbita*, *Ficus*, *Fragaria*, *Malus*, *Mangifera*, *Musa*, *Persea*, *Persica*, *Pirus*, *Prunus*, *Ribes*, *Rubus*, *Sorbus*, *Vaccinium*, *Vitis*)
21625, 21640, B23262, B23264

Fungi (parasitic) 21783, 21910, 21984, 22085, 22283, 22331, 22470, 22655, 22662, 22687, 22752, 22804-5, 22851, 22872, 22963, 23385, 23687, 23713, 23846, 24114, 24216-7, 24452, 24632, 25072, 25128

G

Garlic see *Allium*

Glycine 21538, 21549, 21556, 21590, 21701, 21887, 21925, 22028, 22080, 22084, 22123, 22162-3, 22279, 22352, 22489, 22685, 22698, 22735, 22761-2, 22787, 22806, 22830-1, 22923, 22998, 23002-3, 23007, 23085, 23174-6, 23235, 23250-2, 23282-3, 23339, 23367-8, 23448, 23504, 23608-9, 23645, 23647, 23732, 23735, 23737, 23782, 23784, 23849, 23877, 23883, 23972, 23980, 23999, 24088, 24122-5, 24137, 24259-60, 24337, 24364, 24408, 24424, 24444, 24508, 24537, 24546, 24649, 24705, 24774, 24820, 24824, 24839, 24841, 24861, 24968, 24970, 24985, 25026, 25103, 25109, 25136-8

Gossypium 21705, 21740, 21787, 22024-7, 22043, B22115, 22123, 22266, 22356, 22403, 22484-5, 22509, 22521, 22660, 22734, 22762, 22826, 22870, 22964, 23587, 23602-4, 23717, 23756-7, B23767, 23768, 23979-80, 24001, 24036, 24088, 24143, 24149, 24331, 24424, 24508, 24527, 24716, 24804, 24806, 24839, 25037, 25088

Gourd see *Cucurbita*

Gram chick pea see *Vigna*

Grape fruit see *Citrus*

Grape vine see *Vitis*

Grasses (*cf.* also *Avena, Bromus, Carex, Cynodon, Cyperus, Dactylis, Digitaria, Festuca, Hordeum, Lolium, Oryza, Panicum, Paspalum, Pennisetum, Phleum, Poa, Saccharum, Secale, Setaria, Sorgum, Triticum, Zea*)
21527, 21552-3, 21667, 21684, 21705, 21739, 21760, 21762, 21794-5, 21814, 21847, 21862, 21871, 21942, 21944, 21973, 22041, 22053, 22074-6, 22119-20, 22123, 22156, 22184, 22249, 22251, 22328, 22344, 22366, 22379, 22570, B22572, 22617, 22631, 22635, 22668, 22684, 22700, 22716, B22822, 22823, 22833-6, 22838-40, 22868, 22896, 22916, 22919, 22950, 22983, 23001, 23033, 23061, 23064, 23081, 23136, 23138, 23221, 23231, 23234, 23245, 23313, 23338, 23405, 23447-8, 23457, 23522, 23586, 23620, 23714, 23741, B23767, 23894, 23903-5, 23992, 24006, 24015, 24044, 24088, 24159-65, 24181, 24185, 24214-5, 24231, 24266, B24271, 24276, 24310, 24314, 24419, 24424, 24431, 24446, 24509, 24512, 24563, 24588, 24624, 24759, 24861, 24935, 24968, 25023, 25049, 25060, 25123

Groundnut see *Arachis*

H

Halobacterium 21714-6, 22022, 22045, 22059-60, 22179, 22218, 22753, 22760, 23432, 23458, 23495-6, 24056, 24195, 25110

Halophilous plants (*cf.* also Salt marsh and strand plants)
22251, 22380, 22544, 23902, 24070, 24378, 24755

Hawthorn see *Crataegus*

Heath plants and communities 22034-5

Hedera 21705, 24557

Helianthus 21708, 21729, 21785, 21847, 21897, 21943, 21970, 21973, 21995, 22031, 22197, 22246-7, 22273, 22415, 22651, 22734, 22772, 22968, 23026, 23036, 23052, 23117, 23250-1, 23329, 23339, 23408, 23448, 23547, 23625, 23673, 23703, 23714, B23767, 23814, 24000-1, 24110, 24143, 24196, 24216, 24364, 24441, 24537, 24595, 24598, 24610, 24627, 24716, 24839, 24863, 24928, 25136, 25138

Hemlock see *Tsuga*

Hemp see *Cannabis*

Hevea B21979

Hibiscus 21511-2, 21659, 23023-4, 23066, 24739

Hickory see *Carya*

Holly see *Ilex*

Hop see *Humulus*

Hordeum 21520, 21529, 21532, 21558, 21575, 21627, 21630, 21710, 21801-3, 21825, 21860, 21870, 21984, 22001, 22011, 22028, 22123, 22156, 22202, 22240, 22281, 22348, 22441, 22462, 22470, 22479, 22589, 22596-7, 22600-2, 22618-9, 22627, 22674-5, 22696, 22737, 22800, 22804-5, 22829, 22833, 22836, 22840, 22945, 22952, 22987, 22990, 23009, 23014, 23072, 23115-6, 23221, 23231, 23357, 23398,

23402-3, 23428-9, 23443, 23468, 23493-4, 23529, 23531, 23547, 23555, 23586, 23642, 23681, 23704, B23767, 23773, 23775, 23777-80, 23821-3, 23833, 23859, 23944, 23954, 23960, 23969, 23973-4, 24001, 24016, 24055, 24076, 24079, 24158, 24198-9, 24247, 24275, 24280, 24284, 24328, 24330, 24364, 24397, 24429-30, 24459-65, 24468, 24545, 24590, 24625, 24633, 24716, 24734, 24827, 24889-90, 24912, 24916, 24923, 24933, 24940, 24962, 25015, 25037, 25052, 25072, 25090, 25111, 25134, 25136

Hornbeam see *Carpinus*

Horse chestnut see *Aesculus*

Horsetail see *Equisetum*

Humulus 23557, 23829, 24267, 24532, 24771, 25103

I

Ilex 22302, 22924, 23556, 24557

Ipomoea 22028, 22250, 24468, 24820, 24883

Ivy see *Hedera*

J

Jerusalem artichoke see *Helianthus*

Jointgrass see *Paspalum*

Juglans 22190, 22719, 22761

Jute see *Corchorus*

K

Kalanchoë 21652, 21748, 21885, 22098-9, 22508, 22963, 23011, 23121, 23374, 23379, 23481, 23764, 23928, 24126, 24656-8, 24742, 25118

Kale see *Brassica*

Kenaf see *Hibiscus*

Kohlrabi see *Brassica*

L

Lactuca 21613, 21671, 21744-5, 21871-2, 22091, 22131, 22141-2, 22233, 22372, 22483, 22615, 22647, 22747, 22797, 22996, 23477, 23568, 23577, 23717, 23728-30, 24042, 24130, 24166, 24321, 24416-7, 24468, 24471, 24488-91, 24716, 24750, 24870, 24910, 24952, 25093, 25152

Larch see *Larix*

Larix 21633, 21705, 23091, 23319, 23500, 23607, 23830, 24240, 24716, 24958

Lathyrus 23154, 23156, 23168

Leguminous plants (*cf.* also *Arachis, Cajanus, Cicer, Glycine, Lathyrus, Lens, Lupinus, Medicago, Phaseolus, Pisum, Trifolium, Vicia, Vigna*) 21705, 22631, 22823, 23154-6, 23714, 24012

Lemna see *Lemnaceae*

Lemnaceae 21509, 22472-3, 22644, 22819, 23005, 23189, 23494, 24145-6, 24166, 24234, 24669

Lemon see *Citrus*

Lens 23771-2, 24508

Lentil see *Lens*

Lettuce see *Lactuca*

Lichens 21628, 21759, 21761, 21910, 22053, 22404, 22412, 22665, 22993, 23015, 23027, 23069-70, 23121, 23298, 23305-8, 23342, 23423, 23716, 24173, 24614, 24834-5, 25015, 25025, 25054

Lilac see *Syringa*

Linden see *Tilia*

Linseed see *Linum*

Linum 21880, 21882, 22240, 24075, 24077-8, 24141

Liverworts 24246

Locust see *Robinia*

Lolium 21624, 21692, 22039, 22096, 22112, 22504, 22516, 22676, 22779, 23061, 23063, 23331, 23713, 23897, 23983-4, 24033, 24100, 24201, 24375, 24435-8, 24744-5, 25045-7

Lucerne see *Medicago*

Lupine see *Lupinus*

Lupinus 22285-6, 22593, 22755, 23026, 23562, 23588, 24007

Lycopersicon 21570-1, 21676, 21686, 21860, 21948, 21977, 22028, 22037, 22130, 22188, 22719, 22747, 22818, 22860, 22999, 23075, 23101, 23106, 23323-5, 23467, 23502, 23528, 23565, 23690, 23694, 23716, 23728-9, 23738, 23853, 23878, 24071, 24094, 24129, 24229-30, 24299, 24372, 24468, 24527, 24645, 24716, 24806, 24808, 24883-4, 25117

M

Macereed see *Typha*

Maize see *Zea*

Malus 21625, 21640, 22028, 22085, 22418, 22661, 22678, 22855, 23045, B23262, B23264, 23295, 23341, 23382, 23406, 23846, 23887, 24270, 24336, 24363, 24531, 24602, 24641

Mangifera 24751

Mango see *Mangifera*

Mangold see *Beta*

Manihot 22845, 23448

Manioc see *Manihot*

Maple see *Acer*

Marrow see *Cucurbita*

Meadowgrass see *Poa*

Medicago 21748, 21960, 22030, 22123, 22130, 22279, 22660, 22792, 23194, 23338, B23407, 23448, 23534, 23588, 23895-6, 24080, 24558, 24574, 24716, 24723, 25037, 25137, 25155

Medicinal plants (*cf.* also *Cynodon, Hibiscus, Ricinus, etc.*) 22028, 22209-10, 22380, 22963, 23166, 23386, 23471, 23680, 24559, 24568

Melon see *Colocynthis, Cucumis*

Millet see *Panicum*

Morus 23727, 24325-6, 24528, 24605, 24701, 25095, 25097

Mosses (*cf.* also *Sphagnum*) 21626, 21699, 21737, 21759, 22053, 22056, 22231, 22270, 22543, 22644, 22767, 22992-4, 23189, 23317, 23342, 23460, 23616, 23632, 23789, 23872, 23969, 24017, 24091, 24137, 24554, 24557, 24833, 24961, 24989-90, 25025

Mulberry see *Morus*

Musa 22747, 24468

Musk-melon see *Cucumis*

Mustard see *Sinapis*

N

Napier grass see *Pennisetum*

Nicotiana 21549, 21613, 21665, 21705, 21724, 21771, 21791, 21925, 21961, 21994, 22023, 22050, 22066, 22069-70, 22097, 22296, 22313, 22396, 22484, 22529, 22647, 22719, 22734, 22739-41, 22796, 22835, 22851-2, 22980-1, 23051, 23180, 23231, 23254-6, 23343, 23369, 23373, 23404, 23427, 23512, 23547, 23588, 23618, 23624, 23629, 23643, 23647, 23714, 23717, 23725, B23767, 23788, 23870, 23887, 23920, 23954, 23969, B24024, 24064, 24113, 24136, 24190, 24351, 24364, 24414, 24424, 24442, 24476, 24536-7, 24655, 24683, 24693-4, 24716, 24748, 24754, 24779, 24804, 24853, 25044, 25079, 25108, 25136-8

Nostoc 21628, 22590, 22853, 24709, 25145

O

Oak see *Quercus*

Oat see *Avena*

Oil palm see *Elaeis*

Olea 21705, 22002-3, 22187, 22278, 23243, 24232, 24988

Olive see *Olea*

Onion see *Allium*

Orange see *Citrus*

Orchardgrass see *Dactylis*

Orchids 22175-6, 23787, 23801, 24114

Ornamental plants (*cf.* also *Agave*, *Antirrhinum*, *Asparagus*, Coniferous plants, *Cyperus*, Deciduous trees and shrubs, *Eucalyptus*, *Euphorbia*, *Ficus*, *Hedera*, *Hibiscus*, *Ilex*, *Lathyrus*, *Lupinus*, *Pelargonium*, *Perilla*, *Rosa*, *Tradescantia*, *etc.*) 21748, 21798, 21814, 21840, 21961, 21975, 21998, 22028, 22098, 22130, 22148-9, 22183, 22314-5, 22333-4, 22380, 22508, 22537, 22643, 22647, 22662, 22719, 22725-7, 22770-1, 22925, 22980, 23026, 23036, 23091, 23129, 23132, 23244, 23265-6, 23350, 23434, 23450, 23503, 23618, 23647, 23747, B23767, 23817, 23829, 23859, 23873, 24206, 24217, 24285, 24317, 24369-70, 24372, 24400, 24506, 24672, 24760-1, 24788, 24915, 24919-21, 24948, 24969, 25117

Oryza 21506, 21596, 22012, 22123, 22305, 22700, 22806, 22869, 22880, 22895, 22965-6, 22998, 23002-4, 23059, 23091, 23235, 23250-2, 23352, B23407, 23448, 23657, 23669, 23706, 23714, 23732-4, 23736-7, 23758, 23773, 23814, 23922, 24002, 24142, 24283-4, 24324, 24388, 24452, 24508, 24510, 24609, 24646-7, 24661-2, 24692, 24705, 24769, 24806, 24808, 24820-2, 24841, 24851-2, 24972, 25037, 25086, 25136-7

P

Paddy see *Oryza*

Palms see *Cocos*, *Elaeis*

Panicum 21926-7, 22123, 22249, 22266, 22281, 22397, 22489, 22698, 22700, B22822, 22835-6, 22878-9, 22983, 23001, 23032, 23038, 23072, 23231, B23407, 23465, 23586, 23813, 24040, 24115, 24164-5, 24310, 24769, 24850, 24861, 25048-9

Paprika see *Capsicum*

Para-rubber tree see *Hevea*

Parasitic plants 23652

Parsley see *Petroselinum*

Parsnip see *Pastinaca*

Paspalum 21705, 22366, 22449, 24806

Pastinaca 22110

Pasture plants see Forage plants

Pea see *Pisum*

Peach see *Persica*

Peanut see *Arachis*

Pear see *Pirus*

Peavine see *Lathyrus*

Pecan see *Carya*

Pelargonium 21828, 21849, 22016, 22028, 22098, 23129, 23180, 24442

Pennisetum 21941-2, 22183, 22249, 22670, 22910, 23231, 24139

Pepper see *Capsicum*

Perilla 21702, 25073

Persea 22250, 24468

Persica 22014-5, 22028, 22153, 22747, B23262, B23264, 23687, 23846

Petroselinum 22660, 22747, 23577, 23580, 24304, 24648, 24723, 25070

Phaseolus 21518, 21528, 21533, 21560, 21603, 21606, 21616, 21647, 21676, 21683, 21705, 21730, 21733, 21786, 21842, 21918, 21925, 21961, 21968, 22006, 22028, 22041, 22066, 22113, 22130, 22156-7, 22171, 22200-1, 22240, 22273, 22275-6, 22305-7, 22484-5, 22506, 22594, 22647, 22660, 22663, 22691, 22719, 22734, 22747, 22777, 22780-1, 22789, 22915, 22937-8, 22963, 22975, 23060, 23077, 23108-9, 23170, 23173, 23205-6, 23231, 23242, 23263, 23287, 23291, B23407, 23434, 23450, 23494, 23508, 23513-4, 23523, 23547, 23608, 23716-7, 23750, B23767, 23793, 23836, 23859, 23866, 23870, 23874-5, 23898, 23900-1, 23958-9, 23961, 23969, 23972, 23997, 24001, 24018-9, 24100, 24116, 24127, 24138, 24143, B24250, 24256, 24284, 24291, 24341, 24348, 24364, 24383, 24411, 24468, 24508, 24537, 24542, 24546, 24592-3, 24598, 24626, 24628, 24631, 24636, 24716, 24747, 24750, 24754, 24760-1, 24808, 24885, 24912, 24922, 24941, 25019, 25028, 25136

Phleum 22071, 22279, 23237, 23375, 24277, 24436, 24438

Photosynthetic bacteria see Bacteria, photosynthetic

Phragmites 23910, 25103

Picea 21540, 21629, 21633, 21705, 22041, 22086-7, 22447, 22584, 22734, 23082, 23135, 23257, 23319, 23666, 23791, 23818, 23828, 23834, 23856, 23891, 24057, 24336, 24404, 24556-7, 24604, 24606, 24785, 24837, 24934, 24958, 25115

Pigeon pea see *Cajanus*

Pine see *Pinus*

Pineapple see *Ananas*

Pinus 21633, 21660, 21705, 21831, 21925, 22028, 22189, 22400, 22505, 22523, 22598, 22660, 22681, 22714, 22719, 22734, 22778, 22932, 23091, 23257, 23313, 23500, 23534, 23605, 23666, 23818, 23828, 23830, 23861, 23916-7, 24219, 24232, 24240, 24336, 24432, 24530, 24557, 24670, 24958, 25025, 25089, 25116

Pirus 22468, 22606, B23262, 23263, B23264, 24023, 24468, 24602

Pisum 21530-1, 21545, 21565, 21576, 21579, 21594, 21607-8, 21613, 21662, 21690, 21705, 21727-8, 21747, 21825, 21857, 21943, 21947-8, 21996-7, 22028, 22058, 22067, 22120, 22123, 22156, 22183, 22193, 22237, 22281, 22335, 22338, 22347, 22371, 22405, 22423-4, 22462, 22482, 22512-3, 22540-1, 22547, 22560, 22566, 22595, 22615, 22647, 22656, 22660, 22685, 22698, 22747, 22891, 22894, 22941, 22977, 22980-1, 22988, 23032, 23068, 23118, 23120, 23141, 23150, 23154, 23156, 23159, 23168, 23199, 23205, 23215, 23343, 23461, 23486, 23491, 23630, 23660, 23709, 23716, 23719, 23726, 23742-3, 23748, 23750, B23767, 23847, 23851, 23859, 23870, 23914, 23918, 23934-5, 23949, 23964, 23969-70, 23975, 24001, 24012, 24020, B24024, 24048, 24050, 24084-5, 24104, 24106-7, 24154, 24167-8, B24176, 24188, 24226, 24241, 24284, 24364, 24385, 24415, 24457, 24468, 24478-9, 24484, 24508, 24537, 24545-6, 24716, 24721, 24750, 24812, 24832, 24854,

24861, 24870, 24886, 24910-2, 24973, 24986, 25009, 25070, 25099-101, 25124-5

Plum see *Prunus*

Poa 21762, 22380, 23912, 24314, 24435, 24716, 25031-2

Poplar see *Populus*

Populus 21633, 22138, 22226-9, 22234, 22542, 22614, 22761-2, 23286, 23521, 23670, 23709, 23829, 24045-6, 24105, 24183, 24240, 24336, 24528, 24601, 24739, 24817--8, 24935, 25015, 25018

Porphyridium 21720, 22092, 22308, 22623, 25120

Portulaca 22697-8, 24310

Potato see *Solanum*

Prune see *Prunus*

Prunus (*cf.* also *Armeniaca, Cerasus*)
22284, 22747, B23262, B23264, 23890, B24024, B24176, 24376-8, 24602, 24714

Pseudotsuga 21633, 22087, 23818, 24336, 24372, 24976-7

Pumpkin see *Cucurbita*

Purslane see *Portulaca*

Q

Quercus 21633, 21674, 21705, 21748, 21782, 21808-9, 21854, 22016, 22139, 22165, 22302, 22402, 22573, 22647, 22752, 22762, 22841, 22893, 22917-8, 23091, 23232, 23234, 23342, 23534, 23560, 23679, 23804, 23818, 23829, 23962, 24240, 24567, 24601, 24605, 24670, 24718, 24988

R

Radish see *Raphanus*

Rape see *Brassica*

Raphanus 21719, 21841, 21959, B22052, 22114, 22555, 22571, 22719, 23019, 23125, 23171, 23399-401, 23657, 24032, 24443, 24485, 24557, 24564, 24685, 24806

Raspberry see *Rubus*

Reed see *Phragmites*

Rhodopseudomonas 21649-50, 21821, 21993, 22093-4, 22100-1, 22198, 22230, 22259, 22288-9, 22626, 22733, 22773-4, 22904, 22954, 23047, 23089, 23126-7, 23436, 23825-6, 23886, 23998, 24028, 24095-7, 24109, 24182, 24205, 24305, 24315-6, 24329, 24342, 24696, 25076, 25120

Rhodospirillum 21577, 21651, 21669, 21691, 21777, 21780, 21797, 21855, 21866, 21904, 22088, 22101, 22126-7, 22253, 22305-6, 22354, 22581, 22612-3, 22648, 22759, 22824, 22947, 23362-3, 23436, 23470, 23598, 24034, 24121, 24174, 24189, 24333, 24858-9, 25068, 25104

Ribes 22178, 22747

Rice see *Oryza*

Ricinus 22533, 22734, 23927, 24289, 24386

Robinia 22973, 23551, 23616-7

Rosa 21517, 21809, 22631, 23716, 24063, 24596

Rose see *Rosa*

Rubber tree see *Hevea*

Rubus 22948

Rye see *Secale*

Ryegrass see *Lolium*

S

Saccharum 21950-1, B21979, 21994, 22123, 22249, 22328, 22366, 22532, 22556, 22700, 22713, 22885, 22965, B23407, 23448, 23586, 23714, B23767, 23814, 23835, 23877, 24552, 25137

Safflower see *Carthamus*

Salix 21760-2, 22053, 22778, 23829, 24528, 24605, 24624, 24974

Salt marsh and strand plants (*cf.* also Halophilous plants) 21722, 21903, 21945-6

Sambucus 22614, 24527

Scenedesmus 21678, 21693, 21706, 21738, 21757, 21770, 21805-7, 21888, 22055, 22121, 22180, 22248, 22319, 22442, 22454, 22578, 22642, 22658-60, 22673, 22794-5, 23253, 23272-3, 23311, B23407, 23474, 23614, 23632, 23751, 23824, 23867-8, 23882, 23888-9, 23954, 24053, 24154, 24169, 24273-4, 24371, 24373, 24401-3, 24420, 24483, 24493, 24498-9, 24572, 24579, 24608, 24652, 24700, 24723, 24777--8, 24854, 24881-2, 24944, 24969, 25001, 25121, 25145

Secale 22700, 23713-4, 23905, 24508, 24716, 24890-4, 24912, 24943, 25052, 25061, 25136

Sedge see *Carex*

Sempervirent plants (*cf.* also *Coffea*, Coniferous plants, *Hedera*, *Ilex*, *etc.*) 21705, 22073, 22284, 22492, 23500, 23693, 24604

Service-tree see *Sorbus*

Sesamum 22651-3

Setaria 22630, 24139, 25021

Sinapis 21933, 22447, 22474, 22588, 22902, 23034-5, 23050, 23528, 23564, 23674, 23712, 23714, 23920, 24166, 24368, 24716, 24738, 25029

Sisal see *Agave*

Solanum (*cf.* also *Lycopersicon*) 21564, 21639, 21863, 22039, 22123, 22185, 22250, 22281, 22638, 22701, 22719, 22734, 22829, 22844, 22911, 22999, 23071, 23106, 23226, 23448, 23457, 23544, 23550, 23625, 23695-6, B23767, 23802, 23846, 24228, 24284, 24424, 24428, 24468, 24505, 24940, 25052, 25142

Sorbus 21809, 23859, 24601

Sorghum see *Sorgum*

Sorgum 21613, 21705, 21713, 21739, 21822-3, 22028, 22033, 22042, 22123, 22249, 22328, 22428-30, 22698, 22700, 22762, 22836, 22908-9, 22983, 23007, 23032-3, 23037, 23219-20, 23231, B23407, 23448, 23586, 23714, B23767, 23782-3, 23814, 23877, 23945, 24100, 24244, 24323-4, 24468, 24537, 24546, 24623, 24689, 24725, 24808, 24968, 25037, 25136-7

Soybean see *Glycine*

Sphagnum 23991

Spinach see *Spinacia*

Spinacia 21515-6, 21547-8, 21559, 21567-8, 21573, 21575, 21584, 21588, 21591-2, 21597, 21606, 21611, 21613-4, 21617, 21620, 21631, 21643-6, 21652, 21655, 21662, 21672-3, 21675, 21679, 21687-9, 21698, 21717-8, 21720, 21737-8, 21763, 21766, 21778, 21792, 21816-7, 21819, 21825, 21834, 21839, 21851-2, 21867-8, 21875, 21898, 21900, 21924-5, 21937, 21965, 21983, 21987-9, 21997, 22017-9, 22077, 22081, 22098, 22103, 22107, 22136, 22140-2, 22160-1, 22182, 22194-5, 22203, 22214-7, 22243-4, 22305, 22310, 22320, 22326, 22332, 22345-7, 22355, 22360, 22370-2, 22374-6, 22397, 22406, 22425-7, 22452, 22456, 22463, 22483, 22491, 22514, 22527-8, 22530, 22536, 22539, 22553, 22557, 22574, 22576, 22580, 22583, 22586-7, 22615, 22640, 22644, 22656, 22658-60, 22664, 22669, 22671, 22682, 22689-90, 22693-4, 22698, 22705-9, 22719, 22723-4, 22732, 22736, 22738, 22747-8, 22766, 22801-2, 22812, 22820-1, 22835, 22856, 22862, 22886, 22889, 22905-6, 22915, 22922-3, 22934-6, 22940, 22957, 22967, 22970, 22983, 22995, 23050, 23055, 23057, 23062, 23103, 23122, 23183-4, 23192, 23195-8, 23208-9, 23219, 23227, 23253, 23261, 23276, 23292, 23309-10, 23336-7, 23351, 23359, 23370, 23390, 23394, 23397, 23415-6, 23422, 23473, 23484-6, 23492, 23506-7, 23515, 23517-8, 23538, 23541, 23546, 23549, 23555, 23568, 23573, 23577, 23588, 23591, 23600-1, 23632, 23641, 23656, 23691, 23705, 23715-6, 23721-3, 23725-6, 23728-30, 23750, 23761-2, 23837, 23840-2, 23852, 23858-60, 23868, 23873, 23876, 23906, 23916, 23920, 23927, 23948, 23954-7, 23989, 23993, 24035, 24037, 24049, 24054, 24064, 24066, 24081, 24086, 24097, 24103, 24112, 24116, B24176, 24191--3, 24222-4, 24237-8, 24262-4, 24281, 24298, 24304, 24306-7, 24311-2, 24327, 24335, 24338, 24344, 24351-2, 24354-5, 24360, 24362, 24364, 24374, 24380, 24384, 24395-6, 24442, 24487, 24492-4, 24497, 24501-2, 24549, 24591, 24637, 24648, 24651, 24663-5, 24667, 24682-3, 24720-1, 24723, 24727, 24750, 24757, 24760, 24791, 24828-9, 24840, 24846, 24856, 24860, 24871, 24874-6, 24878-9, 24881, 24888, 24901-4, 24908, 24910, 24917, 24923, 24925, 24944, 24951, 24999, 25002, 25029, 25057, 25062, B25064, 25081-5, 25092, 25094, 25104-5, 25107, 25113-4, 25131, 25137-8, 25145, 25148, 25152, 25159

Spinach beet see *Beta*

Spirodela see *Lemnaceae*

Spruce see *Picea*

Squash see *Cucurbita*

Strawberry see *Fragaria*

Submersed plants see Aquatic macrophytes

Succulents (*cf.* also *Agave*, *Bryophyllum*, Cacti, CAM plants, *etc.*) 23121, 23374, 24297

Sugar beet see *Beta*

Sugar cane see *Saccharum*

Sunflower see *Helianthus*

Sweet potato see *Ipomoea*

Syringa 22071, 22165, 22614, 23859, 24601, 25018, 25102

T

Tamarisk see *Tamarix*

Tamarix 23112

Tapioca see *Manihot*

Taxus 21705, 24674, 24716

Tea see *Thea*

Thea 22647, 24526

Theobroma 21661-3, 23907-8

Thuja 23745

Tilia 22168, 23829, B24024, 24240, 24336, 24601, 24817-8, 25018, 25102

Timothy see *Phleum*

Tobacco see *Nicotiana*

Tomato see *Lycopersicon*

Tradescantia 21840, 22946, 23612, 23873, 25117

Trifolium 21869, 22123, 22563, 22611, 22668, 23063, 23338, 23588, 23594, 23716, 23895, 24001, 24427, 24558, 24968, 25103

Triticum 21529, 21542, 21544, 21561, 21593, 21602, 21634, 21642, 21653-4, 21677, 21703, 21785, 21800, 21844-5, 21847, 21886, 21947-9, 21967, 22011, 22033, 22049, B22052, 22064, 22071, 22118, 22123, 22212, 22221-2, 22260, 22281, 22290, 22295, 22305, 22309, 22331, 22341-2, 22369, 22409-10, 22431, 22441, 22448, 22467, 22519, 22524, 22639, 22698, 22700, 22719, 22734, 22785-6, 22793, 22829, 22833, 22835-6, 22840, 22849, 22856-7, 22882-4, 22960, 22983, 23007, 23049, 23073, 23079-80, 23086, 23095, 23114, 23145, 23149, 23162, 23172, 23193, 23216, 23231, 23277, 23321, 23339, 23346, B23407, 23417, 23435, 23448, 23494, 23561, 23586, 23650, 23655, 23665, 23672, 23713-4, 23717, 23748, 23750, 23763, B23767, 23773, 23778, 23782, 23790, 23805, 23810-2, 23831, 23859, 23862-3, 23870, 23873, 23899, 23903-5, 23950, 23994, 24022, 24026, 24064, 24108, 24124, B24132, B24176, 24190, 24196, 24199, 24207, 24252, 24284, 24286, 24290, 24295, 24309, 24364, 24391, 24424, 24449, 24468, 24504, 24508, 24523, 24534, 24626, 24633, 24653-4, 24657, 24715-6, 24726, 24752-3, 24775-6, 24781, 24806, 24808, 24890, 24940-1, 24949-50, 25015, 25037, 25052, 25059, 25134, 25136-7, 25140, 25143, 25145

Tsuga 22224, 22321, 23616-7, 24213

Tundra plants and ecosystems 22144, 22314-5, 22543, 23654, 23865, 24301, 24535, 24564, 24634, 24758, 25024

Turnip see *Brassica*

Typha 23271, 23684

Zea (continued)

22601, 22608, 22635, 22641, 22660, 22698, 22700, 22712, 22719, 22734, 22754, 22799, 22806, 22813-6, 22829, 22835, 22840, 22869, 22875, 22958, 22974, 22983-5, 22998, 23002-3, 23021, 23030-3, 23054, 23060, 23067, 23091-2, 23094, 23106, 23108, 23134, 23170, 23231, 23235, 23250-2, 23258, 23343, 23351, 23353, 23376, 23392, 23397, B23407, 23408, 23418-9, 23424, 23426, 23448, 23457, 23461, 23485, 23547, 23586, 23592, 23608, 23619, 23647, 23665, 23671, 23676-7, 23689, 23714, 23716-7, 23732, 23735, 23737, 23739-40, 23746, B23767, 23782, 23814, 23851, 23859, 23863, 23870, 23877, 23887, 23893, 23919, 23923, 23954, 23969, 23985, 24000-1, 24003, 24005, 24008, 24018, B24024, 24033, 24039, 24047, 24074, 24083, 24100, B24132, 24143, 24157, 24164-5, 24196, 24236, 24244, 24268, 24284, 24318, 24323-4, 24364, 24424, 24468, 24470, 24474, 24504, 24507-8, 24511, 24538, 24543, 24545-8, 24597, 24607, 24615, 24657, 24705, 24723, 24725, 24734, 24737, 24750, 24769, 24772, 24794, 24805-6, 24808, 24841, 24850, 24854, 24861, 24887, 24912, 24931, 24933, 24959, 25015, 25037, 25052, 25087, 25091, 25099, 25101, 25134, 25136-8, 25156, 25158

Zebrina see *Tradescantia*